The Harvest of a Century

The Harvest of a Century

Discoveries of Modern Physics in 100 Episodes

Siegmund Brandt

Physics Department
University of Siegen

OXFORD
UNIVERSITY PRESS

OXFORD

UNIVERSITY PRESS

Great Clarendon Street, Oxford OX2 6DP

Oxford University Press is a department of the University of Oxford.
It furthers the University's objective of excellence in research, scholarship,
and education by publishing worldwide in

Oxford New York

Auckland Cape Town Dar es Salaam Hong Kong Karachi
Kuala Lumpur Madrid Melbourne Mexico City Nairobi
New Delhi Shanghai Taipei Toronto

With offices in

Argentina Austria Brazil Chile Czech Republic France Greece
Guatemala Hungary Italy Japan Poland Portugal Singapore
South Korea Switzerland Thailand Turkey Ukraine Vietnam

Oxford is a registered trade mark of Oxford University Press
in the UK and in certain other countries

Published in the United States
by Oxford University Press Inc., New York

© Siegmund Brandt 2009

British Library Cataloguing in Publication Data

Data available

Library of Congress Cataloging in Publication Data

Data available

Printed in Great Britain
on acid-free paper by
CPI Antony Rowe, Chippenham, Wilts

ISBN: 978-0-19-954469-1

10 9 8 7 6 5 4 3 2 1

To Martin

Preface

The intention of this book is to give an impression of the advances of physics in the twentieth century by retracing important discoveries. In this period physicists focussed their interest primarily on two topics, the microscopic structure of matter and the structure of space and time. Sometimes an exact date is attributed to the beginning of modern physics, 14 December 1900, when Planck presented his radiation law, featuring the quantum of action. Half a century later Einstein stated that "this discovery [of the quantum of action] became the basis of all twentieth-century physics and has almost entirely conditioned its development ever since". That is still true, now that the whole century has passed. In spite of this, our tale begins a little earlier with important experimental discoveries, those of X rays and the electron, and those of radioactivity, radium, and alpha, beta, and gamma rays before we touch on Planck's work.

The book is conceived as a string of a hundred episodes, covering both experimental and theoretical physics. This structure is meant to allow readers to follow their own particular interests, picking topics of their choice. Still, episodes are arranged in chronological order and readers with no or little previous knowledge in modern physics may want to follow that order, since they will become familiar with the development of the field and its concepts as they go along. Each episode is a short story of the scientists involved, their time, and their work; each is self-contained, with references to others where required. The text is intended for easy reading. Occasionally, a more thorough discussion of experimental set-ups and theoretical concepts is presented in special boxes for readers interested in more detail.

In contrast to science itself, history of science is not well ordered; it develops in twists and turns with ramifications and blind alleys. In our collection of episodes, which can be seen as a mosaic with wide spaces between its stones, many historical details are left unmentioned. The year, given in the title of each episode, is that in which the corresponding discovery was made public, usually by submission of a paper to a learned journal. Often, it took quite some time until discoveries were generally accepted. This, in particular, holds for new theories like Planck's or Einstein's. Again, we scarcely dwell on the reception of the discoveries we describe.

The selection of topics is, of course, to some extent arbitrary. On the other hand, many of those chosen definitely had to be included. It is sincerely hoped that every reader will find at least half of her or his own favourite hundred. The episodes refer to atomic, nuclear, and particle physics, to physics of condensed matter, to relativity, and to physical properties of the universe.

An important starting point for atomic physics was the development of optical spectroscopy in the nineteenth century. Spectroscopic measurements became the basis of Planck's work; when the electron and the nucleus had been found, they allowed the construction and the refinement of atomic models. The culmination point was the formulation of quantum mechanics in the 1920s, a theory which, in extended form, seems to govern all microscopic physics.

Nuclear physics began with the discovery of radioactivity which provided projectiles of rather high energy for scattering and collision experiments. In the early 1930s higher energies, reached by accelerators, and neutrons as new projectiles brought this field to the foreground. A peak was reached with the discovery of nuclear fission and the construction of nuclear reactors in the late 1930s and the 1940s, respectively. Nuclear physics experiments gradually made it clear that there were two additional forces in nature besides the gravitational force (which binds planets to a star) and the electromagnetic force (which binds electrons to a nucleus). These new forces were called strong and weak, respectively. The strong force was needed to keep the nucleus together, the weak one to explain a particular type of its decay. Nuclear physics brought some progress in the understanding of the weak force but no lasting success with the strong.

Particle physics is an off-spring of nuclear physics. It makes use of the very high energies provided by cosmic radiation and modern accelerators and studies collisions of particles simpler than nuclei (except for the proton, the simplest nucleus). Such experiments led to the discovery of many new particles coupled to the strong force (hadrons) and of a few showing only weak and, at most, electromagnetic interactions (leptons). In the 1960s it was found, that hadrons are composed of more fundamental particles, called quarks, and later on that there are as many quarks as there are leptons. On the basis of quarks and leptons, successful theories of the strong and the weak interaction were developed (the latter even was unified with the electromagnetic one) and confirmed by experiment.

From the diversified field of condensed matter physics, episodes are presented on unusual and unexpected phenomena such as superconductivity, superfluidity, and the quantum Hall effects, all understood later on the basis of quantum mechanics. But quantum mechanics also helped to design novel materials and devices. In that context we describe the development of the band model of semiconductors and the construction of transistors and semiconductor laser diodes.

Relativity plays a special role in two respects. It was created essentially by one man, Einstein, and, as a theory of space and time, it pertains to all of physics. Two episodes tell how special and general relativity theory came into being in 1905 and 1915, respectively. The special theory is indispensable in nuclear and particle physics. Astrophysical observations, connected to the general theory, are the subject of several episodes.

In all fields, mentioned above, active research is currently done. Often, interest was re-kindled by new experimental possibilities, such as new detectors for particles and light, powerful systems for data acquisition and handling, and a light source of new quality, the laser.

The book ends with two episodes demonstrating that research requires persistence. One experiment, in 1995, succeeded with the Bose–Einstein condensation of neutral atoms, an effect predicted by Einstein seventy years earlier. The other one, extending into the twenty-first century, showed that neutrinos, particles postulated by Pauli also about seventy years before, possess tiny, non-vanishing masses. Although Pauli himself only stated that the neutrino mass was small and possibly 'equal to zero', most physicists had tacitly assumed that possibility to be the truth.

It may surprise the reader as it has, at first, struck me, that the majority of episodes plays in the first half of the century, whereas my own active life in research, essentially, coincides with the second half. Besides a possible personal bias, such as the veneration of scientists from the previous two generations, I offer the following explanation: Modern physics, in its different branches and as a whole, has evolved from an explorative, often qualitative stage full of surprise discoveries to a stage of synthesis, in which theories were formulated, which was then followed by a stage of precision measurements providing stringent tests of the latter. The subject matter for our episodes was taken from the first two stages. The third one, although very important for science, was simply ignored unless it led to contradictions with theory.

Portraits and photographs, figures, and references to the literature are considered an important part of every episode. I am grateful to the organizations, journal publishers, and individuals who allowed me to reproduce historical photographs and original figures. A particular pleasure was to read or re-read the original publications, while preparing an episode. Many are still captivating and some are written more to the point and clearer than later presentations in textbooks; a prominent example is Einstein's 1905 paper on special relativity. All episodes contain references to the original literature which is now, mostly, accessible electronically. Cited as well are biographies and autobiographies of scientists, reminiscences in various collections, and biographical articles by colleagues. Valuable biographic information is also provided by the Nobel Foundation [1]. From the general literature on the recent history of physics I want to mention here Pais' inspiring tale of nuclear and particle physics [2], Mehra's and Rechenberg's monumental work on the development of quantum theory [3] and the history of solid-state physics by Hoddeson and co-authors [4].

I gratefully acknowledge the help of friends and colleagues at various stages during the preparation of this book. I am obliged to my old friend Andrzej Wróblewski for fruitful conversations on the selection of topics, to Hans Dieter Dahmen for discussions and help concerning theoretical points, to Alexander Khodjamirian for his help with Russian texts, and to Tilo Stroh, who created many of those explanatory figures which were not taken from the original literature and who shared with me his expertise on computer typesetting. Claus Grupen, Gottfried Holzwarth, Alexander Khodjamirian, Diethard Schiller, Tilo Stroh, and Andrzej Wróblewski read the manuscript of some or all episodes and offered helpful suggestions; needless to say that remaining mistakes are mine. Last, but not least, I thank my son Martin, to whom the book is dedicated, for his detailed advice and helpful criticism in matters of solid-state physics.

Siegen, July 2008   Siegmund Brandt

[1] http://nobelprize.org.

[2] Pais, A., *Inward Bound – Of Matter and Forces in the Physical World.* Oxford University Press, Oxford, 1986.

[3] Mehra, J. and Rechenberg, H., *The Historical Development of Quantum Theory, 6 vols.* Springer, New York, 1982–2001.

[4] Hoddeson, L., Braun, E., Teichmann, J., and Weart, S., *Out of the Crystal Maze – Chapters from the History of Solid-State Physics.* Oxford University Press, Oxford, 1992.

Contents

Prologue – The Nineteenth-Century Heritage

In 1874 Max Planck was sixteen years old. He was about to enter the university in Munich and sought the advice of Jolly[1] who was professor of physics there. Jolly told him that he considered the edifice of physics to be essentially completed and that only a few details might still be discovered in this field. In short, Jolly told Planck that there was not much future in the study of physics.

Fortunately Planck was not discouraged. A quarter of a century later, on 14 December 1900, Max Planck delivered a talk at one of the regular meetings of the German Physical Society in Berlin about work with which he laid the foundations of quantum theory which goes far beyond the now so-called *Classical Physics* of the nineteenth century. Because of this the year 1900 is generally considered the birth year of *Modern Physics*. The story of Planck's discovery will be told in one of our first episodes. But before that let us briefly look at the status of physics at the end of the nineteenth century.

Newton

Euler   Lagrange   Hamilton   Jacobi

The theoretical description of *mechanics* was firmly based on the laws of motion and on the law of gravitation set up by Newton[2] in the seventeenth century. It was extended and mathematically developed in the eighteenth century in particular by Euler[3] and Lagrange[4] and in the nineteenth century by Hamilton[5] and Jacobi[6]. Mechanics described accurately planetary motion, the behaviour of rigid and of elastic bodies, the flow of liquids, and the waves of sound in solids, liquids, and gases. It was also used to great advantage by mechanical engineers in the industrial revolution.

The industrial revolution was made possible by the steam engine. Its importance called for the development of a comprehensive *theory of heat* or *thermodynamics*. This was finally based on the two fundamental laws of thermodynamics. The first one, the law of energy conservation, was formulated by

---

[1] Philipp Johannes Gustav (von) Jolly (1809–1884)   [2] (Sir) Isaac Newton (1643–1727)   [3] Leonhard Euler (1707–1783)   [4] Joseph Louis Lagrange (1736–1813)   [5] (Sir) William Rowan Hamilton (1805–1865)   [6] Carl Gustav Jacob Jacobi (1804–1851)

Joule[7], Mayer[8], and Helmholtz[9]. The second law, expressing that entropy cannot decrease, was found by Kelvin[10] and by Clausius[11] who followed prior work done by Carnot[12]. The hypothesis that all matter consists of small constituents, the atoms, made it possible to develop the *kinetic theory of heat*. It

Joule   J. R. Mayer   Helmholtz

Carnot   Kelvin   Clausius

Boltzmann

stated that the temperature of a body is directly connected to the kinetic energy of the microscopic motion of its constituents. This approach culminated in the *statistical mechanics* of Boltzmann[13], who was able to reduce all thermodynamic concepts to mechanical ones, and thus to reduce thermodynamics to a branch of mechanics. In particular, Boltzmann gave a rather simple meaning to the complicated concept of *entropy*, which had been introduced by Clausius. He showed that entropy, which increases with time in all irreversible thermodynamic processes, is directly related to the *probability* with which a state is realized. For instance, for a volume filled with gas it is much more probable that the gas molecules are spread out evenly throughout the volume rather than to be concentrated in one corner. Boltzmann had thus introduced probability as a law of nature. At the end of the nineteenth century there was still a debate

---

[7] James Prescott Joule (1818–1889)   [8] Julius Robert Mayer (1814–1878)   [9] Hermann Ludwig Ferdinand (von) Helmholtz (1821–1894)
[10] William Thomson (Lord Kelvin) (1824–1907)   [11] Rudolf Clausius (1822–1888)   [12] Nicolas Léonard Sadi Carnot (1796–1832)   [13] Ludwig Boltzmann (1844–1906)

going on among scientists whether such an approach to science is acceptable. Moreover, the very existence of atoms and molecules was not established for all critics.

The phenomena of *electricity* and *magnetism* were found to be interconnected through the discoveries of Oersted[14] and Faraday[15] and were subsequently described by Maxwell[16] with his beautiful theory of *electrodynamics*. This theory predicted the existence of *electromagnetic waves*. These were produced and detected with apparatus conceived and realized by Hertz[17], who invented the dipole antenna and developed techniques to use it both as transmitter and as detector of electromagnetic waves.

Maxwell

Oersted

Faraday

Hertz

The physics of light or *optics* had been developed, in particular, by Fresnel[18] into an elaborated mathematical theory of waves, and these light waves were shown through the experiments of Hertz to be electromagnetic waves. Thus optics was reduced to a branch of electrodynamics. Since an electric current is caused by moving electric charge, there was a connection between electrodynamics and mechanics. Electromagnetic waves were considered to be similar to mechanical waves and were therefore thought to need a material medium, the *ether*, to propagate in. Thus it was thought that the whole edifice of physics was built on the laws of mechanics.

Fresnel

Although macroscopic phenomena seemed to be well understood, rather little was known, however, about the microscopic structure of matter. The study of material substances was mainly the subject of chemical research and there was important progress in this field. The concept of chemical elements came to the forefront when Meyer[19] and Mendeleyev[20] independently pointed out a way to organize them systematically in the *Periodic Table*. A chemical element was thought to be composed of *atoms*, i.e., indivisible smallest objects, which were different from the atoms of all other elements. A chemical compound then was composed of *molecules* built up of atoms from one or several elements. But both chemists and physicists were divided over the question whether atoms really exist or whether the *atomic hypothesis* just provide a convenient means to talk about the wealth of chemical information.

---

[14] Hans Christian Oersted (1777–1851)   [15] Michael Faraday (1791–1867)   [16] James Clerk Maxwell (1831–1879)   [17] Heinrich Rudolph Hertz (1857–1894)   [18] Augustin Jean Fresnel (1788–1827)   [19] Julius Lothar Meyer (1830–1895)   [20] Dimitry Iwanovich Mendeleyev (1834–1907)

J. L. Meyer

Mendeleyev

Chemists knew in what proportion different elements form compounds and were thus able to make a table of *atomic weights* of the elements. The atomic weight, now called *atomic mass number*, was the mass of an atom of a particular element expressed as a multiple of the mass of the hydrogen atom. (Later that definition was changed slightly. At present, it is the mass of an atom expressed as the multiple of $1/12$ of an atom of the carbon isotope ^{12}C.) Analogously, one defines the *molecular mass number*. The absolute masses of the atoms were still unknown. In 1811 Avogadro[21], using results by Gay-Lussac[22], conjectured that, for equal pressure and equal temperature, equal volumes of different gases contain the same number of molecules. We now call one *mole* the amount of substance one gets if one multiplies the atomic (or molecular) mass number of a substance by the mass 1 g. And we call *Avogadro's constant* the number of atoms (or molecules) contained in one mole, for instance, the number of atoms contained in 1 g of hydrogen (or rather, more precisely, in 12 g of the carbon isotope ^{12}C). More than fifty years after Avogadro's conjecture that number was determined for the first time in 1865 on the basis of measurements performed by Loschmidt[23]; for a long time it was called *Loschmidt's number*. It was felt that the atomic hypothesis would gain

Gay-Lussac   Avogadro   Loschmidt

---

[21] Lorenzo Romano Amedeo Carlo Avogadro (1776–1856)   [22] Joseph Louis Gay-Lussac (1778–1850)   [23] Jan Joseph Loschmidt (1821–1895)

support if the analysis of very different phenomena would yield the same numerical value of Avogadro's number. When Einstein entered the scientific stage in 1905, he not only presented his special theory of relativity and his light-quantum hypothesis but also several methods to determine Avogadro's number, including his analysis of *Brownian motion*.

In the second half of the nineteenth century, through the work of Kirchhoff[24] and Bunsen[25], it became a well-established fact that elements, when heated, emit light at discrete characteristic wavelengths. This fact was interpreted by some as evidence that the atoms of these elements oscillate at characteristic frequencies. In general, the spectra of the elements were too complicated to allow for a simple analysis. The spectrum of hydrogen, however, did show a very regular structure and in 1885 Balmer[26] found a formula with only one constant describing the spectral lines of hydrogen with very high precision. No physical justification for the *Balmer formula* was found at that time.

Kirchhoff   Bunsen   Balmer

In his doctoral examination in 1879 Max Planck, as customary at the time, had to submit and defend orally a number of theses. His thesis number three was: *The assumption of absolutely indivisible components of matter contradicts the principle of conservation of energy.* Little did he dream that it was his discovery of the quantum of action that later enabled science to explain atoms and, more generally, to a large extent the structure of matter.

---

[24] Gustav Robert Kirchhoff (1824–1887) [25] Robert Wilhelm Bunsen (1811–1899) [26] Johann Jacob Balmer (1825–1898)

Röntgen's X Rays (1895)

In 1833, Michael Faraday did his research on the conduction of electricity through a liquid, which was a solution of some salt in water. Two metal plates, called *electrodes*, were immersed into the liquid and connected to a battery. The electrode connected to the positive pole of the battery was called *anode*, the other *cathode*. In such a set-up an electric current flows and, at the same time, substance is deposited at the electrodes. The latter process is called *electrolysis*. As an example, if the liquid is a solution of silver nitrate, then silver is deposited at the cathode. The mass deposited is proportional to the electric charge passed through the liquid. The transport of electricity and of material obviously is due to charged matter. As the atomic hypothesis gained support, this charged matter was conceived to consist of positively or negatively charged atoms or molecules called *ions*. The smallest possible charge of an ion was obtained by dividing *Faraday's constant*, i.e., the total charge needed to transport one *mole* of chemically monovalent substance, silver in our example, by *Avogadro's constant*, namely, the number of atoms (or molecules) per mole. This charge was called the *atom of electricity* by Helmholtz in 1881 and *electron* by Stoney[1] in 1891. We now call it the *elementary charge*, since the term *electron* was soon used to designate a fundamental particle (see Episode 4).

Hittorf

Goldstein

The conduction of electricity in solids like copper but also in liquid metals like mercury was not really understood because nothing material seemed to be transported together with the electric charge. Also not understood but full of interesting phenomena were electric currents in gases. During the whole nineteenth century the problem had been studied and the techniques used had been slowly improved. Better pumps yielded lower gas pressures. Geissler[2] produced such pumps and also glass tubes with tight metal feed-throughs for the electrodes. The tubes were filled with the desired gas at the desired pressure and sealed so that the low pressure was maintained without the need of pumping. Large induction coils were developed which could provide high voltages although only for short times. The passing of electricity through gas is still today called a *gas discharge* because a capacitor, a coil, or a battery might be discharged by the current. Largely depending on the type of gas, its pressure, the voltage, and the geometrical arrangement of the electrodes, a variety of phenomena was observed, mainly through visual inspection of the tube which showed a more or less intensive glow at one or several places. Of course, the optical spectra of this light, characteristic for the gas filling, were studied. Also, more or less distinct 'rays' were observed by Hittorf[3] in 1869. They seemed to move away from the cathode and were called *cathode rays* by Goldstein[4] in 1876. In 1886 Goldstein constructed a tube with a hole or *canal*

[1] George Johnstone Stoney (1826–1911) [2] Johann Heinrich Wilhelm Geissler (1815–1879) [3] Johann Wilhelm Hittorf (1824–1914) [4] Eugen Goldstein (1850–1930)

in the cathode and observed what he called *canal rays*, penetrating through this hole from the discharge region between anode and cathode. We now know that the cathode rays are electrons (Episode 4) and the canal rays are positive ions (Episode 22). But then the nature of these two types of rays was unknown and many scientists experimented with gas discharges. One could form the discharge tube in such a way that cathode rays did not hit the anode but instead fell on the glass wall of the tube. The glass itself was then observed to fluoresce, i.e., to emit light there. Lenard[5], following a suggestion of his teacher Hertz, had constructed tubes with an opening covered by a thin aluminium foil, the *Lenard window*, through which cathode rays could penetrate into the surrounding air. In 1895 yet another kind of rays was discovered by Röntgen when he experimented with gas discharge tubes.

Röntgen

Röntgen[6] [1] was born in the small town of Lennep in the Rhineland as son of a textile merchant. When he was three years old, his family moved to Appeldoorn in the Netherlands, where his mother had relatives. Röntgen failed to pass the end-of-school exam which opened the way to university studies, but learnt that this exam was not required by Eidgenössische Polytechnikum (Federal Polytechnique) in Zurich and, in 1865, moved there to study mechanical engineering. After obtaining his diploma Röntgen followed the advice of Kundt[7], his physics professor, and took a Ph.D. degree at the University of Zurich. Röntgen was Kundt's assistant at several universities before becoming a professor himself: 1875 in Hohenheim, 1876 in Strasbourg, 1879 in Giessen, and 1888 in Würzburg.

There, on 8 November 1895, Röntgen observed something completely unexpected. In the first lines of his paper *On a new kind of rays* [2], which we quote from the English translation of the time, he reports as follows:

> A Discharge from a large induction coil is passed through a Hittorf's vacuum tube, or through a well-exhausted Crookes' or Lenard's tube. The tube is surrounded by a fairly close-fitting shield of black paper; it is then possible to see, in a completely darkened room, that paper covered on one side with barium platinocyanide lights up with brilliant fluorescence when brought into the neighbourhood of the tube, whether the painted side or the other be turned towards the tube. The fluorescence is still visible at two metres distance. It is easy to show that the origin of the fluorescence lies within the vacuum tube.

Röntgen had the good fortune that such a screen, used to detect ultraviolet light, happened to lie near his discharge tube. Alone and in complete secrecy, for about seven weeks, he studied the properties of the new rays – he called them *X rays* in great detail. Röntgen determined qualitatively the absorbing power of different substances for these rays and found lead to be particularly effective. Besides barium platinocyanide he identified other substances capable to fluoresce and he found photographic plates and films to be sensitive to X rays. Contrary to light, he found X rays to be neither reflected by a surface nor refracted by a prism or a lens and, contrary to cathode rays, he did not find them to be deflected by a magnet. Röntgen localized the geometrical source of the X rays as the spot of the tube wall which was hit by the cathode rays

---

[5] Philipp Eduard Anton Lenard (1862–1947), Nobel Prize 1905   [6] Wilhelm Conrad Röntgen (1845–1923), Nobel Prize 1901   [7] August Eberhard Kundt (1839–1894)

and therefore showed fluorescence, i.e., emitted visible light. He proved this by moving that spot around with a magnet which deflected the cathode rays. Moreover, he recorded that the human eye is insensitive to X rays even if held very near to their source.

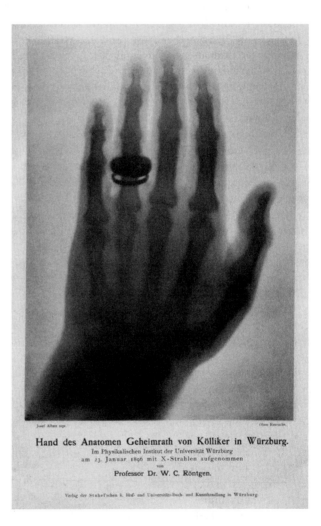

Hand des Anatomen Geheimrath von Kölliker in Würzburg.
Im Physikalischen Institut der Universität Würzburg
am 23. Januar 1896 mit X-Strahlen aufgenommen
von
Professor Dr. W. C. Röntgen.

X-ray photograph of von Kölliker's hand, taken by Röntgen on 23 January 1896.

Spectacular use of X rays is made in medicine. That is possible because of the different absorbing power of different materials and the overall high penetration power of these rays. Already on the second page of his paper Röntgen writes:

> If the hand be held before the fluorescent screen, the shadow shows the bones darkly, with only faint outlines of the surrounding tissues.

He also reports that he produced X-ray photographs of a hand but also of several technical objects revealing their interiors. Just before the end of 1895 Röntgen submitted his paper as a session report to the Würzburg Physico-Medical Society. He had it printed immediately and he sent copies including some of his photographs to a number of colleagues. It is because of these photographs that Röntgen's discovery very soon made it to the newspapers around the world, even before he gave his lecture in Würzburg.

The oral report to the Physico-Medical Society, accompanied by demonstrations, was delivered on 23 January 1896. At the end of his lecture Röntgen asked the anatomist von Kölliker to allow him to take an X-ray photograph of his hand. Von Kölliker agreed, the photograph was made on the spot, and von Kölliker proposed to call the new rays 'Röntgen-Strahlen', i.e., *Röntgen rays*, a term still in use in the German-speaking countries.

Röntgen, who published only two more papers on X rays [3,4], was awarded the first Nobel Prize in physics in 1901. At that time the physical nature of X rays was still unknown. (He himself had conjectured they might be longitudinal ether waves.) Generously, he donated the prize money to the University of Würzburg. Also deliberately he refused to take patents on the production of X rays, which certainly would have made him a wealthy man.

Röntgen presented some of the tubes, used in his early work, to the Deutsches Museum in Munich. It is not known exactly with which type of tube X rays were first observed. A modern X-ray source is rather different from Röntgen's original tubes. It no longer contains a gas but is evacuated. It still contains two electrodes, the cathode and the anode, which, in this particular case, is called *anticathode*. The cathode is heated by a special heating circuit and thus made to emit electrons (the cathode rays). These electrons are accelerated by a constant high voltage between the two electrodes and hit the anticathode, where the X rays are produced. Their quality depends on the voltage and on the material of the anticathode.

X rays have become a most valuable tool not only in medicine and biology but also in revealing the structure of crystals and of the atoms themselves. We shall encounter them again and again.

In 1900 Röntgen accepted the physics chair at the University of Munich. Having spent most of his professional life in small and homely university towns, he did not feel at ease with his recent fame as a world celebrity in the Bavarian capital. Although X-ray research was done in his institute, Röntgen himself more and more stayed away from active scientific work. Laue, who, in 1912, cleared up the nature of X rays (Episode 20) remembers [5] that the only time he could talk to him in peace and quiet was, 'when in a very full train I found the only free place in a third-class compartment opposite to His Excellency. There I formed the impression that one could have communicated with him quite well, if only there had been more opportunity.' Laue continues: 'Often one has asked for reasons why this man, after his epoch-making achievements of 1895/96, had so retracted. Many motives were suggested, some little flattering for Röntgen. I consider them all false. In my opinion, the impression of his discovery had so overpowered him that he, who made it when he was fifty, never recovered. For – and only few think of it – a great feat is a burden for him who achieved it. ... It needed much to compile three papers, which, like Röntgen's from 1895 to 1897 exhausted the subject so much, that for a decade hardly anything new could be said about it. And with what ingenious care were they written! I know only few papers of discoveries which would not contain minor or major mistakes. With Röntgen everything was cast-iron.'

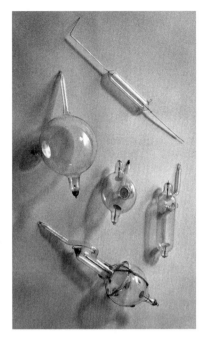

Gas discharge tubes used by Röntgen in 1905/06 (now in the Deutsches Museum, Munich). Photo: Deutsches Museum.

[1] Glasser, O., *Wilhelm Conrad Röntgen und die Geschichte der Röntgenstrahlen.* Springer, Berlin, 1995.

[2] Röntgen, W. C., *Sitzungsberichte der Würzburger Physikalisch-medicinischen Gesellschaft*, (1895). Also in *Annalen der Physik und Chemie* **64** (1898) 1. Engl. transl. *Nature* **53** (1896) 274.

[3] Röntgen, W. C., *Sitzungsberichte der Würzburger Physikalisch-medicinischen Gesellschaft*, (1896). Also in *Annalen der Physik und Chemie* **64** (1898) 12.

[4] Röntgen, W. C., *Sitzungsber. Preuss. Akad. Wiss. (Berlin)*, (1897) 576. Also in *Annalen der Physik und Chemie* **64** (1898) 18.

[5] von Laue, M., *Mein physikalischer Werdegang.* In [6], Vol. III, p. V.

[6] von Laue, M., *Gesammelte Schriften und Vorträge, 3 vols.* Vieweg, Braunschweig, 1961.

2 Becquerel Discovers Radioactivity (1896)

As we have seen, the first observation of X rays was brought about by a fortunate coincidence. The discovery of radioactivity, likewise unexpected, was triggered by the discovery of X rays although the two are completely unrelated. The story is told by Marie Curie in the introduction to her thesis [1] as follows: 'The first tubes producing Röntgen rays were tubes without metallic anticathode. The source of the Röntgen rays was a spot on the glass wall which was hit by the cathode rays; at the same time this spot showed a lively fluorescence. One could therefore ask oneself if the emission of Röntgen rays did not necessarily accompany the production of fluorescence. This idea was pronounced first by M. Henri Poincaré. Shortly afterwards, M. Henry announced that he had obtained photographic prints through black paper with the help of zinc sulphide.' She goes on to report that also two other gentlemen, M. Niewenglowski and M. Troost, obtained the blackening of photographic plates wrapped in black paper when they caused different materials to fluoresce by exposing them to light. But she also reports that none of these results could be reproduced and how then Becquerel made his astonishing discovery.

The story can still be followed in the *Comptes Rendus* of the French Academy of Sciences for the first three months of 1896. It had been triggered by Poincaré's remark 'should one not ask oneself if not all bodies, whose fluorescence is sufficiently strong, emit, apart from visible rays, the X rays of Röntgen, whatever may be the cause of their fluorescence?' Poincaré[1], an eminent mathematician with a strong interest in physics, had been among those who had received Röntgen's paper and photographs and had asked this question in the *Revue générale des Sciences* of 30 January 1896. The note by Henry, communicated to the Academy on 10 February by Poincaré himself, contains the observation that zinc sulphide, stimulated to fluoresce by light, blackened a photographic plate wrapped in paper. Similar reports by Niewenglowski and by Troost, who used different materials, were made on 17 February and 9 March, respectively. Much later one came to the conclusion that the samples used by the three gentlemen, under the influence of humidity in the air, produce traces of hydrogen sulphide, which, by chemical action, blackens the photographic plate.

Becquerel

Becquerel[2] was born in Paris and studied civil engineering at the École Polytechnique. Having obtained his diploma in 1877, he opted for a career as scientist. He did research in optics, took his Ph.D. in 1888, and was elected member of the Academy of Sciences a year later. In 1892 Becquerel became professor at the Natural History Museum in Paris, like his father and grandfather before

[1] Henri Poincaré (1854–1912)   [2] Antoine Henri Becquerel (1852–1908), Nobel Prize 1903

him (and later his son after him). In 1895 he was also appointed professor at the École Polytechnique. His speciality (as that of his father and grandfather) was the study of *fluorescence* and *phosphorescence*. These are properties of materials to emit light if properly stimulated, usually by irradiation with light of shorter wavelength than that of the emitted light. Strictly used, the two terms apply to prompt and somewhat delayed emission, respectively. (The general expression is *luminescence*.) But that is of no importance here. In the *Académie des Science* Röntgen's discovery was discussed in January 1896. It was quite natural for Becquerel to wonder, as Poincaré had suggested, whether X rays were produced by some kind of phosphorescence.

Becquerel began investigations with materials which were phosphorescent if exposed to sunlight. Because of its scientific tradition the Museum possessed many samples of such material. This was a decisive advantage which he had over his competitors. His technique was clear and simple. He wrapped a photographic plate in thick black paper, protecting it from light, placed on it the phosphorescent substance, exposed the arrangement for some time to the direct sunlight, and finally developed the plate. Since X rays were known to traverse the paper, the plate would be blackened if X rays were produced. At first the results were negative. But when he used a sample of potassium uranyl disulphate $K_2UO_2(SO_4)_2 2H_2O$, he observed an effect which he reported to the Academy on 24 February [2].

A week later, on 2 March 1896, Becquerel reported again to the Academy [3]. It is this report that contained the discovery which would make him famous. We first quote from a paragraph dealing with repetitions of his previous experiment:

> A photographic plate, gelatine with silver bromide, was enclosed in an opaque frame in black cloth, closed on one side by a sheet of aluminium; if one exposed the frame in full sunshine, even for a whole day, the plate was not fogged; but, if one had fixed on the aluminium sheet, on the outside, a lamella of uranium salt [...] and if one exposed it for several hours to the sun, one recognised, after one had developed the plate in the usual way, that the silhouette of the crystalline lamella appeared in black on the sensitive plate.

The unexpected discovery is described two paragraphs further down:

> I shall insist in particular on the following fact which seems to me very important and outside the phenomena one could expect to observe: The same crystalline lamellas, placed with respect to photographic plates in the same way [...] but kept in the dark, produce again the same photographic prints. Here is how I have been led to make that observation: Of the previous experiments some had been prepared on Wednesday 26th and on Thursday 27th of February and, since on these days the sun had shown itself only intermittently, I had kept the experiments prepared and placed the frames back in the dark into a drawer, leaving the lamellas of uranium salt in place. Since the sun had also not shown itself the following days, I developed the plates, expecting to find very feeble images. The silhouettes appeared, on the contrary, with great intensity.

Becquerel then went on to describe how he repeated experiments in complete darkness with the same result. Thus besides X rays (or Röntgen rays) there was yet another kind of new rays, which were emitted seemingly without cause

Photographic plate blackened by radioactivity with notes in Becquerel's hand.

by a certain phosphorescent substance although that substance had not been stimulated to phosphoresce.

During the course of 1896 Becquerel continued to investigate these rays. Still in March he found that they discharge an electroscope, i.e., they give some electric conductivity to air [4]. He also performed observations, which later turned out to be erroneous, namely, that the rays could be reflected, refracted, and could possess polarization. These erroneous findings seemed to attribute to Becquerel's rays most of the properties of ordinary light; they seemed to be better understood than Röntgen rays. It can be assumed [5] that for this reason there was little interest in them for the next two years. In May 1896, Becquerel found out that all uranium compounds, phosphorescent or not, which he had studied, emitted rays [6]. He concluded that pure metallic uranium should show the strongest radiation and confirmed that hypothesis by experiment. Becquerel now called his rays *uranic rays*. At the end of 1896, he reported on the absorbing power of different materials with respect to his rays. Moreover, he stressed the fact that the source of energy, which makes uranium emit rays, was completely unknown.

We can summarize the events of 1896 in Paris by stating that Becquerel discovered *radioactivity*. He found it to be a property of the element uranium. It could be detected photographically and electrically. An unknown source of energy had to exist to keep the radiation going. Although Becquerel discovered radioactivity he did not coin the name. That was done by Marie Curie whom we shall meet again in Episode 5.

[1] Curie, M. S., *Recherches sur les substances radioactives*. Gauthier–Villars, Paris, 1903.

[2] Becquerel, H., *Comptes Rendus Acad. Sci.*, **122** (1896) 420.

[3] Becquerel, H., *Comptes Rendus Acad. Sci.*, **122** (1896) 501.

[4] Becquerel, H., *Comptes Rendus Acad. Sci.*, **122** (1896) 559.

[5] Wróblewski, A. K., *Acta Physica Polonica*, **30** (1999) 1179.

[6] Becquerel, H., *Comptes Rendus Acad. Sci.*, **122** (1896) 1086.

Zeeman and Lorentz – A First Glimpse at the Electron (1896)

In 1896 a young Dutch physicist, Pieter Zeeman[1], submitted a paper to the Royal Academy of Sciences in Amsterdam [1]. In the introductory paragraph he writes:

> Several years ago ... it occurred to me whether the light of a flame if submitted to the action of magnetism would perhaps undergo any change. ... With an extemporized apparatus the spectrum of a flame, coloured with sodium, placed between the poles of a Rühmkorff electromagnet, was looked at. The result was negative. Probably I should not have tried this experiment again so soon had not my attention been drawn two years ago to the following quotation from Maxwell's sketch of Faraday's life ...:—"... in 1862 he made the relation between magnetism and light the subject of his very last experimental work. He endeavoured, but in vain, to detect any change in the lines of the spectrum of a flame when the flame was acted on by a powerful magnet."

Zeeman in 1925

Zeeman, in the laboratory of Kamerlingh Onnes at the University of Leiden, had at his disposal a powerful new Rowland[2] grating for the study of spectra. Placing a piece of asbestos, impregnated with kitchen salt and heated by the flame of a Bunsen burner, between the pole pieces of an electromagnet, Zeeman clearly observed the two yellow D lines of sodium. When the current in the magnet was turned on, the two lines 'were distinctly widened'. Once the current was turned off they resumed their original narrow widths. Zeeman varied his experiment in many ways to ascertain that his effect was real.

When he was absolutely sure, Zeeman was looking for a theoretical interpretation of his results. Let us again quote from his paper:

> A real explanation ... seemed to me to follow from Prof. Lorentz's theory.
>
> In this theory it is assumed that in all bodies small electrically charged particles with definite mass are present, that all electric phenomena are dependent upon the configuration and motion of these "ions," and that light-vibrations are vibrations of these ions. The charge, configuration, and motion of the ions completely determine the state of the ether. The said ion, moving in a magnetic field, experiences mechanical forces ... and these must explain the variation of the period [of vibration]. Prof. Lorentz, to whom I communicated these considerations, at once kindly informed me of the manner in which, according to his theory, the motion of an ion in a magnetic field is to be calculated, and pointed out to me that if the explanation following from his theory be true, the edges of the lines of the spectrum ought to be circularly polarized. The amount of widening

---

[1] Pieter Zeeman (1865–1943), Nobel Prize 1902   [2] Henry Augustus Rowland (1848–1901)

13

Lorentz in 1927

Zeeman triplet

Zeeman doublet

no field

Splitting of a spectral line by normal Zeeman effect.

might then be used to determine the ratio between charge and mass, to be attributed in this theory to a particle giving out the vibrations of light.

The above mentioned extremely remarkable conclusion of Prof. Lorentz relating the state of polarization in the magnetically widened lines, I have found fully confirmed by experiment.

We read here, in a single publication, of the ideal case of scientific progress: Experiment stimulates theoretical considerations. These yield predictions about the results of new experiments. Those, once performed, confirm the predictions and thus make the theory plausible.

Lorentz[3] [2], professor at the University of Leiden, had done important work in the theory of electrodynamics. As Zeeman writes in his paper, he had been able to describe the electric and magnetic properties of matter by the assumption that it consists of positively and negatively charged ions. He deduced from Oersted's law about the force, which a magnetic field exerts on an electric current in a wire, the force with which it acts on a single moving charge or ion. This *Lorentz force* became an essential part of electrodynamics besides the *Coulomb force* which describes the action of an electric field on a charge. Moreover, Lorentz made Maxwell's equations compact and easier to handle by introducing the now conventional vector notation for all fields and summarized his theory in a book [3]. Because of these advances Einstein used to refer to the *Maxwell–Lorentz theory*.

The argument of Lorentz with respect to the *Zeeman effect* runs as follows. An ion oscillating about a fixed point, according to Hertz, emits light with the frequency of the oscillation. If one constructs an xyz coordinate system in space, then the oscillation can be considered a superposition of a linear oscillation along the z direction and two circular motions (one clockwise, the other anticlockwise) in the xy plane. All three motions have the same period, hence the same frequency, and light with a single frequency is emitted. If, however, a magnetic field, pointing in the z direction, exists, the Lorentz force has to be taken into account. It vanishes for the motion in z direction, leaving the period unchanged. Its influence on the two circular motions is such that the period of one is increased, that of the other decreased. Therefore light with three different frequencies is emitted. A spectral line is split into three lines. The shift in frequency between the lines depends on the magnetic field and on the charge-to-mass ratio of the ions. If observed along the z direction, only the outer two lines (forming a *Zeeman doublet*) are visible, because an oscillating charge does not radiate along its direction of oscillation. The light of the two lines is polarized circularly in opposite directions. If observed perpendicularly to the z direction, all three lines (forming a *Zeeman triplet*) can be observed. They are linearly polarized, the outer ones in z direction, the central one perpendicular to it.

Zeeman, who, in his first paper, only observed a broadening of spectral lines and not a splitting into doublets or triplets, took the broadening as a measure for the splitting. He found a value for the charge-to-mass ratio of the ions, which was about a factor of thousand higher than for the ions known from electrolysis. In further studies, published early in 1897 [4,5], he observed the

---

[3] Hendrik Antoon Lorentz (1853–1928), Nobel Prize 1902

expected doublets and triplets for a spectral line of cadmium and all the predicted properties concerning their polarization. Cadmium shows the so-called *normal Zeeman effect*, i.e., the one described above. In the paper [4] Zeeman also commented on his value for the charge-to-mass ratio e/m. He wrote: 'It is very probable that these 'ions' differ from the electrolytical. It is true that by means of the latter many phenomena can be interpreted ... but the high value of e/m which I have found makes it extremely improbable that we have to deal with the same mass in the two cases.' Here we have the first appearance of a charged particle much lighter than an atom. Thus, we have a first glimpse at the *electron* (see Episode 4).

In 1897 Zeeman became lecturer and in 1900 professor at the University of Amsterdam where he founded a large laboratory which now bears his name. In 1902 Lorentz and Zeeman shared the second Nobel Prize in physics.

It is typical of Lorentz, who was a very modest man, that he did not write himself on his theory of the Zeeman effect but allowed Zeeman to publish it in his first paper. Lorentz was a native of Arnhem, a town in the South of the Netherlands. He entered the University of Leiden in 1870 and got his B.Sc. in mathematics and physics already in the following year. He returned to Arnhem where he worked as teacher and at the same time prepared his dissertation. He obtained his doctor's degree in 1875 in Leiden where he went only for the ceremony and continued his teaching and private research work in Arnhem. Three years later, at the age of 26, he was appointed Professor of Theoretical Physics in Leiden with a chair especially erected for him. Through his work on electromagnetic theory, Lorentz gradually became an internationally known figure but not until 1900 did he get into personal contact with scientists from abroad. He received offers to chairs from foreign universities but refused all. From their beginning in 1911 to his death he chaired all *Solvay Conferences* (see Episode 12). Prominent participants praised him for his skilful and tactful way in leading and summarizing the discussions equally well in English, French, and German. In his later years, Lorentz chaired a National Committee in charge of the calculation necessary to prepare the closing of the *Zuiderzee*, a large bay in the North of the Netherlands by a dyke. The calculations, in which Lorentz took an active part, had to take into account the changes in sea currents and the resulting changes in the tide effected by that dyke. They took many years and were entirely successful. The dyke now stands for far more than half a century. We have mentioned already that Einstein was a great admirer of Lorentz. He was often in Leiden, meeting with Lorentz and with his successor to the Chair of Theoretical Physics, Paul Ehrenfest.

The story of the Zeeman effect does not end with Lorentz's theory. In fact sodium, with which the first indication of the effect was observed, behaves in a more difficult way. The two sodium D lines together split up into ten lines. Sodium displays the so-called *anomalous Zeeman effect*. That, for nearly thirty years, remained a puzzle which was solved only when the concept of electron *spin* was introduced by two young physicists who then also worked at the University of Leiden (see Episode 36).

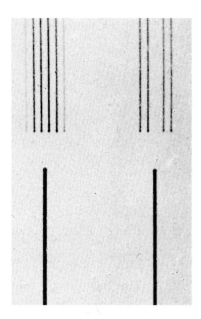

Anomalous Zeeman effect of sodium D lines. Top: with field, all polarizations. Bottom: without field. The wavelength increases towards the right. From [6], Plate I, Fig. 2, © 1925 by Julius Springer in Berlin. With kind permission from Springer Science+Business Media.

[1] Zeeman, P., *Vers. Kon. Akad. Wetensch. Amsterdam*, **5** (1896) 181, 242. Engl. transl. in *Philosophical Magazine* 43 (1897) 226.

[2] de Haas-Lorentz, G. (ed.), *H. A. Lorentz – Impressions of his Life and Work*. North-Holland, Amsterdam, 1957.

[3] Lorentz, H. A., *Versuch einer Theorie der electrischen und optischen Erscheinungen in bewegten Körpern*. Brill, Leiden, 1895.

[4] Zeeman, P., *Vers. Kon. Akad. Wetensch. Amsterdam*, **6** (1897) 13. Engl. transl. in *Philosophical Magazine* 44 (1897) 55.

[5] Zeeman, P., *Vers. Kon. Akad. Wetensch. Amsterdam*, **6** (1897) 99. Engl. transl. in *Philosophical Magazine* 44 (1897) 255.

[6] Back, E. and Landé, A., *Zeemaneffekt und Multiplettstruktur der Spektrallinien*. Springer, Berlin, 1925.

4 The Discovery of the Electron (1897)

In 1897 cathode rays were known for nearly three decades. They were produced in a gas discharge and gave rise to X rays when they fell on matter, e.g., the glass wall of a discharge tube. But their nature was unknown. There was contradictory experimental evidence. Some researchers took them for charged particles, in fact for ions which were familiar from electrolysis. Perrin had determined the sign of their charge to be negative. Others, among them Heinrich Hertz, who was unable to influence the rays by an electric field and to detect a magnetic field which should be caused by the rays if they were moving charges, took them to be electromagnetic waves.

A Charged Particle in an Electric or Magnetic Field

We use the following symbols: q charge, m mass, $\mathbf{v}$ velocity, $\mathbf{p} = m\mathbf{v}$ momentum of a particle, $\mathbf{E}$ electric field, $\mathbf{B}$ magnetic field (more exactly: field of magnetic induction), $\mathbf{F}_e = q\mathbf{E}$ electric force, and $\mathbf{F}_m = q(\mathbf{v} \times \mathbf{B})$ magnetic (Lorentz) force acting on the charged particle.

Trajectories in homogeneous fields: The trajectory of a charged particle in a homogeneous electric field is a parabola like that of a stone in a homogeneous gravitational field. The trajectory of a charged particle in a homogeneous magnetic field with a velocity perpendicular to the field is a circle of radius $R = (q/m)(v/B)$.

Energy gained in an electric field: If a particle is originally at rest and is accelerated in a constant electric field over the distance ℓ, it acquires the kinetic energy $E_{\text{kin}} = mv^2/2 = qE\ell = qU$, where $U = E\ell$ is the electric potential over the distance ℓ. The velocity gained is $v = \sqrt{2(q/m)U}$.

Electric deflection: If a particle of momentum $\mathbf{p}$ traverses a region of length ℓ with an electric field $\mathbf{E}$ in the direction perpendicular to $\mathbf{p}$, it suffers a change $\Delta\mathbf{p} = \mathbf{F}_e \Delta t$ with $\Delta t = \ell/v$. For $\Delta p \ll p$ the angle of deflection is $\Theta_e = \Delta p/p = |\mathbf{F}_e| \Delta t/p = (q/m)(E/v^2)\ell$.

Magnetic deflection: If, instead, the region traversed contains a magnetic field, perpendicular to $\mathbf{p}$, the angle of deflection is $\Theta_m = \Delta p/p = |\mathbf{F}_m| \Delta t/p = (q/m)(B/v)\ell$.

In 1897 three physicists, Wiechert[1], Kaufmann[2], and Thomson[3] [1–3], in independent experiments, found that cathode rays are indeed negatively charged particles having the peculiar property that the ratio of the mass and the charge of these particles is on the order of 1000 times smaller than for the lightest ion, the ion of hydrogen. The term *electron*, which had previously been used to denote the *elementary charge*, was soon adopted as the name of the new particle. The electron was the first of the later so-called *elementary particles* to be dis-

---

[1] Emil Wiechert (1861–1928) [2] Walter Kaufmann (1871–1947) [3] (Sir) Joseph John Thomson (1856–1940), Nobel Prize 1906

covered. Unlike others, e.g., the proton or the neutron which were discovered later, it is still considered to be elementary in the sense that it is not composed of other more fundamental particles. Electrons are responsible for the transport of electricity in solids. Scientists and engineers have learnt to influence and steer this transport in the most subtle ways thus creating the technical field of *electronics* and those wonderful machines, the *electronic computers*. The discovery of the electron therefore was not only important for fundamental physics but also laid the basis of an enormous industrial and social revolution.

Before looking at the three experiments, it is helpful to state the most basic facts about gas discharges in modern terms. A gas consists of electrically neutral atoms or molecules and therefore normally does not allow the flow of an electric current which needs movable charge carriers. The atoms contain electrons. Through an *ionization* process an electron can be removed from an atom. Then, instead of a neutral atom, one has two charge carriers, a negatively charged electron and a positively charged *ion*. In the presence of an electric field between two electrodes, the electrons move towards the anode and the ions towards the cathode. The discharge remains stable only if the carriers lost at the electrodes (and possibly also in the gas and at the walls) are continuously replaced. This can be done in several ways, e.g., by electrons, which after acquiring energy from the electric field, are able to ionize gas atoms or molecules. Also, the ions, hitting the cathode with considerable energy, can liberate electrons from the cathode material. In the experiments discussed here most of the electrons were produced in this way. In these experiments, performed at rather low gas pressure and rather high voltage, a stable gas discharge takes the form of a *glow discharge*. At atmospheric or higher pressures and low voltage it is an *electric arc*. An arc of very short duration, caused, for instance, by the discharge of a capacitor, is called a *spark*. Wiechert, Kaufmann, and Thomson, as Röntgen and many others before them, were working with a glow discharge. In such a discharge, the electric field between the electrodes is not at all uniform. It is high in a small region near the cathode, the *cathode fall*, and comparatively low everywhere else.

Let us now return to the experiments by Wiechert, Kaufmann, and Thomson. All three used a glow discharge at low pressure to produce cathode rays. A fraction of these traversed an opening in the anode (or, in the case of Kaufmann, passed by the anode which was a thin wire). After some distance they were detected through the fluorescence they caused on a piece of glass or on the glass wall of the apparatus. The influences of an electric and a magnetic field on the rays were studied. These influences, through two separate equations, are related to the velocity and the charge-to-mass ratio of the particles. These two quantities could thus be determined from the observations. It must be pointed out here that this analysis was possible only because in 1895 Lorentz, by introducing the *Lorentz force*, had been able to explain how charged particles move in a magnetic field. In detail, of course, the experiments differed from one another, as did the conclusions drawn by the authors.

On 7 January 1897, Emil Wiechert presented a talk [4] including a demonstration of his experiment to the Physico-Economic Society in Königsberg, East Prussia (now Kaliningrad, Russia). He measured the deflection of the cathode rays by a magnetic field. As to the influence of an electric field, he

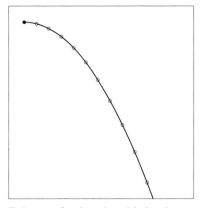

Trajectory of a charged particle in a homogeneous electric field pointing downwards in the plane of the figure. The full little circle indicates the initial position. Later positions, equidistant in time, are marked by open circles.

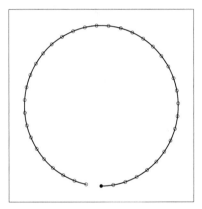

Trajectory of a charged particle in a homogeneous magnetic field pointing perpendicular the plane of the figure.

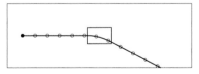

Deviation of a charged particle by an electric field in a small region.

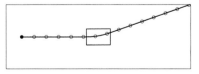

Deviation of a charged particle by a magnetic field in a small region.

Wiechert

Kaufmann

argued that the kinetic energy of the rays is at most the energy which a charged particle gains if it traverses the complete distance between cathode and anode but at least a certain fraction of it. Assuming further, as he wrote, 'the charge [to be] 1 electron' (i.e. one elementary charge), he concluded from his experiment that the mass of the particles, which constitute the cathode rays, was less than $1/400$ of the mass of the hydrogen atom. He wrote: 'This upper bound for [the mass] shows unambiguously that cathode rays cannot be the ordinary chemical atoms.'

In May 1897, Walter Kaufmann, working at the Physics Institute of Berlin University, submitted his paper [5] for publication. He computed the velocity of the rays by assuming that they had traversed the complete distance between cathode and anode. Like Wiechert he measured the magnetic deflection. For the charge-to-mass ratio he found a value which was a factor 1000 larger than that of hydrogen ions. He pointed out that this ratio did not depend on the gas filling of his apparatus as might be expected if ions were formed from the gas atoms or molecules. Kaufmann worded his conclusions very carefully. Unlike Wiechert he made no assumption as to the charge of the cathode rays and therefore gave no definite mass value.

In August 1897, Joseph John Thomson, Cavendish Professor of Experimental Physics at Cambridge University, submitted his paper [6]. It is the most complete of the three. He did not rely on the electric field between cathode and anode for the velocity determination. Instead, he used a second electric field perpendicular to the particle motion, which the particles had to traverse once they had passed the anode, and measured the deflection caused by that field. At first, like Hertz before him, he could not detect an effect of that field on the cathode rays. But when he decreased the gas pressure and with it the number of ions in the gas, which made the gas conduct electricity, he measured reliably the deflection caused by that field. Measuring also a magnetic deflection, he was able to determine the velocity and charge-to-mass ratio from these two deflections. Like Kaufmann he found it to be about 1000 times as large as that of hydrogen ions and independent of the gas used.

But, unlike Kaufmann, he drew a number of conclusions. He wrote: 'The smallness of m/e [the mass-to-charge ratio] may be due to the smallness of m or the largeness of e, or to a combination of these two. That the carriers of charges in cathode rays are small compared to ordinary molecules is shown, I think, by Lenard's results ...' He thus inferred from the small mass of the new particle that they also are small in size, in particular, smaller than atoms. Lenard had found that cathode rays can traverse quite an amount of material, the *Lenard window*, and thus must be able to traverse many atoms without being stopped. Thomson explained that by their smallness. He called the new particles *corpuscles* (the name *electron* came somewhat later) and declared them to be the *primordial atoms*, i.e., the smallest constituents of ordinary chemical atoms. He went on to speculate that the charge of his corpuscles was larger than the elementary charge and how atoms were built up of them and how molecules were formed of such atoms. These latter speculations did not stand the test of time. While all atoms contain electrons, the bulk of their mass is contained in the *atomic nucleus*, see Episode 16.

If a piece of metal, placed on an insulator, is irradiated by ultraviolet light,

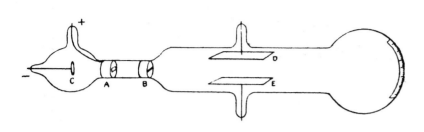

Thomson's tube. A gas discharge is maintained between the cathode C and the anode A. Electrons passing the slit in the anode are collimated by the diaphragm B and deflected by an electric field between the two plates D and E and a magnetic field perpendicular to it. The electron beam produces a luminescent spot on the far end of the glass vessel where its position is recorded on a scale. From [6]. Reprinted by permission from Taylor & Francis Ltd, `http://www.inform` `aworld.com`.

then it acquires a positive electric charge. In other words, it loses negative electric charge, for instance, in the form of negatively charged particles. This is called the *photoelectric effect*. In 1899 Thomson showed [7] that these particles have the same charge-to-mass ratio as the cathode rays. They also are electrons. The photoelectric effect can also liberate an electron from a single atom and turn that into an ion. In the same paper [7], Thomson used this to measure the charge of the ions and thus the charge of the electrons with the help of the *cloud chamber*, which had just been invented by his student Wilson. This charge, within experimental errors, turned out to be the same as the elementary charge known from electrolysis. If we call that (positive) charge e, then electrons have the charge $-e$. From the known charge-to-mass ratio of the electron its mass could now be found. As already said in the beginning of this episode, it is about 2000 times smaller than that of the lightest atom, the hydrogen atom.

J. J. Thomson in his laboratory.

[1] Thomson, J. J., *Recollections and Reflections*. Bell, London, 1936.

[2] Thomson, G. P., *J. J. Thomson and the Cavendish Laboratory in his Day*. Nelson, London, 1964.

[3] Lord Rayleigh, R. J., *The Life of Sir J. J. Thomson*. Cambridge University Press, Cambridge, 1942.

[4] Wiechert, E., *Schriften der Physikalisch–Ökonomischen Gesellschaft zu Königsberg*, **38** (1897) 3.

[5] Kaufmann, W., *Annalen der Physik und Chemie*, **61** (1897) 544.

[6] Thomson, J. J., *Philosophical Magazine*, **44** (1897) 293.

[7] Thomson, J. J., *Philosophical Magazine*, **48** (1899) 547.

5 Marie and Pierre Curie – Polonium and Radium (1898)

In 1867, Maria Salomee Skłodowska was born in Warsaw which then was part of the Russian Empire. Under the name Marie Curie[1] [1–3] she was to become the most famous woman scientist of all time and the only person to receive the Nobel Prize both in physics and in chemistry. After her school years she worked for some time as governess before, in 1891, she began to study physics in Paris where her sister lived by that time. After passing the exam in physics as best of her class in 1893, she went on to study mathematics. Besides, she took on experimental work on magnetic properties of steels. Looking for a laboratory in which to perform that work, she was introduced, early in 1894, to Pierre Curie[2] [4], who worked at the *École municipale de Physique et de Chimie industrielles*. He had already done important research. Together with his brother Paul-Jacques Curie he had discovered *piezoelectricity*, which later became important in many technical applications. That same year Pierre Curie found relations on the temperature dependence of magnetism in matter, which still bear his name. Marie and Pierre married in 1895. Their first daughter, Irène, whom we shall meet again in Episode 54, was born in September 1897. Shortly thereafter Marie Curie, having heard of Becquerel's discoveries, began to work on radioactivity.

She began by introducing a quantitative method into that young field of research. To do that she made use of the property of radioactivity to give the air some electric conductivity. Two circular metal plates, 8 cm in diameter, were oriented horizontally, one 3 cm above the other. A thin layer of a pulverized substance was spread evenly over the lower plate. The system of two plates was charged by connecting the plates for a short time to the poles of a battery. Then the capacitor, formed by the two plates, was allowed to discharge. The discharge current was a measure for the conductivity of the air and thus for the radioactivity of the substance. To accurately measure this tiny current, Marie Curie used a clever technique based on the piezoelectric properties of crystals, discovered by her husband. For the discharge of a capacitor not only the current but also the change of voltage with time is related in a well-determined way to the conductivity. Marie Curie measured the voltage by compensating it with that produced by a piezoelectric crystal. The latter generates a voltage if a weight is hung on it. She measured a time and a weight with high accuracy and from these she computed the change of voltage and finally the current. In this way she was able to measure tiny currents of a few pico amperes (10^{-12} A).

With this simple but ingenious apparatus she obtained a measure for the radioactivity of, as she wrote herself, a great number of metals, salts, oxides,

---

[1] Marie Skłodowska Curie (1867–1934), Nobel Prize 1903 and 1911   [2] Pierre Curie (1859–1906), Nobel Prize 1903

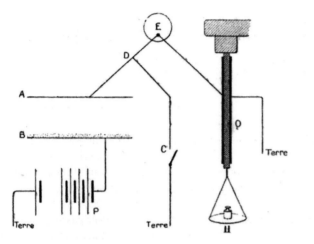

The set-up Marie Curie used to determine the radioactivity of a sample (sketch from her thesis [5]). The sample is spread evenly on the plate B which forms a capacitor with plate A. The capacitor is charged to the potential P provided by a battery by closing the switch C and allowed to discharge by opening that switch. The potential of plate A is measured with high accuracy by producing an equal potential of opposite sign with the weight Π on the faces of the quartz Q so that there is no voltage across the electrometer E.

and minerals. In April 1898, in her first publication of only three pages [6] she presented these results and from them drew important conclusions:

1. *All compounds of uranium are radioactive. They are the more radioactive, the more uranium they contain.* Shortly afterwards, Marie Curie used this finding to declare radioactivity an *atomic phenomenon*: The more atoms of a given radioactive element in a substance, the more radioactive it is.

2. *Also, the compounds of thorium are radioactive.* She thus discovered radioactivity of an element other than uranium. The same discovery had shortly before been made by Schmidt[3]. But she did not know of his result.

3. *There are two particular minerals containing uranium, which, in contrast to point 1, show a much higher activity than can be explained by their uranium content. These are pitchblende (a uranium oxide) and chalcolite (uranyl copper phosphate).* Marie Curie explained this finding by the assumption that these minerals contained an element much more radioactive than uranium. As a first test of this explanation, she produced chalcolite synthetically from pure substances in the laboratory and found it no more active than other uranium salts.

Marie Curie had declared a physical property of some atoms, radioactivity, to be a signature which could be used to find new chemical elements. Such a step had been done once before by Kirchhoff and Bunsen with their spectral analysis, when they used the emission of light of particular frequencies as a signature for new elements. Spectral analysis had its greatest success when the element helium was discovered by analysis of light from the sun. Building on this bold idea, Marie Curie, together with her husband, founded the field of *radiochemistry*. Beginning with pitchblende, they used the techniques of chemistry like dissolving, precipitating, filtering, and crystallization to enrich the new element in their samples, and the physical property of radioactivity to

---

[3] Gerhard Carl Schmidt (1865–1949)

detect it. Already in July 1898, they obtained a sample with a radioactivity 400 times as large as that of pure uranium. They proposed to name the new element *polonium* 'after the name of the land of origin of one of us' [7].

Pierre and Marie Curie at work.

In analytic chemistry there is a well-established procedure in which a sample of unknown composition is divided into sub-samples, each containing only a small group of elements or some of their compounds. Polonium was found in one of these groups. In quite another group, containing the elements barium, potassium, and strontium, the Curies also found radioactivity. Working together with the chemist Bémont[4], they were able to prepare samples with activities between 60 and 900 times the strength of pure uranium [8]. In the optical spectrum of these samples, Demarçay[5] found a so far unknown spectral line. Its intensity increased with the radioactivity of the sample. The Curies concluded that they had found yet another new element which they called *radium*. Their strongest sample not only blackened a photographic plate much faster than uranium but, like Röntgen's X rays, it even caused visible light to be emitted if brought near to barium platinocyanide. The Curies made the important statement that there was no source of energy for that light in apparent contradiction to the principle of Carnot, i.e., the principle of conservation of energy.

It was now that the truly heroic period in the life of Marie Curie began. The Curies did not have enough pitchblende to go on enriching the radioactivity of their samples, let alone to isolate the new elements polonium and radium in their pure form. That, however, was needed to study the chemical properties of the new elements and to determine their atomic mass numbers. After all,

---

[4] Gustave Bémont (1857–1932)   [5] Eugène Anatole Demarçay (1852–1904)

the very concept of an element was a chemical one. Pitchblende was mined in *Sankt Joachimsthal*, Bohemia, then a part of Austria. In fact, mining there had a long and glorious tradition. Originally, there were rich silver mines in Sankt Joachimsthal and the place gave its name to a popular silver coin, the *Thaler*. That name still finds itself in all currencies called *dollar*. Pitchblende was mined to obtain uranium for the Bohemian glass industry. Fortunately, the waste material after uranium extraction still was useful for the Curies and they were donated first a hundred kilogrammes and later several tons of that material. Working for several years under miserable conditions in a shed by the *École de Physique*, Marie Curie with her own hands performed the chemical processing of large quantities of material while her husband by physical measurements supervised the process of enrichment of radioactivity. She was able to produce a small amount of nearly pure radium salt and to study the chemical properties of radium, which turned out to be an alkaline earth element situated in the Periodic Table below barium and chemically very similar to it.

In the later stages of that work, Marie Curie got some help from a chemical factory in the form of a first preprocessing of pitchblende. She suggested to the young scientist Debierne[6], who worked in the factory, to look for yet another radioactive element. Indeed Debierne discovered *actinium* in 1899.

In the summer of 1903 Marie Curie obtained her Ph.D. with a thesis [5] summarizing her important work. At the end of that year, Henri Becquerel and Marie and Pierre Curie were awarded the Nobel Prize in physics. In 1904 Pierre Curie was finally made professor at the *Sorbonne*, the University of Paris. Rutherford remembered a visit to Paris in 1903 in the following words [9]: 'During the summer, I visited Professor and Madame Curie in Paris and found the latter was taking her degree of D.Sc. on the day of my arrival. In the evening, my old friend, Professor Langevin[7], invited my wife and myself and the Curies and Perrin to dinner. After a very lively evening, we retired about 11 o'clock in the garden, where Professor Curie brought out a tube coated in part with zinc sulphide and containing a large quantity of radium in solution. The luminosity was brilliant in the darkness and it was a splendid finale to an unforgettable day. At that time we could not help observing that the hands of Professor Curie were in a very inflamed and painful state due to the exposure to radium rays. This was the first and last occasion I saw Curie. His premature death in a street accident in 1906 was a great loss to science and particularly to the rapidly developing science of radioactivity.'

Marie Curie succeeded to her husband's chair and thus became the first woman professor at the Sorbonne. In 1911 she was awarded the Nobel Prize in chemistry. She continued her work in radioactivity and, in particular, radiochemistry until shortly before her death in 1934.

In 1995 the remains of Pierre and Marie Curie were transferred to the *Panthéon* in Paris, where France lays to rest her most honoured citizens. The solemn ceremony was attended by the presidents of France and Poland and also by Ève Curie, the younger of the two daughters of the couple, author of a most charming biography of her mother [1].

[1] Curie, È., *Madame Curie*. Gallimard, Paris, 1938.
[2] Reid, R., *Marie Curie*. Paladin, London, 1978.
[3] Ksoll, P. and Vögtle, F., *Marie Curie*. Rowohlt, Reinbek, 1988.
[4] Curie, M., *Pierre Curie*. Payot, Paris, 1924.
[5] Curie, M. S., *Recherches sur les substances radioactives*. Gauthier–Villars, Paris, 1903.
[6] Curie, Mme. S., *Comptes Rendus Acad. Sci.*, **126** (1898) 1101.
[7] Curie, P. and Curie, Mme. S., *Comptes Rendus Acad. Sci.*, **127** (1898) 175.
[8] Curie, P., Curie, Mme. P., and Bémont, G., *Comptes Rendus Acad. Sci.*, **127** (1898) 1215.
[9] Eve, A. S., *Rutherford – Being the Life and Letters of the Rt Hon. Lord Rutherford, O.M.* Cambridge University Press, Cambridge, 1939.

---

[6] André-Louis Debierne (1874–1949) [7] Paul Langevin (1872–1946)

<table>
<tr><td>**6**</td><td># Alpha, Beta, and Gamma Rays (1899)</td></tr>
</table>

Rutherford

In 1899 a young New Zealander, Ernest Rutherford[1] [1,2], appeared on the scene of radioactivity research, which he was to dominate for the next decades. He had studied at Canterbury College, Christchurch, which was part of the University of New Zealand and performed important research work on the detection of electromagnetic waves. In 1894, he won the prestigious 1851 Exhibition Science Scholarship and went to Cambridge to work under J. J. Thomson. Together with Thomson he began to study the effect of X rays on gases. They described in detail how X rays in a gas produce pairs of ions with opposite charge and thus make the gas a conductor. They found quite different behaviours for the positive and negative ions. After the electron had been detected, Thomson quickly made it clear that the negative ions were, in general, electrons. This pioneering work on the *ionization* of gases helped Rutherford to become an expert in the detection of radioactivity in which ionization plays a key role.

In 1898, at the age of 27 years, Rutherford became professor at McGill University, Montreal. His first work on radioactivity [3], written still at Cambridge, was published in 1899. The paper carries the title *Uranium Radiation and the Electrical Conduction Produced by It*. But it is by no means a mere extension of his work with X rays to radioactivity. Rutherford's experimental set-up was very similar to that used by Marie Curie. A horizontal metal plate was covered by a thin layer of powdered uranium or uranium compound. Another plate, parallel to the first one, was placed 4 cm above it. The two plates were connected by an external electric circuit containing an electrometer and a battery. The current observed in the electrometer was a measure for the number of ions between the two plates and thus for the capability of the radiation, emitted by the uranium, to ionize the air between the plates. Rutherford covered

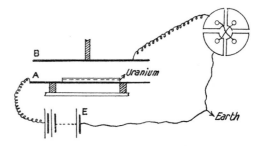

The apparatus with which Rutherford distinguished α and β rays. From [3].

the radiating powder with layers of thin aluminium foil, each having a thickness of $5\,\mu m$, and found that the current fell steeply with each added layer

---

[1] Ernest (Lord) Rutherford (of Nelson) (1871–1937), Nobel Prize 1908

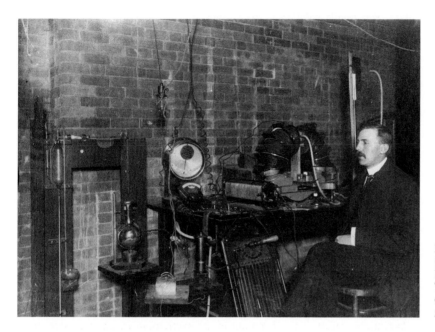

Rutherford 1906 in Montreal. Otto Hahn tells [5] that for this photograph he was asked to lend his detachable white cuffs to Rutherford. The photographer, who worked for the science journal *Nature*, wanted to add some elegance to the rather drab cellar laboratory.

for the first few layers and only very slowly for the following many layers. He concluded 'that the uranium radiation is complex, and there are present at least two distinct types of radiation – one that is very readily absorbed, which will be termed for convenience α radiation, and the other of a more penetrative character, which will be termed the β radiation.' In 1900, Paul Villard[2] in Paris found a third and even more penetrating type of rays in the radiation of radioactive materials [4] which Rutherford named γ radiation.

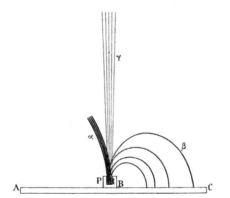

The paths taken by α, β, and γ rays in a magnetic field perpendicular to the plane of the figure. From [6], the doctoral dissertation of Marie Curie.

It was soon found out that β rays could be deflected by a magnetic field in the same ways as cathode rays. More detailed experiments showed that, indeed, β rays were electrons [7]. The properties of γ rays were found to be very similar to Röntgen rays. That, just like the latter, they are electromagnetic

---

[2] Paul Ulrich Villard (1860–1934)

waves, was finally established by Rutherford and Andrade[3] [8] in 1914 after methods to measure their very short wavelengths had become available.

Therefore, although that was not known at the time, of the three types of radiation, only the α rays were something completely new. Rutherford not only determined their properties in a series of beautiful experiments but also used them to make new discoveries, among them the discovery of the atomic nucleus. Otto Hahn, one of the first young scientists coming from abroad to work with Rutherford at McGill, recalled later [5]: 'At the time Rutherford was chiefly occupied with α particles, which indeed remained his first love throughout his life.' In this episode, we shall sketch Rutherford's experiments which established the properties of α particles, although the last one was done 10 years after α radiation was discovered. There were three major steps: (i) α rays can be deflected in a magnetic field, they have positive charge, (ii) the charge of an α particle is $2e$, where e is the elementary charge, (iii) α particles are doubly charged ions of helium.

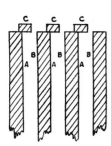

The system of plates and slits used by Rutherford to detect the magnetic deflection of α rays. From [9].

Magnetic Deflection of α Particles

For some time α rays were thought to be neutral because, contrary to β rays, no deflection by a magnetic field could be observed. Rutherford then devised a method [9] to detect very small deflections. The radiation was made to traverse between a system of parallel metal plates 37 mm long, placed 0.42 mm apart from each other, i.e., it had to run through long, narrow slits. If a magnetic field was applied perpendicular to the direction of the radiation and in the plane of the plates, then, if the radiation consisted of charged particles, even a small deflection would make the particle hit the plates and reduce the number of particles traversing the whole system of plates. Placing this system between a source of α radiation and a detector, Rutherford indeed found a rapid decrease of radiation in the detector when he increased the magnetic field. Having thus established that α particles are charged, Rutherford determined the sign of their charge. To that end he covered in part the exit side of each slit, making the exit sides asymmetric. If the deviation is towards the covered part, then the radiation in the detector will be smaller than if it is towards the uncovered part. In this way, Rutherford found that α particles carry positive electric charge. His results also allowed a coarse quantitative analysis, from which he concluded that the charge-to-mass ratio of α particles is typical of ions, not electrons. He concluded: 'The α rays from radium are thus very similar to the *Canal Strahlen* [canal rays] observed by Goldstein which have been shown by Wien to be positively charged bodies moving with high velocity. The velocity of the α rays is, however, considerably greater than that observed for *Canal Strahlen*.'

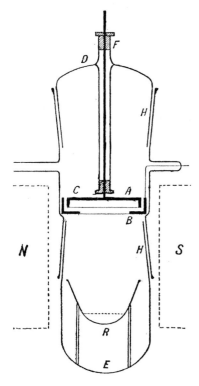

The apparatus with which Rutherford and Geiger determined the charge of the α particle. From [10], Fig. 1. Reprinted with permission.

The Charge of α Particles

In 1908 Rutherford, now professor in Manchester, together with his collaborator Hans Geiger published a paper entitled *An Electrical Method of Counting the Number of α-Particles from Radio-active Substances* [11]. In the introduction to their paper they write: 'If the number of α-particles expelled from a

---

[3] Edward Neville da Costa Andrade (1887–1971)

definite quantity of radio-active matter could be determined by a direct method, the charge carried by each particle could be at once known by measuring the total positive charge of the α-particles.' Because of its many applications we shall discuss this counting method in detail in Episode 13. Here, we turn directly to the next paper by Rutherford and Geiger [10], entitled *The Charge and Nature of the α Particle*. The apparatus was an exhaustible glass tube H with two stoppers. The lower stopper E contained a radium source R emitting a known number of α particles per second. A certain fraction of these, known through the geometry of the apparatus, could enter a capacitor formed by the plate AC and the plate B, mounted in the upper stopper D, and deposit their charge there. A voltage developed between the capacitor plates from which the total charge Q on the capacitor could be computed. From the number N of α particles deposited, the charge $q = Q/N$ of a single α particle was then easily found. In practice, of course, the experiment was not easy at all. The α particles could ionize the residual gas or other matter on their way and thus create additional charge. Care had to be taken to avoid these from reaching the capacitor. Particularly dangerous were electrons with a longer range. They were called δ rays by J. J. Thomson, although they were not part of radioactivity but of secondary nature. By a strong magnetic field (between the poles N and S) the trajectories of these electrons were made to curl up around their place of origin. Rutherford and Geiger stated their result like this: '... we may conclude with some certainty that the α particle carries a charge $2e$ ...'.

α Particles are Doubly Ionized Helium Atoms

A proof of this statement was published by Rutherford and Royds[4] in 1909 [12]. They introduced radium emanation, a radioactive gas (see Episode 8), into a glass capillary with a wall thickness of only 0.01 mm. The capillary was placed inside another tube. α particles emitted by the emanation could traverse the wall of the capillary, lose their energy, and turn into uncharged helium atoms in the outer tube. From time to time this tiny amount of helium was compressed into a very small volume and examined spectroscopically. After 2 days the most prominent yellow line of helium was seen and after 6 days all the stronger lines characteristic for that element. Thus, the authors wrote: 'We can conclude with certainty from these experiments that the α particle after losing its charge is a helium atom.' We still want to add a few lines on the discovery of the element helium. In 1860 Kirchhoff and Bunsen developed spectral analysis, a method to identify chemical elements by their characteristic patterns of spectral lines, which are narrow regions in the wavelength of the light emitted by elements when heated, see Episode 7. In 1868 Janssen[5] and also Lockyer[6] observed a yellow line, hitherto unknown, in the light of the sun. It was concluded that the line was caused by an element unknown on earth, which was called *helium*. In 1895 Ramsay[7] detected helium on earth as an inert gas given off by an uranium-bearing mineral when heated. We now understand that the proximity of helium, found in the earth's crust, to uranium is caused by the radioactivity of the latter.

[1] Eve, A. S., *Rutherford – Being the Life and Letters of the Rt Hon. Lord Rutherford, O.M.* Cambridge University Press, Cambridge, 1939.

[2] Rutherford, E., *The Collected Papers of Lord Ernest Rutherford of Nelson (Edited by J. Chadwick)*, 3 vols. Allen and Unwin, London, 1962–1964.

[3] Rutherford, E., *Philosophical Magazine*, **47** (1899) 109. Also in [2], Vol. 1, p. 169.

[4] Villard, P., *Comptes Rendus Acad. Sci.*, **130** (1900) 1010, 1178.

[5] Hahn, O. in [2], Vol. 1, p. 164.

[6] Curie, M. S., *Recherches sur les substances radioactives*. Gauthier–Villars, Paris, 1903.

[7] Kaufmann, W., *Physikalische Zeitschrift*, **4** (1902) 54.

[8] Rutherford, E. and da Andrade, E. N., *Philosophical Magazine*, **27** (1914) 854. Also in [2], Vol. 2, p. 432.

[9] Rutherford, E., *Philosophical Magazine*, **5** (1903) 177. Also in [2], Vol. 1, p. 549.

[10] Rutherford, E. and Geiger, H., *Proceedings of the Royal Society*, **A81** (1908) 162. Also in [2], Vol. 2, p. 109.

[11] Rutherford, E. and Geiger, H., *Proceedings of the Royal Society*, **A81** (1908) 141. Also in [2], Vol. 2, p. 89.

[12] Rutherford, E. and Royds, T., *Philosophical Magazine*, **17** (1909) 281. Also in [2], Vol. 2, p. 163.

---

[4] Thomas Royds (1884–1955)   [5] Pierre Jules César Janssen (1824–1907)   [6] (Sir) Norman Lockyer (1836–1920)   [7] (Sir) William Ramsay (1852–1916), Nobel Prize 1904

7 Max Planck and the Quantum of Action (1900)

Planck in the last decade of the nineteenth century.

It is often said that Planck's discovery, or rather his postulate of the quantum of action in December 1900, marks the beginning of quantum physics, the dominant subject of twentieth-century physics. Somewhat less known is that what Planck aimed at and achieved in 1900 was the answer to a problem pointed out by Kirchhoff forty years earlier.

We met Planck[1] [1,2] briefly in the Prologue when he decided to study physics in spite of advice to the contrary. His father, at that time, was professor of law in Munich. His grandfather and his great-grandfather had been professors of theology in Göttingen. Planck studied in Munich and, for one year, in Berlin under Helmholtz and Kirchhoff. He was fascinated with the elegance of the theory of thermodynamics and, in particular, with the second heat law as formulated by Clausius. Therefore, he chose work on that theorem as subject for his Ph.D. thesis submitted to the Philosophical Faculty in Munich in 1879. In 1880 Planck became Privatdozent in Munich and in 1885 professor in Kiel. In 1889 he succeeded Kirchhoff as professor of theoretical physics at the University of Berlin.

In 1860 Kirchhoff had been professor of physics in Heidelberg where Bunsen was professor of chemistry. Working together, they discovered *spectral analysis*. It was known that, if brought into a flame, some chemical substances coloured the flame characteristically. Kirchhoff and Bunsen studied the spectrum of the emitted light and found that it contained sharp *spectral lines*, i.e., light emitted only in very narrow regions of wavelength, characteristic for the elements in the substance. They predicted and demonstrated that new elements, so far missing in the Periodic Table, could be found with the help of spectral analysis. The spectra of the elements would later yield most important information about the structure of the atoms and would help to explain the Periodic Table.

Kirchhoff observed that light of the spectral lines, which is emitted by a substance, is also absorbed by that substance. For example, sodium vapour can both emit and absorb yellow light corresponding to a particular spectral line. This finding led him to study the emittance and absorption of radiation in the context of thermodynamics. He developed the concept of a *black body*, a hollow body with reflecting walls and containing some bodies of the same temperature. Later a black body was realized as a cavity with thick, heatable walls and a small opening. When heated, radiation would be emitted by the inside wall of the body, be absorbed by other parts of the wall, re-emitted, etc. The radiation would be in *thermal equilibrium* with the material of the walls and only a small part would leave the opening to be analysed. Kirchhoff

---

[1] Max Karl Ernst Ludwig Planck (1858–1947), Nobel Prize 1918

found, that the spectral intensity density of this *blackbody radiation*, i.e., the intensity per unit wavelength and unit solid angle, is a function depending only on the wavelength and the temperature but independent of the dimensions or the material of the body. In his publication [3] he stressed the importance of determining this universal function, finally found by Planck forty years later, after much experimental and theoretical work had been done by others.

Early Radiation Laws

We use the following symbols: λ wavelength and ν frequency of radiation, $c = \lambda\nu$ velocity of light, T absolute temperature, $u_\lambda(T)$ spectral energy density per unit wavelength and unit volume, $u_\nu(T) = cu_\lambda(T)/\nu^2$ same but per unit frequency, $u(T) = \int_0^\infty u_\lambda(T)\,\mathrm{d}\lambda = \int_0^\infty u_\nu(T)\,\mathrm{d}\nu$ integrated energy density; A, a, b constants; $f(\nu/T)$ some unknown function of the argument ν/T. We can then write:

$$u(T) = AT^4 \quad \text{Stefan–Boltzmann law} \quad ,$$

$$u_\nu(T) = \nu^3 f(\nu/T) \quad \text{Wien's displacement law} \quad ,$$

$$u_\nu(T) = \frac{8\pi b\nu^3}{c^3} \exp(-a\nu/T) \quad \text{Wien's radiation law} \quad .$$

Planck in 1906

The spectral intensity density of blackbody radiation is connected in a simple way to the spectral energy density (per unit wavelength and unit volume) of the radiation within the cavity of the black body. Thus the latter is obtained experimentally by measuring the former. At the time of Kirchhoff's publication it was not possible to measure the spectral intensity density. In fact, it took nearly 20 years until the integrated intensity, the intensity of the whole spectrum without knowledge of the wavelength dependence, was measured. That was finally achieved by Stefan[2] in Vienna in 1879 [4]. He found that the integrated energy density is proportional to the fourth power of the temperature of the black body. Five years later Boltzmann, introducing the concept of *radiation pressure*, showed that Stefan's empirical law could be obtained theoretically from the second law of thermodynamics [5]. This was the first firm step towards finding Kirchhoff's function.

The second step was done in 1893 by Wien[3] in Berlin. He was working in the *Physikalisch-Technische Reichsanstalt* (National Institute for Physics and Technology), which had recently been founded following a proposal and a donation by the industrialist Siemens[4]. The institute was meant to do scientific research of particular importance to the growing technical industry and was headed by Helmholtz. At that time, artificial illumination by electric arc and by gas lamps was already an important issue. These lamps, essentially, are highly heated pieces of material and emit a spectrum very similar to that of the idealized black body. This is why blackbody radiation was studied at the Reichsanstalt experimentally and theoretically. Wien performed the following thought experiment. He considered a cavity at temperature T in thermal equilibrium with the radiation in it. Then the temperature was increased to

Wien during his years at the Physikalisch-Technische Reichsanstalt in Berlin (1889–1896).

---

[2] Josef Stefan (1835–1893)   [3] Wilhelm Karl Werner Wien (1864–1928), Nobel Prize 1911   [4] Werner (von) Siemens (1816–1892)

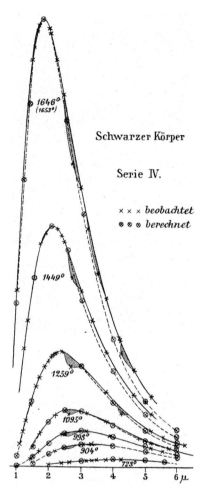

Schwarzer Körper

Serie IV.

× × × beobachtet
⊗ ⊗ ⊗ berechnet

The findings of Lummer and Pringsheim for the intensity of the blackbody radiation as a function of its wavelength. The measurements (× symbols connected by full lines) deviate from values computed according to Wien's law (⊗ symbols connected by broken lines). From [9].

$T + \Delta T$ and, of course, the spectral energy density of the radiation changed. Next, the cavity was expanded adiabatically (i.e., without heat exchange with its surroundings) until it reached the temperature T again. Since the temperature was the same, the spectral energy density also had to be the same as in the beginning. Since, during the expansion process the walls of the cavity moved outside, all wavelengths were increased in a known way by *Doppler effect*. Although Wien had not found Kirchhoff's function, he had unravelled how that function changed with the wavelength if the temperature was changed. This was *Wien's displacement law* [6].

In 1896 Wien derived an explicit formula for Kirchhoff's function, later called *Wien's radiation law* [7]. To arrive at that formula he had, however, to make assumptions, which were not easily justified. Nearly at the same time, Paschen[5] at the Technical University in Hanover for the first time measured the blackbody spectrum [8]. It coincided, within the measurement accuracy, with Wien's law.

Planck had become interested in the interaction of radiation with matter in 1895. In the cavity of a *black body* radiation is continuously absorbed and re-emitted by the walls. In this way thermal equilibrium is reached and Kirchhoff's spectral energy density of the radiation is established in the cavity. Planck called this process *radiation damping* and realized that it is similar to an irreversible process in thermodynamics in which the entropy increases. The concept of *entropy* had been introduced by Clausius. Like temperature and volume it is a quantity describing the state of a thermodynamic system.

Planck tried to extend the concept of entropy from a material system to radiation. As simplest system he considered a cavity with reflecting walls containing one *resonator*, i.e., a Hertzian dipole antenna with a fixed eigenfrequency, which would absorb and emit radiation. Later, he let the cavity contain a very large number of resonators, radiating at all possible wavelengths (or frequencies). In other words, the material walls of the black body were considered to consist of resonators. Planck found a formula relating the energy of a resonator with a given eigenfrequency to the spectral energy density at that frequency. Next, he made a plausible *ansatz* for the entropy of the resonator. Since there is a thermodynamic equation, connecting entropy, energy, and temperature, he now knew the resonator energy as a function of temperature and, since he had related that already to the spectral energy density, he knew the latter. It turned out to be the same as in Wien's radiation law [10]. Planck was satisfied to have derived it from thermodynamic considerations only.

Wien's radiation law contains two constants (a and b in the notation used in our Box). Since he had identified the function proposed by Wien with Kirchhoff's universal function, Planck became convinced that a and b were fundamental constants of nature, on the same footing with the speed of light c in vacuum and with the gravitational constant G_N in Newton's law of gravitation. He proposed to make these four constants the basis of a system of units for length, mass, time, and temperature based on natural constants. The constant b is, essentially, the quantum of action h introduced by Planck in 1900. The natural unit of mass, in the spirit of Planck's proposal, is therefore

---

$m_{\text{Planck}} = \sqrt{hc/G_{\text{N}}}$. Under the name *Planck mass*, it recently gained importance in theoretical elementary particle physics, astrophysics, and cosmology.

In 1899, by the time Planck reported his results, experiments, however, began to show deviations from Wien's radiation law. Lummer[6] and Pringsheim[7] had overcome difficulties to measure the blackbody spectrum in the far infrared, i.e., for wavelengths much larger than that of visible light. They found the spectrum to be in good agreement with Wien's radiation law for short wavelengths but to deviate from it for long ones [9]. There were also theoretical developments. Lord Rayleigh[8] argued [11] that, at least for long wavelengths, the spectral energy density should rise linearly with temperature for fixed wavelength whereas Wien's law predicted a much slower rise with temperature. At the *Technische Hochschule* (Technical University) in Berlin, Rubens[9] and Kurlbaum[10] measured the intensity of blackbody radiation with high precision for a wavelength of 51 microns. This was possible because Rubens, together with Nichols[11], had developed the method of *residual rays* by which infrared radiation of a fixed wavelength is filtered out of incoming radiation by repeated reflection off a particular crystal. They found a linear increase with temperature [12].

Planck learned of this result on Sunday, 7 October, 1900 when Mr. and Mrs. Rubens were visiting Mr. and Mrs. Planck at their home. After the visitors had left, Planck found a way to change his *ansatz* for the entropy. With this change he obtained a radiation law, which, for small wavelengths, practically coincided with Wien's law but, at large wavelengths, showed the required linear rise. By the same evening he informed Rubens by a postcard and, on 19 October, he presented his new law in a session of the Physical Society in Berlin [13]. Although Planck now had a formula which was in full agreement with experiment, he also knew that he still had to formulate his entropy in an unambiguous way. He achieved this within less than two months. On 14 December, again at a session of the Physical Society, he reported on his derivation [14].

Planck had found that he could advance only if he accepted Boltzmann's probabilistic definition of entropy. Moreover, to compute probabilities, he assumed that a resonator with a particular frequency ν could only possess the fixed energy values $E_\nu = 0, h\nu, 2h\nu, \ldots$. Here h was a constant of nature, now called *Planck's constant* or *Planck's quantum of action*. That meant that a resonator could contain only an integer number of *energy quanta* $\varepsilon = h\nu$. This was the decisive break with classical physics in which every positive energy value was allowed. After this revolutionary step the rest was (more or less) simple calculation. Planck considered a subsystem of a large number of resonators of frequency ν, which together carried the energy E. He then computed the number of possible ways (in Boltzmann's language called the number of *complexions*) to distribute the total of $P = E/\varepsilon = E/h\nu$ energy quanta on these resonators. To illustrate this he even gave a numerical example for one such complexion: 7 quanta on the first resonator, 38 on the second, 11 on the third, etc. (Note that he distinguished between resonators but *not* between quanta. He did not consider the case *quantum 1 on resonator 1, quan-*

Statue of Planck in front of the Humboldt University in Berlin (created by Bernhard Heiliger in 1949).

---

[6] Otto Richard Lummer (1860–1925)   [7] Ernst Pringsheim (1859–1917)   [8] John William Strutt (Lord Rayleigh) (1842–1919), Nobel Prize 1904   [9] Heinrich Leopold Rubens (1865–1922)   [10] Ferdinand Kurlbaum (1857–1927)   [11] Ernest Fox Nichols (1869–1924)

Planck's Radiation Law

can be written in the two equivalent forms for frequency and wavelength, respectively

$$u_\nu = \frac{8\pi h\nu^3}{c^3}\left[\exp\left\{\frac{h\nu}{kT}\right\}-1\right]^{-1} \quad , \quad u_\lambda = \frac{8\pi hc}{\lambda^5}\left[\exp\left\{\frac{hc}{\lambda kT}\right\}-1\right]^{-1} \quad .$$

Planck's Constant h

The numerical value of Planck's constant is

$$h = 6.626\,069\,3\,(11)\times 10^{-34}\,\mathrm{J\,s} \quad .$$

The number in parentheses is the experimental error in the last digits.

Boltzmann's Constant k

has the numerical value

$$k = 1.380\,650\,4\,(24)\times 10^{-23}\,\mathrm{J\,K^{-1}} \quad .$$

Dirac's Symbol $\hbar$

We shall often use the shorthand introduced by Dirac for a frequently appearing fraction,

$$\hbar = \frac{h}{2\pi} \quad .$$

tum 2 on resonator 2 to be different from *quantum 2 on resonator 1, quantum 1 on resonator 2*.) Next, he added to his subsystem another subsystem of N' resonators of frequency ν' carrying the energy E', and so on, finally forming a complete system of resonators of all frequencies. Of course, there were many ways to distribute the total energy E_0 of the system on the partial energies $E, E', E'', \ldots$. Of all these, Planck, following Boltzmann's reasoning, chose that one which yielded the largest number of complexions for the complete system, that is, the one which was most probable. In this way he found for every frequency the energy contained in the resonators. From this he already knew to compute the spectral energy density of the radiation. It was the one he had found two months earlier.

Now also the two constants this law contained had a clear meaning. One was *Planck's constant h*. The other was *Boltzmann's constant k*. This way to write the latter was actually introduced by Planck. (Boltzmann himself had al-

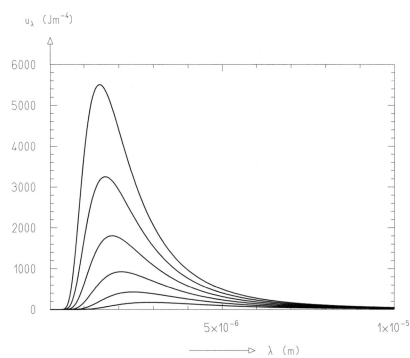

Illustration of Planck's radiation law. The temperature rises from the lowest curve ($T = 1000\,\mathrm{K}$) in steps of $200\,\mathrm{K}$ to the highest curve ($T = 2000\,\mathrm{K}$).

10-Euro coin, issued in 2008 in commemoration of Planck's 150th birthday and displaying a graph of his radiation law.

ways written it as ratio R/N_A of the *gas constant R* and *Avogadro's constant N_A*.) Comparing his formula to the recent measurements of the blackbody radiation, Planck was able to get rather precise numerical values for these constants. Immediately, he made good use of this knowledge. Since, at the time, the gas constant was known quite well but Avogadro's constant only rather poorly, Planck computed the latter as $N_A = R/k$. As a consequence, he was able to also improve the numerical value of the elementary electric charge e, which was known to be connected to *Faraday's constant of electrolysis F* and Avogadro's constant through $e = F/N_A$.

To summarize Planck's achievements we repeat that, within two months late in 1900, he found his radiation law, which was Kirchhoff's long-sought universal function, he introduced the concept of *energy quantization* (restricted, for the time being, to his resonators, which, essentially, were microscopic Hertzian dipoles), he found a new constant of nature, and he improved the numerical values of two other important constants. All these achievements remain valid to this day. The details of his derivation, however, have been criticized and improved upon several times. The first one to do so was Einstein (see Episode 10).

Besides his scientific successes, Planck's life reflects in a sad way the drama of Europe in the second half of the nineteenth and the first half of the twentieth century: His brother Hermann fell in the war of 1870/71 with France when Planck was still at school. His son Karl fell in 1916 at Verdun. His son Erwin was executed early in 1945 because of his involvement in the attempt of July 1944 to kill Hitler. Moreover, Planck's daughter Grete died in 1917 giving birth to her first child and her fate was shared by her twin sister Emma in 1919.

Einstein wrote about Planck's radiation law and the discovery of the quantum of action [15]:

> This discovery became the basis of all twentieth-century physics and has almost entirely conditioned its development ever since. Without this discovery it would not have been possible to establish a workable theory of molecules and atoms and the energy processes that govern their transformations. Moreover, it shattered the whole framework of classical mechanics and electrodynamics and set science a fresh task: that of finding a new conceptual basis for all physics. Despite remarkable partial gains the problem is still far from a satisfactory solution.

[1] Planck, M., *Wissenschaftliche Selbstbiographie*. Barth, Leipzig, 1948.

[2] Heilbron, J. L., *The Dilemmas of an Upright Man*. University of California Press, Berkeley, 1986.

[3] Kirchhoff, G., *Annalen der Physik und Chemie*, **109** (1860) 275.

[4] Stefan, J., *Sitzungsber. Akad. Wiss. (Wien)*, **79** (1879) 391.

[5] Boltzmann, L., *Annalen der Physik*, **22** (1884) 291.

[6] Wien, W., *Sitzungsber. Preuss. Akad. Wiss. (Berlin)*, (1893) 55.

[7] Wien, W., *Annalen der Physik*, **58** (1896) 662.

[8] Paschen, F., *Annalen der Physik*, **58** (1896) 455.

[9] Lummer, O. and Pringsheim, E., *Verh. d. Dt. Phys. Gesellschaft*, **1** (1899) 215.

[10] Planck, M., *Sitzungsber. Preuss. Akad. Wiss. (Berlin)*, (1899) 440.

[11] Lord Rayleigh, J. W. S., *Philosophical Magazine*, **49** (1900) 439.

[12] Rubens, H. and Kurlbaum, F., *Sitzungsber. Preuss. Akad. Wiss. (Berlin)*, (1900) 929.

[13] Planck, M., *Verh. d. Dt. Phys. Gesellschaft*, **2** (1900) 202.

[14] Planck, M., *Verh. d. Dt. Phys. Gesellschaft*, **2** (1900) 237.

[15] Einstein, A., *Out of my Later Years*, chap. In Memoriam Max Planck. Philosophical Library, New York, 1950.

8 Rutherford Finds the Law of Radioactive Decay (1900)

In his first research work done at McGill, Rutherford made two discoveries at a time: He found a new element, now called *radon*. Moreover, he discovered the exponential law describing the decay of radioactive substances.

The story of these discoveries began when Rutherford was asked by a young colleague, Robert B. Owens, who was Professor of Electrical Engineering, to suggest a research subject. Rutherford proposed to investigate the radiation emitted by thorium as he had done before with the radiation from uranium (see Episode 6). A very similar apparatus was used, in which a current through an electrometer indicated the intensity of the radiation, and qualitatively similar results were obtained. But there were also unexpected difficulties. Rutherford and Owens, in a short joint paper [1] wrote that the radiation was not constant, 'but varied in the most capricious manner.' They traced the reason for that effect and stated: 'The sensitiveness of thorium oxide to slight currents of air is very remarkable and made it difficult to work with.'

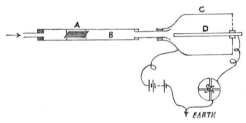

Some emanation was blown from its point of production A into the vessel on the right, where its radioactivity was measured by registering the current between the electrodes C and D. From [2].

At this stage Owens left off and Rutherford continued alone [2]. By covering, in his apparatus, a layer of thorium oxide with successive layers of paper, Rutherford found that the intensity of the radiation dropped for one and two layers because the α rays were absorbed. But if he added many more layers, the radiation stayed practically constant. He concluded that thorium gave off some radioactive substance, which he called *emanation* and which could pass through the paper. The emanation consisted of neutral particles because it was not influenced by the electric field in the apparatus and therefore was easily blown away by a slight air current. Rutherford used this property to separate the emanation from the thorium by blowing it in a controlled way into a separate detection vessel, which consisted of a metal cylinder with an insulated metal rod along its axis. By connecting cylinder and rod in the familiar way over a battery and an electrometer he could study the radioactivity of the emanation. He blew emanation into the detector, then stopped the air current and observed the electrometer. He found its reading to fall relatively fast with time. Rutherford presented his measurements in tabular and graphical form

34

Exponential Law of Radioactive Decay

At a given time t there are n radioactive particles. Some of them decay so that the number changes by dn in the time dt. The number dn is negative (because n is reduced) and it is proportional to dt (which, of course, is assumed to be small). Rutherford also assumed it to be proportional to n itself, that is, he assumed that *the number of particles decaying is proportional to the number of particles present* or, in other words, that *in a given time a certain fraction of particles decays*. He therefore had $dn = -\lambda n \, dt$ where λ is a constant. This is one of the simplest differential equations. Its solution is

$$n = n_0 \, e^{-\lambda t} \quad .$$

Here n_0 is the number of particles at the time $t = 0$. The equation states that the number n of radioactive particles falls exponentially with time. By differentiating the equation with respect to time one finds that also the number dn/dt of decays per unit time decreases exponentially, as found in the experiment.

The constant $\tau = 1/\lambda$ is called *mean-life time*. The *half-life time* or simply *half-life* is $t_{1/2} = \tau \ln 2 \approx 0.6931 \, \tau$.

---

and wrote: 'The result shows that the intensity of the radiation has fallen to one-half its value after an interval of about *one minute* ... [It] was too small for measurement after an interval of ten minutes.'

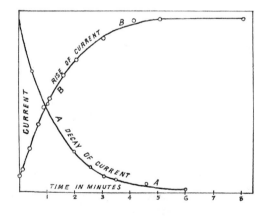

Rutherford's decay and rise curves. From [2].

With this statement Rutherford created the concept of the *half-life* of a radioactive substance. But he did more. He gave the correct formula describing the time dependence of radioactive decay. It was obvious from the graphical representation of his measurements that the current in the electrometer dropped off exponentially with time. That means it decreased in a smooth way such that it was one half of its original value after one minute, one quarter after two minutes, etc. The current was carried by ions created in the air of the detection vessel because of the decay of emanation particles. The number of ions and thus the current is proportional to the frequency of decays, that is, to the number of decays per unit time. Rutherford found a way to understand that this number in fact drops exponentially (see Box).

Lord Rutherford's coat of arms. From [4]. Reprinted with permission from Cambridge University Press.

The discovery of radioactive decay and the exponential law governing it was most remarkable in several respects. For the radioactive substances known until then no decrease of radioactivity with time had been observed. They seemed to perpetually radiate with undiminished intensity. Their half-lives, determined later to be on the order of a thousand years or more, were too large to have been observed. But even if the idea of decay was accepted there was an unprecedented question: How did a particle (let us now call it an atom) know when to decay? It seemed that the decay was determined purely by chance. The law of decay states that in each second a certain fraction of the atoms, then existing, decays. This is true even after many half-lives have elapsed. The few remaining atoms have by no means grown older in the sense that they would decay more easily. The decay law even holds for a single atom. As long as it exists, the probability for it to decay in the next second is a constant. The atoms decay quite independently of each other. In 1910 Rutherford and Geiger showed this independence in a very elegant experiment [3].

Rutherford also made a first step to study the creation of emanation. He placed thorium oxide, covered by many layers of paper, in a detector and blew away the emanation with a current of air. After shutting off the air, the radioactivity increased. It reached half its final intensity after one minute, three quarters after two minutes, etc. Rutherford presented these results in the form of a *rise curve*, which he drew in the same plot as the *decay curve* of the emanation. He found that they were mirror images of one another. The rise is, of course, due to the production of emanation by the thorium. Rutherford explained the curve in detail. The final intensity is reached when, per unit time, the number of emanation particles decaying is the same as the number produced. In the beginning the decay number is smaller, because less emanation particles exist. Hence the intensity rises.

We shall see the pair of decay and rise curves again in Episode 9. When Rutherford was made a baron and became *Lord Rutherford of Nelson*, the pair of curves adorned his coat of arms. With time the nature of the emanation was revealed. It is an inert or noble gas like helium or neon. Still in 1900 Marie and Pierre Curie found emanation emitted from radium which, however, had a different half-life. In 1903 a third type of emanation was found, which was produced by actinium, the radioactive element discovered by Debierne in 1899. As we know now, the different emanations are different *isotopes* – we shall see in Episode 9 how the concept of isotopes was introduced – of the same element, later called *radon*. For a long time they were called *thorium emanation*, *radium emanation*, and *actinium emanation*.

Later in 1900 Rutherford travelled to New Zealand where he was married to his fiancée, Miss Mary Georgina Newton, whom he knew since his student days at Christchurch. Here are few lines from a letter [4] he wrote to her on 2 December 1899: 'In your last letter you mentioned that I hadn't mentioned if I was doing any research work. My dear girl, I keep going steadily turning up at the Lab. five nights out of seven, till 11 or 12 o'clock, and generally make things buzz along. I sent off on Thursday another long paper for the press which is a very good one, even though I say so ...' That was the very paper [2] on the discoveries of radon and half-life.

[1] Rutherford, E. and Owens, R. B., *Transactions of the Royal Society of Canada*, **2** (1899) 9. Also in [5], Vol. 1, p. 216.

[2] Rutherford, E., *Philosophical Magazine*, **49** (1900) 1. Also in [5], Vol. 1, p. 220.

[3] Rutherford, E. and Geiger, H., *Philosophical Magazine*, **20** (1910) 698. Also in [5], Vol. 2, p. 203.

[4] Eve, A. S., *Rutherford – Being the Life and Letters of the Rt Hon. Lord Rutherford, O.M.* Cambridge University Press, Cambridge, 1939.

[5] Rutherford, E., *The Collected Papers of Lord Ernest Rutherford of Nelson (Edited by J. Chadwick)*, *3 vols.* Allen and Unwin, London, 1962–1964.

The Transmutation of Elements (1902)

9

Ernest Rutherford had found that thorium produces a radioactive gas, which he called *emanation* and which decays with a half-life of about one minute (see Episode 8). In his further studies he discovered that the wall of a vessel, which had contained emanation, showed radioactivity even after the emanation had decayed [1]. He called this phenomenon *excited radioactivity* and the new substance an *active deposit*. Rutherford measured eleven hours as half-life of the active deposit, to be compared to one minute for the emanation. In the summer of 1900 a young scientist, Frederick Soddy[1], came from Oxford to McGill University to become Demonstrator in Chemistry there. Rutherford, who, with his new gas and deposit, could do with the help of a chemist, was glad that Soddy agreed to do research work with him.

They set themselves the task to find out how thorium produces emanation and began with chemical treatment of thorium compounds. As Marie Curie had done with pitchblende on her way to the discovery of polonium and radium, they subjected thorium compounds to the standard methods of chemical separation, hoping to find some new substance. Knowing that the presence of emanation is accompanied by radioactivity, they used the latter as an indicator for any substance separated from thorium. Shortly before Christmas 1901 they were able to prepare samples, which showed a much higher radioactivity than thorium. Just as Marie Curie had done, Rutherford and Soddy assumed that a new radioactive element was contained in these samples. They called it *thorium X*, abbreviated ThX [2]. The thorium from which it had been separated, on the other hand, had lost most of its radioactivity. Besides the chemical method to enrich ThX also a physical one, simply washing thorium with water, was found. Soddy later remembered [3]: 'how, in grim determination not to be outdone by a mere chemist, and to show him that physical methods were just as good, Rutherford shook up some thoria with gallons of water till he was tired, then boiled all the water down triumphantly and produced from the residue a minute quantity of the new body.'

When they resumed work after the Christmas holidays, they found that the samples with ThX had completely lost their activity. They also found that the thorium, from which it had been abstracted, had nearly regained completely its original activity. Rutherford and Soddy studied the effect in more detail and came up with a revolutionary explanation [4]. First, they found that the half-life of ThX was about four days. In the same time, the activity of the thorium, from which the ThX had been removed, rose to half its original value.

Rutherford was already familiar with this kind of behaviour from when he discovered emanation. Therefore it was clear that thorium somehow produced ThX. It was also found by experiment that the emanation was given off not

Soddy

---

[1] Frederick Soddy (1877–1956), Nobel Prize 1921

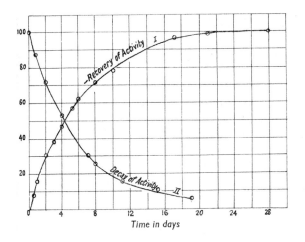

Decay and recovery of thorium X measured by Rutherford and Soddy. From [4], DOI: 10.1039/CT9028100837 – Reproduced by permission of The Royal Society of Chemistry.

by the original thorium but by the ThX. Therefore, there was a chain of substances:

$$thorium \rightarrow ThX \rightarrow emanation \rightarrow active\ deposit.$$

Having observed that the four substances in this chain each had very distinct chemical properties and also half-lives, Rutherford and Soddy came to the conclusion that they are different chemical elements. They stated that a chemical element can change into another, that into yet another, etc. This statement was in direct opposition to the very definition of an element, that says that its atoms are absolutely stable. To defend their point of view they wrote:

> There is not the least evidence for assuming that uranium and thorium are not as homogeneous as any other chemical element, in the ordinary sense of the word, so far as the action of *known* forces is concerned. The idea of the chemical atom in certain cases spontaneously breaking up with evolution of energy is not of itself contrary to anything that is known of the properties of atoms, for the causes that bring about the disruption are not among those that are under our control, whereas the universally accepted idea of the stability of the chemical atom is based solely on the knowledge we possess of the forces at our disposal.

The discovery of Rutherford and Soddy showed that the *transmutation* of elements was possible. That had be sought in vain by the alchemists who had hoped to make gold from lesser matter. Rutherford therefore preferred the non-committal term 'transformation'.

Let us now look at the very last sentence of the paper [4]:

> Nothing can yet be stated of the mechanism of the changes involved, but whatever view is ultimately adopted it seems not unreasonable to hope that radioactivity affords the means of obtaining information of processes occurring within the chemical atom.

As we know now, these were prophetic words and we want to give a short sketch of how they were fulfilled, although we shall come back to the developments in several episodes. Rutherford discovered (Episode 16) that each atom consists of a *nucleus* carrying most of the mass of the atom and a number Z of positive elementary charges and a total of Z electrons each with one negative elementary charge. The nucleus of the lightest atom was later, in 1920, given the name *proton* by Rutherford. It was found that elements consist of atoms, which are chemically indistinguishable but may have different masses and, if radioactive, different half-lives. Soddy gave them the name *isotopes*, because they occupy the same position in the Periodic Table. Besides the proton an uncharged particle, the *neutron*, was found in the nucleus by Chadwick, one of Rutherford's brilliant students (Episode 48). In general, a nucleus consists of Z protons and N neutrons, i.e., a total of $A = Z + N$ *nucleons*. An isotope is uniquely described by two of these numbers, usually Z and A. Rutherford's thorium emanation, for instance, has $Z = 86, A = 220$. It is an isotope of the element radon and referred to by the symbol $^{220}_{86}\text{Rn}$ or simply by ^{220}Rn. The symbol of the element implies the number Z. Transmutation and radioactive decay are connected by a law formulated by Soddy [5] and by Fajans[2] [6].

We want to formulate this law in modern words. It is based on the properties of α particles (helium nuclei ^4_2He), β particles (electrons which we could denote as $^0_{-1}\text{e}$) and γ particles (electromagnetic radiation carrying neither mass nor charge). Therefore, if an isotope emits an α particle, its proton and nucleon numbers are changed by the loss of two protons and four nucleons, $Z \to Z - 2$, $A \to A - 4$. Correspondingly, the emission of a β particle brings about the change $Z \to Z + 1$, $A \to A$. The emission of a γ particle leaves both Z and A unchanged.

The complete decay chain of thorium is rather complicated and has several branches. The main branch reads

$$^{232}_{90}\text{Th} \xrightarrow{\alpha} {}^{228}_{88}\text{Ra} \xrightarrow{\beta} {}^{228}_{89}\text{Ac} \xrightarrow{\beta} {}^{228}_{90}\text{Th} \xrightarrow{\alpha} \mathbf{^{224}_{88}\text{Ra}} \xrightarrow{\alpha} \mathbf{^{220}_{86}\text{Rn}} \xrightarrow{\alpha}$$

$$^{216}_{84}\text{Po} \xrightarrow{\alpha} {}^{212}_{82}\text{Pb} \xrightarrow{\beta} {}^{212}_{83}\text{Bi} \xrightarrow{\beta} {}^{212}_{84}\text{Po} \xrightarrow{\alpha} {}^{208}_{82}\text{Pb} \quad .$$

We see that two isotopes appear in it for the elements thorium (Th), radium (Ra) and lead (Pb). Rutherford's thorium X (the radium isotope $^{224}_{88}\text{Ra}$) and his thorium emanation (the radon isotope $^{220}_{86}\text{Rn}$) are shown in fat print.

In 1908 Rutherford was awarded the Nobel Prize in chemistry 'for his investigations into the disintegration of the elements, and the chemistry of radioactive substances'. In the evening of the presentation of the prizes by the King of Sweden there was a banquet, where Rutherford in his speech 'declared that he had dealt with many transformations with various periods of time, but that the quickest he had met was his own transformation in one moment from a physicist to a chemist' [3]. Abraham Pais points out [7] that in his opinion Rutherford is the only scientist who made his greatest discovery after he had already won the Nobel Prize. Indeed, his most important work was yet to come, the discovery of the atomic nucleus. Soddy, too, became a Nobel laureate. In 1921 he got the prize in chemistry 'for his contributions to our knowledge of the chemistry of radioactive substances, and his investigations into the origin and nature of isotopes'.

[1]   Rutherford, E., *Philosophical Magazine*, **49** (1900) 161. Also in [8], Vol. 1, p. 232.

[2]   Rutherford, E. and Soddy, F., *J. Chem. Soc., Trans.*, **81** (1902) 321. Also in [8], Vol. 1, p. 376; extended version published as [9].

[3]   Eve, A. S., *Rutherford – Being the Life and Letters of the Rt Hon. Lord Rutherford, O.M.* Cambridge University Press, Cambridge, 1939.

[4]   Rutherford, E. and Soddy, F., *J. Chem. Soc., Trans.*, **81** (1902) 837. Also in [8], Vol. 1, p. 435; extended version published as [10].

[5]   Soddy, F., *Chemical News*, **107** (1913) 97.

[6]   Fajans, K., *Physikalische Zeitschrift*, **14** (1913) 131, 136.

[7]   Pais, A., *Inward Bound – Of Matter and Forces in the Physical World.* Oxford University Press, Oxford, 1986.

[8]   Rutherford, E., *The Collected Papers of Lord Ernest Rutherford of Nelson (Edited by J. Chadwick), 3 vols.* Allen and Unwin, London, 1962–1964.

[9]   Rutherford, E. and Soddy, F., *Philosophical Magazine*, **4** (1902) 370. Also in [8], Vol. 1, p. 472.

[10] Rutherford, E. and Soddy, F., *Philosophical Magazine*, **4** (1902) 569. Also in [8], Vol. 1, p. 495.

---

[2] Kasimir Fajans (1887–1975)

10 Einstein's Light-Quantum Hypothesis (1905)

Einstein in his younger days.

There can be no doubt that Einstein[1] [1–3] is still the most widely known scientist of all time. He entered the scientific stage in 1905, the year in which he published several papers of utmost importance and the year in which he obtained his doctor's degree. In the first of these papers, he proposed the *light quantum*, which for his whole life he considered to be his most daring idea.

Einstein was born in Ulm. When he was one year old the family moved to Munich, where his father and his uncle set up a small company for the production of electric machinery. He attended primary and secondary school in Munich. In 1894 the factory, which wasn't doing well, was re-established in the North of Italy and the family moved there as well, leaving young Albert in Munich to finish school. Although – contrary to some tales – he was an excellent pupil, he disliked school because of its authoritarian style and left in 1895 without exam to join his family in Italy. In October of that year he sat for the entrance examination of the Eidgenössische Polytechnikum (Federal Polytechnique) in Zurich. He failed although he got good marks in mathematics and the sciences and took the advice to obtain the Swiss *Matura*, which would entitle him to enter the Polytechnique. To this end he attended for one year the cantonal school in Aarau. He boarded with the family of one of his teachers and enjoyed the liberal spirit of the school. In the fall of 1896, now aged 17, he became a student at the Polytechnique.

Among Einstein's teachers were the physicist Weber, who had discovered an anomaly in the specific heat of diamond (see Episode 12), and the mathematician Minkowski[2], who later devised an elegant formulation of Einstein's special theory of relativity (Episode 11). Einstein preferred to study by himself, reading, among others, books and papers by Kirchhoff, Hertz, Helmholtz, Lorentz, and Boltzmann. Many lectures he did not attend regularly. What concerned mathematics he relied on the scrupulous notes taken by his fellow student and friend Grossmann[3]. In 1900, after not quite four years, he passed the final exam and was qualified as *Fachlehrer*, i.e., as high-school teacher of science. The Polytechnique, contrary to the University of Zurich, then did not have the right to award a doctor's degree.

In vain Einstein now looked for a paid assistantship at a Swiss or a foreign university. Finally in 1902, recommended by the father of Grossmann, he got a position as *technical expert third class* at the Swiss Federal Patent Office in Bern. He liked his work there, married, became the father of a son, founded the *Akademie Olympia*, in which he and two friends read books of literary or philosophical contents, and still made immortal contributions to science.

---

[1] Albert Einstein (1879–1955), Nobel Prize 1921   [2] Hermann Minkowski (1864–1909)   [3] Marcel Grossmann (1878–1936)

Planck's radiation law of 1900 (Episode 7) was taken by most physicists simply as a formula describing experimental facts. It seems that, apart from Planck himself, Einstein was the first who critically looked at its derivation. Much later he would write [1]: 'Planck actually did find a derivation, the imperfections of which remained at first hidden, which latter fact was most fortunate for the development of physics.'

We recall that Planck had considered three different quantities, the spectral energy density of the radiation, the energy of his material resonators for a given frequency and the entropy of the ensemble of a very large number of resonators covering all frequencies. Earlier already he had related the second quantity with the first. By his bold way to construct the entropy he got equations for the second and thus the first quantity.

Einstein's first paper of 1905 [4] carries the title *On a heuristic point of view concerning the production and conversion of light*. In Section 1, *On a difficulty concerning the theory of 'black radiation'*, he pointed out that, while the mean energy of the resonators indeed has the value expected for an ensemble at a given energy, resonators for very low frequencies have unreasonably high energies. To avoid that difficulty he continued using Wien's radiation law rather than Planck's, knowing that the two are equivalent to each other at high frequencies. He computed the entropy of radiation in a given frequency interval in a fixed volume at fixed temperature, compared it to the entropy of a gas of n molecules in the same volume at the same temperature and found a striking similarity between the two formulae. They were identical if one assumed that the radiation also consisted of n 'molecules' or rather *light quanta* each carrying the energy $E = h\nu$ with h being Planck's constant and ν the frequency of the radiation. He wrote:

> Monochromatic radiation of low density (within the region of validity of Wien's radiation formula) behaves in thermodynamic respect in such a way as if it consists of mutually independent energy quanta of magnitude $h\nu$.[4]

Planck had introduced energy quantization but he had restricted it to the energy of his material resonators. Einstein, in whose reasoning resonators did not appear, quantized the energy of radiation, i.e., of the electromagnetic field.

At the end of the paper Einstein showed that there was already indirect experimental evidence for the existence of light quanta and he described an experimental way to ascertain it. Light quanta would naturally explain an empirical rule found by Stokes. This rule states that in *luminescence* the incident light has to have at least the frequency of the emitted light. With light quanta this rule was simply explained: An incident light quantum of energy $E_1 = h\nu_1$ can cause a light quantum $E_2 = h\nu_2$ to be emitted only if ν_2 is not larger than ν_1.

The last section of Einstein's paper is *On the production of cathode rays by exposing solid bodies to light*. This process is called the *photoelectric effect* or simply *photoeffect*. The effect was first found in 1887 by Hertz [5] as a by-product to the famous experiments in which he demonstrated electromagnetic waves. In both his transmitter and receiver he had spark gaps and he observed that the production of a spark in the receiver was facilitated if the electrodes

---

[4] Einstein used different symbols in this formula

of that spark gap were exposed to the light of the transmitter spark. A year later Hallwachs[5] [6] observed that a zinc plate, carrying negative charge with respect to its environment, is discharged when exposed to ultraviolet light. No discharge occurs if it carries positive charge. After cathode rays had been identified as (the negatively charged) electrons, it became clear that the emission of electrons was the reason for the discharge. Lenard, whose work [7] Einstein quotes and describes as pioneering, made the first semi-quantitative analysis in 1902. Lenard worked with an evacuated vessel containing two electrodes.

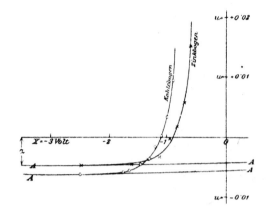

Lenard's observation of different threshold potentials for arc light produced with carbon electrodes (*Kohlebogen*) and zinc electrodes (*Zinkbogen*). From [7]. Copyright Wiley-VCH Verlag GmbH & Co. KGaA. Reproduced with permission.

One, the *photocathode* made of aluminium, was illuminated. The other, the *anode*, was put on an electric potential U with respect to the cathode. The value of U was kept slightly negative. Electrons carry the negative charge $-e$. If they could overcome the potential difference between the two electrodes, they gave rise to a current in the external circuit. That meant that the kinetic energy E_{kin}, which the electrons have when they leave the cathode, had to be larger than $-eU$. Lenard observed that there was a steep rise of current if the absolute value of the potential fell below a certain threshold. That threshold depended on the light source he used, an electric arc with carbon electrodes or one with zinc electrodes. These sources do not produce monochromatic light but it was known that the average wavelength was smaller (and therefore the frequency higher) for the carbon arc. That result, although qualitative, supported the light-quantum hypothesis. On the average, electrons produced by light from a carbon arc had a higher kinetic energy than in the case of a zinc arc.

Einstein assumed that electrons were initially at rest in the cathode. They acquired the energy $h\nu$ of the light quantum but needed a certain amount of energy W (called *work function*) to leave the metal. Thus their kinetic energy E_{kin} in the vacuum was $h\nu - W$. With this he was able to give a formula[4] for the threshold potential U_0 beyond which a current would be measured if the cathode was illuminated by monochromatic light of frequency ν,

$$-eU_0 = h\nu - W \quad ,$$

of which he wrote:

---
[5] Wilhelm Hallwachs (1859–1922)

If the derived formula is right, then U_0, plotted in Cartesian coordinates as a function of the frequency of the exciting light, has to be a straight line, the inclination of which is independent of the nature of the substance under study.

The slope is $-h/e$, the negative quotient of Planck's constant h and the elementary charge e. In 1916, 11 years after the publication of Einstein's paper, Millikan published his results [8,9] in which he verified this formula with great precision. He had developed photocathodes from lithium, sodium, and potassium, which had low work functions allowing for the photoelectric effect with visible light and had used monochromatic light from the various lines of the mercury spectrum.

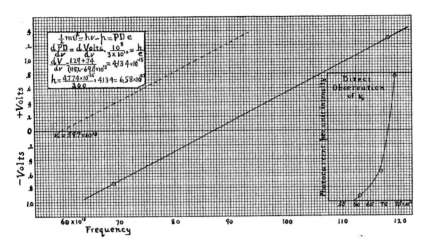

Millikan's measurement of the threshold voltage U_0 as a function of the light frequency ν and his computation of Planck's constant h from that measurement. Figure reprinted with permission from [8]. Copyright 1916 by the American Physical Society.

Millikan, as well as many physicists at the time, accepted the formula he had verified but he did not accept the light-quantum hypothesis. In fact he writes in the very first sentence of his paper [8]: 'Einstein's photoelectric equation ... cannot in my judgement be looked upon at present as resting upon any sort of satisfactory theoretical foundation.'

The Swedish Royal Academy of Sciences, when it awarded the Nobel Prize in Physics to Einstein in 1921, did so 'for his services to Theoretical Physics, and especially for his discovery of the law of the photoelectric effect'. No direct mention was made of the light quantum. The theory of relativity, which by then had made Einstein world-famous, was not mentioned at all.

With his light-quantum hypothesis Einstein attributed particle properties to light, which until then was considered to have only wave properties. Eventually this was also recognized in scientific language when in 1926 the physical chemist Lewis[6] introduced the term *photon*, which has essentially replaced the name *light quantum*. In Episode 32, we shall see how de Broglie generalized Einstein's idea by giving wave properties to particles.

---

[6] Gilbert Lewis (1875–1946)

[1] Einstein, A., *Autobiographical Notes*. In Schilpp, P. A. (ed.): *Albert Einstein: Philosopher-Scientist*. Open Court, La Salle, Ill., 1949.

[2] Pais, A., *'Subtle is the Lord ...'* – *The Science and Life of Albert Einstein*. Oxford University Press, Oxford, 1982.

[3] Seelig, C., *Albert Einstein – Leben und Werk eines Genies unserer Zeit*. Europa Verlag, Zürich, 1960.

[4] Einstein, A., *Annalen der Physik*, **17** (1905) 132.

[5] Hertz, H., *Annalen der Physik*, **33** (1887) 983.

[6] Hallwachs, W., *Annalen der Physik*, **33** (1888) 301.

[7] Lenard, P., *Annalen der Physik*, **313** (1902) 149.

[8] Millikan, R. A., *Physical Review*, **7** (1916) 18.

[9] Millikan, R. A., *Physical Review*, **7** (1916) 355.

11

Einstein Creates the Special Theory of Relativity (1905)

Einstein, ca. 1910

Einstein's paper on the light quantum (Episode 10) was received by the *Annalen der Physik* on 18 March 1905. At the end of April he completed his Ph.D. thesis on a new method to determine Avogadro's constant, which was accepted by the University of Zurich in July. On 11 May his paper on Brownian motion (see Episode 14) was received by the *Annalen* and on 30 June the paper *On the Electrodynamics of Moving Bodies* [1], in which Einstein created what he later called the *Special Theory of Relativity*. About this paper Abraham Pais, Einstein's unsurpassed biographer, writes [2]: 'Einstein was driven to the special theory of relativity mostly by aesthetic arguments, that is, by arguments of simplicity. The same magnificent obsession would stay with him for the rest of his life. It was to lead to his greatest achievement, general relativity, and to his noble failure, unified field theory.'

A number of problems had accumulated in connection with Maxwell's electrodynamics and the electromagnetic waves found by Hertz. We mention but one, directly connected to the classical concept of a *wave*. A wave needs a material *carrier*. For a sound wave the carrier can be the air. A sound wave is a pattern of regions of higher or lower air pressure and it is this pattern that moves to transport sound, not the air itself, which stays at rest. The speed of sound with respect to the carrier is a property of the carrier. The speed of sound, measured by an observer who moves with respect to the carrier, is different from it. No carrier for electromagnetic waves was known but by common belief it had to exist and was called the *ether*. It had to have curious properties: it had to be elastic and solid (like rubber or steel) because it had to carry transverse waves; it had to fill all space or one would not see the stars; it had to be practically massless and allow essentially frictionless motion of material bodies through it or Newton's celestial mechanics would not hold. It was assumed that the ether was at rest with respect to the fixed stars. In a famous experiment performed in 1887, Michelson[1] and Morley[2] looked in vain for a difference between the speed of light in the direction of the earth's motion around the sun and the direction perpendicular to it, although they should have been able to observe it.

In the introduction to his paper [1] Einstein stresses that Maxwell's electrodynamics, 'as it is usually interpreted at present', leads to asymmetries not shown by the phenomena. As an example he mentions magnetic induction. If a conductor and a magnet are in relative motion to each other, then a current flows in the conductor. That is the phenomenon. But there are two theoretical descriptions. If the conductor is at rest and the magnet moves, then an electric

---

[1] Albert Abraham Michelson (1852–1931), Nobel Prize 1907   [2] Edward William Morley (1838–1923)

field is induced giving rise to the current. If the magnet is at rest and the conductor is moved, then a force, the *Lorentz force*, acts on the electrons which are moved with the conductor through the magnetic field and this force causes the current. He mentions the failure to observe a relative motion of the earth with respect to the ether and states that this failure suggests that the concept of a reference frame at 'absolute rest' has no physical meaning. It is not clear to what extent Einstein at that time was familiar with the discussion of these problems because he gives not a single reference to the literature. But he solved them and he did so by critical examination of *space* and *time*.

Before discussing Einstein's work let us look for a moment at the concept of relativity in classical mechanics. Newton's second law states that a body stays in a state of motion with constant velocity if no force acts on it. It is also called the *law of inertia*. A coordinate system, in which this law (and also Newton's third law) is valid, is called an *inertial frame*. If another coordinate system moves with constant velocity with respect to an inertial frame, then Newton's laws hold in that system, too. It also is an inertial frame. If the coordinates of a point are denoted by x, y, z in one inertial frame and by x', y', z' in another, the latter can be computed from the former. The process is called *Galilei transformation*. Of course, the time was assumed to be the same in both systems, $t' = t$. It was this assumption which was corrected by Einstein.

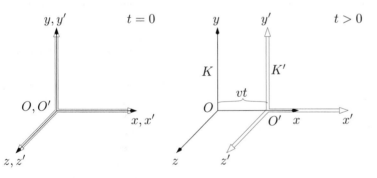

The two coordinate systems K and K' have parallel axes. K' moves with respect to K with the velocity v in x direction. Both systems coincide at the time $t = 0$ (left). At an arbitrary time t their origins have the distance vt.

Einstein defined the usual Cartesian coordinate system with x, y, z axes. The coordinates of a point are measured by yard sticks at a time shown by a clock positioned very near that point. He assumed that yard sticks and clocks are spread all over the coordinate system and gave the following prescription for the *synchronization* of the clocks. A light signal is sent from a point A at time t_A (measured with the clock at A) and arrives at a point B at time t_B (measured with the clock at B). It is reflected there and arrives back at A at time t'_A. Since the signal takes the same time for both ways, the clocks are synchronized if the two differences $t_B - t_A$ and $t'_A - t_B$ are equal. A system in which space and time is measured in this way is now called an *inertial frame*.

Next, Einstein postulated two principles:

1. *Relativity Principle:* If two inertial frames move relative to one another with constant velocity, then the laws of physics have the same form in both frames. (This principle was formulated as in classical mechanics, but was now complemented by another.)

Box I. Galilei Transformation and Lorentz Transformation

The coordinate system K' moves with velocity v in x direction with respect to the system K. At the time $t = t' = 0$ the two systems coincide.

Galilei Transformation	*Lorentz Transformation*
$x' = x - vt$,	$x' = \gamma(x - \beta ct)$,
$y' = y$,	$y' = y$,
$z' = z$,	$z' = z$,
$t' = t$.	$t' = \gamma(t - (\beta/c)x)$.

In the Lorentz transformation $\beta = v/c$ is the ratio of the relative velocity v and the speed of light c and the *Lorentz factor* $\gamma = 1/\sqrt{1 - v^2/c^2}$ is derived from it.

---

2. *Constancy of the Speed of Light:* The speed of light c is the same in all inertial frames and independent of the motion of the light source in that frame.

Einstein considered a frame K in which space and time coordinates are denoted by x, y, z, t and a frame K' in which they are called x', y', z', t'. The coordinate axes of these frames are parallel to each other and coincide at time $t = t' = 0$. The frame K' moves with respect to K with velocity v in x direction. He found formulae which allow to compute space and time coordinates in the frame K' as functions of those in K and vice versa. They are now called *Lorentz transformations* (see Box I).

The essential point here is that the time also is transformed because, in general, it is not the same in the two frames. Consider an *event*, i.e., a set x, y, z, t of coordinates. That may be the changing of a traffic light at a position x, y, z at time t. The traffic light is mounted in the frame K and the event is observed by a pedestrian standing near it. But the event can also be observed by the driver of a car passing the traffic light who carries with him his own reference frame. Even if properly synchronized before, the clocks of the pedestrian and the driver show different times. The difference is tiny for the speed of a car but sizeable if the relative velocity of the two frames gets near the speed of light.

FitzGerald[3] and, independently, Lorentz had explained the negative result of the Michelson–Morley experiment in the following way. The length of the instrument (in fact of all matter) was assumed to be shortened in the direction of its motion through the ether in such a way that the change of the speed of light would go unnoticed. Lorentz thought that this *Lorentz contraction* was brought about by action of the ether on the molecules of the material. The name *Lorentz transformations* recalls this work of Lorentz but the transformations were derived by Einstein directly from his two postulates. He showed that the contraction follows directly from them and therefore, provided they are true, is a property not of matter but of space itself. If $\Delta x = x_B - x_A$ is the difference between the x coordinates of two points which are at rest in the frame K, and if the coordinates x'_A, x'_B of these points in frame K' are measured at the same

---

[3] George Francis FitzGerald (1851–1901)

time t', and their difference $\Delta x' = x'_B - x'_A$ is formed, then

$$\Delta x' = \Delta x/\gamma \quad , \qquad \gamma = 1/\sqrt{1 - v^2/c^2} \quad ,$$

where the *Lorentz factor* γ depends only on the relative velocity v between the two frames and the velocity c of light. Obviously $\gamma = 1$ for $v = 0$. If, on the other hand, v approaches c, then γ approaches infinity. We may describe the *space contraction* by saying that the difference in x between two points is smaller in all reference frames compared to the difference in a frame in which the x coordinates of these points are at rest. Einstein described that a sphere at rest in one reference frame is a somewhat flattened ellipsoid in all other frames and that it becomes completely flat as the relative velocity with respect to the frame of the sphere approaches the speed of light c. Because of this, he recognized c as the highest possible velocity. He derived a formula for the addition of velocities in relativity theory. A new formula was, of course, needed because if a frame K' moves with velocity c in x direction with respect to K and a third frame K'' does the same with respect to K', then the velocity of K'' with respect to K cannot be $2c$ but has to be c.

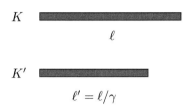

Lorentz contraction: A rod of length ℓ, oriented in x direction and at rest in the system K, is reduced to length ℓ' if described in system K'.

The converse of *space contraction* is *time dilatation*. The time difference Δt between two readings of a clock which is at rest in K, if transformed to the frame K', is multiplied by the Lorentz factor,

$$\Delta t' = \gamma \, \Delta t \quad .$$

Time runs faster in all reference frames except those in which the clock stays at rest. Let us look at a clock mounted in a space ship in fast motion relative to the earth. Then, since the earth moves fast relative to the space ship, time on the earth (as observed from the ship) runs faster and thus it runs differently for a pair of twins, one staying on the earth, the other being on board the ship. This is the now so-called *twin paradox* mentioned by Einstein in more sober words (no twins, no space ship, just clocks and reference frames).

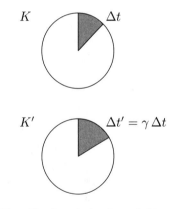

Time dilatation: A time interval Δt, measured with a clock at rest in system K, is lengthened to $\Delta t'$ if described in system K'.

After this discussion of space and time, which he called the *kinematic part* of his paper, Einstein passed to the *electrodynamic part*. He showed that the equations of electrodynamics obey his relativity principle; we now say that they show *Lorentz covariance*. One consequence is an intimate relation between electric and magnetic field which do not transform as separate vectors in a change of reference frame. Particularly beautiful is his derivation of the *Lorentz force*. That force, introduced by Lorentz (see Episode 3) acts on a charged particle if that is in motion in a magnetic field. Einstein showed that the force results from the transformation of the electric field from the frame in which the particle rests to the frame moving along with the particle.

Whereas Maxwell's electrodynamics obeys Einstein's relativity principle Newton's mechanics does not. Einstein computed the kinetic energy of a particle moving with velocity v and found it to be

$$E_{\text{kin}} = mc^2(\gamma - 1) \quad .$$

Here m is the *rest mass* of the particle, i.e., the mass determined in a reference frame in which the particle is at rest, c the speed of light and γ the Lorentz factor computed with the velocity v of the particle. It is obvious that this energy approaches infinity if v approaches c. In this equation the product mc^2 of rest mass and the square of the speed of light appeared for the first time.

In September 1905, Einstein completed a paper of only three pages [3] in which he declared:

> The mass of a body is a measure for its energy content.

He had come to this conclusion by considering a body which emitted the energy E_{rad} in the form of electromagnetic radiation and found that its mass was reduced by E_{rad}/c^2. Therefore the total energy content of a particle at rest with mass m is

$$E = mc^2 \quad .$$

(This famous formula does not appear explicitly in the paper.)

We can now add the last two formulae, that is, compute the total energy E_{tot} of a particle by adding its rest energy E to its kinetic energy E_{kin} and obtain

$$E_{\mathrm{tot}} = \gamma mc^2 = Mc^2 \quad , \qquad M = \gamma m \quad .$$

Here $M = \gamma m$ is the mass of a particle with velocity v which for $v = 0$ becomes equal to its rest mass.

The first scientist to take an interest in Einstein's work was Planck, who also was the first after Einstein to publish on relativity. Slowly Einstein's name became known. In February 1908 he became Privatdozent at the University of Berne while continuing to work at the patent office.

In September 1908, Minkowski, who had been one of Einstein's mathematics professors in Zurich and then was at the University of Göttingen, gave a talk at a scientific conference in Cologne. It was published [4] only after Minkowski's death in 1909 and contained an important mathematical analysis of the concepts of space and time as introduced by Einstein. The main idea is the following. In a given coordinate system in three-dimensional space a point has the coordinates x, y, z. A *(three-)vector* representing the point can be constructed. It has the three components (x, y, z) and is visualized by an arrow drawn from the coordinate origin to the point. By Pythagoras' theorem the square of its length is

$$x^2 + y^2 + z^2 \quad .$$

Minkowski

Now, if the coordinate system is rotated about the origin and the vector stays in place, then the components of the vector change but its length, of course, does not, the above expression is *invariant under rotations*. Since the speed of light c is a constant independent of the reference frame, the time t can be replaced by a quantity $w = ct$ which has the dimension of a length like a space coordinate but represents the time. Minkowski constructed a four-dimensional space spanned by x, y, z, and w and represented every *event* by a *four-vector* in this space. He found that the relation

$$x^2 + y^2 + z^2 - w^2$$

was independent of the reference frame, i.e., *invariant under Lorentz transformations*. This four-dimensional space is now called *Minkowski space*. Of course, because of the curious minus sign, Pythagoras' theorem has a somewhat different form there. One speaks of *Minkowski geometry* as opposed to normal *Euclidean geometry*. Minkowski's four-vectors not only are useful to simplify calculations but also make clear how intimately space and time coordinates are related to each other. They vary with a change of reference frame in much the same way as the coordinates of a point change if the coordinate system is rotated.

Minkowski, in his Cologne talk [4] expressed this relation as follows:

> The views on space and time I would like to develop are grown on the soil of experimental physics. Therein lies their strength. Their tendency is a radical one. From now on space by itself and time by itself shall entirely collapse to shadows and only a sort of union of both shall retain its independence.

This union was later given the name *Minkowski space* (see Box II); Minkowski himself called it the *world*.

Box II. Minkowski Space

We give the space coordinates x, y, z of a point the more convenient names (x^1, x^2, x^3) and add a fourth coordinate $x^0 = ct$, which is the time t multiplied by the speed of light c. We form a *four-vector* in Minkowski space with the components x^0, x^1, x^2, x^3 and find that its length, defined as $s^2 = (x^0)^2 - (x^1)^2 - (x^2)^2 - (x^3)^2$, is unchanged under Lorentz transformations. The minus signs can be taken into account by writing

$$s^2 = \sum_{\mu=0}^{3} \sum_{\nu=0}^{3} \eta_{\mu\nu} x^\mu x^\nu = \eta_{\mu\nu} x^\mu x^\nu \quad .$$

The shorter expression on the right holds if, as is always done in relativity theory, one uses the *Einstein summing convention*, which implies summation over identical indices. The 16 symbols $\eta_{\mu\nu}$ form the *metric tensor*. They have the values

$$\eta_{00} = 1 \quad , \qquad \eta_{11} = \eta_{22} = \eta_{33} = -1 \quad , \qquad \eta_{\mu\nu} = 0 \quad \text{for} \quad \mu \neq \nu \quad .$$

Also the distance between two points in Minkowski space is an invariant and, in particular, the distance between a point and another one in the very close neighbourhood, a *line element*

$$\mathrm{d}s^2 = c^2 \, \mathrm{d}\tau^2 = \eta_{\mu\nu} \, \mathrm{d}x^\mu \mathrm{d}x^\nu \quad .$$

Let the $\mathrm{d}x^\mu$ be the coordinate differences of two points on the trajectory of a particle. In a reference frame, in which the particle is at rest, obviously $\mathrm{d}s^2 = \mathrm{d}(x^0)^2 = c^2 \, \mathrm{d}t^2$. The invariant quantity $\mathrm{d}\tau$ is called the *proper time* of the particle.

In normal three-dimensional space Newton's equation of motion for particle under the action of a force with components F^1, F^2, F^3 is

$$\frac{\mathrm{d}^2 x^i}{\mathrm{d}t^2} = F^i \quad , \qquad i = 1, 2, 3 \quad .$$

If the *Newton force* is generalized to a *Minkowski force* with four components (K^0, K^1, K^2, K^3), one can write down the relativistic equations of motion,

$$\frac{\mathrm{d}^2 x^\mu}{\mathrm{d}\tau^2} = K^\mu \quad , \qquad \mu = 0, 1, 2, 3 \quad .$$

---

[1] Einstein, A., *Annalen der Physik*, **17** (1905) 891. Also in [5], p. 26, Engl. transl. in [6].

[2] Pais, A., *'Subtle is the Lord ...'* – *The Science and Life of Albert Einstein*. Oxford University Press, Oxford, 1982.

[3] Einstein, A., *Annalen der Physik*, **18** (1905) 639. Also in [5], p. 51, Engl. transl. in [6].

[4] Minkowski, H., *Physikalische Zeitschrift*, **10** (1909) 104. Also in [5], p. 54, Engl. transl. in [6].

[5] Lorentz, H. A., Einstein, A., and Minkowski, H., *Das Relativitätsprinzip*. Teubner, Berlin, 5th ed., 1923.

[6] Sommerfeld, A. (ed.), *The Principle of Relativity*. Dover, New York, 1923.

12 Nernst and the Third Theorem of Thermodynamics (1905)

Nernst

Walther Nernst[1] [1] belongs to the small group of scientists, who, in the twentieth century, excelled in both experiment and theory. In 1905 he postulated his *new heat law*, a theorem in thermodynamics, i.e., in classical physics. It turned out soon, however, that there is a close relation between that theorem and quantum theory. When Nernst discovered his heat law he had just become professor of physical chemistry at the University of Berlin. He had moved there from Göttingen where he had held the same position. Being particularly successful in applying the laws of thermodynamics to chemical and electrochemical reactions he was one of the founders of his field. Generations of students and young researchers grew up with his textbook *Theoretical Chemistry, From the Point of View of Avogadro's Rule and of Thermodynamics* [2]. He had also made an invention by which he had become a wealthy man. His electric lamp, the *Nernst lamp*, produced bright white light. Unlike the light bulb of Edison[2] it glowed in normal air and needed no vacuum.

The theory of heat or, as it is also called, thermodynamics, is different from other fields of physics in the sense that it allows to make statements about very complex systems whose composition need not even be known in detail. It began to be successful when the difference between heat and temperature was clearly understood and additional quantities useful to describe a thermodynamic system were introduced. The theory was then based on two fundamental laws, called the first and the second *heat law* or *theorem of thermodynamics* which had to be accepted as empirical facts.

The *first heat law* is the *law of conservation of energy*, which was known in mechanics since long but was established as a general law of nature only in the nineteenth century. Expressed in the language of thermodynamics it states that the change of the *inner energy* of a system is simply the difference between the *heat* absorbed and the mechanical or electrical *work* delivered by the system. Heat and work are just forms of energy. Energy cannot be created or destroyed. Thus there cannot be a machine yielding work indefinitely without needing an input of energy. Such an impossible machine is called a *perpetuum mobile of the first kind*. The first heat law becomes obvious in the framework of the *kinetic theory of heat* because there the inner energy of a substance is simply the mechanical energy of its microscopic constituents. Therefore the first heat law is nothing else but the energy-conservation law of mechanics. The *molar heat* of a substance is the heat needed to raise the temperature of one mole of the substance by one degree. In the kinetic theory the molar heat of a monoatomic substance (containing only one sort of atoms) is independent of the substance and of the temperature. This rule had been established

[1] Walther Nernst (1864–1941), Nobel Prize 1920   [2] Thomas Alva Edison (1847–1931)

experimentally by Dulong[3] and Petit[4] in 1819 and explained in the theoretical framework of statistical mechanics by Boltzmann in 1871. But already in 1872 Weber[5] observed for the first time that it was seriously violated in the case of diamond which shows the 'right' molar heat at very high temperature but a significantly lower value at room temperature. Here experiment contradicted the kinetic theory which was based on the assumption that the atoms behave according to the laws of mechanics.

The *second heat law* introduces further restrictions. It states that certain processes do not happen although they are allowed by the first law. The second heat law is most commonly known in the two equivalent forms it was given by Kelvin and Clausius, respectively.

Kelvin stated that there is no process producing no other effect than to extract heat from a reservoir and to transform it completely into work. It excludes the existence of a machine, called a *perpetuum mobile of the second kind*, which would produce work by only cooling off a large heat reservoir like the oceans. In any process only part of the heat absorbed can be transformed into work. The rest is 'wasted'. Kelvin was able to use the second heat law to construct an *absolute temperature scale*. Its unit is now called the *Kelvin*. The temperature difference 1 Kelvin (1 K) is equal to the temperature difference 1°C. But in the absolute scale negative temperatures do not occur. The *absolute zero* is at $-273.15°C$.

Clausius stated that there is no process producing no other effect than to extract heat from a colder reservoir and to transfer it to a warmer one. He introduced the concept of *entropy* and, with it, could express the second heat law in a more abstract form: In an isolated system the entropy never diminishes. The definition of entropy is somewhat complicated. The change of entropy of a system is the amount of heat absorbed by the system divided by the absolute temperature. This definition holds only if the absorption happens in a *reversible* way, i.e., if it is possible to reverse the absorption in such a way that the initial state is reached again and no effect of the whole process remains. Moreover, in general, the amount of heat has to be tiny. If it is not, the change of entropy has to be computed as the sum – or rather the integral – over many tiny changes. Since only the change of entropy is defined in this way, the absolute value of entropy is only known up to an additive constant. We already mentioned in the Prologue of this book that Boltzmann was able to give a probability interpretation to entropy. He was then able to derive the second heat law in his *statistical mechanics*.

Helmholtz introduced the concept of *free energy*, the energy available in a system which can be transformed into work (in a process in which the temperature is kept constant). With this new concept he was able to express the second heat law in yet another way, now called the *Helmholtz equation*. The left-hand side of the equation is the difference between the change in free energy and the change in inner (total) energy. The right-hand side is the absolute temperature multiplied with the temperature derivative of the change in free energy.

Nernst connected the Helmholtz equation with the puzzle of molar heats. He knew from experiments that the left-hand side of that equation becomes

---

[3] Pierre Louis Dulong (1785–1838)   [4] Alexis Thérèse Petit (1791–1820)   [5] Heinrich Friedrich Weber (1843–1912)

A Way to Nernst's Heat Law

With T temperature, Q heat, U (inner) energy, and A free energy the Helmholtz equation reads

$$-\Delta Q = \Delta A - \Delta U = T\frac{\mathrm{d}\Delta A}{\mathrm{d}T} \quad .$$

Nernst postulated that in the limit $T \to 0$ the quantities in the Helmholtz equation show the following behaviour:

$$\frac{\mathrm{d}\Delta A}{\mathrm{d}T} \to 0 \quad , \quad \text{i.e.,} \quad \Delta A = \Delta U \quad \text{and therefore also} \quad \frac{\mathrm{d}\Delta U}{\mathrm{d}T} \to 0 \quad .$$

Nernst lecturing

very small for low temperatures and he postulated that it vanishes altogether if the temperature approaches absolute zero. By a careful discussion of this approach he concluded further that the molar heat or, generally, the specific heat of all substances vanishes at zero absolute temperature. This means that the more one approaches absolute zero the more difficult it becomes to extract heat from a substance and that the zero point of the absolute temperature scale itself cannot be reached. This is *Nernst's theorem*, also called the *third heat law* or the *third theorem of thermodynamics*. It can also be expressed by saying that the entropy of any system at absolute zero is equal to zero. Thus, Nernst's theorem provides the constant which was still open in the definition of entropy. For reasons of symmetry it is sometimes stated as follows: There is no machine (*perpetuum mobile of the third kind*) with which the temperature of a system can be lowered to absolute zero. The model substance of thermodynamics and statistical mechanics is the *ideal gas* formed by a large number of atoms which have no internal structure and no interaction. The ideal gas does not liquefy or freeze and does not obey the new heat law. Nernst was hoping for the development of a theory with *gas degeneracy* which would comply with his law.

Nernst liked to tell the story that he first had the idea for his new law during one of his usual lectures to students in Berlin. He first reported his work [3] in a meeting of the Royal Society of Sciences at Göttingen and later published it together with many subsequent measurements in a small book [4]. He did not hide the pride about his discovery and in the later editions of his famous textbook [2] he referred to it in the subject index under the letter 'm' – *mein Wärmesatz* (my heat theorem). Peierls, who sat at his feet as a first-year student, called him a 'great physicist of rather small stature and an even smaller sense of humility' [5].

With his co-workers, notably with Eucken[6] and Lindemann[7], Nernst undertook a successful programme to measure specific heats at very low temperatures. After the First World War Lindemann, in the spirit of Nernst, made the Clarendon Laboratory in Oxford into one of the leading centres of low-temperature research. In 1933 he started a successful programme to welcome and help scientists fleeing Germany because of the Nazis. One of them, Kurt

---

[6] Arnold Eucken (1884–1950)   [7] Frederick Alexander Lindemann (Viscount Cherwell) (1886–1957)

Mendelssohn[8], himself a student of Nernst, in his biography [1] gives a touching account of the atmosphere in Nernst's institute and with high praise tells about higher education and scientific research in the Kaiser's Germany. Lindemann, in the Second World War, became Churchill's scientific adviser and was made Lord and later Viscount Cherwell.

Nernst and Lindemann in Oxford, 1937.

Whereas the first and second heat laws found their explanations in a deeper-lying theory of classical physics, the third one did not. On the contrary, it conflicted with the kinetic theory based on classical statistical mechanics. Already in 1906 Einstein had applied Planck's new quantum theory to the question of specific heats of solid bodies [6,7]. He constructed a model in which one particular resonance frequency influences the probability of emission and absorption of quanta of different frequencies. With the resonance frequency in the far infrared he found drastic discrepancies from the Dulong–Petit rule. When Nernst, in 1909, found out that the measurements of his group were well described by Einstein's formula, he travelled to Zurich, where Einstein had moved from Berne by that time, to discuss with him. Nernst was hoping that the ideal gas would obey his heat law or, as he called it, show *gas degeneracy*, if treated in the framework of the developing quantum theory. His hope was eventually fulfilled in two different ways (see Episodes 33 and 41).

Nernst was a man of many contacts and great influence. In 1910, he convinced the Belgian industrialist Solvay[9] to call (and finance) an international scientific conference on *The Theory of Radiation and the Quanta*. The first *Solvay Conference* [8], held in autumn 1911 in Brussels, was at the same time the first conference on quantum theory. It is interesting to study the photograph of its participants. Lorentz presided over the proceedings. Lindemann was one of the scientific secretaries. The founding fathers of the theory, Planck and Einstein, were present and also Sommerfeld, who would soon make important contributions to it. Of course, one looks in vain for those who completed the theory in 1925, because in 1911 they were just about to enter their teens.

[1] Mendelssohn, K., *The World of Walther Nernst – The Rise and Fall of German Science*. Macmillan, London, 1973.

[2] Nernst, W., *Theoretische Chemie, vom Standpunkte der Avogadroschen Regel und der Thermodynamik*. Enke, Stuttgart, 1st ed., 1893.

[3] Nernst, W., *Nachr. Kgl. Ges. Wiss. Göttingen*, **No. 1** (1906) 1.

[4] Nernst, W., *Die theoretischen und experimentellen Grundlagen des neuen Wärmesatzes*. Knapp, Halle, 1st ed., 1919.

[5] Peierls, R., *Bird of Passage – Recollections of a Physicist*. University Press, Princeton, 1985.

[6] Einstein, A., *Annalen der Physik*, **20** (1906) 199.

[7] Einstein, A., *Annalen der Physik*, **22** (1906) 180.

[8] Mehra, J., *The Solvay Conferences on Physics*. Reidel, Dordrecht, 1975.

---

[8] Kurt Mendelssohn (1906–1982)   [9] Ernest Solvay (1838–1922)

<table>
<tr><td>

13

</td><td>

Observing a Single Particle – The Rutherford–Geiger Counter and Later Electronic Detectors (1908)

</td></tr>
</table>

While he was professor in Montreal, Rutherford received several offers for professorships from Universities in the United States, among others from Columbia and Yale, and refused them. But when he was offered a chair at the University of Manchester as successor of Schuster[1], he accepted. He wrote to the principal of McGill University that [1] 'the determining factor in deciding to go to Manchester was my feeling that it is necessary to be in closer contact with European science than is possible on this side of the Atlantic.' *Tempora mutantur!*

Geiger (left) and Rutherford in Manchester, 1912.

In Manchester, where Rutherford took up his duties in October 1907, he found a well-equipped and well-functioning laboratory and, more important, a new co-worker. Already in January 1908, in a letter to Hahn, he was able to report important new results [1]:

> I forget whether I told you that we have detected a single alpha particle by an electric method. ... I am working with Dr Geiger – one of your countrymen who was here last year. He is a very excellent experimenter and a great assistance to me. The method is to fire an alpha particle into a small hole covered with mica into a cylinder about 60 cm long where the air pressure is about 30 cm [about 40 % of normal atmospheric

[1] Arthur Schuster (1851–1934)

pressure which corresponds to a mercury column of 76 cm]. There is a thin central wire and a voltage is applied of about 1000 [Volts] until a discharge almost passes. Under such conditions, the ionization produced in the gas by the alpha particle is magnified 2000 times by collisions. The effect of each alpha particle is marked enough to show an audience.

Before the work of Rutherford and Geiger[2] all electric detectors for radiation had been *ionization chambers*. In such instruments the ion pairs produced by the radiation directly carry the electric current through the gas of the chamber. It is the same current that is measured by an instrument in an external circuit. The number of ion pairs produced by a single α particle is relatively large (typically 100 000 in air). But the resulting current is nevertheless so small that it could not be detected at that time when electronic amplifiers for the external circuit did not exist. Rutherford and Geiger were the first to use an internal amplification directly in the gas volume. The electric field between the outer cylinder and the thin central wire is *inhomogeneous*, it is small near the cylinder but rises steeply in the immediate vicinity of the wire. If the wire is kept on positive voltage with respect to the cylinder, the electrons (the negative partners of the ion pairs) *drift* towards the wire. In the low field their energy is insufficient to ionize gas atoms when they collide with them. Very near the wire, however, in the high field the electrons acquire appreciable energies between two collisions such that they create new ion pairs. The electrons from the new pairs create further pairs, etc. In this way an avalanche of ion pairs is formed. If the voltage gets higher than a certain critical value, then other processes come into play. Ultraviolet light, which is emitted by gas atoms hit by electrons, can lead to the emission of electrons by the outer cylinder. These form new avalanches such that the discharge continues even after the original α particle lost all its energy. One speaks of an *autonomous discharge*. If the voltage is kept below the critical value, then the total number of ion pairs is proportional to the original number created by the α particle. The detector works in the *proportional mode*. It is now called a *proportional counter*.

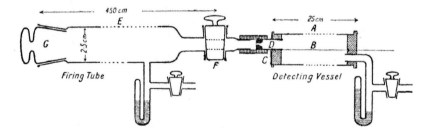

Firing Tube · Detecting Vessel

Rutherford's and Geiger's first apparatus for counting α particles. The radium source is placed in the glass tube E. α particles travelling to the right could pass the stopcock F (if open), the tube D and enter the counter consisting of the brass tube A and the wire B, electrically insulated from each other by ebonite corks C. From [2], Fig. 1. Reprinted with permission.

The experimental arrangement of Rutherford and Geiger was simple [2]. A weak radium source (typically 0.1 mg of radium) was placed at one end of an evacuated tube of 4.5 m length. At the other end was the counting tube with a small opening covered by a thin mica window which the α particles could traverse.

From the distance between the source and the opening and from the size of

----

[2] Hans Geiger (1882–1945)

A record on film of the electric signals obtained by Rutherford and Geiger. From [4]. Reprinted by permission from Taylor & Francis Ltd, http://www.inform aworld.com.

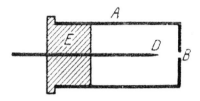

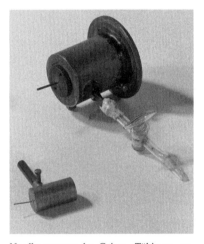

Cut through Geiger's needle counter. From [3], Fig. 3, © 1937 by Julius Springer in Berlin. With kind permission from Springer Science+Business Media.

Needle counters by Geiger, Tübingen, ca. 1930. Photo: Deutsches Museum.

Geiger–Müller counters by Geiger, Tübingen, ca. 1933. Photo: Deutsches Museum.

the opening it was known which fraction of all α particles emitted by the source would enter the detector. The electrometer indicating the current through the detector was observed and every time it was 'thrown', was counted as a detection of an α particle. Since the counting had to be done by eye (and by hand), the numbers had to be kept small, about 5 per minute. Hence the weak source and the long distance. As the final result Rutherford and Geiger stated that 1 g of radium emits 3.18×10^{10} α particles per second.

It is difficult today to realize the impression which the detection of single particles made on the scientific world. This was an important step to make obvious the existence of smallest constituents of matter. It thus came that Rutherford, who had already shown that atoms decay, also demonstrated that they really exist as individual entities.

Rutherford and, in particular, Geiger kept on to improve counting techniques. In 1912, they published a method [4] in which the electrometer throws were recorded on a fast-moving photographic film. In this way on the order of 1000 α particles could be registered per minute. The actual counting was done later by inspecting the film.

In October 1912, Geiger moved to the Physikalisch-Technische Reichsanstalt in Berlin to direct the newly established laboratory for radium research. There, in 1913, he developed a counter with an amplification of 10^6 and more [5]. Such a high amplification enabled to register single β particles (electrons), which produce a much smaller ionization in gas compared to α particles. The amplification was reached by the use of a needle with a sharply pointed tip instead of the central wire. Near the tip the electric field was still higher than near a wire. The detector was called the Geiger *needle counter* or *Spitzenzähler*.

In 1925 Bothe and Geiger made in important experiment in which two particles appear (Episode 34). They used two needle counters and selected events in which both counters gave a signal at the same time. This *coincidence method*, later much refined, is now a key technique in modern elementary-particle experiments.

The instrument which is now familiar under the name *Geiger counter* or, more exactly, *Geiger–Müller counter* was developed in 1928 [6] when Geiger was professor in Kiel and Walther Müller[3] his Ph.D. student. The detector itself is of the same construction as the original Rutherford–Geiger counter but the voltage was chosen so high that an autonomous discharge is triggered by each ionizing particle which enters the detector. The discharge is then automatically 'quenched' by special techniques. (Originally the central wire was covered by a thin layer of a very poor conductor. But also additions to the gas or special elements in the external circuit can be used for quenching.) The GM counter yields a sturdy signal which, however, is no longer proportional to the original ionization left by the detected particle.

Rutherford and Geiger also pioneered another line of particle detectors. It

---

[3] Walther Müller (1905–1979)

Meeting of the Bunsen Society in Münster, 1932. From the left: Chadwick, Hevesy, Mrs. Geiger, Geiger, Lise Meitner, Rutherford, Hahn, Stefan Meyer, Przibram.

had been found by Crookes[4] and, independently, by Elster[5] and Geitel[6], that minute flashes of light, called scintillations, occur when α particles fall onto a zinc sulphide screen, i.e., a surface covered with a layer of tiny crystals of zinc sulphide. Crookes developed a small instrument, called *spinthariscope*, which was produced commercially. It is a small cylinder with a zinc sulphide screen, in which a little radium is embedded, on one end and a magnifying lens on the other. Regener[7] had suggested that each individual scintillation was caused by a single α particle. Rutherford and Geiger were able to confirm this by comparing the scintillation method of counting with their original electric method. Observing and counting scintillations was tedious. The eye had to be accommodated in total darkness for some time and to be rested after about half an hour of observations. Strong coffee (and even a pinch of strychnine) was said to help its sensitiveness. On the other hand, the *spatial resolution* of the scintillation method was very high, especially as the scintillations could be observed by microscope. This is why the famous α-ray scattering experiment of Geiger and Marsden (Episode 16) was done by that method.

Even today all particle detectors use the ionization or the emission of light caused by the particle to be detected. The active medium, in which ionization is produced, is often a gas like in the original Rutherford–Geiger counter. But it can also be a semiconductor. If light is produced, it is no longer detected by eye but by light-sensitive electronic devices. The small primary electric signals are amplified by gas amplification and/or by external amplifiers. Modern microelectronic and computing techniques make it possible to build up complex detector systems with hundreds of thousands or even millions of active detectors. The coincidence method can then be implemented in computer programmes to detect special classes of events.

[1] Eve, A. S., *Rutherford – Being the Life and Letters of the Rt Hon. Lord Rutherford, O.M.* Cambridge University Press, Cambridge, 1939.

[2] Rutherford, E. and Geiger, H., *Proceedings of the Royal Society*, **A81** (1908) 141. Also in [7], Vol. 2, p. 89.

[3] Geiger, H., *Durchgang von α-Strahlen durch Materie*. In Geiger, H. and Scheel, K. (eds.): *Handbuch der Physik, Vol. XXIV*, p. 137. Springer, Berlin, 1937.

[4] Geiger, H. and Rutherford, E., *Philosophical Magazine*, **24** (1912) 618. Also in [7], Vol. 2, p. 288.

[5] Geiger, H., *Verh. d. Dt. Phys. Gesellschaft*, **15** (1913) 534.

[6] Geiger, H. and Müller, W., *Physikalische Zeitschrift*, **29** (1928) 839.

[7] Rutherford, E., *The Collected Papers of Lord Ernest Rutherford of Nelson (Edited by J. Chadwick), 3 vols.* Allen and Unwin, London, 1962–1964.

---

[4] (Sir) William Crookes (1832–1919) [5] Julius Elster (1854–1920) [6] Hans Friedrich Karl Geitel (1855–1923) [7] Erich Regener (1881–1955)

14 Jean Perrin and Molecular Reality (1909)

Perrin

By the end of the nineteenth century, the existence of atoms and molecules was accepted by most scientists. However, atoms were still *invisible* to the human eye and remained so for decades. It was Perrin[1] [1,2], who, through careful experiments and quantitative analysis of *visible* phenomena, not only overcame the resistance of the last doubting scientists but also convinced the general public of, as he called it, 'la réalité moléculaire'.

In 1827 the botanist Brown[2] described that tiny particles, like pieces of pollen, dust, or soot, which he observed with his microscope while they were suspended in water, performed a perpetual random motion. Near the end of the nineteenth century, when the *kinetic theory of heat* was well developed, it became more and more accepted that this *Brownian motion* was caused by collisions of the tiny and therefore invisible water molecules with the much larger visible particles. It was taken to be a visual display of the intimate relation between motion and heat. In 1904 Poincaré described it in the words: 'We see under our eyes now motion transformed into heat by friction, now heat changed inversely into motion.' But still then no quantitative exploitation of Brownian motion was possible.

We have mentioned in the Prologue the hypothesis of Avogadro proposed in 1811, that, for equal pressure and equal temperature, equal volumes of different gases contain the same number of molecules. If a volume of 22.4 litres at atmospheric pressure and at a temperature of 0°C contains a gas with only one kind of molecules, then the mass of that gas is equal to the molecular mass number multiplied by 1 g, e.g., 4 g for helium (He), 2 g for hydrogen (H_2), and 32 g for oxygen (O_2). In each case the amount of substance is called 1 *mole*. The number of molecules in 1 mole is called *Avogadro's constant* N_A. One way to demonstrate that molecules exist, is to determine values of N_A from experiment by different methods and show that they coincide.

Once N_A is known, the *size* of molecules and even of atoms can be estimated. In a liquid or a solid the molecules can be assumed to be closely packed. Therefore, the volume taken by one mole, if divided by N_A, is a good estimate for the volume taken by a single molecule.

Experiments showed that for 1 mole of gas the product of the pressure p and the volume V of the gas is proportional to the absolute temperature T, $pV = RT$. The proportionality constant R is independent of the gas and is called (universal) *gas constant* and is easily measured. In the *kinetic theory* of gases, the proportionality law was understood, if one assumed that the N_A molecules in the volume V were in irregular motion, colliding with each other and with the walls, the collisions with the walls providing the pressure p and

[1] Jean Baptiste Perrin (1870–1942), Nobel Prize 1926   [2] Robert Brown (1773–1858)

if, on the average, each molecule had a kinetic energy $\langle E_{\text{kin}} \rangle$ proportional to the temperature T. For a gas consisting of molecules with only one atom, like helium, this proportionality takes the simple form $\langle E_{\text{kin}} \rangle = (3/2)kT$ and one obtains $pV = N_A kT$. Thus there is a very simple relation between the new constant k, called *Boltzmann's constant*, the gas constant R, and *Avogadro's constant* N_A, $N_A k = R$. Since R is known, a measurement of k directly yields N_A. Another result of kinetic theory, due to Boltzmann, is a statement not about the average value but about the distribution of energies over the molecules. The number of molecules with energy E is proportional to $\text{e}^{-E/kT}$, i.e., the number drops fast with increasing energy. This law holds for all kinds of energy. In the gravitational field of the earth the potential energy is proportional to its height above the surface of the earth. Consequently, the average number of molecules per unit volume, i.e., the average density of molecules in the air, drops with height – the air gets 'thinner' with increasing height.

Einstein's Formula for Brownian Motion (1905)

A small particle of radius a suspended in a liquid of viscosity η at absolute temperature T is observed with a microscope from above for a time interval τ. In this interval it is displaced in the plane of vision by a distance r from its initial position. The average of r^2 is proportional to τ,

$$\langle r^2 \rangle = \frac{2kT}{3\pi\eta a}\tau \quad .$$

The proportionality constant is $2kT/(3\pi\eta a)$. If η, a, τ, and T are known and the average $\langle r^2 \rangle$ is determined from many measurements, then Boltzmann's constant k can be computed from the formula.

Does knowledge about the behaviour of molecules in a gas help us to understand the motion of tiny but macroscopic particles in a liquid? It does if we use the law which van't Hoff[3] proposed on the basis of experiments by Pfeffer[4] on *osmosis*. The law is valid for dilute solutions, e.g., the solution of sugar (the solute) in water (the solvent) and states that the molecules of the solute behave very much like those of an ideal gas. In particular, they have the same energy distribution and average energy. It was only a small step to assume that what holds for larger molecules like those of sugar holds also for the much larger particles of Brownian motion.

It was now clear that one had only to measure the average kinetic energy of Brownian particles to determine k and thus N_A. The kinetic energy of a particle of mass m and velocity v is $E_{\text{kin}} = mv^2/2$. One had to determine mass and velocity; but it turned out that the latter was by orders of magnitude smaller than expected. After some discussion it was realized that the true velocity of a Brownian particle could not be observed with the microscope. The particle suffers collisions with individual molecules at a very high rate and it is the velocity between one such collision and the next which should be known. That is

---

[3] Jacobus Henricus van't Hoff (1852–1911), Nobel Prize 1901   [4] Wilhelm Pfeffer (1845–1920)

Einstein's Formula for the Particle Density as a Function of Height (1906)

If n_0 and n are the number of Brownian particles at height 0 and h, respectively, then

$$\frac{n}{n_0} = \exp\left\{-\frac{V(\rho - \rho_0)gh}{kT}\right\} \ .$$

Here V is the volume of a particle, ρ is its density, ρ_0 the density of the liquid and g the acceleration due to the earth's attraction.

---

much higher than the velocity observed because here the effects of many collisions with molecules coming from different directions counteract each other to a large extent.

In his 'miraculous year' 1905 Einstein gave a theory of Brownian motion [3]. He considered a Brownian particle to be a small sphere and computed the average of the square of the distance by which a Brownian particle is displaced in a given time, taking into account the friction of the sphere in water expressed by *Stokes' law* in terms of the viscosity of the water and the radius of the sphere. The distance is taken as the length of a straight line between the initial and final positions and has nothing to do with the details of the particle's path. This distance can be easily observed in the microscope, squared, and averaged over many observations. Early in 1906 Einstein also gave a formula [4] describing the density of Brownian particles as a function of height very similar to the density of air described above. Later in 1906 the Polish physicist Smoluchowski[5] [5] in Lemberg (then Austria, now Ukraine) published a comprehensive paper [6] on Brownian motion which contained Einstein's results and more but which had not been written earlier because Smoluchowski originally had planned also to do experiments.

The experiments were done by Perrin to whom we now, finally, return. He had come by himself to the conclusion that the density of Brownian particles should change with height similar to the density of air molecules and that he could measure Avogadro's constant from that change, provided he was able to produce Brownian particles that were very similar to each other. Working at the *Sorbonne*, the University of Paris, he set out to manufacture tiny spheres of *gamboge*. This substance was used for water colour and originates from a rubber tree. In alcohol and water it forms an emulsion consisting of tiny, perfectly spherical bodies suspended in the liquid. In a centrifuge, particles of different mass and thus of different diameter are deposited in different layers. By repeated, so-called fractional centrifugation, Perrin could select granules of very similar diameter. He developed methods to determine the density of the substance and the diameter and thus the volume of his granules with high precision.

Perrin then prepared a suspension of the selected granules in a very small hermetically closed glass vessel and placed it under the microscope, taking care that top and bottom of the vessel were exactly horizontal. The microscope offered clear vision only in a very thin layer. This layer could be moved up and down very precisely by turning a screw by which the microscope was lifted or

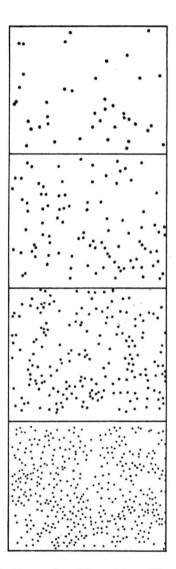

Perrin's recording of his particles at different heights. From [7].

---

[5] Marian Ritter von Smolan-Smoluchowski (1872–1917)

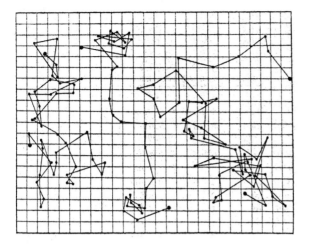

Perrin's recording of the Brownian motion of three of his particles. The dots correspond to positions measured in time intervals of 30 seconds. Dots corresponding to consecutive times are connected by straight lines. From [7].

lowered with respect to the vessel. In this way, the number of granules could be counted in different layers, i.e., at different height. Due to the Brownian motion this number, however, changed constantly and that made counting difficult. Perrin overcame this difficulty with two different methods. One way was to take a photograph and then count the granules on the picture. The other was to observe only a very small field of vision through a small hole and to make that observation possible only for a short time using the shutter of a camera. The hole was chosen so small that at most five granules were visible at the same time so that the number was immediately obvious to the observer. The average density was then obtained by very many such observations. Perrin made his measurement in 1908 and 1909 and published a very detailed account in 1909 [7].

During the course of his experiments he became aware of Einstein's and Smoluchowski's work and also studied the horizontal displacement of Brownian particles in a given time interval. With both methods he obtained values for Avogadro's constant compatible within experimental errors with each other and with the one obtained by Planck from blackbody radiation.

In 1910 Perrin became professor at the Sorbonne and in 1912 he published his book *Les Atomes* [8], in which he explained to the interested public all evidence, available at the time, why atoms exist. It made him a very popular figure. He created the *CNRS* (Caisse (now Centre) National de la Recherche Scientifique) which still today finances and organizes most of extra-university research in France. In 1940 after the occupation of France he fled to the United States where he died in 1942. In 1948 Perrin's remains returned to France on board the battleship *Jeanne d'Arc* and were buried in the Panthéon.

[1] Lot, F., *Jean Perrin et les atomes*. Éditions Seghers, Paris, 1963.

[2] Nye, M. J., *Molecular Reality – A Perspective of the Scientific Work of Jean Perrin*. MacDonald, London, 1972.

[3] Einstein, A., *Annalen der Physik*, **17** (1905) 549.

[4] Einstein, A., *Annalen der Physik*, **19** (1906) 371.

[5] Chandrasekhar, S., Kac, M., and Smoluchowski, R., *Marian Smoluchowski – His Life and Scientific Work*. PWN, Warsaw, 1999.

[6] v. Smoluchowski, M., *Annalen der Physik*, **21** (1906) 756.

[7] Perrin, J., *Annales de Chimie et de Physique*, **18** (1909) 1.

[8] Perrin, J., *Les Atomes*. Felix Alcan, Paris, 1912.

15   Millikan's Oil-Drop Experiment (1910)

Millikan

Millikan[1] [1,2] was the first American-born Noble laureate. He also was the best known and most influential American physicist between the two world wars. In text books, his name is mostly related to the oil-drop experiment for measuring the elementary electric charge e. Millikan was born in Iowa as son of a minister. He studied at Oberlin College and Columbia University where he obtained a Ph.D. in physics in 1895. After a year in Europe, where he attended lectures in Jena, Paris and Berlin and did research work in Göttingen under Nernst, Millikan became assistant and later professor at the Ryerson Laboratory of the University of Chicago which was led by the great Michelson.

In the next decade, Millikan did some research on radioactivity and also on the photoelectric effect. But his main activity was teaching and writing textbooks for both university and high-school students. In 1907, together with his graduate student Begeman[2], he began with measurements of the elementary charge. Let us shortly review the state of the art for the determination of that quantity at the time.

By studying the behaviour of a particle with charge q and mass m in an electric or magnetic field only the ratio q/m, also called the *specific charge* of the particle, can be obtained. We told in Episode 4 how J. J. Thomson achieved this measurement for the electron. To determine q additional information is needed; the mass m has to be known. A direct measurement of the mass of the electron or a charged atom was impossible. Therefore much larger objects were used: tiny water droplets. Thomson used the *cloud chamber* invented by C. T. R. Wilson in its original form, which did not yet produce tracks of charged particles. But it did produce water droplets around ions (see Episode 17). Depending on the amount of expansion, droplets would be formed only for negative ions or for ions of both charges. The ionization could be achieved with X rays or with radiation from radioactive sources. In his experiment [3] Thomson produced a cloud with negative ions. The mass of a water droplet is given by its radius. But if the droplet is small, the radius cannot be measured with a microscope because the image size is given by the resolution of the microscope which is limited by the wavelength of visible light. This is how Thomson determined the size and thus the mass of a droplet: A body, moving under the influence of a constant external force and a frictional force through a gas or liquid (like a parachutist through air), reaches a constant velocity. For a spherical body that velocity can by computed from its radius, from the *viscosity* of the medium and from the external force. Conversely, the radius can be computed if the velocity is measured and the other quantities are known. Thomson measured the velocity with which the cloud fell and therefore deter-

---

[1] Robert Andrews Millikan (1868–1953), Nobel Prize 1923   [2] Louis Begeman (1865–1958)

mined the average droplet size. He also measured the total charge Q of the cloud by letting it discharge a capacitor and he determined the number N of droplets in the cloud by dividing the total water volume of the cloud through the volume of one droplet. The average charge per droplet was then Q/N and he took that quantity as a measurement for the elementary charge.

Thomson's method was improved [4] by H. A. Wilson[3], who measured not one but two velocities, namely the speed of a droplet under its weight alone and the speed under its weight and under the influence of an electric field. From the two measurements one can determine both the mass and the charge of the droplet. There is no need any more to care about the total amount of water or the total charge in the cloud. Wilson tried to measure the speed by observing the sharp upper boundary of the cloud. He found that in the presence of an electric field there were two or even three boundaries moving with different velocities and he concluded that there were droplets with one, two, and more elementary charges.

This was the situation when Millikan entered the field. He and Begeman built an apparatus similar to Wilson's but since they had a larger battery they could produce higher fields which helped them to observe *individual droplets*. In 1909 Millikan published a rather good measurement based on the observation of single charged droplets of water [5]. He claimed to have eliminated 'every possible error due to evaporation' by obtaining drops in equilibrium with the surrounding gas for which evaporation compensated condensation. But it soon became clear that such droplets could not reliably be found. This then was the moment when the oil-drop idea suggested itself.

In 1910 Millikan published the first results obtained with oil drops [6] and in 1911 a comprehensive paper [7]. The experiment was much simplified by the use of oil since there was no need any more for the expansion system of a cloud chamber. Oil drops were obtained by using an *atomizer*, an instrument conceived by J. Y. Lee two years earlier at the Ryerson Laboratory for the study of Brownian motion. It was soon put to more profane use: The atomizer is still found sitting on top of many perfume flasks. The experimental setup consisted essentially of two horizontal metal disks, one 16 mm above the other, which could be connected to a high-voltage source (3000...8000 V). Droplets could enter through a pinhole in the upper plate; they were illuminated and could be observed with a microscope. Most droplets were charged when leaving the atomizer. They could be made to hover if the electric field exactly compensated gravity. As described above, charge and mass of a droplet were obtained by measuring its speed without electric field and also with a field switched on. The latter speed changed if a droplet changed its charge. That happened spontaneously but could also be provoked by the presence of a source of radioactivity. Millikan describes that he could follow a single drop indefinitely and measure individually the different charges it carried.

In his paper [6] he writes:

> Mr. Harvey Fletcher[4] and myself, who have worked together on these experiments since December, 1909, studied in this way between December and May from one to two hundred drops which had initial charges

$$F_e = qE$$

m, q

$$F_g = mg$$

A droplet of mass m and charge q under the influence of the gravitational force F_g and the electrostatic force F_e.

---

[3] Harold Albert Wilson (1874–1964)   [4] Harvey Fletcher (1884–1981)

between 1 and 150, and which were upon as diverse substances as oil, mercury and glycerin, and found in every case *the original charge of the drop an exact multiple of the smallest charge which we found that the drop caught from the air.*

This smallest charge was the elementary charge e, which Millikan determined with remarkable precision. Measurements with improved apparatus were published in 1913 and in 1917 when an accuracy of one in ten thousand was reached [8].

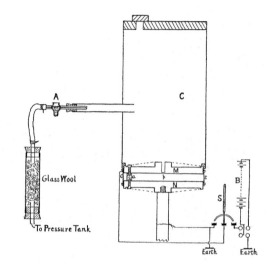

Millikan's original apparatus. A cloud of oil droplets is sprayed into the dust-free chamber C. One or more fall through the pinhole p into the space between the two capacitor plates M and N to which a voltage from the battery B can be applied by the switch S. Figure reprinted with permission from [7]. Copyright 1911 by the American Physical Society.

In the laboratory Millikan used to call the charged drops *ions* although that term is reserved for charged atoms or molecules and in his memoirs [2] he tells a delightful little story:

> One night Mrs. Millikan and I had invited guests for dinner. When six o'clock came I was only half through with the needed data on a particular drop. So I had Mrs. Millikan apprised by phone that "I watched an ion for an hour and a half and had to finish the job," but asked her to please go ahead with dinner without me. The guests later complimented me on my domesticity because what they said Mrs. Millikan had told them was that Mr. Millikan had "washed and ironed for an hour and a half and had to finish the job."

In 1916 Millikan published his other piece of precision measurement, the determination of Planck's constant on the basis of Einstein's theory of the photoelectric effect, which we mentioned already in Episode 10. When it became likely that the United States would enter the First World War the *National Research Council* (NRC) was formed. Millikan became vice-chairman and led the subcommittee on submarine detection. In 1917 he joined the Army as head of the Signal Corps's Science and Research Division, beginning as major and rising to the rank of lieutenant colonel. The NRC played an important role coordinating science and finding funds during and also after the war. When the war was over Millikan returned to his work in Chicago but in 1921, urged by Hale[5], another NRC member and director of the Mount Wilson Observatory,

---

[5]   George Ellery Hale (1868–1938)

he became the first leader of the California Institute of Technology in Pasadena then abbreviated C.I.T. but now best known as *Caltech*. (His title was *Chairman of the Executive Council*; only his successors are called *president*.) This institution rose amazingly fast to the forefront of research by attracting leading scientists and, not least, because of Hale's and Millikan's ability to raise large funds from wealthy Californians.

Millikan with (from the left) Nernst, Einstein, Planck, and von Laue in Berlin, 1928.

Millikan also was director of the Norman Bridge Laboratory of Physics of Caltech. His interest had shifted to *cosmic radiation*, a name he coined. He performed experiments and also published rather speculative papers on the nature and the origin of these rays. In 1932 Anderson in Millikan's group in Pasadena discovered the positron in a cosmic-ray experiment (see Episode 49).

In 1982 a charmingly written article by Harvey Fletcher [9], who had died the previous year, was published under the title *My work with Millikan on the oil-drop experiment*. In 1909 he told Millikan that he would like to work under him on a Ph.D. thesis and Millikan arranged a meeting with him and Begeman. In that meeting it was agreed that Fletcher should work on the measurement of *e* with droplets of non-volatile liquids such as oil or mercury. Fletcher bought an atomizer and some watch oil in a drugstore and set up a provisional apparatus with which he demonstrated that the method would allow the measurement of *e* and the study of Brownian motion. He had worked with Millikan on both subjects for about a year when one day Millikan looked him up at home, told him that for his Ph.D. he would need a paper of which he was the only author and suggested that Fletcher be the sole author of papers on Brownian motion and he, Millikan, the sole author of papers on the elementary charge. Fletcher, although disappointed, agreed. But he stressed that he bore no grudge against Millikan and that he wrote down the story only because some of his classmates had spread the rumour that he had been unfairly treated. (For this and another incident concerning Millikan and the oil-drop experiments see also [10].)

[1] Kargon, R. H., *The Rise of Robert Millikan*. Cornell University Press, Ithaca, 1982.

[2] Millikan, R. A., *The Autobiography of Robert A. Millikan*. Prentice-Hall, New York, 1950.

[3] Thomson, J. J., *Philosophical Magazine*, **48** (1899) 547.

[4] Wilson, H. A., *Philosophical Magazine*, **5** (1903) 429.

[5] Millikan, R. A., *Physical Review*, **29** (1909) 560.

[6] Millikan, R. A., *Science*, **32** (1910) 436.

[7] Millikan, R. A., *Physical Review*, **32** (1911) 349.

[8] Millikan, R. A., *Philosophical Magazine*, **34** (1917) 1.

[9] Fletcher, H., *Physics Today*, **June** (1982) 43.

[10] Goodstein, D., *American Scientist*, **Jan.–Feb.** (2001) 54.

16

The Atomic Nucleus (1911)

In the first decade of the twentieth century atoms were a reality to nearly every scientist. It was also known that they were *not* indivisible since, by ionization, one or more electrons could be removed from an atom. That meant that an atom contains one or more electrons. As we have seen (Episode 14), the size of atoms was known. But nothing was known about their inner structure. Several models of the atom were proposed but could not be subjected to severe experimental tests. The most popular model was one by J. J. Thomson, sometimes called the *plum-pudding model* [1]. It described the atom as a sphere filled out evenly with positive charge which contained the point-like electrons.

In 1903 Lenard studied in detail the absorption of cathode rays, i.e., electrons, by different materials [2]. He found that they can traverse quite thick layers of solid matter and concluded that the cathode rays must be able to traverse the atoms themselves. He therefore assumed that the atoms are composed of 'finer constituents', which he called *dynamides*, 'with many free spaces between them' such that the cathode rays could pass through these free spaces. Lenard was the first to probe the structure of the atom by shooting particles at it. Such *scattering experiments* with atoms or its constituents as *targets* are done to this day.

The true structure of the atom was revealed by Rutherford and his collaborators by scattering α particles on atoms and, in particular, by Rutherford's theoretical analysis of the scattering process. Rutherford discovered α scattering in 1906 [3] when he was still in Montreal. In an evacuated tube he let a beam of collimated α particles pass through a slit and then fall onto a photographic plate on which he could observe the image of the slit. When he placed a thin foil of mica between slit and plate, the image was blurred.

In Manchester, Rutherford chose the scattering of α particles as one of the first research subjects in his new laboratory and asked Geiger and a young research student, Marsden[1], to study it in detail. To measure the scattering angle, i.e., the angle by which an α particle was deviated from its original direction, instead of a photographic plate a zinc-sulphide screen was used. On it the position, where each α particle hit the screen, could thus be observed under the microscope. It was found that the average scattering angle was small but that there were exceptions. Geiger wrote in 1910 [4]:

> It is also of interest to refer here to experiments made by E. Marsden and myself (see [5]) on the diffuse reflection of the α-particles. It was found that some of the α-particles falling upon a metal plate appear to be reflected, *i.e.* they are scattered to such an extent that they emerge again on the side of incidence. It was shown that from gold 1 in about 8000 of the incident α-particles suffers reflection, and that this reflection takes

---

[1] (Sir) Ernest Marsden (1889–1970)

Box I. Interaction Cross Section, Differential Cross Section

We are interested in the probability that a *projectile particle* interacts with a *target particle*. The projectile particles form a *beam* of, for instance, electrons or α particles. The target particles are contained in a piece of matter, the *target*. They are, for instance, the atoms in the target. We define the *impact parameter b* as the distance of closest approach between the centre of the projectile particle and centre of the target particle if there was no interaction at all. Suppose for the moment that projectile and target particles behave like hard spheres with the radii r_P and r_T, respectively. Then an interaction takes place only if b is less than the sum of the two radii, $b < r_P + r_T$. We call the *geometric cross section*

$$\sigma_{geom} = \pi(r_P + r_T)^2$$

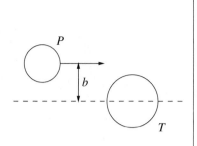

Initial position of projectile P and target particle T in a plane containing the initial projectile momentum, b is the impact parameter.

the cross section of a sphere with a radius, which is the sum of the two radii. An interaction takes place if the centre of the projectile particle would hit a circular disk with the area σ_{geom} surrounding the centre of the target particle and oriented perpendicular to the direction of incidence.

As seen from the projectile particle, a certain fraction of the target is covered by such disks. This fraction is

$$P = \sigma n \, \Delta x \quad,$$

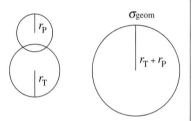

Initial position of P and T as seen in the initial direction of P (left). The geometrical cross section is the surface of a disc of radius $r_P + r_T$ (right).

where n is the density of target particles inside the target and Δx is the target thickness. At the same time, P is the *interaction probability* for a projectile particle with the target. Thus the cross section is related in a simple way to the interaction probability. In general, this relation defines the concept of an interaction cross section, since the naive geometrical interpretations usually do not hold true.

Often one is interested in details of the interaction, for instance, in the probability that the angle of scattering, i.e., the angle between the initial and the final direction of the projectile, is between ϑ and $\vartheta + d\vartheta$. In analogy this probability is given by

$$dP(\vartheta) = d\sigma(\vartheta) \, n \, \Delta x = n \, \Delta x \, \frac{d\sigma(\vartheta)}{d\vartheta} \, d\vartheta \quad.$$

Here $d\sigma(\vartheta) = 2\pi b \, db$ is the area of a circular ring of radius b and thickness db, since a given impact parameter leads to a certain scattering angle, $b = b(\vartheta)$. Usually one is interested in the *differential cross section* with respect to the *solid angle* $d\Omega = 2\pi \sin \vartheta \, d\vartheta$, i.e., the scattering into the region between the two cones with opening angles ϑ and $\vartheta + d\vartheta$. It is

$$\frac{d\sigma(\vartheta)}{d\Omega} = \frac{d\sigma(\vartheta)}{db} \frac{db}{d\vartheta} \frac{d\vartheta}{d\Omega} = \frac{b}{\sin \vartheta} \frac{db}{d\vartheta} \quad.$$

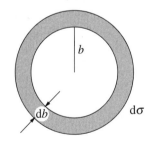

The surface $d\sigma(\vartheta)$ of a ring with radii b and $b + db$ is the cross section for scattering by an angle between θ and $\theta + d\theta$.

From this relation it is clear that the differential cross section, and hence the probability for a certain scattering angle, are known, if the relation $b = b(\vartheta)$ between impact parameter and scattering angle is known.

place within a relatively thin surface layer equivalent to about 5 mm. of air.

Rutherford was very much intrigued by these large scattering angles. Until then it had been thought that larger angles came about by multiple scattering off electrons while the α particles traversed one or several atoms. But Rutherford realized that larger forces than those of a single electron were needed. He

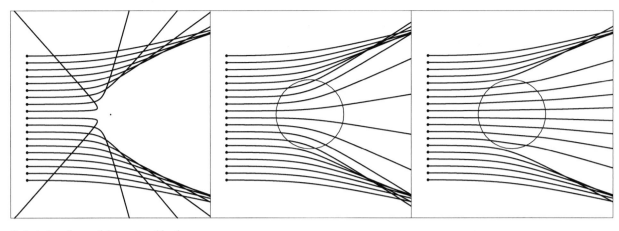

Trajectories of α particles scattered by three different positive charge configurations, a point charge (left), a uniformly charged sphere (middle), and a sphere carrying a uniform surface charge (right). Only for the point charge very large scattering angles occur.

therefore assumed a large *central charge* within the atom. Only later he used the word *nucleus*. In addition to its charge the nucleus was to contain essentially the complete mass of the atom. He thought at first that the central charge must be negative so that it attracted an α particle, which could swing around it like a comet coming from far outside is swung around by the sun. But he realized that the α particle could just as well be deflected by a large positive central charge. He then wrote a paper [6] in which he worked out the scattering theory in all detail. Rutherford did not use the term *scattering cross section* but *probability of scattering* which is essentially the same. We have sketched the theory here in two boxes, using modern nomenclature. With this theory Rutherford was able to completely predict the angular distribution of the scattered α particles as a function of the energy of the α particles, the number Z of elementary charges in the nucleus of a target atom and of the target thickness.

In the theoretical part of his paper, Rutherford made it clear that by measuring the distribution of scattering angles it could not be decided whether the central charge was positive or negative. But in the concluding section, 'for concreteness', he considered 'an atom having a positive central charge Ne, and surrounded by a compensation charge of N electrons.' After all, it was well known that electrons carry the charge $-e$. (The nuclear charge number, called N by Rutherford, is now generally denoted by Z.) This was Rutherford's nuclear model of the atom.

Geiger wrote later [7]:

> Rutherford must have thought much about this peculiarity [the large scattering angles]: one day (in 1911) he came, obviously in the best of moods, into my room and told me he now knew what the atom looked like and how the large scatterings were to be understood. Probably still that same day I began with an experiment to verify the relation, predicted by Rutherford, between the particle number and scattering angle.

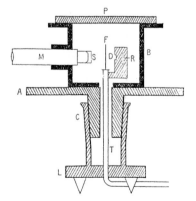

Cut through the apparatus of Geiger and Marsden. From [8]. Reprinted by permission from Taylor & Francis Ltd, http://www.informaworld.com.

In a classic experiment Geiger and Marsden checked Rutherford's predictions [8]. Their apparatus consisted of a stationary part, mounted on the plate L, which carried a hollow cone C, and a scattering chamber B covered by the plate P and mounted on the plate A, which sat on a cone and could be rotated in C. In the fixed part, situated in the axis of rotation of the moving part, was

---

Box II. The Rutherford Scattering Cross Section

The electric force between a fixed charge Q and another charge q is, except for a factor, mathematically the same as the gravitational force between a fixed mass M and outer mass m. The only difference is that, depending on the signs of the charges, the electric force can be repulsive or attractive, whereas the gravitation is always attractive. If the moving particle comes in from infinity, its trajectory is a hyperbola. In the case of an electric force the scattering angle ϑ, i.e., the angle between the initial and the final asymptote of the hyperbola, is related to the impact parameter b by

$$b = \frac{qQ}{4\pi\varepsilon_0} \frac{1}{E_{\text{kin}}} \cot\frac{\vartheta}{2} \quad .$$

Here E_{kin} is the kinetic energy of the projectile. The factor $4\pi\varepsilon_0$ appears, if one writes down the force in SI units (ε_0 is the electric field constant). The scattering angle is the same for a repulsive or an attractive force. By introducing the charges of the α particle $q = 2e$ and of the nucleus $Q = Ze$ and using the relation between differential cross section and impact parameter (see bottom of Box I) one gets the differential *Rutherford cross section*

$$\frac{\mathrm{d}\sigma(\vartheta)}{\mathrm{d}\Omega} = \frac{Z^2 e^4}{64\pi^2\varepsilon_0^2} \frac{1}{E_{\text{kin}}^2} \frac{1}{\sin^4\frac{\vartheta}{2}} \quad .$$

---

the target foil F. It was irradiated by a beam of α particles emitted by the source R and collimated by the diaphragm D. Source and diaphragm were also fixed. The α particles scattered in F and hitting the screen S gave rise to scintillations which were observed with the microscope M on which the screen was mounted. The scattering angle was easily varied by turning the movable part to which the microscope belonged. The whole assembly was constructed as a single vessel, which was evacuated by connecting the tube T to a pump. In this way, it was assured that the scattering took place in the foil and not in air.

Geiger and Marsden were able to verify Rutherford's predictions with rather high precision. Moreover, by using different target materials, from carbon to gold, they showed that the number Z of elementary charges in the central charge, i.e., the nucleus, was approximately half the atomic mass number A. They stated that, for elements with an atomic mass number greater than that of aluminium, this approximation, $Z \approx A/2$, was good to about 20%. From that Rutherford computed that, if an α particle with the energy used in the experiment was to be scattered directly backwards ($\vartheta = 180°$) by a gold nucleus, it had to approach the central charge up to the very small distance of 3×10^{-12} cm. Thus, while the atom had a radius of about 10^{-8} cm, the nucleus, which contained practically all its mass, had a radius of at most 3×10^{-12} cm, because otherwise deviations from Rutherford's scattering formula would have shown up in the experiment. Only a tiny part of the atom's volume was used up by the nucleus. Here were Lenard's free spaces.

There is a letter of 2 September 1932 by Rutherford, now in Cambridge, to Geiger who was then professor in Tübingen. The second sentence reads [9]:

> They were happy days in Manchester and we wrought better than we knew.

[1] Thomson, J. J., *Philosophical Magazine*, **6** (1903) 673.

[2] Lenard, P., *Annalen der Physik*, **12** (1903) 714.

[3] Rutherford, E., *Philosophical Magazine*, **12** (1906) 134. Also in [10], Vol. 1, p. 859.

[4] Geiger, H., *Proceedings of the Royal Society*, **A83** (1910) 492.

[5] Geiger, H. and Marsden, E., *Proceedings of the Royal Society*, **A82** (1909) 495.

[6] Rutherford, E., *Philosophical Magazine*, **21** (1911) 669. Also in [10], Vol. 2, p. 238.

[7] Geiger, H. in [10], Vol. 2, p. 296.

[8] Geiger, H. and Marsden, E., *Philosophical Magazine*, **25** (1913) 604.

[9] Eve, A. S., *Rutherford – Being the Life and Letters of the Rt Hon. Lord Rutherford, O.M.* Cambridge University Press, Cambridge, 1939.

[10] Rutherford, E., *The Collected Papers of Lord Ernest Rutherford of Nelson (Edited by J. Chadwick)*, *3 vols.* Allen and Unwin, London, 1962–1964.

17 Tracks of Single Particles in Wilson's Cloud Chamber (1911)

A fitting introduction to this episode is the following quotation from the speech of C. T. R. Wilson[1] at the Nobel Banquet in Stockholm on December 10, 1927:

> I feel that I owe what success I have achieved in my scientific work mainly to my good fortune in beginning my experiments on clouds just at the right time – a few months before Röntgen made his great discovery. And my choice of a subject to work upon was not due to any forethought on my own part nor to any good advice received, but just to the fact that in the autumn of 1894 I spent a few weeks on a cloudy Scottish hill-top – the top of Ben Nevis. Morning after morning I saw the sun rise above a sea of clouds and the shadow of the hill on the clouds below surrounded by gorgeous coloured rings. The beauty of what I saw made me fall in love with clouds and I made up my mind to make experiments to learn more about them. Working in J. J. Thomson's laboratory during the years when X-rays and radio-activity were discovered, I could not help being interested in ions – and with ions and clouds I have worked ever since.

A cloud consists of tiny water droplets. The creation of a cloud in clear sky means that (invisible) water vapour is turned into (visible) liquid water. Whether water, mixed with air, is gaseous or liquid depends on the ratio of water to air, on the temperature and on the pressure. By changing one or more of these parameters, vapour can be changed into liquid or vice versa. Air containing vapour and in a state that a tiny parameter change would turn that into liquid is said to be *saturated*. It is then easy to bring it into a *supersaturated* state in which droplets can be formed. It was known that condensation of vapour into liquid takes place first at *condensation nuclei* such as small dust particles.

Wilson began to make clouds in his own *cloud chamber* in the Cavendish Laboratory in Cambridge in the beginning of 1895. He enclosed air, which was not quite saturated, in a small chamber. When he suddenly lowered the pressure, beautiful clouds appeared. The pressure was lowered by allowing the gas mixture to expand into a larger volume. When he cleaned the air of all dust particles he was able to reach a rather high degree of supersaturation. Wilson could expand the volume of his chamber by one fourth of its original volume and no droplets were formed. Only for larger expansion there was condensation in the form of larger drops like a light rain.

In the beginning of 1896, when the news of Röntgen's discovery reached Cambridge, J. J. Thomson at once began to study ionization caused by X rays and also Wilson exposed his chamber to the new rays and already at his first trial found that 'a fog which took minutes to fall' was formed [1,2]. He proved

C. T. R. Wilson

[1] Charles Thomson Rees Wilson (1869–1959), Nobel Prize 1927

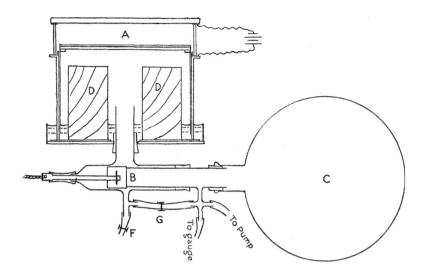

Wilson's first track chamber. The tracks are formed in the flat volume A. Below is the expansion mechanism. From [3], Fig. 1. Reprinted with permission.

that the condensation nuclei for the small droplets in the fog really were ions and he was able to distinguish drop formation by positive and by negative ions. Drops formed around an ion could be influenced together with the ion itself by an electric field. J. J. Thomson in 1899, assuming that each droplet contained a singly charged ion, was able to measure the elementary charge. That was the first application of the cloud-chamber method in fundamental physics. Later Millikan refined the method of charge measurement. But he used oil drops for which there was no need of condensation and which, moreover, did not evaporate (see Episode 15).

Only more than ten years later it occurred to Wilson, 'that the track of an ionizing particle might be made visible and photographed by condensing water on the ions which it liberated' [4]. He built a new apparatus with faster expansion including suitable means of illumination and photography. The first trials were done in spring 1911. In his Nobel Lecture, Wilson reported [4]:

> The first tests were made with X-rays with little expectation of success, and in making an expansion of the proper magnitude for condensation on the ions while the air was exposed to the rays. I was delighted to see the cloud chamber filled with little wisps and threads of clouds – the tracks of the electrons ejected by the action of the rays. The radium-tipped metal tongue of a spinthariscope was then placed inside the cloud chamber and the very beautiful sight of the clouds condensed along the tracks of the α-particles was seen for the first time. The long thread-like tracks of fast β-particles were also seen when a suitable source was brought near the cloud chamber.

After a first publication [5] with a few photographs in 1911, Wilson published a comprehensive paper [3] in 1912 which already contained a photograph of the track of an α particle which, in two distinct places, changes its direction because the particle suffers Rutherford scattering. The cloud chamber which showed the tracks of individual charged (and thus ionizing) particles became an extremely valuable instrument of fundamental research. If placed in a magnetic field the trajectory of a particle became curved and the radius

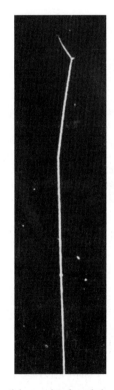

An α particle, coming from below, is scattered twice by nuclei from the gas filling of the chamber. The recoil nucleus of the second scattering itself leaves a track. From [3], Plate 6.3. Reprinted with permission.

of curvature allowed to determine the momentum of the particle. Complicated *events*, for instance, the interaction of an incoming projectile particle with a target particle in the chamber, could be analysed in all detail, especially if a pair of stereoscopic photographs was taken from which the directions (and momenta) of all charged particles involved could be reconstructed. The number of ions produced per centimetre along the trajectory depends on the charge of the particle and its velocity. It is large for high charge and low velocity. Since α particles are doubly charged and, usually, because of their much higher mass, have a lower velocity than β particles, the droplet density in the tracks of α particles, generally, is much higher. Also slow β particles produce tracks with a higher droplet density than fast ones. Because of its low mass, a slow β particle is easily deflected by the much heavier atomic nuclei. Its path is not straight but appears quite irregular. By all these features α and β particles, and also others, can be told apart in the cloud chamber.

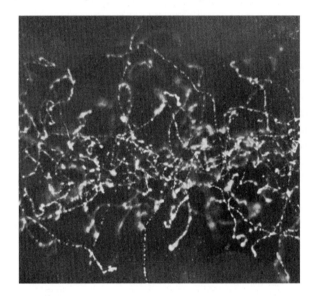

Tracks of electrons liberated by a horizontal beam of X rays. From [3], Plate 9.1. Reprinted with permission.

[1] Wilson, C. T. R., *Proceedings of the Royal Society*, **A59** (1896) 338.
[2] Wilson, C. T. R., *Proceedings of the Royal Society*, **A61** (1896) 240.
[3] Wilson, C. T. R., *Proceedings of the Royal Society*, **A87** (1912) 277.
[4] Wilson, C. T. R., *Nobel Lectures, Physics 1922–1941*, p. 194. Elsevier, Amsterdam, 1965.
[5] Wilson, C. T. R., *Proceedings of the Royal Society*, **A85** (1911) 285.
[6] Blackett, P. M. S. and Ochialini, G., *Nature*, **130** (1932) 363.
[7] Gentner, W., Maier-Leibnitz, H., and Bothe, W., *Atlas typischer Nebelkammerbilder*. Springer, Berlin, 1940. Engl. transl. [9].
[8] Rochester, G. D. and Wilson, J. G., *Cloud Chamber Photographs of the Cosmic Radiation*. Pergamon, London, 1952.
[9] Gentner, W., Maier-Leibnitz, H., and Bothe, W., *An atlas of typical expansion chamber photographs*. Pergamon, London, 1954.

In 1932 Blackett and Occhialini, also working at the Cavendish Laboratory, combined the cloud chamber with Geiger counters [6]. A set of counters gave a signal if one or several particles from the cosmic radiation had traversed the chamber. The ionization along the trajectories in the chamber was preserved long enough for the chamber to be expanded and for the tracks of droplets to be formed so that a photograph could be taken.

Important discoveries were made with the help of Wilson's cloud chamber, some of them completely unexpected, like that of the *positron* (Episode 49) and that of the so-called *strange particles* (Episode 68). Two beautiful collections of photographs from many laboratories give an impressive overview of the research with cloud chambers in radioactivity [7] and in cosmic rays [8].

Kamerlingh Onnes – Liquid Helium and Superconductivity (1911)

Superconductivity is the property of some conductors to completely lose their electric resistance below some, usually very low, temperature. It was discovered in mercury in 1911 by Kamerlingh Onnes[1] [1,2], who, in 1908, had succeeded in liquefying helium and thus in producing the 'the lowest temperature on earth' at the time.

Some gases can be liquefied simply by putting them under sufficient pressure. An example is the gas in a cigarette lighter. Gases for which that method does not work were called *permanent*. In 1869, studying the properties of carbon dioxide, Andrews Thomas[2] found that for a given gas there is a *critical temperature* above which liquefaction by pressure is impossible. The pressure needed at the critical temperature is the *critical pressure*. Inspired by these concepts van der Waals[3], in his Ph.D. thesis of 1879, published an equation which describes the transition from gaseous to liquid state for a *real gas*. In addition to the gas constant of the *ideal gas*, known from the kinetic theory of heat, this equation contains two constants describing the properties of the particular gas. In 1880, when van der Waals was the first professor of physics at the recently founded University of Amsterdam, for his theory of *corresponding states* he rewrote this equation. It still contained two constants but these were now the critical temperature and the critical pressure.

In 1882 Onnes, a native of Groningen in the north of the Netherlands, who had studied in Groningen and also in Heidelberg under Kirchhoff and Bunsen, became professor of experimental physics at the University of Leiden, a position he was to hold for 42 years. In his inaugural lecture [3] he described with admiration the work of his compatriot van der Waals and stressed the necessity of reaching lower temperatures. We quote a famous phrase from that lecture:

> *Door meten tot weten* [through measuring to knowing] is the motto which
> I would like to write over every physics laboratory.

It was Onnes' aim not only to produce liquid gases, as cold as possible, but also to produce them in relatively large quantities. He then wanted to use them as baths to cool down materials whose properties he wanted to study. To achieve that he set up a large laboratory with quasi-industrial aspects quite unusual for the time. The method by which he wanted to reach low temperatures was a *cascade* of cooling circuits. In the first circuit a suitable gas (Onnes chose chloromethane) is compressed at room temperature to become liquid, which in a large volume vaporizes (at least in part) and as a result of vaporization cools off (in the case of chloromethane to $-90°C$). The cool vapour is brought back to the compressor so that the circuit is closed. On the way to

---

[1] Heike Kamerlingh Onnes (1853–1926), Nobel Prize 1913   [2] Andrews Thomas (1813–1885)   [3] Johannes Diderik van der Waals (1837–1923), Nobel Prize 1910

the compressor the vapour can be used to cool a second circuit containing a gas that can be liquefied by compression at the temperature reached by the first circuit, and so on. For the second circuit ($-140°$C) Onnes used ethylene and for the third ($-180°$C) oxygen.

In 1892 after 10 years of work, Onnes produced liquid oxygen. But by that time it was already available commercially through a different process invented by Linde[4], who used the *Joule–Thomson effect*. If a gas is expanded from high pressure though a fine throttle into a low-pressure vessel, then the gas cools because the molecules have to perform work against intermolecular forces when moving away from each other. For oxygen the cooling is $0.25°$C per atmosphere of pressure drop. The cooled gas is compressed again but on its way to the compressor it is made to cool the gas streaming to the expansion valve. In this way oxygen can be liquefied in a single circuit. The *Linde process* was not only more elegant but it also helped to reach really low temperatures, since the temperature of liquid oxygen is still higher than the critical temperature of hydrogen. Therefore hydrogen could not have been liquefied in a simple cascade with a hydrogen circuit following an oxygen circuit. In 1896 Dewar[5], using the Linde process on hydrogen gas cooled to liquid-oxygen temperature, produced $20\,\text{cm}^3$ of liquid hydrogen. Dewar was also the inventor of the *Dewar flask*, a very well-insulating container for the storage of liquid gases, now known to everybody as *thermos flask* and used to keep coffee warm or lemonade cold.

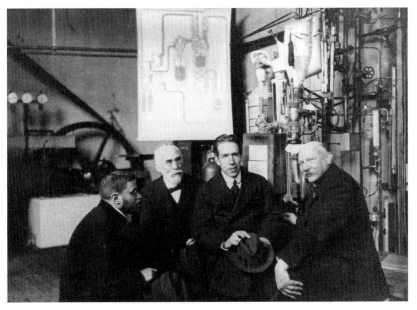

From the left: Ehrenfest, Lorentz, and Bohr with Kamerlingh Onnes in his laboratory in Leiden, 1919.

It seemed that Onnes was beaten. But he won the final round by liquefying helium. He added two more circuits to his cascade, one in which he produced liquid hydrogen the way Dewar had done and a second which should work in a similar way starting from gaseous helium at liquid-hydrogen temperature ($-250°$C). He also had to get hold of a sufficient supply of helium gas, which

[4] Carl (von) Linde (1842–1934)   [5] (Sir) James Dewar (1842–1923)

then was not commercially available (for the discovery of helium see the end of Episode 6). With the help of his brother, director of the department of foreign trade relations in Amsterdam, he got a supply of monazite from North Carolina, which contained between 1 and $2\,\mathrm{cm}^3$ of helium gas per gram. It took four chemists and three years to extract and purify 360 litres.

Finally on 10 July 1908, a quarter of a century after Onnes began to set up his laboratory, liquid helium was produced in Leiden. To collect the liquid and to keep it as cold as possible, he had constructed a set of four Dewar flasks, one nested in the other. The innermost was to contain liquid helium, the next hydrogen, the third oxygen and the outermost ethylene. Onnes had created his own platform for publications, the *Communications from the Physical Laboratory of the University of Leiden*, in which *The Liquefaction of Helium* was published as number 108 [4]. He gave a particularly vivid report in his Nobel Lecture [5] from which we quote a short passage:

> It was a wonderful sight when the liquid, which looked almost unreal, was seen for the first time. It was not noticed when it flowed in. Its presence could not be confirmed until it had already filled up the vessel. Its surface stood sharply against the vessel like the edge of a knife. How happy I was to be able to show condensed helium to my distinguished friend van der Waals, whose theory had guided me to the end of my work on the liquefaction of gases.

The boiling temperature of liquid helium under atmospheric pressure is $4.2\,\mathrm{K}$ or $-269°\mathrm{C}$. It can be raised somewhat by increasing the pressure or lowered by decreasing it, i.e., by pumping off the vapour rapidly. Onnes began to measure the electric resistance of metals in the new temperature range. He soon concentrated on mercury because it could be made particularly pure by distillation. Mercury is liquid at room temperature and becomes solid when cooled. A system of glass capillaries was constructed which, when cold, contained a 'wire' of solid mercury. Mercury conductors leading to the outside were connected to the ends of the wire and to intermediate positions. If an electric current was passed through the wire, a potential drop due to the electric resistance could be measured between the intermediate leads.

By the time Onnes began his measurements, there was already a rather developed theory of electric conduction in metals, mainly due to Drude, who had introduced the idea of a *free electron gas*. It predicted a proportionality of resistivity and absolute temperature, i.e., the electric resistance was expected to reach zero at zero absolute temperature. However, deviations from this law had been observed and predictions, based on measurements, offered three possibilities at zero temperature: vanishing resistance, a finite value, and even infinite resistance. In all cases a smooth approach to zero temperature was predicted.

The resistance measurements, initiated by Onnes, were actually performed by his assistant Holst[6], whose contribution is acknowledged in the publications. What Onnes reported in April [8] and May [9] and with more details in November [7], 1911, had not been foreseen. When the mercury was cooled below the temperature of $4.2\,\mathrm{K}$, its resistance dropped *abruptly* to an immeasurably low value so that only an upper limit could be given.

In his Nobel Lecture, given two years later, he said:

---

[6] Gilles Holst (1886–1968)

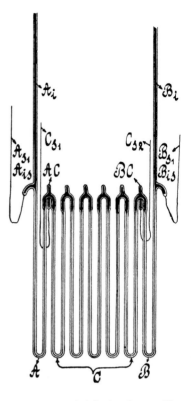

The mercury 'wire' in its glass capillary. From [6].

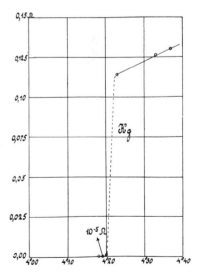

Kamerlingh Onnes' graph of measurements showing that near $T = 4.2\,\mathrm{K}$ the electric resistance of mercury drops abruptly. From [7].

[1]   van Helden, A. C., *The Coldest Spot on Earth – Kamerlingh Onnes and Low Temperature Research 1882–1923*. Museum Boerhaave, Leiden, 1989.

[2]   Dahl, P. F., *Superconductivity – Its Historical Roots and Developments from Mercury to Ceramic Oxides*. American Institute of Physics, New York, 1992.

[3]   Laesecke, A., *Journal of Research of the National Institute of Standards and Technology*, **107** (2002) 261.

[4]   Kamerlingh Onnes, H., *Leiden Communications*, **No. 108** (1908). Also in [10] p. 164.

[5]   Kamerlingh Onnes, H., *Nobel Lectures, Physics 1901–1921*, p. 27. Elsevier, Amsterdam, 1967.

[6]   Kamerlingh Onnes, H., *Leiden Communications*, **No. 133a,b,c** (1913). Also in [10] p. 273.

[7]   Kamerlingh Onnes, H., *Leiden Communications*, **No. 124c** (1911). Also in [10] p. 267.

[8]   Kamerlingh Onnes, H., *Leiden Communications*, **No. 120b** (1911). Also in [10] p. 261.

[9]   Kamerlingh Onnes, H., *Leiden Communications*, **No. 122b** (1911). Also in [10] p. 264.

[10] Gavroglu, K. and Goudaroulis, Y. (eds.), *Through Measurement to Knowledge – The Selected Papers of Heike Kamerlingh Onnes 1853–1926*. Kluwer, Dordrecht, 1991.

Thus the mercury at $4.2\,\mathrm{K}$ has entered a new state, which, owing to its particular electrical properties, can be called the state of *superconductivity*.

It is very important to the study of the properties of this state that tin and lead can also become superconductive. The transition point of tin lies at $3.8\,\mathrm{K}$, that of lead probably at $6\,\mathrm{K}$. Now that we are able to use these metals, which are easy to work, all types of electrical experiments with resistance free apparatus have become possible. To take one example: a self-contained coil, cooled in the magnetic field, should, when the field is removed, be able to simulate for some time an Ampere molecular current. There is also the question as to whether the absence of Joule heat makes feasible the production of strong magnetic fields using coils without iron, for a current of very great density can be sent through very fine, closely wound wire spirals.

In this short passage, Onnes not only summarized his discovery but also proposed two applications of it, the construction of a coil in which a *persistent current* flows as long as desired and the construction of an electromagnet with a current and therefore a magnetic field as high as desired. In two notes added in proof to the printed version of the lecture he reported his own results concerning these two points.

He had, in April–June 1914, achieved a persistent current. A short-circuited coil had been placed in a Dewar flask. Before it became superconducting, a large electromagnet was brought near the apparatus so that the flask was between its pole shoes. In this way there was a changing magnetic field which induced a current in the coil. Because of the coil's resistance that current quickly came to a halt. When the coil was superconducting the magnet was moved away. Again the field changed but the induced current persisted. It could be observed with the help of a compass needle placed near the superconducting coil.

In 1932, when liquid helium was not yet available in Britain, a persistent current was demonstrated to the public in style during a lecture at the Royal Institution in London. The current was started in a simple lead ring, immersed in liquid helium in a Dewar vessel, at 3 p.m. in Leiden. The vessel was brought to the Airport near Amsterdam, piloted by a Scottish lord in his private plane to London and driven by car to the lecture. There the current was still found to be running at 9 p.m. The amusing story with excerpts of letters and telegrams exchanged between the gentlemen involved in the enterprise is told in [2].

What concerned the possibility of creating a very large magnetic field Onnes, in the second note, added in proof to his Nobel Lecture, stated that at a given temperature there is a threshold value of the field beyond which the magnetic field itself destroys the superconductivity of the material.

Superconductivity, found by Kamerlingh Onnes in 1911 as a surprise to him and everybody else, has been an active field of research ever since. It took decades until a satisfactory explanation of the astonishing phenomenon was given (see Episode 78).

Hess Finds Cosmic Radiation (1912)

As early as 1785 Coulomb[1], to whom we owe the law governing electrostatic forces, found that an electrically charged body gradually loses its charge even if it is very well insulated. It was found out that this effect is due to a small electric conductivity of the air. After Becquerel had observed that this conductivity increased when radioactive material was present, it became generally accepted that the radiation emitted by that material created pairs of positive and negative ions in the air and that the conductivity of the air was caused by radioactive substances in the earth and possibly in the air itself. If that was so, then the conductivity of atmospheric air should decrease with rising distance from the surface of the earth.

In 1912, Hess[2] [1] discovered that the contrary is true and that therefore there existed an extraterrestrial source of radiation, later named *cosmic radiation*. As turned out much later, it consists of protons and heavier atomic nuclei with a wide range of energies. Some have an energy very much larger than can be achieved by modern particle accelerators. Because of that, many important discoveries could be made with cosmic rays. The discoverer of cosmic radiation itself, however, remains little known.

Hess

Hess was born in Styria, one of the most beautiful regions of Austria. He went to the gymnasium and also studied physics in Graz, the regional capital, obtaining his doctor's degree in 1906. Hess then moved to the University of Vienna where he worked in the institute of Exner[3]. Here he became familiar with the problem of *Luftelektrizität* (*air electricity*), as it was called then, a field in which also the young Schrödinger did experiments. Austria was the only country in which radium was produced in a semi-industrial style. This activity had been triggered by the discovery of radium in Bohemian pitchblende donated to Pierre and Marie Curie by Austria. In 1910 the *Radium-Institut* in Vienna was founded and Hess, already a Privatdozent, became assistant to its director, Stefan Meyer[4].

The most sensitive ionization chamber of the time had been constructed by Wulf[5]. Having become a Jesuit father at age twenty, Wulf had studied under Nernst in Göttingen and become physics professor at the Jesuit University in Valkenburg, Netherlands. His chamber, the *Wulf electrometer*, was a cylindrical vessel containing about 2 litres of air. Inside it, a tiny electroscope, consisting of a pair of threads, was mounted on insulators. It could be charged from the outside so that it was on a different voltage compared to the vessel. The voltage was read with a microscope from outside. Because of the very small electric capacity of the electroscope, a loss of voltage of 1 Volt per hour corresponded to an ionization of about 1 ion per cubic centimetre and second.

---

[1] Charles Augustin Coulomb (1736–1806)   [2] Victor Franz Hess (1883–1964), Nobel Prize 1936   [3] Franz Serafin Exner (1849–1926)   [4] Stefan Meyer (1872–1949)   [5] Theodor Wulf (1868–1946)

Wulf had measured the conductivity of air not only on the ground but also considerably above it, namely on the upper platform of the Eiffel tower [2]. He found no significant difference and concluded that the decrease of radiation from the earth was compensated by radiation from some other source which increased with height. He called for further experiments to be performed at larger heights with the help of balloons. It was suggested that the additional radiation might originate from the material of the tower. Still, the idea of a new radiation, possibly from an extraterrestrial source, was born.

Hess (with nautical cap) in the gondola of his balloon, admired by the local youth.

Hess was not the first to perform measurements on board a balloon but he was the first to do it in a well-planned way over an extended period of time after he had obtained funds from the Austrian Academy of Sciences. He was able to charter a balloon with a pilot and to make a total of nine balloon rides. On the last seven he reports in the paper which contains his discovery [3]. To the basket of the balloon he fitted a board on which three ionization chambers of the Wulf type were mounted. Two had walls of 2 mm thickness and were sealed so that the air inside was kept at the same pressure independent of the balloon's height. These would register only the rather penetrating γ rays. The third had a very thin wall, was kept open, and would allow also some β rays to enter it. For the first six rides the balloon was filled with the local town gas which did not carry it very high. Hess reached a maximum height of 1600 m above sea level (Vienna is at 156 m) and he used these flights to perform long-time measurements at about 350 m. He observed small but significant changes of the ionization over times of about an hour and learnt to average over them.

He found no significant change between day and night and also found no drop during a solar eclipse. Also he developed a routine of careful readings before starting and after landing (if possible with the balloon still over his equipment) to take into account radiation from the balloon itself or from the ballast.

On 7 August 1912, Hess started with a balloon filled with hydrogen from Aussig (Ústí nad Labem) on the river Elbe in Bohemia aiming to reach large heights. He drifted north, crossed the border to Germany near Dresden and after six hours landed not far from Berlin, having reached a maximum height of 5300 m. Hess had observed before that the ionization drops a little in the first few hundred metres but then rises slowly. Now he found that above 1500 m this rise became steeper and steeper. Near 5000 m the ionization was nearly three times as large as measured on the ground. He concluded [3],

Hess in later years, demonstrating the use of a Wulf electrometer.

> ... that a radiation of very high penetrating power enters our atmosphere from above, and still in it's [the atmosphere's] lowest layers gives rise to ionization which is observed in closed vessels. The intensity of this radiation seems to be subject to time variations which can still be observed if readings are taken in hourly intervals.

Hess believed his radiation to consist of γ rays, since of all radiation known at the time these had the highest penetrating power. Consequently, he called it *Ultra Gamma Strahlung* and later *Ultrastrahlung*. The name *cosmic radiation* is due to Millikan.

It has been noted that Hess was the only Austrian Nobel Laureate in Physics who did his work in Austria. Tongue-in-cheek, one might say that this is not true by the letter, since the decisive readings of his instruments were made while he was flying over Germany towards Berlin. At any rate it was in Berlin where Bothe and Kolhörster in 1929 made the next important step to clear up the nature of cosmic rays (see Episode 34). Already in the summer of 1913, Kolhörster had confirmed the findings by Hess in three balloon flights, reaching a maximum height of 6000 m [4].

At first there was little belief in the new source of radiation. Hess continued to work as assistant in the Radium-Institut. Only in 1920 was he offered a position as associate professor in Graz; in 1925 he became full professor. In 1931 he moved to the University of Innsbruck because there he had the possibility to establish a laboratory on a mountain top, the *Hafelekar*, at a height of 2300 m. However, he missed the opportunity to switch to more modern detectors and thus he, the discoverer and pioneer of cosmic rays, did not exploit the new radiation as much as he might have done. In 1932 Anderson discovered the positron in a cloud chamber exposed to cosmic radiation on a mountain (Episode 49) and in 1936 Hess and Anderson shared the Nobel Prize.

After the annexation of Austria in 1938 Hess was dismissed from his position. Warned of his imminent arrest, he fled to Switzerland and then accepted a professorship at Fordham University in New York where he retired in 1958.

[1] Federmann, G., *Victor Hess und die Entdeckung der Kosmischen Strahlung*. Master's thesis, Universität Wien, 2003.
[2] Wulf, T., *Physikalische Zeitschrift*, **11** (1910) 811.
[3] Hess, V. F., *Physikalische Zeitschrift*, **13** (1912) 1084.
[4] Kolhörster, W., *Physikalische Zeitschrift*, **14** (1913) 1153.

20 Max von Laue – The Nature of X Rays and the Atomic Structure of Crystals (1912)

von Laue

The nature of the X rays discovered by Röntgen in 1895 was not known. Röntgen himself conjectured that they might be longitudinal ether waves as opposed to the transverse ones, the electromagnetic waves found by Hertz. Since in Röntgen's original experiment the X rays originated from the point where cathode rays, i.e., electrons, hit matter, Wiechert and also Stokes[1] suggested already in 1896 that X rays were emitted by electrons while the latter were decelerated. In Maxwell's theory an electric charge with a velocity, which is not constant, emits electromagnetic waves. In the Hertzian dipole antenna the charges oscillate to and fro. In a Röntgen tube electrons lose their velocity hitting a piece of matter. The fact that interference effects, characteristic of all waves, in particular, light and Hertzian waves, were not observed for X rays, did not preclude that they were electromagnetic waves. Their wavelength might be too small for the detection of interference.

In 1903 Barkla[2] found that X rays are *heterogeneous* as he called them. He let them traverse material, measured the intensity with an ionization chamber behind the material and recorded how the intensity dropped with the thickness of the absorber. He expected the intensity simply to decrease exponentially with thickness. That would have meant that if a certain thickness of absorber decreased the intensity by some factor, say fifty per cent, then adding again the same thickness would decrease the remaining intensity also by fifty per cent. Since Barkla found a more complicated absorption behaviour, he concluded that the X rays are composed of different kinds each following an exponential absorption law but with a different absorption length.

Barkla also found that the absorber is turned into a source of X rays. He analysed this secondary radiation choosing a direction perpendicular to the direction of the incoming radiation (not to be influenced by that) and found it to be *homogeneous*, i.e., to follow the simple absorption law. This homogeneous radiation was considered to be a kind of fluorescent radiation, typical of the material of the absorber brought about by irradiating it with X rays. It was therefore also called *characteristic radiation*. Looking for details Barkla and his group found that each element was capable of emitting two different types of characteristic radiation, which they called *K radiation* and *L radiation*, leaving letters M, N, ... for possible later discoveries.

The nature of X rays, homogeneous or heterogeneous, remained a mystery. They could be understood as electromagnetic waves of short wavelengths or as new neutral particles. The former standpoint was taken, for instance, by

---

[1] (Sir) George Gabriel Stokes (1819–1903)   [2] Charles Glover Barkla (1877–1944), Nobel Prize 1917

Barkla, the latter by William Henry Bragg. One of Bragg's arguments ran like this: X rays, produced by electrons falling on matter, fly more or less in the same direction as the incident electrons. That is easily understood if one assumes them to be particles. The production of X rays can then be seen as a collision process, just as one billiard ball hitting another. For some time the two scientists fought out the *Barkla–Bragg controversy* in the columns of *Nature*. Sommerfeld [1] showed that, contrary to the expectations of Bragg and others, electromagnetic radiation is emitted mostly in forward direction if a fast electron suffers a sudden deceleration. The German term *bremsstrahlung* [breaking radiation] is still used commonly in the literature.

It was the work of Laue[3] and the experiment done by Friedrich[4] and Knipping[5] on his suggestion that cleared up the nature of X rays once and for all and that, moreover, beautifully demonstrated that crystals are composed of atoms arranged in a regular lattice. Laue [2] had studied mathematics and physics in Strasbourg, Göttingen, Munich, and Berlin, where in 1903 he took his Ph.D. with a thesis under Planck. Feeling that he still had to continue his studies he went for another two years to Göttingen. In 1905 Planck offered him a position as his assistant. Laue worked with Planck on the latter's speciality, the entropy of radiation. In the autumn of 1905 Planck gave a talk in the Berlin Physics Colloquium on Einstein's first paper *The Electrodynamics of Moving Bodies*. Laue was deeply impressed. In 1906, when on a mountaineering trip in Switzerland, as one of the first (possibly the very first) visitor from abroad, he looked up Einstein in the patent office in Bern. In 1907 he published a paper in which he showed that classic experiment by Fizeau[6], who had measured the velocity of light in a moving liquid, was in accordance with Einstein's theory. Laue became a Privatdozent in Berlin and, also in that capacity, moved to Munich University in 1909. In 1910 he wrote the first book on the theory of relativity. The writing was done in a little boathouse constructed on piles above the water of Lake Starnberg with a beautiful view on the Alps. 'So well I never hit it again', Laue later remembered [2].

The towering figures in Munich were Röntgen (who, however, took only little part in research any more) and Sommerfeld. Early in 1912 Ewald[7], then a Ph.D. student of Sommerfeld, looked up Laue in his flat. Sommerfeld had asked him to study theoretically the behaviour of light waves in a spatial lattice of polarizable atoms. Laue could not help with the theory but it occurred to him that possibly a similar problem could be studied experimentally if one assumed that a crystal was a regular spatial lattice of atoms and if one passed X rays through a crystal. (It had been conjectured in the nineteenth century, in particular by Bravais[8], that the regular shape of a crystal is due to the underlying regular lattice of atoms of which the crystal is composed. But that was a mere hypothesis and not widely discussed.) Assuming the wavelength of X rays to be on the order of the distance between neighbouring atoms, one might be able to see interference effects. He mentioned this idea to Ewald and it soon got around in the closely knit group of young physicists in Munich. Soon Friedrich, who had just obtained his Ph.D. with Röntgen and now was

Laue in 1898 during his military service.

[3] Max Theodor Felix (von) Laue (1879–1960), Nobel Prize 1914   [4] Walter Friedrich (1883–1968)   [5] Paul Knipping (1883–1935)   [6] Armand Hippolyte Louis Fizeau (1819–1896)   [7] Peter Paul Ewald (1888–1985)   [8] Auguste Bravais (1811–1863)

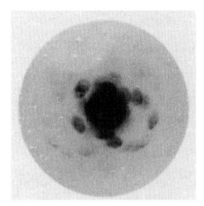

The first exposure by Friedrich and Knipping showing an indication of X-ray diffraction by a crystal of copper sulphate. From [3]. Copyright Wiley-VCH Verlag GmbH & Co. KGaA. Reproduced with permission.

The improved set-up of Friedrich and Knipping. On the left of a lead screen S is the X-ray tube with the anticathode A. A beam of X rays falls through an opening in the screen and is collimated by the diaphragms B_1, B_2, B_3, B_4. Diffraction takes place in the crystal Kr, mounted on the goniometer G. The photographic plate is normally at P_5; alternative positions are P_1, P_2, P_3, P_4. The set-up is enclosed in the lead box K. The undiffracted beam leaves through the lead tube R. From [3]. Copyright Wiley-VCH Verlag GmbH & Co. KGaA. Reproduced with permission.

Sommerfeld's assistant, became interested. Sommerfeld had to be convinced that the experiment was important enough for Friedrich to start it right away in spite of other assignments. Knipping, a Ph.D. student of Röntgen, joined in the effort.

Friedrich and Knipping sent a collimated beam of X rays, 3 mm wide, onto a crystal of copper sulphate and placed a photographic plate at some distance behind the crystal. They observed a dark spot, where the undiffracted beam hit the plate. That spot was surrounded by a more or less regular pattern of further spots, which they attributed to diffraction by the crystal lattice. Only after Laue saw the plate did he start in earnest to analyse the problem of diffraction by a spatial lattice. Later he reminisced: 'In deep thought I went home by way of Leopoldstrasse after Friedrich had shown me the picture. And already near my flat, Bismarckstrasse 22, in front of the house Siegfriedstrasse 10, the thought came for the mathematical theory of the phenomenon.'

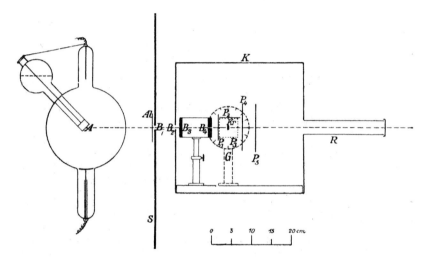

Once the idea had come, the rest was simple. Laue advanced in three steps. First he recalled the laws of diffraction of light by a one-dimensional lattice or grid. If white light shines perpendicularly onto a grid made of fine wires, whose distance is on the order of the wavelength of this light, then on a screen behind the grid there will be a line of white light in the forward direction and a repetition of rainbow-like spectra to the left and to the right. If monochromatic light is used, there will be a series of regularly placed sharp lines. This is due to the fact that for a given wavelength only under certain angles light from equivalent points in the grid interferes constructively (i.e., has a phase difference corresponding to an integer number of wavelengths). He then turned to a two-dimensional lattice by replacing the grid by a mesh. Now two such conditions have to be met. Instead of a line pattern only a point pattern is observed for monochromatic light. For white light it is a pattern of patches each displaying side by side the colours of the rainbow. (The reader can easily do the experiment by looking at a street light through the cloth of her or his umbrella at night.) Up to here all was well known. But now Laue found that for a three-dimensional lattice a third condition had to be met. The spots are characterized by two angles (diffraction in horizontal and vertical direction). A third condition means that only a small number of spots is formed and only for certain wavelengths. Laue, of course, wrote his conditions in mathematical form, later called the *Laue equations*.

Friedrich and Knipping improved their apparatus. They reduced the beam diameter, shielded their set-up by a lead box from stray radiation and, for the same reason, allowed the undiffracted beam to leave it through a lead tube. Instead of the original copper sulphate they used a carefully polished crystal of zinc blende. Also, they took great care to make the symmetry axis of the crystal coincide with the beam axis. As a result, they obtained photographs with beautifully symmetric sharp spots fitting Laue's theory. Such a photograph is now called a *Laue diagram* and the process leading to it is called *Laue diffraction*. A joint paper was written [3] with the theoretical part signed by Laue and the experimental part signed by Friedrich and Knipping and communicated by Sommerfeld on 8 June 1912 to the Bavarian Academy of Science. That same day Laue gave a talk on a session of the German Physical Society in Berlin 'at the same spot', as he proudly remembered, 'where in December 1900 Planck had first talked about his radiation law and the quantum theory'. Laue, Friedrich and Knipping assumed that the zinc blende crystal had a simple cubic lattice with zinc and sulphur atoms alternating on the corners of cubes of side length a. According to Laue's theory, the radiation responsible for one point in the diagram is monochromatic. All the points in the Laue diagram of zinc blende corresponded to just five different wavelengths. In July 1912 an extended numerical analysis of the diagram [4] was published by Laue alone. He found that all the points in the diagram were explained by his theory but that not in every position in which his theory allowed for a point such a point was observed. He therefore assumed that the radiation emitted by the X-ray tube was a mixture of five monochromatic radiations, each with its own wavelength.

As in the case of Röntgen's original discovery, the photographs were extremely convincing. Other researchers immediately were attracted by the new field of X-ray spectroscopy and the discoveries by the Braggs (Episode 21) and Moseley (Episode 24) soon followed.

In 1913 Laue's father was raised to hereditary nobility which implied a change of name for his son. As Max von Laue, he became one of the figures shaping physics research in Germany and, because of his upright character and his love for relativity, a staunch defender of Einstein against his adversaries in dark times. Already in 1914 von Laue received the Nobel Prize in physics.

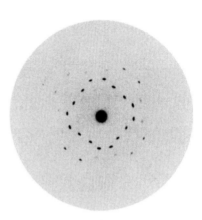

Laue diagram of zinc blende by Friedrich and Knipping. From [3]. Copyright Wiley-VCH Verlag GmbH & Co. KGaA. Reproduced with permission.

[1] Sommerfeld, A., *Physikalische Zeitschrift*, **10** (1909) 969. Also in [5], Vol. IV, p. 369.

[2] von Laue, M., *Mein physikalischer Werdegang*. In [6], Vol. III, p. V.

[3] Friedrich, W., Knipping, P., and Laue, M., *Münchner Sitzungsberichte*, (1912) 303. Also in *Annalen der Physik*, **346** (1913) 971 and in [6], Vol. I, p. 183.

[4] Laue, M., *Münchner Sitzungsberichte*, (1912) 363. Also in [6], Vol. I, p. 208.

[5] Sommerfeld, A., *Gesammelte Schriften, 4 vols.* Vieweg, Braunschweig, 1968.

[6] von Laue, M., *Gesammelte Schriften und Vorträge, 3 vols.* Vieweg, Braunschweig, 1961.

21   Bragg Scattering (1912)

William Henry Bragg

William Lawrence Bragg

One day at the end of 1885 William Henry Bragg[1] [1], a twenty-three year old mathematician who had graduated the previous year, was on his way to attend a lecture in Cambridge. He met the lecturer, J. J. Thomson, in King's Parade and they talked about a vacant professorship in Australia. Bragg was advised by Thomson to apply and shortly afterwards got the position since the electors of the new professor in the distant colony did their work in London. Later Bragg wrote [1]:

> I was Professor of Mathematics *and* Physics. I had never learnt any of the latter, nor worked at the Cavendish except for a couple of terms after I had taken my degree, it was supposed by the electors that I would probably pick up enough as I went along to perform my duties at the Adelaide University ...

Bragg lectured very successfully. In 1889 he married Gwen Todd, daughter of Postmaster General, Inspector of Telegraphs, and Government Astronomer Charles Todd[2] [2] and in 1890 their first child William Lawrence Bragg[3] [2] was born. The boy was known as Willie or Bill and only after being knighted called himself Sir Lawrence not to be confused with his father, Sir William.

Only when he was past forty, it occurred to William Bragg that he might do some research of his own. In 1904, while preparing an address to the Australian Association for the Advancement of Science, he read up about recent results on radioactivity and he had the idea to measure the range of α particles, i.e., the distance they travel in air. After setting up a simple apparatus, he found that the range depended on the substance which emitted the α particles. Rutherford, with whom Bragg began a correspondence, encouraged him to go on and now Bragg worked and published regularly on radioactivity and also on X rays. Bragg thought of X rays and γ rays as massive particles or, more precisely, as electrically neutral pairs of a positive and a negative particle in order to account for experimental findings, e.g., for the fact that electrons, produced by ionization when X rays fall on material, predominantly move along the direction of the original X rays [3].

In 1909 Bragg and his family moved to England, where he had accepted a professorship at Leeds. His son William Lawrence, who had already studied physics at Adelaide, continued his studies at Trinity College in Cambridge, where Bragg senior had been a student of mathematics. In the summer of 1912 both heard of Laue's theory of X-ray diffraction (Episode 20) and of the experiment by Friedrich and Knipping and both set out to analyse the Laue diagram of zinc blende published in [4]. Already in October Bragg senior published a rather curious observation [5]. If the crystal is cubic, with atoms sitting in the

[1] (Sir) William Henry Bragg (1862–1942), Nobel Prize 1915   [2] (Sir) Charles Todd (1826–1910)   [3] (Sir) William Lawrence Bragg (1890–1971), Nobel Prize 1915

corners of cubes with edges of length a, and if the incident X rays pass along one axis (we call it the x axis) of the crystal through one atom serving as origin of the coordinate system, then each spot of the Laue diagram is reached by a straight line going through the origin and another atom at the distance na, with n being a whole number. The position of that other atom is given by its three coordinates $n_x a, n_y a, n_z a$, where n_x, n_y, n_z are, of course, integer numbers. According to Pythagoras' theorem then $n^2 = n_x^2 + n_y^2 + n_z^2$. It follows that, if x and y are the coordinates of a spot on the Laue diagram, z is the distance of the photographic plate from the crystal, and if the numerical values of x, y, z are given as integer numbers, then also the sum of their squares must be an integer. This fact had been observed by Bragg. He spoke of 'avenues' in the crystal open to the outgoing radiation, which, it seems, for him might still be his neutral pairs.

Meanwhile, at Trinity College, Bragg junior gave a most lucid and comprehensive explanation of Laue diffraction. He stressed the existence of planes of atoms in the crystal, in particular, the fact that, in a given crystal, there are planes oriented in different ways. He then pointed out that when radiation falls onto a single plane of atoms, part of the radiation is reflected as if it had struck a mirror. That could be shown by assuming every atom in the plane to be the source of an elementary spherical wave and using a venerable construction, due to Huygens[4], which ensured that the waves from all atoms interfered constructively. Now the crystal is made up of many parallel planes, a distance d apart from each other. Bragg junior therefore required that the radiation reflected from parallel planes also had to interfere constructively and obtained the *Bragg equation*, also called the *Bragg condition*,

$$n\lambda = 2d \sin \varphi \quad ,$$

where n is a whole number $(1, 2, 3, \ldots,$ the *order* of diffraction), λ the wavelength, and φ the angle of incidence and reflection, measured with respect to the plane. He had found, just as Laue had, that radiation leaves the crystal only in definite directions and for each direction with a fixed wavelength. In fact he showed that his equation and also his father's law of squares could be obtained from the Laue equations. In spite of that his reasoning brought an important advance.

William Lawrence Bragg explained the mystery of the missing spots in the Laue diagrams. He pointed out that planes with many atoms per unit area (as exposed to the incident beam) would reflect more radiation than those sparsely populated. Also he mentioned reasons why zinc blende might not be a simple cubic but a face-centred cubic crystal with atoms not only in the corners but also in the middle of each face of the elementary cube. With these two assumptions he was able to explain the different intensities of the spots in the Laue diagram. For many spots the intensity was so weak that they were not seen at all. Moreover, he gave an explanation for the shape of the spots. These were not strictly point-like but elliptic with the shorter axis of the ellipse pointing towards the centre of the diagram. It had been found that these ellipses became more longish if the distance of the photographic plate from the crystal

Cut through a simple crystal with three out of the very many different families of parallel planes containing its atoms.

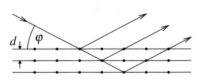

Illustration from which the Bragg equation can easily be derived. Condition for the constructive interference of radiation reflected from the planes of the crystal with a spacing d is that the path difference of neighbouring reflected rays is an integer multiple of the wavelength λ.

---

[4] Chistiaan Huygens (1629–1695)

was increased. Every beam of rays collimated by circular diaphragms has a slightly conical form. If the rays in the cone are reflected by parallel planes in the crystal and then fall onto the photographic plate, they leave an elliptical image, the exact shape of which depends on the distance between crystal and plate just as observed. Bragg junior, who was only 22 years old and was still studying for his M.A. degree, presented his beautiful analysis on 11 November 1912 to the Cambridge Philosophical Society [6].

Bragg junior also tested his theory performing an experiment. C. T. R. Wilson had suggested to him that *Bragg reflection* should be easily observed if the cleavage planes of crystals were used for reflection. Bragg let a pencil beam of X rays fall onto mica under $10°$ with respect to the crystal plane. On a photographic plate placed at right angles behind the crystal he observed not only rays which had traversed the crystal but also some which had been reflected by it [7]. He needed an exposure time of only minutes compared to hours required for Laue diagrams.

There was a discussion whether the 'diffracted' radiation might be produced when the incident X rays hit the crystal and be of a different nature altogether. This idea appealed to Bragg senior because it could save his corpuscular interpretation of X rays. He studied Bragg-reflected radiation with an ionization chamber and found early in 1913 that it ionized gas just as ordinary X rays did [8]. The experiment could only be explained if one assumed that X rays are diffracted as waves are, spreading out in space. But in the process of ionization they behave like well-localized particles hitting individual atoms.

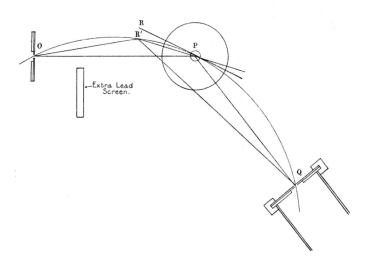

The sketch used by the Braggs to describe their first spectrograph. From [9], Fig. 5. Reprinted with permission.

The two Braggs now began to work together on the construction of the first *X-ray spectrograph* [9]. X rays, emitted by a platinum anticathode, fell through the slit O onto the face of a crystal R of rock salt, were reflected by it at P, and entered the ionization chamber through its entry slit Q. Of course, radiation of wavelength λ reached the detector on the path OPQ only if the crystal formed an angle φ with the direction OP and also an angle φ with the direction PQ and if λ and φ were connected by the Bragg equation. The crystal could be rotated about an axis through P. If it was in a different position the Bragg condition could be met for another path $OR'Q$ of the radiation. For practical use there were collimators allowing radiation to travel only along OPQ. Crystal and chamber were mounted on the turn table and on an arm of an optical spectroscope, respectively, allowing the angular positions φ of the crystal and 2φ of the chamber to be varied. For any particular pair of these angles the current in the ionization chamber was then a measure for the intensity of X radiation with a wavelength λ given by the Bragg equation.

The Braggs discovered distinct maxima for three different wavelengths and were able to observe these maxima in first, second, and third order of diffraction. They found these maxima also if they used other crystals. Having determined by the analysis of Laue diagrams that rock salt, i.e., ordinary kitchen salt, forms simple cubic crystals, they could compute the distance d between neighbouring atoms and thus also the wavelengths of some X-ray lines characteristic for the platinum used as anticathode in the X-ray tube. The lines are part of the spectrum of L radiation of platinum.

Sir William Lawrence Bragg (left) in later years with Max von Laue and the crystallographer Isidor Fankuchen.

At first most of the credit for the discoveries was attributed to Bragg senior, although Lawrence Bragg, who had worked alone in Cambridge, had published his beautiful analysis by himself. In 1915 father and son together were awarded the Nobel Prize. Being only 25, Lawrence Bragg was (and still is) the youngest-ever laureate. To celebrate the golden jubilee in 1965 he gave a Nobel Guest Lecture in Stockholm. In 1919 he succeeded Rutherford as professor in Manchester and in 1938, after Rutherford's death, he succeeded him as Cavendish Professor in Cambridge. He devoted his life to the X-ray analysis not only of simple crystals but also of complex structures in metallurgy, organic chemistry, and biochemistry. His group was outstandingly successful. Two of its members, Perutz[5] and Kendrew[6], revealed the structure of *haemoglobin*, the carrier of oxygen in the blood.

[1] Caroe, G. M., *William Henry Bragg 1862–1942 – Man and Scientist*. Cambridge University Press, Cambridge, 1978.

[2] Hunter, G. K., *Light is a messenger – the life and science of William Lawrence Bragg*. Oxford University Press, Oxford, 2004.

[3] Bragg, W. H., *Nature*, **77** (1908) 270.

[4] Friedrich, W., Knipping, P., and Laue, M., *Münchner Sitzungsberichte*, (1912) 303. Also in *Annalen der Physik*, **346** (1913) 971 and in [10], Vol. I, p. 183.

[5] Bragg, W. H., *Nature*, **90** (1912) 219.

[6] Bragg, W. L., *Proceedings of the Cambridge Philosophical Society*, **17** (1912) 43.

[7] Bragg, W. L., *Nature*, **90** (1912) 410.

[8] Bragg, W. H., *Nature*, **90** (1913) 572.

[9] Bragg, W. H. and Bragg, W. L., *Proceedings of the Royal Society*, **A 88** (1913) 428.

[10] von Laue, M., *Gesammelte Schriften und Vorträge, 3 vols.* Vieweg, Braunschweig, 1961.

---

[5] Max Ferdinand Perutz (1914–2002), Nobel Prize 1962   [6] John Cowdery Kendrew (1917–1997), Nobel Prize 1962

22

J. J. Thomson Identifies Isotopes (1912)

The idea of atoms as final constituents goes back to antiquity. In 1803 Dalton[1] formulated it in a precise scientific form and thus, at the same time, created the framework of modern chemistry. One of his five postulates reads: *Atoms of the same element are similar to one another and equal in weight.* This statement could be taken as a definition of an element. But the name *element* was, and is, used for a substance that cannot be decomposed into others by chemical means. Prout[2] noted the fact that the atomic mass numbers of those elements for which these numbers had been determined with some accuracy were multiples of the atomic mass number of hydrogen. (The latter can be taken to be equal to one.) Therefore, in 1815, he pronounced the hypothesis that the atoms of all elements were made up of hydrogen atoms. In the course of the nineteenth century atomic mass numbers were obtained with increasing accuracy and many elements were found for which that number was far from an integer (i.e., far from a whole number). Prout's hypothesis was therefore rejected but there was no reason to doubt Dalton's postulate. In the twentieth century the situation was reversed: The second half of Dalton's postulate was shown to be wrong and Prout's hypothesis was shown to carry some truth.

J. J. Thomson

In 1907 J. J. Thomson, professor at the Cavendish Laboratory, Cambridge, began to develop a method to measure the mass of atoms and molecules [1]. Thomson wanted to study in detail the positive ions in a gas discharge. It had been observed by Goldstein that *canal rays* would traverse a hole in the cathode of a gas-discharge tube. These rays would produce a glow along their path. At first they were thought to be neutral, since, contrary to *cathode rays*, i.e., the negatively charged electrons, they seemed not to react to a magnetic field. But Wien showed that strong fields deflected them and that they carried positive charge. Thomson replaced the simple hole in the cathode by a long fine tube; he used a hypodermic needle and later even finer tubes. The canal rays, which he preferred to call *positive rays*, were thus made to form a thin and highly parallel beam. Let us call the direction of the beam the x direction. The space containing the gas discharge was separated from the space in which the beam could propagate. By continuous pumping this latter space was kept at a much lower pressure. The fine tube presented a large resistance for the flow of gas into it. Here the ray was made to pass through a region containing an electric field and a magnetic field both parallel to one another and perpendicular to the beam. We call the field direction the y direction. The ions were deflected in y by the electric field and in z, i.e., perpendicular to x and y, by the magnetic field. After that region the ions were allowed to propagate freely for a certain distance and then fall onto a fluorescent screen or a photographic

---

[1] John Dalton (1766–1844) [2] William Prout (1785–1850)

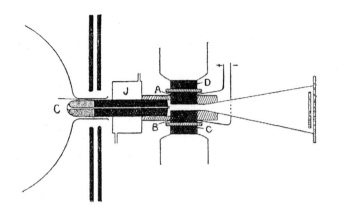

Positive-ray apparatus. In a large spherical glass vessel, part of which is indicated on the left of the figure, a gas discharge is maintained between an anode the cathode C. Positive ions can pass the long thin canal in the cathode and are deflected by a magnetic field between the pole pieces C, D of a magnet. The pole pieces carry the plates A, B of a capacitor providing an electric field. The deflected ions travel through the conical vessel on the right and, at its end, fall onto a photographic plate. The water jacket J carries away the heat produced by ions falling onto the walls of the canal. From [2]. Reprinted by permission from Taylor & Francis Ltd, http://www.informaworld.com.

plate oriented perpendicular to the x direction, i.e., parallel to the yz plane. In this arrangement ions of a fixed charge-to-mass ratio fall onto a parabola on the screen. Different points on the parabola correspond to different velocities of the ions. A fast ion suffers less deflection than a slow one. Ions with different charge-to-mass ratio form different parabolae.

Parabolae registered by Thomson. Positive ions of different charge-to-mass ratio form pieces of parabolae in one quadrant of the photographic plate. The white halo at the centre is caused by atoms and molecules which carry no charge when passing through the electric and magnetic fields and, therefore, are not deflected. By changing the gas filling and inverting the directions of one or both fields all four quadrants of the plate can be used successively for registrations. The two parabolae on the upper right are caused by the ions of ^{20}Ne and ^{22}Ne. From [3].

Enlarged section of the preceding figure. Arrows indicate the parabola of ^{20}Ne and that of ^{22}Ne which is weaker and partly overlaps with the former. From [3].

After some development work on his apparatus Thomson indeed observed clearly separated parabolae of different atoms or molecules which carried one elementary charge e (sometimes $2e$ or even a higher multiple of e) with the discharge tube filled with gases which contained light elements. Such elements have practically integer atomic mass numbers. When, in 1912, he filled the discharge tube with neon, which has an atomic mass number of 20.2, he observed two parabolae, a strong one corresponding to the atomic mass number 20 and a weaker one corresponding to 22. He concluded that the element neon contains two sorts of atoms and that the atomic mass number of neon, measured in the conventional way, is the average of the atomic mass number of the two kinds weighted with the abundances of these kinds.

That there are several different atoms corresponding to the same element had been suspected, and to some extent, demonstrated before. It was found by studying the many new radioactive 'elements' or 'bodies', that there were groups which, although differing in half-life, where chemically inseparable (see Episode 9). Such inseparable 'elements' were called *isotopes* by Soddy [4], because they occupy the same position in the Periodic Table. Now Thomson had shown that there are isotopes also for non-radioactive elements and that the atoms of different isotopes of the same element have different masses. Since the mass of an atom is essentially equal to the mass of its nucleus, isotopes of the same element have nuclei of different mass. The nuclei have, however, the same charge because the number of positive elementary charges of the nucleus is equal to the number of electrons in the atomic shell and that number determines the chemical properties. Soddy would later say the atoms of a chemical element 'have identical outsides but different insides'. Of all the different researches that Thomson undertook the work with positive rays is the most extended in time. In 1912 he published a comprehensive paper [2] and in 1913 the results obtained up to that time were presented in a book [5].

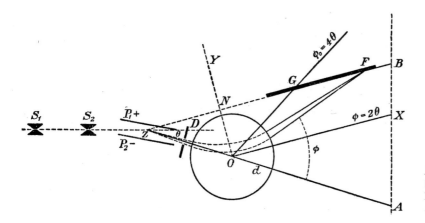

Aston's mass spectrograph of 1919. From [6]. Reprinted by permission from Taylor & Francis Ltd, http://www.informaworld.com.

The work on isotopes at the Cavendish Laboratory was continued and extended by Aston[3], who began his research there in 1909. Aston was the first to try to separate macroscopic quantities of isotopes. After trials to separate the isotopes of neon by fractional distillations had failed, he met with some success by trying the fractional diffusion of neon through pipe clay since the diffusion velocity depends on the mass of the diffusing atoms. Like most research Aston's work was interrupted by war work but was resumed in 1919.

Aston conceived and built an apparatus, which he called the *mass spectrograph* [6]. As the name implies, it produces a spectrum of atomic or molecular masses. Aston realized that by the fine tube, characteristic of the Thomson method, much intensity was lost. Instead by a tube, he collimated the beam by two slits S_1, S_2 (see figure) as one does in the optical spectroscopy of light. Since he wanted to produce one image of the last slit on the photographic plate for all ions of the same charge-to-mass ratio, electric and magnetic deflections had to take place in the same plane, which had to be perpendicular to the slit. The direction of the magnetic field, therefore, was no longer parallel to that of the electric field but perpendicular to it. Also the two fields were separated from each other in position. Let us follow the path of ions of a fixed charge-to-mass ratio. They first passed the electric field by which they were deflected downwards. The slow ions were, of course, deflected more than the fast ones. After passing through a rather wide diaphragm D they entered the region of the magnetic field, where they were deflected upwards, again the slower more than the fast. By a clever choice of the fields and the position of the photographic plate all ions could be focussed onto a single line F on the plate even though they had different velocities. Aston had achieved *velocity focussing*.

With this apparatus and later with even more improved spectrometers, Aston found isotopes for nearly every element in the Periodic Table and he found a strikingly simple regularity which he called the *whole-number rule* [7]. If not the mass of the hydrogen atom but $1/16$ of the atomic mass of oxygen was taken as unit[4], then the atomic mass numbers of all isotopes were, within the

Aston

---

[3] Francis William Aston (1877–1945), Nobel Prize 1922   [4] At present $1/12$ the atomic mass of the isotope ^{12}C of carbon is taken.

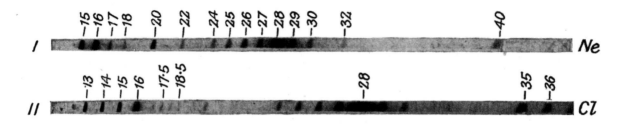

Two of Aston's early mass spectra. The lines marked 20 and 22 are due to singly charged ^{20}Ne and ^{22}Ne ions, those marked 35 and 36 to singly charged ^{35}Cl and ^{36}Cl ions, and those marked 17.5 and 18.5 to doubly charged ^{35}Cl and ^{37}Cl. Most of the remaining lines are caused by ions of other atoms or molecules and serve as calibration. From [7]. Reprinted by permission from Taylor & Francis Ltd, http://www.inform aworld.com.

accuracy of the experiment, which was about one part in a thousand, integer numbers. There was one exception: hydrogen. Its atomic mass number, being 1.008, deviated from 1 by about a per cent.

It was found plausible that all nuclei with nuclear charge number Z and atomic mass number A were made up of A hydrogen nuclei (protons) and $A - Z$ electrons. Since the mass of the electrons was negligible, the whole-number rule was explained. It was still intriguing that a nucleus containing A protons is lighter than are A separate protons by about one per cent. But that could be explained by a first rough theory of nuclear binding. Protons are bound in the nucleus and it takes energy to separate them. By Einstein's famous relation $E = mc^2$ the binding energy E is related to a mass m and it is this mass by which the mass of the constituents of a nucleus is increased when the nucleus is broken up. These considerations were published first in quantitative form by Lenz[5] [8]. After the discovery of the *neutron* it became clear that the nucleus consists of Z protons and $A - Z$ neutrons. But since the neutron has only a slightly higher mass than the proton that does not change the simple reasoning given above.

The development of mass spectroscopy and for quite some time also its exploitation was very much a matter of the Cavendish Laboratory and, to some extent, also of Trinity College, Cambridge. J. J. Thomson had been a student and later a Fellow of Trinity and, after his retirement as Cavendish Professor in 1919, he became Master of Trinity College. Aston was elected a Fellow of Trinity in 1920 and, being a bachelor, lived in the college for the rest of his life.

J. J. Thomson was an extraordinarily successful teacher. In an appendix to his memoirs [9], titled *List of my Cavendish Students who became Fellows of the Royal Society*, we find 27 names, among them the names of seven Nobel Laureates (C. T. R. Wilson, Rutherford, Barkla, Richardson, Aston, Appleton, and G. P. Thomson, J. J. Thomson's son). Thomson was deeply convinced of the importance of teaching for research. He wrote [9]:

> There is no better way of getting a good grasp of your subject, or one likely to start more ideas for research, than teaching it or lecturing about it, especially if your hearers know very little of it, and it is all to the good if they are rather stupid. You have then to keep looking at your subject from different angles until you find the one which gives the simplest outline, and this may give you new ideas and lead to further investigations.

[1] Thomson, J. J., *Philosophical Magazine*, **13** (1907) 561.

[2] Thomson, J. J., *Philosophical Magazine*, **24** (1912) 209.

[3] Thomson, J. J., *Rays of Positive Electricity*. Longmans, Green, London, Second ed., 1921.

[4] Soddy, F., *Nature*, **92** (1913) 400.

[5] Thomson, J. J., *Rays of Positive Electricity*. Longmans, Green, London, 1913.

[6] Aston, F. W., *Philosophical Magazine*, **38** (1919) 707.

[7] Aston, F. W., *Philosophical Magazine*, **39** (1920) 611.

[8] Lenz, W., *Naturwissenschaften*, **8** (1920) 181.

[9] Thomson, J. J., *Recollections and Reflections*. Bell, London, 1936.

---

[5] Wilhelm Lenz (1888–1957)

Bohr's Model of the Atom (1913)

Bohr[1] [1,2] ranks among those few physicists which exert the strongest scientific influence on others. Many important theoreticians of the twentieth century regarded him their teacher, although, like Einstein, he did not have a single Ph.D. student. Bohr was the son of a professor of medicine and physiology at the University of Copenhagen; the Bohr family, for generations, had been one of teaching and learning. His mother was the daughter of a banker and philanthropist. His younger brother Harald, Niels' closest friend, became an eminent mathematician and, as a young man, was an excellent soccer player and member of the Danish National team. Bohr began to study physics in Copenhagen in 1903. He solved a prize problem, requiring extensive experiments which he performed in his father's laboratory (there was none for the physics chair) and won the prize in 1907. In 1909 he obtained a master's degree with work on the electron theory of metals and in 1911 his Ph.D. with a thesis on the same subject.

In the autumn of 1911, Bohr, with a stipend for one-year postdoctoral studies, joined the group of J. J. Thomson in Cambridge. But he did not feel at ease there. After meeting Rutherford he decided to work under him in Manchester for the second half of his stay in England and in this way became familiar with the concept of the atomic nucleus developed by Rutherford early in 1911 (Episode 16). Rutherford had also put forward a model of the atom. It consisted of a nucleus with an integer number Z of positive elementary charges e surrounded by Z electrons, each, of course, carrying the charge $-e$.

After his return to Denmark, Bohr set himself the task to elaborate this model theoretically. It was already well known that the Rutherford model was inconsistent with classical physics. It did not explain why atoms are stable. It had both a *static* (all particles in the atom at rest) and a *dynamic* instability (electrons orbiting the nucleus like planets the sun in the solar system). The latter came about because electrons on an orbit are accelerated and because accelerated charges lose energy by radiation, as Maxwell had predicted and Hertz had demonstrated. Bohr therefore knew that 'new physics' had to be used on the problem. This new physics was the theory of Planck's quantum. His first idea was to restrict the number of possible orbits of an electron by allowing it only to have certain discrete, *quantized* energy values just as Planck had done with his oscillators. When discussing his idea with an acquaintance Bohr was asked: 'But how does it do with the spectral formulae?'. It turned out that Bohr at the time was not familiar with the *Balmer formula*.

Already in 1885, Balmer had found a simple but very accurate formula, which allowed to compute the frequencies (and wavelengths) of the light for the spectral lines emitted by atomic hydrogen. Each of the infinitely many

Bohr on his way back from England in 1912.

---

[1] Niels Henrik David Bohr (1885–1962), Nobel Prize 1922

Bohr in 1916

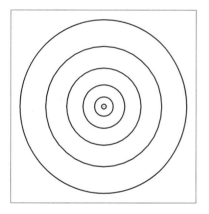

Sketch of the Bohr orbits for values of the principal quantum number from $n = 1$ (centre) to $n = 6$.

frequencies is determined by two natural numbers n and m, i.e., numbers $1, 2, 3, \ldots$, and by a constant R which had to be found from experiment. In Balmer's formula each frequency is given by the difference of two *terms*, one depending on n, the other on m, only. Later Rydberg[2] and Ritz[3] found formulae, more complicated but of similar structure, for the spectra of other atoms. The constant R, already appearing in the Balmer formula, is now called the *Rydberg constant*. When Bohr had looked up the Balmer formula it took him only a few weeks to complete his model and to write a publication. He sent the paper to Rutherford, asking him for criticism and also asking him to forward it to a British scientific journal. Rutherford was impressed with Bohr's results but not with his writing and suggested to shorten the paper. In a letter to Bohr he wrote: 'As you know it is the custom in England to put things very shortly and tersely in contrast to the Germanic method, where it appears to be a virtue to be as long-winded as possible.' Bohr, however, refused and Rutherford did not insist. The paper [3] appeared in July 1913. Long-winded or not, it was to gain world fame for Bohr.

Bohr's theory was built on the two *Bohr postulates*, which we paraphrase in modern language:

1. An atom has different *stationary states*, each of a definite energy. The motion of the electrons in these states can be discussed using classical mechanics. The state of lowest energy is called the *ground state*.
2. There can be *transitions* between stationary states. A transition from a higher to a lower energy is accompanied by the emission of a light quantum corresponding to the energy difference. A transition from a lower to a higher energy is possible if a light quantum corresponding to the energy difference is absorbed by the atom.

Because of its simple spectrum Bohr declared that the hydrogen atom is the simplest atom (which, at the time, was by no means clear) with only one electron and a nucleus of charge number $Z = 1$. By a simple calculation, in which he assumed the electron mass to be negligibly small compared to the mass of the nucleus, he was now able to obtain the Balmer formula for the possible frequencies of the light emitted or absorbed by the hydrogen atom and to compute the energies of the stationary states. These energies E_n could be enumerated by a natural number, $n = 1$ for the ground state, $n = 2$ for the *first excited state*, $n = 3$ for the *second excited state*, etc. A number like n is now called a *quantum number*. By his second postulate the frequencies ν_{nm} of the light quanta emitted or absorbed were connected to the difference of two of these energies by $E_n - E_m = h\nu_{nm}$. Thus the two *terms* of the spectral formulae were directly connected to the energies of two stationary states.

Assuming the orbits of the electron to be circles, Bohr computed their radii. They, of course, depended on n. For the ground state, which has $n = 1$, he found the value of $a_B = 0.55 \times 10^{-8}$ cm, now called the *Bohr radius*. For a state with quantum number n the radius is $a_n = n^2 a_B$. For the Rydberg constant Bohr obtained an expression in terms of the fundamental constants e (elementary charge), m (electron mass), and h (Planck's constant). This

[2] Johannes Robert Rydberg (1854–1919)   [3] Walter Ritz (1878–1909)

Bohr's Spectral Formula

The frequency ν of light, emitted when a single electron bound to a nucleus of charge Ze changes from a stationary state (with principal quantum number) m to a state n, is

$$\nu = \frac{E_n - E_m}{h} = cZ^2 R_\infty \left(\frac{1}{n^2} - \frac{1}{m^2} \right) \quad,$$

where h is Planck's constant, c is the velocity of light, and R_∞ is Rydberg's constant. For the hydrogen atom ($Z = 1$) this is Balmer's formula. Bohr was able to write Rydberg's constant in terms of fundamental constants,

$$R_\infty = \frac{e^2 m}{8 \varepsilon_0^2 h^3 c} \quad.$$

(ε_0 is the electric field constant.) If the mass M of the nucleus is taken into account, the mass m of the electron is replaced by its reduced mass $\mu = mM/(m + M)$.

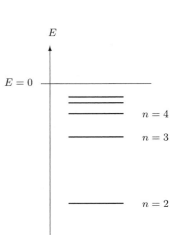

expression was verified within the precision by which the latter three constants were known.

Bohr also considered nuclei with a higher charge number Z surrounded by one electron. His spectral formula for such systems was similar to that of hydrogen, the only difference being a factor Z^2 in front of the Rydberg constant. Rutherford had shown that for helium the nuclear charge number is $Z = 2$. Therefore, the simplest system of this type is the helium ion He^+, i.e., a helium atom from which one of its two electrons is removed by ionization. Bohr suggested that a spectrum, first observed in 1896 in the light of some stars by Pickering[4] and in 1912 by Fowler[5] in a gas discharge of hydrogen with an admixture of helium, was emitted by ionized helium. As one of the first reactions to Bohr's paper Fowler [4] pointed out that, if one followed Bohr's interpretation, there was a still slight difference between the Rydberg constant R_H obtained experimentally from the hydrogen spectrum and the corresponding constant R_{He}. Bohr [5] was able to show that this difference was fully explained if the masses of the hydrogen and the helium nucleus were taken into account. This success convinced many scientists that Bohr's theory should be taken seriously. In October 1913, Hevesy[6] wrote to Rutherford about a congress which took place in Vienna [6]:

> Speaking with Einstein on different topics we came to speak on Bohr's theory, he told me that he had once similar ideas but he did not dare to publish them. "Should Bohr's theory be right, it is of the greatest importance." When I told him about the Fowler Spectrum the big eyes of Einstein looked still bigger and he told me "Then it is one of the greatest discoveries". I felt happy hearing Einstein saying so.

Later in 1913 Bohr wrote two follow-up papers [7,8] in which he applied his theory to other atoms and even to molecules. Of this work many qualitative and some quantitative results still remain valid. Bohr began with accepting the recent suggestion by van den Broek[7], who, by the way, was an amateur scientist,

Energy values of stationary states in Bohr's model of the hydrogen atom. All are negative; they approach zero as n approaches infinity.

---

[4] Edward Charles Pickering (1846–1919)   [5] Alfred Fowler (1868–1940)   [6] Georg von Hevesy (1885–1966), Nobel Prize 1943   [7] Antonius Johannes van den Broek (1870–1826)

that the number of charges in the nucleus is equal to the position number of the corresponding element in the Periodic Table. He then argued that the electrons orbiting the nucleus are arranged in separate rings (later called *shells*). Every shell can hold a certain fixed number of electrons. If that number is exceeded, a new shell is begun. The chemical properties of the atom are determined by the outermost shell. The energy needed to remove an electron from that shell is the *ionization energy* of the atom. Bohr also explained the *characteristic radiation*, which is part of the X-ray spectrum and to which we shall come back in Episode 24.

At the end of 1913, Bohr presented a summary of his work to the Danish Physical Society [9]. This report also contained the first version of what was later called Bohr's *correspondence principle*. It states that in some limit the results computed in the new theory have to converge to those obtained in the classical theory. Bohr found that, for very large quantum numbers n, the frequency $\nu_{n,n-1}$ of light emitted in the transition from a stationary state n to the state $n-1$ is equal to the one expected classically, i.e., to $\omega_n/2\pi$ with ω_n being the angular frequency of the orbiting electron.

Bohr was the first who applied the quantum theory of Planck and Einstein to the atom. Because of the precision with which he could compute simple spectra it was soon clear that there had to be some truth is his theory, although a deeper reason for his postulates could not be given at the time. In 1913 Bohr was assistant at the University of Copenhagen, the only one in Denmark. There was only one professorship for physics. This made it quite difficult for him do work independently. Eventually, an Institute of Theoretical Physics was created for Bohr and officially opened in 1921 to which he attracted many brilliant young scientists. We shall meet quite a few of them later. Copenhagen became an exceptional centre of modern physics presided over by Niels Bohr.

[1] Pais, A., *Niels Bohr's Times, In Physics, Philosophy, and Polity.* Clarendon Press, Oxford, 1991.

[2] Bohr, N., *Collected Works, 10 vols.* North Holland, Amsterdam, 1972–1999.

[3] Bohr, N., *Philosophical Magazine*, **26** (1913) 1. Also in [2], Vol. 2, p. 161.

[4] Fowler, A., *Nature*, **92** (1913) 95.

[5] Bohr, N., *Nature*, **92** (1913) 231. Also in [2], Vol. 2, p. 274.

[6] Eve, A. S., *Rutherford – Being the Life and Letters of the Rt Hon. Lord Rutherford, O.M.* Cambridge University Press, Cambridge, 1939.

[7] Bohr, N., *Philosophical Magazine*, **26** (1913) 476. Also in [2], Vol. 2, p. 188.

[8] Bohr, N., *Philosophical Magazine*, **26** (1913) 857. Also in [2], Vol. 2, p. 215.

[9] Bohr, N., *Fysisk Tidsskr.*, **12** (1914) 97. Engl. transl. in [2], Vol. 2, p. 281.

Planck (right) on a visit in Copenhagen with Bohr, 1930.

Moseley and the Periodic Table of Elements (1913)

In 1913 and 1914, respectively, Moseley[1] [1] published two papers which, once and for all, established a firm connection of the Periodic Table, which was based on empirical chemistry, to the physical structure of atoms. In 1915 at the age of 27, he was killed in the ill-fated campaign of the British against the Turkish army on the Gallipoli peninsula.

Moseley in the Balliol–Trinity laboratory in Oxford around 1910.

Moseley's family background and education were exceptional. His father, Henry Nottidge Moseley, and both his grandfathers, Henry Moseley and John Gwynn Jeffreys, were Fellows of the Royal Society. His father, who had been professor of zoology at Oxford, died when Moseley was only four years old. From then on his mother saw to it that he got the best education available. In 1901 he won a King's scholarship for the prestigious Public School of Eton and in 1906, again with a scholarship, he entered Trinity College, Oxford. He studied physics under Townsend[2] and, after graduating in 1910, joined Rutherford's outstandingly successful group in Manchester. He did some work on radioactivity but, immediately after learning of Laue's theory of X-ray diffraction and the experiment by Friedrich and Knipping in the summer of 1912, he became focussed on X rays.

---

[1] Henry Gwyn Jeffreys Moseley (1887–1915)   [2] (Sir) John Sealy Edward Townsend (1868–1957)

Just like Lawrence Bragg he wanted to understand Friedrich's and Knipping's results in detail. To achieve that he joined forces with Darwin[3] who also worked in Rutherford's group and who was the son of an Astronomer Royal and the grandson of Charles Darwin. Together they developed essentially the same theory as Lawrence Bragg, and Moseley reported their results in November in the physics colloquium in Manchester. William Henry Bragg, who was present, told them that his son had already come to the same conclusion. The two young men decided not to publish a paper on theory but to set up their own experiment. They looked for a possible ionizing power of Bragg-reflected X rays, observed it, but published [2] this result slightly later than William Henry Bragg (see Episode 21). As a next step they constructed their own X-ray spectrograph competing with the Braggs. Again they were only second, but the angular resolution of their apparatus was much better and the sensitivity of their ionization chamber much higher.

In their publication of mid-1913 Moseley and Darwin not only reported on high-precision measurements of the three X-ray lines found by the Braggs but also showed that two of these lines were in fact pairs of lines with very similar wavelengths [3]. Moreover, they were the first to identify these lines with the characteristic *L radiation* of Barkla (see Episode 20). In addition they presented the first quantitative determination of a continuous X-ray spectrum.

Moseley was intrigued by the characteristic radiation and wanted to measure it for elements other than platinum. The use of the precision spectrometer was cumbersome, since the lines were very sharp and could easily be missed, if the detector was not swept in very small steps of $1'$, i.e., $\frac{1}{60}^\circ$. To overcome this difficulty he designed the first X-ray spectrometer with photographic registration by replacing the ionization chamber with its entrance slit at one angle by a photographic plate extending over a certain angular region. The exposure time needed for the photographic plate to yield a line spectrum was only five minutes and thus, in a very short time, Moseley produced the first of his two famous papers [4] in which he showed the spectra of K radiation of ten different substances. Nine of these (calcium (20), titanium (22), vanadium (23), chromium (24), manganese (25), iron (26), cobalt (27), nickel (28), and copper (29)) were chemical elements, the tenth, brass, was an alloy, i.e., a mixture of two elements, namely copper (29) and zinc (30). The number given in parentheses is the position number of the particular element in the Periodic Table, which begins with hydrogen (1), helium (2), ... Moseley arranged the spectra, one below the other in a step-like fashion, in such a way that a given wavelength was in the same position for all spectra. It then became clear by simple inspection of this 'step ladder' that the spectrum of K radiation of each element contains two strong lines (which Moseley called K_α (for the longer wavelength) and K_β (for the shorter) and that this pair of lines moves to shorter and shorter wavelengths in a monotonic fashion if one moves step by step from calcium to zinc in the order indicated above.

Only a few months before Moseley's work, Bohr had published his model of the atom (Episode 23) with Z electrons, each of electric charge $-e$ circling an atomic nucleus of charge Ze. Bohr had taken the nuclear charge number Z

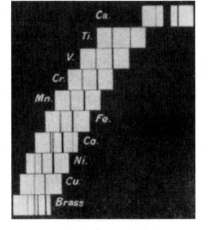

Moseley's 'step ladder'. For most samples there are only two prominent lines, the stronger one (on the right) being K_α. From [4].

---

[3] (Sir) Charles Galton Darwin (1887–1962)

Moseley's Law

Inspired by Bohr's spectral formula (Episode 23) Moseley wrote down a formula of the form

$$\nu(K_\alpha) = c(Z - b_K)^2 R_\infty \left(\frac{1}{1^2} - \frac{1}{2^2} \right) \quad , \qquad b_K \approx 1 \quad ,$$

for the K_α lines of atoms with nuclear charge number Z. Here c is the velocity of light and R_∞ is Rydberg's constant. For the L_α radiation he found

$$\nu(L_\alpha) = c(Z - b_L)^2 R_\infty \left(\frac{1}{2^2} - \frac{1}{3^2} \right) \quad , \qquad b_L \approx 7.4 \quad .$$

---

to be identical with the position number of the corresponding element in the Periodic Table. His theory could explain the visible spectra of the hydrogen atom ($Z = 1$) and the positive ion of helium ($Z = 2$) with only one electron. But he could not make calculations for atoms with more electrons. Moseley realized that, in contrast to visible spectra, the characteristic X-ray spectra, in particular the spectrum of K radiation, was simple also for atoms of high Z. Since Bohr had conjectured that the electrons in an atom are arranged in separate rings and since in his model transitions to the innermost ring correspond to the highest energies, i.e., the shortest wavelengths, Moseley wrote:

> The very close similarity between the X-ray spectra of the different elements shows that these radiations originate inside the atom, and have no direct connexion with the complicated light-spectra and chemical properties governed by the structure of its surface.

Moseley also gave a formula describing the frequency of the K_α radiation for all elements which he had studied and predicting it for all others and, on the basis of very sparse data, even gave a similar formula for the L_α radiation. The formulae later were called *Moseley's law*.

Moseley's work made it clear once and for all that indeed the position number in the Periodic Table is equal to the number Z of positive elementary charges in the nucleus of an atom. It also showed that Z is more important for the spectroscopic and chemical properties of an atom than the atomic mass number A. This is evident in the case of the elements cobalt ($Z = 27, A = 58.9$) and nickel ($Z = 28, A = 58.7$), where even the order in A differs from that in Z.

At this stage of his work Moseley decided to leave Manchester and to move back to Oxford, although Rutherford had offered him a fellowship for the academic year 1913/14, and although he got no paid position in Oxford. His motives are not entirely clear but it seems he thought that it would be easier eventually to obtain a professorship in Oxford if he was on the spot. With a grant of 1000 Belgian Francs from the Solvay Foundation he set up new equipment in Townsend's laboratory, where he was allowed to work as a guest. For his X-ray tube he used a design originally due to Kaye. In this tube anticathodes of different material were mounted on a tiny truck. A given anticathode could be moved to be exposed to the cathode rays without breaking the vacuum of the tube.

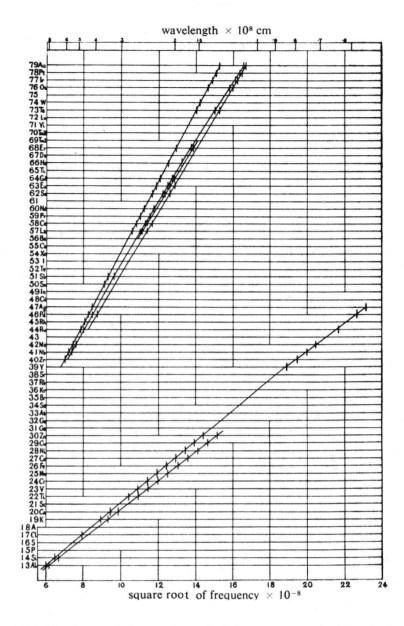

Moseley diagram: frequencies of K lines (bottom) and L lines (top) as measured by Moseley for most elements. From [5].

With Moseley's technique and Moseley's law it was easy to determine the number Z for virtually any known element. For elements with higher values of Z the L radiation had to be used, since the voltage available for X-ray tubes was not high enough to produce the K radiation with its shorter wavelength. Already in April 1914 Moseley published his results [5]; one comprehensive diagram contains the frequencies of K or L lines for most elements between aluminium ($Z = 13$) and gold ($Z = 79$). In the conclusions he wrote:

> Known elements correspond with all numbers between 13 and 79 except three. There are here three possible elements still undiscovered.

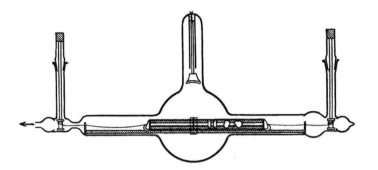

Moseley's X-ray tube with targets, i.e., anticathodes, mounted on an aluminium truck which could be 'drawn to and fro by means of a silk-fishing line wound on brass bobbins'. From [5].

These were the elements with $Z = 43, 61$, and 75. In fact, also the element with $Z = 72$, taken to be a rare earth, was missing. Moseley had assumed its existence, because it was reported in the chemical literature, but could not get a sample of it to use in his measurements. All four elements were found between 1922 and 1945, two in terrestrial material (hafnium, $Z = 72$, and rhenium, $Z = 75$). The other two had to be produced by nuclear reactions (technetium, $Z = 43$, and promethium, $Z = 61$) since these radioactive elements do not seem to exist in the earth's crust.

Moseley was invited to report on his work at the meeting of the British Association for the Advancement of Science held in Australia in August 1914. It was the month in which the First World War began. Immediately after his talk Moseley travelled back to England by the next steamer to volunteer for the army. He even 'pulled private strings' and became a lieutenant in the Royal Engineers. On 10 August 1915, he perished in the Battle of Sari Bair. Also William Henry Bragg's younger son Robert was killed in the Gallipoli campaign.

A simple explanation for Moseley's law, sought for in vain by Moseley himself, was given by Kossel[4] in 1914 [6,7]. The mystery was why $(Z - b)^2$ appeared in the law rather than Z^2 as in Bohr's formula. Kossel, who knew not only Moseley's emission spectra but also absorption spectra of X rays, argued that characteristic radiation is a two-step process. If the electrons of an atom are arranged in rings or, as they are now called, shells, then the incident cathode rays, as a first step, can remove an electron from the innermost shell, the K shell with principal quantum number 1. In a second step an electron from the next shell, the L shell with principal quantum number 2, can make a transition to the K shell and, doing so, emit K_α radiation. This is very similar to the transition $2 \rightarrow 1$ in the hydrogen atom described by Bohr's formula, except that the K shell is not empty. The remaining electron or electrons have negative charge and thus effectively reduce the positive charge Ze of the nucleus. Since Moseley found $b_K \approx 1$ there was one remaining electron in the K shell. It follows that all atoms in their ground state possess two electrons in the K shell. Similarly K_β radiation is produced by a transition from the M shell to a (previously emptied) place in the K shell, L_α radiation is a transition from the M shell to the L shell and so forth.

---

[4] Walther Kossel (1888–1956)

[1] Heilbron, J., *H. G. J. Moseley – The Life and Letters of an English Physicist 1887–1915*. University of California Press, Berkeley, 1975.
[2] Moseley, H. G. J. and Darwin, C. G., *Nature*, **90** (1913) 594.
[3] Moseley, H. G. J. and Darwin, C. G., *Philosophical Magazine*, **26** (1913) 210.
[4] Moseley, H. G. J., *Philosophical Magazine*, **26** (1913) 1024.
[5] Moseley, H. G. J., *Philosophical Magazine*, **27** (1914) 703.
[6] Kossel, W., *Verh. d. Dt. Phys. Gesellschaft*, **16** (1914) 898.
[7] Kossel, W., *Verh. d. Dt. Phys. Gesellschaft*, **16** (1914) 953.

<table>
<tr><td>

25

</td><td>

The Franck–Hertz Experiment (1914)

</td></tr>
</table>

Franck in 1938

Hertz

In 1914 James Franck[1] [1] and Gustav Hertz[2] [2], a nephew of Heinrich Hertz, were working together in the Physics Institute of Berlin University led by Rubens. Both were from Hamburg and knew each other since their student days. At that time, Hertz was assistant and Franck *Privatdozent*, i.e., a scientist with the right to give lectures but not having a professorship.

They were particularly interested in *ionization*, i.e., in the process in which an electron is removed from an atom. The minimum energy which is required to ionize an atom is called the *ionization energy*. To measure it, Franck and Hertz constructed an apparatus in which they wanted to study the ionization brought about in the atoms of a gas or vapour by electrons, which were emitted from a hot wire by thermionic emission and then accelerated in an electric field so that their energy was well known. For an electron energy smaller than the ionization energy they expected no energy exchange between electrons and atoms. In contrast, for higher energies, they expected the electrons to lose exactly the ionization energy.

The apparatus was a glass tube containing the gas or vapour and an arrangement of three electrodes, a thin, electrically heated platinum wire emitting the electrons in the centre of the tube, a cylindrical platinum grid of 4 cm radius, surrounding the wire, and an outer cylindrical electrode (also from platinum) of 4.2 cm radius. The grid was placed on a variable positive voltage U with respect to the wire and the outer electrode on a negative voltage, 0.5 V, with respect to the grid. They were able to work with mercury vapour by putting a drop of mercury into the tube, putting the tube in a bath of hot paraffin, and connecting it to a vacuum pump which was operating continuously.

Recording the electric current between the wire and the outer electrode, which was carried by the electrons reaching the latter, they found results just as they had expected [3]. The current rose with the voltage U until a certain value U_0 was reached, then dropped sharply but rose again up to a maximum at $2U_0$ to be followed by another drop and a third maximum at $3U_0$, etc. For the voltage U_0 they found the value $U_0 = 4.9$ V. Now, since an electron, accelerated by an electric field over the voltage U, gains the energy $E = eU$ (where e is the elementary charge) the results are easily interpreted. As long as the electron energy is smaller than $E_0 = eU_0$, the electrons cannot lose energy in collisions with the mercury atoms. When they reach the grid their energy is large enough to overcome the small repelling field between grid and outer electrode. At a voltage slightly above U_0, however, the electrons reach the energy E_0 a little distance before the grid. They lose their energy there in collisions and, before reaching the grid, cannot gain enough energy from

[1] James Franck (1882–1964), Nobel Prize 1925   [2] Gustav Ludwig Hertz (1887–1975), Nobel Prize 1925

the field to traverse the repelling field outside the grid. Therefore the current drops. As the voltage increases, the collision zone moves inward towards the wire. After the collision, the electrons can again gain enough energy to reach the outer electrode. The current rises again until, at a voltage just above $2U_0$ a second collision with energy loss happens near the grid. The current drops again, etc. Franck and Hertz believed E_0 to be the ionization energy of the mercury atoms.

It was well known that, in gas discharges, mercury emits radiation, in particular, at the wavelength $\lambda_0 = 253.6\,\text{nm}$ corresponding to the frequency $\nu_0 = c/\lambda_0$, where c is the velocity of light. Franck and Hertz pointed out that the energy E_0 and the frequency ν_0 were connected by the relation $E_0 = h\nu_0$, where h is *Planck's constant*. Thus they had not only shown for the first time that electrons lose their kinetic energy to mercury atoms in energy quanta E_0, but they had also shown that these energy quanta are equal to the energy of the light emitted by the same atoms if interpreted with Einstein's light-quantum hypothesis. Of course, they could also turn the argument around and, from the known experimental values of E_0 and ν_0 compute h. This was the first experimental determination of Planck's constant not using blackbody radiation. In a second experiment [4] Franck and Hertz showed that they could indeed excite the emission of a spectrum with a single line of frequency ν_0 using electrons with energies just above E_0.

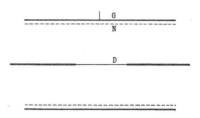

Cut through the cylindrical arrangement of electrodes. In the centre is a thin platinum wire D heated by an electric current, which is passed through it. It is surrounded by a platinum net N on the outside of which is the solid platinum electrode G. From [3]. Reprinted with kind permission of the Deutsche Physikalische Gesellschaft.

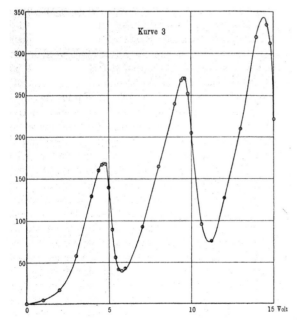

The electron current, reaching the outer electrode G, as a function of the accelerating voltage between the wire D and the net N. There are three sharp maxima, separated by $4.9\,\text{V}$ from one another. From [3]. Reprinted with kind permission of the Deutsche Physikalische Gesellschaft.

When they wrote their papers early in 1914, Franck and Hertz were not familiar with Bohr's new theory of 1913. It was Bohr who pointed out in 1915 [5] that their experiments 'may possibly be consistent with the assumption' that E_0 is not the ionization energy but only the energy needed to produce an excited state of the mercury atom. By emission of a light quantum of frequency

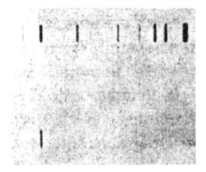

The spectrum of mercury obtained from a gas discharge in mercury vapour (top) shows many lines; Franck and Hertz obtained a single-line spectrum (bottom) by exciting the atoms of mercury vapour by electrons with an energy slightly above the excitation energy. From [4]. Reprinted with kind permission of the Deutsche Physikalische Gesellschaft.

[1] Lemmerich, J., *Max Born und James Franck, der Luxus des Gewissens*. Reichert, Wiesbaden, 1982.

[2] Kuczera, J., *Gustav Hertz*. Teubner, Leipzig, 1985.

[3] Franck, J. and Hertz, G., *Verh. d. Dt. Phys. Gesellschaft*, **16** (1914) 457.

[4] Franck, J. and Hertz, G., *Verh. d. Dt. Phys. Gesellschaft*, **16** (1914) 512.

[5] Bohr, N., *Philosophical Magazine*, **30** (1915) 384.

[6] Hoffmann, D., *Physikalische Blätter*, **51** (1995) F–157.

ν_0 the atom could then return to the ground state. Thus, the experiments by Franck and Hertz were later considered as new and independent support not only of Planck's quantum theory and Einstein's light-quantum hypothesis but also of Bohr's theory of the atom with stationary states of discrete energies. In 1926 Franck and Hertz were awarded the Nobel Prize for 1925. It is typical of Franck that he concluded his Nobel Lecture with the words: 'We know only too well that we owe the wide recognition that our work has received to contact with the great concepts and ideas of M. Planck and in particular of N. Bohr.'

As is well known, the twentieth century was full of dramatic and tragic events. In a particular way these are reflected in the further careers of Franck and Hertz.

In 1922 Franck, together with Max Born, became professor in Göttingen which, under their leadership, was soon to be a world centre in modern physics. When the Nazis took power in Germany in January 1933, they immediately issued a law which dismissed Jews and also socialists and communists from public office such as a professorship. Persons who had served in the First World War were at first exempt from this inhuman rule. Nevertheless, Franck, who had been a lieutenant and had been awarded the iron cross, first class, for bravery, resigned his office already in April 1933. The news were carried by many German papers including parts of his resignation letter. Franck moved to the United States where he became professor in Baltimore in 1935 and in Chicago in 1938. During the Second World War, like many scientists, he worked on the atomic bomb. Within his Chicago group passionate discussions took place about the use of the bomb. As a result, Franck drafted the 'Franck Report' which he handed in Washington to the Deputy Secretary of Defence on 12 July 1945. It argued against the use of the bomb against the Japanese population and for a demonstration on uninhabited land and urged for an international treaty against the use of atomic weapons in wars. In 1953 Franck accepted the honorary citizenship of the town of Göttingen. He died there, during a visit, in 1964.

Hertz, who had become professor at the Technical University in Berlin, resigned his office in 1935. He chose to change into industry where he felt less exposed to the regime. There he led a group working on isotope separation. In 1945, together with many scientists, he was brought to the Soviet Union to work there. In 1954 he became professor in Leipzig, where, as the only Nobel Prize winner in East Germany, he was able to use his influence for the promotion of research. Born, who met him in 1955, wrote in a letter to Franck, quoted in [6], that Hertz told him: 'What I say publicly has nothing to do with what I think. For instance, I say, I don't go to the West because I detest the conditions there. The truth is, I don't go, not to make myself suspicious and thus not to endanger the return of many physicists which are still held back in Russia.' Hertz died in Berlin in 1975.

Einstein Completes the General Theory of Relativity (1915)

In Episode 11 we witnessed the creation of special relativity. The word *special*, later given to that theory by Einstein, signifies that the relative motion between two reference frames is rectilinear with constant velocity.

In 1907, still working at the patent office in Bern, Einstein began to study the laws of physics in reference frames with an accelerated relative motion. When he completed this work in 1915 he called it the *General Theory of Relativity*. At various occasions Einstein recalled his starting point in this project. It struck him that a man falling from the top of a roof, he said, did not feel his own weight. In the reference frame of the building it is the weight or *gravitational force* which make the man fall but in a reference frame moving with the man there is another force, exactly counteracting the weight so that there is no net force. In that frame the man stays at rest. Einstein realized that acceleration and gravitation are equivalent to each other. That was later called the *equivalence principle*. If he would be able to extend his theory of relativity to accelerated reference frames, he would be able to do for the theory of gravitation what he had done for electrodynamics with special relativity. He gave a first glimpse at his new topic in a review article on special relativity written in 1907.

In 1909 Einstein was appointed associate professor at the University of Zurich and in 1911 full professor at the German (spoken) University of Prague, the capital of Bohemia, then part of Austria. It was here that he wrote the first paper completely devoted to general relativity, which bears the title *On the Influence of Gravity on the Propagation of Light* [1]. He considered a coordinate system K at rest in a homogeneous gravitational field. Let us imagine it on the surface of the earth with the z axis pointing upwards. There is a body S_1 in the origin and a body S_2 above it at height h on the z axis. An amount of energy is sent off by S_2 in the form of electromagnetic radiation and, after travelling down the z axis, is transferred to S_1. The question was, has that energy increased because the light *fell* for a distance h in the gravitational field, and, if so, by how much? To answer that question Einstein used another coordinate system K', which is accelerated with respect to K in z direction and in which no gravitational field exists. Using the equivalence principle he had replaced gravitation by acceleration. Let us call the acceleration g. A third coordinate system K_0 was used to study the energy transfer to S_1. At the moment of transfer it has the same velocity as K' but is not accelerated with respect to K. Therefore in this system Einstein could use results from special relativity. He found that the Energy E_1 absorbed by S_1 was indeed larger than the energy E_2 emitted by S_2 by an amount $\Delta E_2 = E_2 gh/c^2$ provided that ΔE_2 was much smaller than E_2. The product $gh = \Delta\varphi$ is the difference in the potential φ of the gravitational field in the positions of S_2 and S_1. (One knows from school that the energy needed to lift a body of mass m from the lower of these positions to the upper is $mgh = m\Delta\varphi$.) Thus Einstein found that an amount E of electromagnetic energy falling a distance h in a gravitational field increases by $\Delta E = Egh/c^2 = E\Delta\varphi/c^2$. On the other hand, if it climbs against the field, it decreases by that energy.

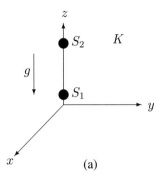 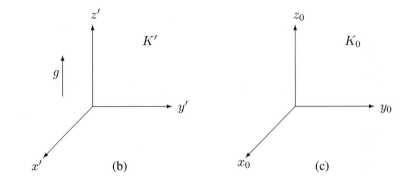

The three coordinate systems used by Einstein in 1911. (a) The system K is not accelerated; in it there acts a constant gravitational force $F_\mathrm{g} = mg$, pointing downwards, on a body of mass m. (b) The system K', related to K by the equivalence principle, is accelerated (with acceleration g) in z direction. Although there is no gravitational force in K' its equivalence is provided by the acceleration. (c) The system K_0 is not accelerated but moves with constant velocity with respect to K. At a fixed moment in time it has the same velocity as K'.

It is characteristic for Einstein that in the same paper he proposed a way to verify his predictions experimentally. He computed the influence of gravity on the frequency ν of light and found a frequency shift of $\Delta\nu = \nu\Delta\varphi/c^2$. From the above, that is obvious if one considers a light quantum of energy $E = h\nu$ (Episode 10). However, here Einstein did not use his still unproven light-quantum hypothesis. From this result he computed that the frequency of a spectral line, emitted by an atom on the surface of the sun, would be reduced by two parts in a million when that light reached the earth. Thus a line spectrum originating from the sun is shifted to lower frequencies and therefore to longer wavelengths compared to a spectrum emitted in the laboratory. This *redshift* is difficult to measure because the surface of the sun is a nasty environment with high pressure, storms, and magnetic fields, all influencing spectral lines. But with modern techniques it has been well established even in the laboratory with radiation climbing against the earth's gravitation for only a few metres (see Episode 81).

Einstein also computed the *bending of light* by the gravitation. If the light of a star passes near the surface of the sun and is then observed by an astronomer on the earth, the star appears to be in a slightly different position because the light was attracted by the sun and thus the ray was bent on its way from star to the earth. Einstein found a bending angle of 0.83 seconds of an arc. This number was too small by a factor of two but nobody knew because the effect had not been measured. However, Einstein himself realized that his theory had to be refined. For a homogeneous gravitational field, a field that is constant everywhere, he could replace gravitation by a single transformation to an accelerated coordinate system. For a more complicated field like that of the sun or that of all stars the transformation would have to be different for every point in space. That seemed a formidable problem.

In 1912 Einstein returned to Zurich, this time as full professor at the *ETH* (Eidgenössische Technische Hochschule), as the Polytechnikum had been

---

Riemann Space

There is a global reference frame with coordinates x^0, x^1, x^2, x^3 and at every point a local system with coordinates $\xi^0, \xi^1, \xi^2, \xi^3$. In this latter system, special relativity theory holds in a small vicinity of the point, i.e., the coordinates $\xi^0, \xi^1, \xi^2, \xi^3$ span a Minkowski space, so that the invariant line element in the local system is

$$ds^2 = \eta_{\alpha\beta}\, d\xi^\alpha d\xi^\beta$$

(see Box II in Episode 11). A general transformation to the global system yields

$$ds^2 = g_{\mu\nu}\, dx^\mu dx^\nu \quad, \qquad g_{\mu\nu}(x^0, x^1, x^2, x^3) = \eta_{\alpha\beta}\frac{\partial\xi^\alpha}{\partial x^\mu}\frac{\partial\xi^\beta}{\partial x^\nu} \quad.$$

The *metric tensor* $g_{\mu\nu}$ of the Riemann space depends on the coordinates of that space and therefore is more complicated than the metric tensor $\eta_{\alpha\beta}$ of Minkowski space. Transformation of the equation of motion from the local to the global frame yields

$$\frac{d^2 x^\kappa}{d\tau^2} = -\Gamma^\kappa_{\mu\nu}\frac{dx^\mu}{d\tau}\frac{dx^\nu}{d\tau} \quad,$$

where the *Christoffel symbols*

$$\Gamma^\kappa_{\mu\nu} = \frac{\partial x^\kappa}{\partial \xi^\alpha}\frac{\partial^2\xi^\alpha}{\partial x^\mu \partial x^\nu} = \frac{g_{\kappa\lambda}}{2}\left(\frac{\partial g_{\mu\lambda}}{\partial x^\nu} + \frac{\partial g_{\nu\lambda}}{\partial x^\mu} - \frac{\partial g_{\mu\nu}}{\partial x^\lambda}\right)$$

contain the metric tensor and its derivatives. Thus, the acceleration on the left-hand side of the equation of motion and therefore motion itself is determined by the structure of space.

---

called the year before when it had been raised to the rank of a university. There, one of his colleagues in mathematics was his old friend Grossmann whom he told of his new problem and from whom he learned that a theory of very general transformations had been developed in the nineteenth century by Gauss[1], Riemann[2], Christoffel[3], Ricci[4], and Levi-Civita[5]. In collaboration with Grossmann, Einstein began to adapt this new part of mathematics which can be called *differential geometry of the Riemann space* to his problem. He continued by himself when he left Zurich.

A peculiarity of Riemann space is that space itself, in general, is curved. What does that mean? Consider a plane in normal (Euclidean) space and a triangle on that plane. A triangle is a figure composed of pieces of three straight lines. The sum of its three angles is 180°. The equivalent of a straight line on a sphere is a *great circle*, i.e., a circle which has the centre of the sphere as its own centre. Now consider the surface of a sphere and a triangle, composed of three pieces of great circles, on that surface. The sum of its angles is not 180°. (One easily sees that the particular triangle one obtains by drawing an arc from the north pole to the equator, an arc along one quarter of the equator, and an arc back to the pole has three angles of 90°.) The reason is that the two-dimensional spherical surface is curved in three-dimensional space. But

---

[1] Carl Friedrich Gauss (1777–1855)   [2] Bernhard Riemann (1826–1866)   [3] Elwin Bruno Christoffel (1829–1900)   [4] Gregorio Ricci-Curbastro (1853–1925)   [5] Tullio Levi-Civita (1873–1941)

Einstein in his flat in Berlin, Haberlandstrasse.

also a space of three or more dimensions can be curved. The sum of the angles of a triangle in that space is different from 180°. Expressed more formally, this is the case if the *metric tensor* of the space is not constant throughout the space but changes from point to point (see Box). Einstein himself gave a simple example for curvature appearing in general relativity. Consider a circle on a plane and decompose a radius and the circumference into small elements. Let the circle rotate continuously and perform a Lorentz transformation of each element to the reference frame of an observer at rest. Since the relative motion is perpendicular to the radius the elements of the radius stay unchanged but those of the circumference are Lorentz contracted. Therefore, the rule that the circumference is 2π times the radius no longer holds.

In 1913 Planck and Nernst came to Zurich and convinced Einstein to move to Berlin. They offered him the membership in the Prussian Academy of Sciences, a position paid by the Society, as well as a professorship at the University with the right but not the obligation to teach. Einstein, who liked teaching but found that it left him less time for research than he had at the patent office, accepted. In March 1914 he moved to Berlin, where he lived and worked for more than 18 years [2] and where he was very productive. In the war years 1914–1918 alone, he published about 50 scientific papers. Moreover, he was director of the Kaiser-Wilhelm Institute (now Max-Planck Institute) of Physics, and he served as adviser to the Phyikalisch-Technische Reichsanstalt, and as president of the German Physical Society.

Einstein with friends and colleagues in Berlin. Standing (from the left): Walter Grotrian, Wilhelm Westphal, Otto von Bayer, Peter Pringsheim, Gustav Hertz. Sitting a little elevated: Einstein and Fritz Haber, the 1918 Nobel Laureate in chemistry. Sitting: Hertha Sponer, Ingrid Franck, James Franck, Lise Meitner, Otto Hahn. The photograph was taken in 1920 on a farewell party for Franck, who moved to Göttingen.

In hindsight the development of general relativity can be seen to consist of three steps,

1. use of the equivalence principle, i.e., replacement of gravitation by transformation to an accelerated reference frame,
2. formulation of the equations of motion in Riemann space, which connects the acceleration of a particle to the properties of space itself as

expressed by the metric tensor $g_{\mu\nu}$,

3. finding of the *field equations of gravitation*, which connect the metric tensor with the distribution of masses in space.

The physics contents of (2) and (3) are expressed beautifully in the introductory parable of the monumental book [3]:

Space acts on matter, telling it how to move. In turn, matter reacts back on space, telling it how to curve.

The first step was taken by Einstein when he was still at Bern and reconsidered in Prague. The second step was essentially performed by Einstein and Grossmann in Zurich (for a few more details see Box). After several unsatisfactory attempts Einstein found his *field equations* in Berlin, where he presented them on 25 November 1915 in a session of the Prussian Academy [4]. We do not give here the field equations explicitly but only mention that mathematically they are quite different from Newton's law of gravitation. The potential of Newton's gravitational force can be expressed as a single quantity, a *scalar*, depending on the space coordinates. Einstein's potential is a symmetric *tensor* consisting of ten independent elements.

A comprehensive article entitled *The Foundation of the General Theory of Relativity* [5] appeared a little later and in 1917 Einstein published a small book intended for a broader public [6]. For the special case in which the field equations are determined by only one star, the sun, Einstein derived three important results.

1. *Redshift of spectral lines:* It remained as he had found it in 1911 in Prague.
2. *Bending of light:* The angle of bending for starlight just grazing the sun was now twice the Prague value, 1.7 seconds of an arc.
3. *Rotation of the Perihelion of Mercury:* Here Einstein solved a problem known since more than 60 years.

We know from letters written by Einstein at the time that the last point gave him the greatest joy [7]. In Newton's mechanics a planet moves about the sun on an elliptical orbit and that ellipse stays fixed in space (with respect to the fixed stars). The point on the orbit which is nearest to the sun is called *perihelion*. It was known since 1859 that the orbit of Mercury, the innermost planet, is not fixed. Its perihelion rotates very slowly about the sun. The angle of rotation was known already in 1883 to be 43 seconds of an arc per century. The observation was in obvious discrepancy with the laws of classical mechanics. In his general theory of relativity, the new theory of gravitation, Einstein was able to compute the rotation of the perihelion of a planet and for Mercury he found exactly the measured value. In Episode 29 we will learn how the bending of light by the sun was measured. A laboratory experiment verifying the predicted redshift succeeded in 1960, see end of Episode 81.

[1] Einstein, A., *Annalen der Physik*, **35** (1911) 898. Also in [8], p. 72, Engl. transl. in [9].

[2] Kirsten, C. and Treder, H.-J. (eds.), *Albert Einstein in Berlin 1913–1933, Teil I Darstellung und Dokumente*. Akademie-Verlag, Berlin, 1979.

[3] Misner, C. W., Thorne, K. S., and Wheeler, J. A., *Gravitation*. Freeman, San Francisco, 1973.

[4] Einstein, A., *Sitzungsber. Preuss. Akad. Wiss. (Berlin)*, (1915) 844.

[5] Einstein, A., *Annalen der Physik*, **49** (1916) 284. Also in [8], p. 81, Engl. transl. in [9].

[6] Einstein, A., *Über die spezielle und die allgemeine Relativitätstherie, gemeinverständlich*. Vieweg, Braunschweig, 1917. Engl. transl. [10].

[7] Pais, A., *'Subtle is the Lord . . .' – The Science and Life of Albert Einstein*. Oxford University Press, Oxford, 1982.

[8] Lorentz, H. A., Einstein, A., and Minkowski, H., *Das Relativitätsprinzip*. Teubner, Berlin, 5th ed., 1923.

[9] Sommerfeld, A. (ed.), *The Principle of Relativity*. Dover, New York, 1923.

[10] Einstein, A., *On the Special and the General Relativity Theory, a Popular Exposition*. Methuen, London, 1920.

27 Sommerfeld – Spatial Quantization and Fine Structure (1916)

Sommerfeld in Göttingen, 1906.

In 1906 Röntgen, who was then professor of physics at the University of Munich, saw to it that the chair of Theoretical Physics, left vacant by Boltzmann who had moved to Vienna, was offered to Sommerfeld[1]. (Before, it had been offered to Lorentz, who, however, had preferred to stay in Leiden.) Sommerfeld [1–3] was an accomplished mathematician. Earlier in his career he had been assistant to Klein[2], the eminent mathematician in Göttingen. Besides the main lecture courses Sommerfeld used to offer lectures on special topics on which he was currently working. In the winter semester 1914/15 his topic was the Bohr model. He extended it considerably and was able to find a number of new results which were published [4,5] early in 1916.

The gist of Bohr's theory was the fact that it allowed only certain electron orbits. Sommerfeld tried to find the *quantum conditions* by which these orbits could be singled out from the classically allowed ones. Such a condition was already known for a very simple system, the *harmonic oscillator*. If a mass point can move along a fixed line (the x axis) and is pulled to a fixed point $(x = 0)$ by a harmonic force (the force of a spring), then it oscillates about that point with a fixed frequency ν. This is a mechanical model for an oscillator à la Planck. The system has one *degree of freedom* since motion is possible only along the x axis. Now for a Planckian oscillator only certain fixed energies are allowed, the smallest energy difference being $h\nu$. A quantum condition which makes this spacing possible can be expressed by demanding that a certain integral be an integer multiple of Planck's constant, $\int p_x \, \mathrm{d}x = nh$. In this *action integral* the momentum p_x in x direction is integrated over the position x for a complete period of the oscillation. The quantum condition thus gives rise to the *quantum number* n, which can take the values $n = 1, 2, \ldots$. Sommerfeld now demanded that such a condition be fulfilled for every degree of freedom.

In the hydrogen atom the electron has three degrees of freedom corresponding to the three space coordinates. Using as coordinates the distance r from the nucleus and two angles φ and ψ he found three quantum numbers which specify each allowed orbit. The rules for two of these quantum numbers, n_φ and n_ψ, do not depend on the exact form of the force field felt by the electron as long as it is spherically symmetric. Their effects can be summarized as follows:

- *Angular-Momentum Quantization:* The quantum number n_φ determines the absolute value L of the angular momentum of the electron, $L = n_\varphi h/2\pi$. For a circular orbit the absolute value L is the product $L = rp = rmv$ of the radius r of the orbit and the momentum p of the elec-

---

[1] Arnold Sommerfeld (1868–1951)   [2] Felix Klein (1849–1925)

tron, the latter being the product of the mass m and the velocity v of the electron. Not only circular but also elliptical orbits are possible. Even the case $L = 0$, i.e., $n_\varphi = 0$, for which the orbit reduces to an oscillation along a line, can be considered. But since then the electron would oscillate on a straight line right through the nucleus Sommerfeld excluded it.

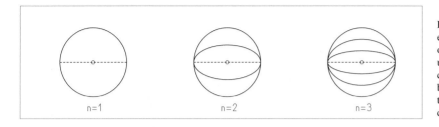

Form of Sommerfeld's elliptic orbits of the electron in the hydrogen atom for three values of the principal quantum number n. The figures are in different scales. The radius of each circular orbit is n^2 times the Bohr radius. The broken line, corresponding to an oscillatory trajectory, was excluded by Sommerfeld because it passes through the nucleus.

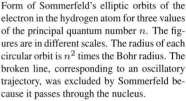

Arrangement of Sommerfeld's elliptic orbits in space, one focus being the nucleus.

- *Spatial Quantization:* The angular momentum is a *vector*, i.e., besides its absolute value, it has a direction. This direction is perpendicular to the plane of the orbit. If the orbit appears clockwise on a blackboard, then the direction is right into the blackboard. If we chose a direction in space, which we call the z direction, the angular momentum has a component L_z with respect to that direction. It is given by $L_z = n_\psi h/2\pi$. The quantum number n_ψ can take the values $n_\psi = \pm1, \pm2, \cdots, \pm n_\varphi$. (At first Sommerfeld admitted $n_\psi = 0$, i.e., $L_z = 0$, but later excluded it for reasons similar to those for which he excluded $n_\varphi = 0$.) The quantization of L_z means that only certain orientations of angular momentum and therefore of the plane of motion in space are allowed. If the force field is spherically symmetric, then no direction is preferred over others. In that case all states, differing only in the quantum number n_ψ, have the same energy, they are said to be *degenerate*. If the spherical symmetry is broken, this *degeneracy* is lifted. The states may have different energies.

If there is only one electron, it is only under the influence of the electric attraction of the nucleus and there are no electric fields of other electrons to be considered. Then, if there is also no external field, there is another degeneracy. The sum of Sommerfeld's third quantum number n_r and of n_φ is equal to Bohr's quantum number n, now called the *principal quantum number*, $n = n_r + n_\varphi$. A stationary state with energy E_n can have a circular orbit ($n_\varphi = n$) but also various elliptical orbits ($n_\varphi = 1, 2, \ldots, n-1$). Because of the

Sommerfeld lecturing on his model.

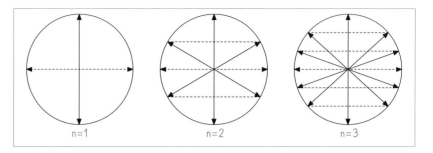

Sommerfeld's illustration of spatial quantization for angular-momentum values $\hbar, 2\hbar, 3\hbar$. The arrows indicate the allowed orientations of the angular-momentum vector with respect to the z axis which points upwards. A figure like this appeared in later editions of [6] which was published after Sommerfeld had excluded the case $L_z = 0$ indicated by the broken horizontal arrows.

degeneracy the energies E_n and therefore the spectra for hydrogen and for the helium ion are the same as obtained by Bohr. But with his elaborate theory Sommerfeld was able to describe other phenomena. Here we mention only the alkali spectra and the fine structure of the hydrogen spectrum.

Let us first consider *alkali atoms*, i.e., lithium, sodium, etc. Here we have one electron in an outer shell in the field of the nucleon and of the other electrons, which are in one or more inner shells. The spectrum is caused by transitions of the outer electron between different stationary states. On orbits far outside the inner shells the force field is that of a nucleus with a single positive charge, because the number of elementary charges Z of the nucleus is effectively reduced to 1 by the $Z - 1$ electrons on the inner shells. One can say that there is a *screening* of the nucleus by the other electrons. Thus, for states with large values of n and circular orbits $(n_\varphi = n)$ the energies are very similar to the ones of hydrogen. The way of counting the quantum number n for alkali atoms is adjusted here to the counting for hydrogen. For small values of n the screening is less effective, the attraction of the nucleus is stronger than in the case of hydrogen – the energy decreases. For non-circular orbits, with n_φ smaller than n, this effect is enhanced, since the electron passes through regions still nearer to the nucleus. Therefore, for the alkali atoms the energies E_{n,n_φ} of the stationary states depend on two quantum numbers. As a consequence, there are many more spectral lines than for hydrogen. The alkali spectra were described quite satisfactorily in this way except for one fact. If observed with high-resolution instruments, the spectral lines are resolved into several neighbouring lines. This is called the *fine structure* of the spectrum. The alkali spectra suggested a splitting of the energies E_{n,n_φ} (for n_φ larger than 1) into two. That called for the existence of one more quantum number. We shall learn about it in Episode 36.

Also the hydrogen spectrum shows a fine structure, although the line splitting is tiny compared to that in alkali spectra. Sommerfeld was able to give a good, quantitative explanation for it. He applied Einstein's special relativity to the hydrogen atom. It is easy to compute the velocity of an electron, which circles the hydrogen nucleus in its ground state. It is a small fraction α of the velocity of light. The fraction can be expressed by fundamental constants. (In the units used by Sommerfeld $\alpha = e^2/\hbar c$, in SI units $\alpha = e^2/(4\pi\epsilon_0\hbar c)$ with e being the elementary charge, h Planck's constant $(\hbar = h/(2\pi))$, c the velocity of light, and ϵ_0 the permittivity of free space). It is now called *Sommerfeld's fine-structure constant*. It turns out that α is very nearly equal to $1/137$. This

Sommerfeld's Fine-Structure Constant α

in SI units is given by

$$\alpha = \frac{e^2}{4\pi\,\epsilon_0\hbar c} \quad .$$

Its numerical value is

$$\alpha = \frac{1}{137.035\,999\,11(46)} \quad .$$

The number in parentheses is the experimental error in the last digits.

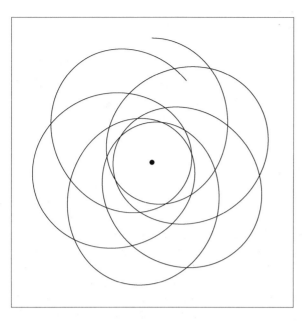

If special relativity is taken into account, the ellipse rotates and the orbit becomes a rosette.

is a small but not completely negligible number. Accordingly, the relativistic dependence of the mass of the electron on its velocity has to be taken into account. On an elliptic orbit the electron is fastest where it is closest to the nucleus, i.e., at the *perihelion*, and slowest where it is farthest away. Its mass changes along the orbit. This brings about that the orbit is no longer an ellipse but a rosette. The electron moves along an ellipse which itself rotates about the nucleus. One speaks about a *perihelion rotation*. It is now obvious that even for hydrogen the energy levels for fixed n but different n_φ are no longer degenerate. Sommerfeld computed the energies E_{n,n_φ} and expressed them as a sum of infinitely many terms, called a *power series*. The first term is the ordinary Balmer (or Bohr) result. The next term is proportional to α^2 (quite small), the next proportional to α^4 (still smaller), etc. It is clear that only the first few terms need be considered, the others being negligibly small. Such expansions in powers of Sommerfeld's fine-structure constant α are still used today in modern quantum electrodynamics.

Bohr, who at the time was in Manchester with Rutherford, in the middle of the First World War, wrote in a letter to Sommerfeld [7]:

> I thank you so much for your most interesting and beautiful papers. I do not think that I have ever enjoyed the reading of anything more than I enjoyed the study of them, and I need not say that not only I but everybody here has taken the greatest interest in your important and beautiful results
> ...
> I do not know how to express, how I wish that the present terribly sad state of the world may change soon. I am hoping very much to meet you soon again and send the kindest regards to you and all the other physicists in your laboratory not only from myself but from all here.

In 1919 the first edition of Sommerfeld's famous book *Atomic Structure and Spectral Lines* [6] appeared. It is from this book that the younger generation of physicists learned atomic theory. Some of the most brilliant were Sommerfeld's Ph.D. students, among them the future Nobel laureates Pauli, Heisenberg, and Bethe.

ATOMBAU

UND

SPEKTRALLINIEN

Von

Arnold Sommerfeld

Professor der Theoretischen Physik an der Universität München

Mit 103 Abbildungen

Braunschweig 1919

Druck und Verlag von Friedr. Vieweg & Sohn

[1] Sommerfeld, A., *Autobiographische Skizze*. In *Geist und Gestalt, Beiträge zur Geschichte der Bayerischen Akademie der Wissenschaften*, vol. 2, p. 100. München, 1959. Also in [3], Vol. IV, p. 673.

[2] Benz, U., *Arnold Sommerfeld*. Wissenschaftl. Verlagsbuchh., Stuttgart, 1975.

[3] Sommerfeld, A., *Gesammelte Schriften, 4 vols.* Vieweg, Braunschweig, 1968.

[4] Sommerfeld, A., *Annalen der Physik*, **51** (1916) 1. Also in [3], Vol. III, p. 172.

[5] Sommerfeld, A., *Annalen der Physik*, **51** (1916) 125. Also in [3], Vol. III, p. 266.

[6] Sommerfeld, A., *Atombau und Spektrallinien*. Vieweg, Braunschweig, 1st ed., 1919.

[7] Bohr, N., *Letter to Sommerfeld, March 19, 1916*. In *Niels Bohr, Collected Works*, vol. 2, p. 603. North Holland, Amsterdam, 1981.

Title page of the first edition of Sommerfeld's book *Atomic Structure and Spectral Lines*. As shown by the signature this copy was originally owned by Walther Gerlach. In the possession of Prof. Fritz Bopp, Siegen; reproduced with his permission.

Nitrogen is Turned into Oxygen (1919)

In June 1919, a paper by Rutherford was published in which he announces another of his great discoveries, the transmutation of an element in a planned experiment. He had already discovered the spontaneous transmutation of radioactive elements (see Episode 9), which could not be influenced by man. Now he showed that transmutation could be achieved, more or less, at will.

When bombarding air, of which 80% is nitrogen, and also pure nitrogen gas with α particles, he found that fast hydrogen nuclei were produced. In the concluding section of his paper [1] he wrote:

> From the results so far obtained it is difficult to avoid the conclusion that the long-range atoms arising from the collision of α particles with nitrogen are not nitrogen atoms but probably atoms of hydrogen, or atoms of mass 2. If this is the case, we must conclude that the nitrogen atom is disintegrated under the intense forces developed in close collision with a swift α particle, and that the hydrogen atom which is liberated formed a constituent part of the nitrogen nucleus.

The discovery resulted from the detailed study of the scattering of α particles by nuclei. The scattering of one nucleus by another is not unlike the collision between one billiard ball and another. In both cases, the fundamental laws of *conservation* of *energy*, *momentum*, and *mass* are observed. That means that the total energy, the total momentum, and the total mass of the two particles are not changed by the collision. Usually the two particles do not exchange mass in the collision; each one keeps its original mass. In that case simple, but interesting properties of the collisions can be inferred:

Rutherford

- *Case 1:* If the relatively light α particle collides with a heavy gold nucleus, the latter stays essentially at rest, the α particle changes direction but not energy or speed. The direction may even be completely reversed.
- *Case 2:* If the target particle is not very much heavier than the projectile, it *recoils* from the projectile and gains some energy. In the cloud chamber then both particles, after the collision, leave tracks. Still the projectile can be reflected backwards.
- *Case 3:* This is no longer true if projectile and target have the same mass. Then, after the collision the direction of projectile and target, in general, form a right angle with each other. In the special case of a *central collision* the projectile transfers all its energy to the target and comes to rest.
- *Case 4:* Finally, if the target particle is lighter than the projectile, then the projectile is always scattered through an angle of less than $90°$. In a central collision the projectile does not come to rest but the target gains a velocity higher than that held originally by the projectile, just as a tennis ball hit by a racket is faster than the racket was.

Taking advantage of these simple facts one can, by studying the collisions between particles, learn something about their masses.

The story of Rutherford's discovery is told by Feather [2]. The original Geiger–Marsden experiment was done with heavy nuclei as targets. But, of course, Rutherford was also interested in the scattering by light ones. In 1914 Marsden, indeed, found fast hydrogen nuclei, when he let α particles pass through hydrogen gas and thus verified case (4) above [3]. Hydrogen nuclei or protons, as Rutherford called them later, because of their smaller charge are less ionizing than α particles. The scintillations they cause in a zinc-sulphide screen are therefore less bright. On the other hand, they have a longer range in matter. Marsden who let α particles traverse hydrogen gas, observed protons at distances from the source which no α particles could reach any more. In 1915, by now professor in New Zealand, he observed fast hydrogen nuclei also when the target gas was air, which consists of nitrogen and oxygen atoms, both considerably heavier than α particles [4]. That did not fit any of the cases above. It was checked that the effect was not due to water vapour in the air and was attributed to the material of the source from which the α particles were emitted.

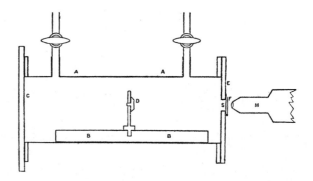

The apparatus with which Rutherford observed the transmutation of nitrogen. From [1].

In 1915 practically all young scientists were at war or working on war research. Also Rutherford did so by working on sound ranging of submarines. But in the second half of 1917 he took time again to do fundamental research, by himself aided only by Mr. William Kay, the laboratory steward. His apparatus, typical of Rutherford, was astonishingly simple. In the middle of an oblong brass box a source D of α particles was mounted. The box could be exhausted or filled with the desired gas at a desired pressure. One end wall had an opening, the outside of which was covered with a zinc-sulphide screen S, which could be observed with a microscope M. On the inside of the opening one or several thin metal sheets could be mounted. These sheets had an absorbing power for α particles equivalent to some centimetres of air. Early in September 1917 Rutherford observed scintillations, when the box was filled with air and when the absorbing power of the air and the metal sheets far exceeded the range of α particles from the source. He soon found that the effect was due to nitrogen because when he replaced the air by pure nitrogen it was higher by a factor $5/4$. (Only $4/5$ of the air is nitrogen.) To rule out other interpretations of his effect, he performed a careful study of α-particle scattering in various gases and only in 1919 published a series of four papers under the title *Collision of α Particles with Light Atoms*. The last paper with the subtitle *An Anomalous Effect in Nitrogen*, only six pages long, from which the quotation above is taken, contains his discovery.

It was now clear that a nitrogen nucleus was changed in the collision with an α particle but it was not clear into what. Rutherford realized that, to answer that question, detailed information about all particles existing after the collision was needed. After he had moved from Manchester to Cambridge, where, in 1919 he succeeded J. J. Thomson as Cavendish Professor, he asked a Japanese collaborator, T. Shimizu, to set up a Wilson cloud chamber in which the tracks of all charged particles taking part in a collision could be studied. When Shimuzu had to return to Japan the work was assigned to Blackett[1].

Blackett, originally a professional naval officer and veteran of the battles of the Falklands and Jutland, after the First World War, had decided to study physics under Rutherford. In his Nobel Lecture of 1948 Blackett wrote: 'Provided by Rutherford with so fine a problem, by C. T. R. Wilson with so powerful a method and by Nature with a liking for mechanical gadgets, I fell with a will to the problem of photographing some half million alpha-ray tracks.' Most photographs [5] were taken with the chamber filled with nitrogen (and water vapour), but for comparison also hydrogen and helium were used as chamber filling. Nearly all nuclear interactions, which Blackett observed, had the properties described above. They correspond to the cases 2, 3, and 4 (with nitrogen, helium, and hydrogen, respectively).

Blackett

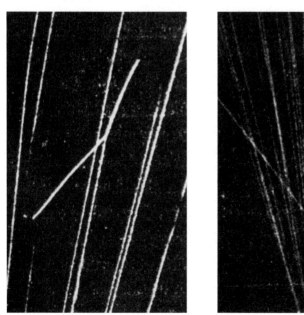

But in nitrogen there were eight events which were quite different. Like the other events they showed two tracks leaving the point of collision, but none of them was an α particle. One, identified by its light ionization, i.e.,

Cloud-chamber photographs showing the elastic scattering of an α particle by a nitrogen nucleus (left), a helium nucleus (middle), and a hydrogen nucleus (right). From [6], Plate 11.3 (left), [5], Plate 6.1 (middle), and [5], Plate 6.4 (right). Reprinted with permission.

---

[1] Patrick Maynard Stuart (Lord) Blackett (1897–1974), Nobel Prize 1948

Cloud-chamber photograph of the transmutation of nitrogen into oxygen [5]. One of the α particles, coming from below, interacts with a nitrogen nucleus from the chamber filling near the top of the picture. The track of the oxygen nucleus points towards the top and a little to the right. The much longer track of the proton points towards the lower left. From [7], Plate 11.3. Reprinted with permission.

[1] Rutherford, E., *Philosophical Magazine*, **37** (1919) 537, 562, 571, 581. Also in [8], Vol. 2, p. 547, 568, 577, 585.

[2] Feather, N. In [8], Vol. 2, p. 15.

[3] Marsden, E., *Philosophical Magazine*, **27** (1914) 824.

[4] Marsden, E. and Lantberry, W. C., *Philosophical Magazine*, **30** (1915) 240.

[5] Blackett, P. M. S., *Proceedings of the Royal Society*, **A107** (1925) 349.

[6] Blackett, P. M. S. and Lees, D. S., *Proceedings of the Royal Society*, **A134** (1931) 658.

[7] Blackett, P. M. S. and Lees, D. S., *Proceedings of the Royal Society*, **A136** (1932) 325.

[8] Rutherford, E., *The Collected Papers of Lord Ernest Rutherford of Nelson (Edited by J. Chadwick), 3 vols.* Allen and Unwin, London, 1962–1964.

by its small droplet density, was a hydrogen nucleus, a proton. The other one was a rather heavy nucleus. From measurements of the angles and ranges of the tracks, Blackett concluded that it had to be an oxygen nucleus with an atomic mass number 17, $^{17}_{8}$O. That isotope of oxygen was not known at the time but found soon after by optical spectroscopy. Blackett had therefore shown that "the assumed 'disintegration' of nitrogen by alpha particles was in reality an 'integration' process". Most of the α particle, upon collision, was integrated into the nitrogen nucleus and only a proton left. One can write the transmutation of nitrogen into oxygen as a chemical formula,

$$^{4}_{2}\text{He} + {}^{14}_{7}\text{N} \rightarrow {}^{17}_{8}\text{O} + {}^{1}_{1}\text{H} \quad .$$

Later Bothe introduced a compact notation in which the absorption of the α particle and the emission of the proton are contained in parentheses between the original and transmuted element,

$$^{14}_{7}\text{N}\,(\alpha, p)\,^{17}_{8}\text{O} \quad .$$

Transmutation reactions like the one above were later called *nuclear reactions*. They were used to produce unknown isotopes. But also elements which do not exist on earth were made, in particular, the element *plutonium* which was and still is produced in an industrial style. For the transmutation of light elements charged particles like the α particle can be used. If it has sufficiently high energy it can enter a nucleus although there is a strong repelling force between the positively charged α particle and the nucleus. For heavier nuclei with very large positive charge this force becomes very strong. But even then a neutral particle, the *neutron*, can bring about a transmutation. For its discovery see Episode 48.

Astronomers Verify General Relativity (1919)

29

In 1704 Newton published his book *Opticks*. In his foreword to a reprint of 1931 Einstein wrote [1]:

> Fortunate Newton, happy childhood of science! He who has time and tranquillity can by reading this book live the wonderful events which the great Newton experienced in his young days. Nature to him was an open book, whose letters he could read without effort ... In one person he combined the experimenter, the theorist, the mechanic and, not least, the artist in exposition. He stands before us strong, certain and alone; his joy in creation and his minute precision are evident in every word and in every figure.

The edition of 1704 ends with 16 *Queries*. In later editions Newton added another 15. The first query reads: *Do not Bodies act upon Light at a distance and by their action bend its Rays?* We have seen that Einstein, in his General Theory of Relativity, answered this question in 1915. His answer was confirmed by the observations of British astronomers in 1919.

One can speculate that Newton may have been inclined to answer his first query in the affirmative. After all, he was convinced that light consisted of tiny particles and particles are attracted by other masses through gravitation. Later the wave picture of light, advocated by Huygens, was accepted by most scientists. It did not allow for a bending of light.

Let us now rephrase Newton's query in the form 'How large is the angle α of bending for the light of a distant star passing near the surface of the sun?' and consider three answers:

1. $\alpha = 0$ – the answer of classical physics if light is assumed as a wave phenomenon with no mass associated to it.
2. $\alpha = 0.87''$ (the symbol $''$ stands for seconds of an arc) – the answer of classical physics if light is assumed to consist of particles with mass. According to Newtonian mechanics the angle of deflection depends only on the velocity of the particles and not on their mass.
3. $\alpha = 1.75''$ – the answer of Einstein's General Relativity in which no assumptions about the nature of light are needed.

The angle α is best measured during a solar eclipse when the sun itself is hidden behind the moon so that the stars near the sun can be observed. Photographs of the stars around the sun are taken through a suitable telescope and the relative positions of the stars are compared to those on photographs taken of the same group of stars in a convenient night. If there is a bending, then the stars seem to be a little pushed away from the centre of the sun on photographs taken during an eclipse. The effect, however, is tiny. To an observer on the earth the diameter of the sun appears as an angle of about half a degree. One second of an arc is $1/3600$ of a degree, $1'' = 1°/3600$. The displacement expected by Einstein is therefore on the order of one thousandth of the diameter of the sun. The accuracy of observation had to be considerably better, of course.

Photographic plate of the 1919 solar eclipse taken by the British expedition at Sobral. The prominent feature is the solar corona which appears dark in the photographic negative. The seven stars used for measurement appear as tiny dark spots between pairs of horizontal bars drawn later on the plate. To the left of these bars I have added the numbers used to denote these stars in the original publication. (In the planning stage it had seemed possible to observe 13 stars which were given the numbers from 1 to 13; the position of seven stars was actually measured.) From [2], Plate 1. Reprinted with permission.

Even before Einstein moved to Berlin he found there a young astronomer, Freundlich[1], then assistant at the Berlin observatory, who was enthusiastic about testing Einstein's prediction. In December 1913 Freundlich obtained the permission of the Prussian Academy [3] to participate in an Argentinean expedition to the Crimea to observe the eclipse due on 21 August 1914. Early in August the First World War broke out. Freundlich was interned in Russia but was allowed to return to Germany in 1915. Einstein later wrote that 'war and weather' prevented the success of the expedition. A success at the time

---

[1] Erwin Finlay Freundlich (1885–1964)

would certainly have puzzled Einstein, because, as we have seen in Episode 26, he found the correct value for the bending angle only in 1915. Before, he had only half that value, the same as in answer 2 above.

In 1917 British astronomers studied the prospects of observations during the eclipse of 29 May 1919. They found them to be exceedingly good since a total of 13 stars would be in the vicinity of the sun. The government was asked for financial support of 1100 £ and a committee was formed under Dyson[2], the Astronomer Royal, to prepare two expeditions. Taking into account possible weather conditions, two places were selected for the observations, *Sobral* in North Brazil and the island of *Principe* off the West African coast. The two teams left Liverpool together on 8 March 1919, and sailed together to Madeira. From there Crommelin[3] and Davidson went on to Sobral and Eddington[4] and Cottingham[5] to Principe. Both teams reached their objectives well in time, set up and tested their instruments and, at the time of the eclipse, enjoyed nearly perfect conditions to take photographs. The photographic plates were measured and the measurements were analysed after the return to Britain and the results were announced in a *Joint Eclipse Meeting of the Royal Society and the Royal Astronomical Society* [4]. This meeting of 6 November 1919, with *Sir Joseph Thomson, O.M., P.R.S., in the Chair* was called by Pais [5] 'the canonisation of Einstein'. The Astronomer Royal himself presented the results of the Sobral expedition and Eddington those obtained with the photographs taken at Principe. Both, within the limits of their errors, supported Einstein's answer 3 and excluded answers 1 and 2.

Eddington

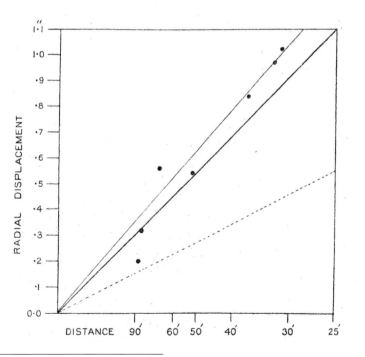

From the observations at Sobral: The radial displacement (in seconds of an arc) of the seven stars plotted against the inverse of the radial distance (in minutes of an arc) from the centre of the sun. The dark line corresponds to Einstein's prediction (answer 3). The dotted line corresponds to answer 2. The thin line is a 'best fit' to the measurements. From [2], Diagram 2. Reprinted with permission.

[2] (Sir) Frank Watson Dyson (1868–1939)   [3] Andrew C. D. Crommelin (1865–1939)   [4] (Sir) Arthur Stanley Eddington (1882–1944)   [5] E. T. Cottingham (1869–1940)

J. J. Thomson summarized as follows:

> It is difficult for the audience to weigh fully the meaning of the figures that have been put before us, but the Astronomer Royal and Prof. Eddington have studied the material carefully, and they regard the evidence as decisively in favour of the larger value of displacement. This is the most important result in the connection with the theory of gravitation since Newton's day and it is fitting that it should be announced at a meeting of the Society so closely connected with him.

Newton had presented many of his discoveries to the Royal Society. He had also been its President and thus Thomson's most distinguished predecessor.

An extensive report [2] by the two expeditions was published a little later. In its conclusions it contains two numbers. The angle of deviation, measured at Sobral was given as $\alpha = (1.98 \pm 0.12)''$ and that measured at Principe as $\alpha = (1.61 \pm 0.30)''$ The numbers were obtained by averaging the displacements found for individual stars. None of the stars, as seen from the earth, appeared to be directly at the rim of the sun. That fact was taken into account in the averaging procedure. In a diagram the deviation for individual stars was shown to decrease with the apparent distance of the star from the centre of the sun in just the way as predicted by Einstein.

Of the astronomers involved in the eclipse observations of 1919 Eddington [6,7], professor of astronomy in the University of Cambridge and director of the Cambridge Observatory, is the best known. He was a highly gifted mathematician, published himself on relativity, and was the author of successful popular books. The measurements of 1919 have been repeated with increasing accuracy with visible and infrared light and also with radio waves from distant stars.

The results of the 1919 eclipse expedition made Einstein the outstanding public figure he still is. The world's newspapers were full about the Theory of Relativity and he himself was asked to contribute. In an article for the London *Times*, which appeared on 28 November 1919, he began by expressing his 'feelings of joy and gratitude toward the astronomers and physicists of England' and his appreciation for testing 'the implications of a theory which was perfected and published during the war in the land of your enemies.' He went on giving a popular exposition of General Relativity which ended with the three predictions to be tested, the perihelion rotation of Mercury, the bending of light, and the redshift. Since the *Times*, on 8 November, had introduced him to the readers as a 'Swiss Jew' he could not resist to add the following to his article:

> A final comment. The description of me and my circumstances in *The Times* shows an amusing feat of imagination on the pert of the writer. By an application of the theory of relativity to the taste of readers, today in Germany I am called a German man of science, and in England I am represented as a Swiss Jew. If I come to be regarded as a *bête noire*, the descriptions will be reversed, and I shall become a Swiss Jew for the Germans and a German man of science for the English!

Although in 1919 Einstein considered his *note* to be nothing more than a joke, such time was to come as will be told at the end of Episode 33.

[1] Newton, I., *Opticks, or, A treatise of the reflections, refractions, inflections & colours of light.* Bell, London, 1931. Reprinted from the fourth edition with a foreword by A. Einstein and an introduction by E. T. Whittaker.

[2] Dyson, F. W., Eddington, A. S., and Davidson, C., *Philosophical Transactions of the Royal Society*, **A220** (1920) 291.

[3] Kirsten, C. and Treder, H.-J. (eds.), *Albert Einstein in Berlin 1913–1933, Teil I Darstellung und Dokumente.* Akademie-Verlag, Berlin, 1979.

[4] Thomson, J. J., *The Observatory*, **42** (1919) 389. Joint Eclipse Meeting of the Royal Society and the Royal Astronomical Society.

[5] Pais, A., *'Subtle is the Lord ...'* *– The Science and Life of Albert Einstein.* Oxford University Press, Oxford, 1982.

[6] Douglas, A. V., *The Life of Arthur Stanley Eddington.* Nelson, London, 1956.

[7] Chandrasekar, S., *Eddington: The Most Distinguished Astrophysicist of His Time.* Cambridge University Press, Cambridge, 1983.

Stern and Gerlach Observe Spatial Quantization (1922)

The *Stern–Gerlach experiment* is described or at least mentioned in practically every textbook on modern physics and its result is presented as clear proof for the *spin* of the electron. However, spin was unknown at the time of experiment and also not discovered on the basis of its outcome. The reason for Stern[1] to propose and undertake the experiment was the wish to test Sommerfeld's idea of *spatial quantization.*

In Sommerfeld's theory, the plane in which an electron orbits an atom can have only certain discrete orientations in space. The same holds for the vector **L** of the orbital angular momentum of the electron which is perpendicular to the plane. We have described in Episode 27 that for an electron in the ground state of an atom in the Bohr–Sommerfeld model the z component L_z of the angular-momentum vector was allowed to assume only two possible values, $L_z = \pm\hbar$, where $\hbar$ is Planck's constant h divided by 2π. (At first Sommerfeld had also admitted an orbit with $L_z = 0$. But he found that in the presence of an electric field, however small, such an orbit would eventually end up in the nucleus, and therefore he excluded it.) Let us take the z direction with respect to which the z component is defined to be the direction of a magnetic field. An orbiting electric charge is equivalent to a circular current and represents a magnetic dipole, i.e., a small magnet whose strength is characterized by its *magnetic moment* $\boldsymbol{\mu} = -\mu_\mathrm{B}\mathbf{L}/\hbar$. Here the constant $\mu_\mathrm{B} = e\hbar/(2m)$ is the *Bohr magneton* and m the electron mass. Thus the magnetic moment of an electron in the ground state was allowed only two possible z components, $\mu_z = \pm\mu_\mathrm{B}$. In 1921 Stern proposed an experiment to test this statement.

Stern called himself an *experimenting theoretician.* He had studied in Freiburg, Munich, and Breslau, where he obtained his Ph.D. in physical chemistry in 1912. The family fortune provided him with the means to follow a scientific career without having to look for paid positions. Stern joined Einstein in Prague and accompanied him when he moved back to Zurich in 1913. As he later said, he had been attracted primarily by Einstein's work on statistical physics. He became a Privatdozent in Zurich and in 1914 moved to the young University of Frankfurt in the same capacity. For the whole of the First World War he served in the German Army. After the war, in Frankfurt, he began to work experimentally on *atomic beams.*

Rays or beams (as they were called later) of charged particles had been known for decades in the form of cathode rays (electrons) or canal rays (ionized atoms or molecules) but beams of neutral atoms were first produced by Dunoyer[2] who had shown that atoms or molecules in high vacuum travel along straight lines. In 1919 Stern used the method to measure for the first time the

Stern

---

[1] Otto Stern (1888–1969), Nobel Prize 1943   [2] Louis Dunoyer (1880–1963)

velocity of atoms due to their thermal motion, which was predicted by Maxwell and Boltzmann and which was at the heart of the kinetic theory of heat. He produced a vapour of silver atoms by heating a platinum wire coated with silver. A beam of silver atoms was produced by slit collimators. The atoms fell on a glass plate placed beyond the collimators and after some time produced a deposit which could be seen under a microscope. When the whole apparatus was made to rotate about the wire, there was a displacement of the deposit from which the mean velocity of the atoms could be obtained. The Institute of Theoretical Physics to which Stern was attached was led by Born since early 1919. It had a workshop and an excellent mechanic who constructed and ran the apparatus under the instructions of Stern.

A silver atom has one electron in its outermost shell. The angular momenta (and the magnetic moments) of the electrons in the inner shells compensate each other. Therefore the outermost electron determines the magnetic moment of the whole atom (a possible contribution from the atomic nucleus is smaller by three orders of magnitude). Stern realized that an inhomogeneous magnetic field would exert a force on the magnetic moment and thus on each atom. The reason is simple. A magnetic moment is like a tiny magnetic needle, which has a north pole and a south pole. If the needle is placed in a magnetic field, the north pole is pulled in the direction of the field lines, the south pole in the opposite direction. If the field has the same strength everywhere, i.e., if it is homogeneous, the two forces compensate each other and there is no net force on the needle. But if it is inhomogeneous the force on one pole can be larger than that on the other and the needle is pulled in one direction. The idea for the experiment proposed by Stern [1] was simple enough. A beam of silver atoms is produced by letting silver evaporate in an oven with a small opening. The beam is collimated and travels in x direction until it falls on a glass plate. Between collimators and plate an inhomogeneous magnetic field is produced. It points in y direction and also changes its strength as a function of y. If the atoms possess a magnetic moment, the field pulls them away from the x axis. If the moments are oriented at random, there will be a broadening of the beam. But if spatial quantization exists with just two possible orientations, then the beam will be split in two. Half the atoms are pulled in the positive and half in the negative y direction. (In fact, the orbiting charge is more complicated than a magnetic needle. It should be treated as a spinning top with a magnetic needle in its axis. But that does not change the simple argument given above.)

Although simple in principle, the experiment was ambitious in practice. Stern asked Gerlach[3] [2] from the Institute of Experimental Physics to join forces with him. Gerlach later told that until then he had never heard about spatial quantization. He had obtained his Ph.D. with spectroscopic work under Paschen in Tübingen in 1912. He, too, served in the army during the war except for a few months when he was dismissed because of sickness. In that time he became Privatdozent in Tübingen. After the war he led the physical laboratory of a large chemical company before he came to Frankfurt in October 1920. Money for research was tight in post-war Germany. Einstein helped with funds from the Kaiser-Wilhelm Institute, Born contributed the entrance fees earned

Gerlach

---

[3] Walther Gerlach (1889–1979)

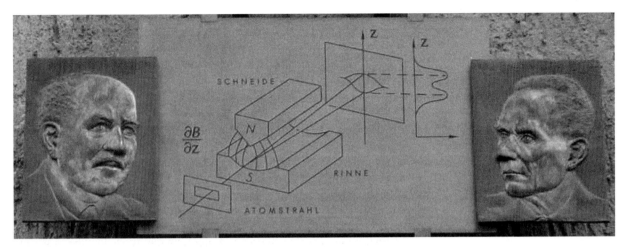

for public lectures he gave on relativity, and there were donations from private persons and from industry. The main experimental problems were mechanical precision (the deflection in the field was only about 0.1 mm) and reliable vacuum (the mercury pumps, made of glass, tended to break). To get a deposit of silver that could be made visible by chemical treatment, stable conditions had to be maintained for hours. The inhomogeneous magnetic field was produced by an electromagnet with pole shoes of rather different shapes. One had the form of a knife edge, the one opposite carried a groove. The direction of the field was from the edge to the groove, the field strength being higher near the edge.

Plaque commemorating the Stern–Gerlach experiment mounted on the building in Frankfurt, where it was performed. It shows portraits of Stern (left) and Gerlach and sketches the splitting of a beam of silver atoms in an inhomogeneous magnetic field, see also the right-hand photograph on the postcard below.

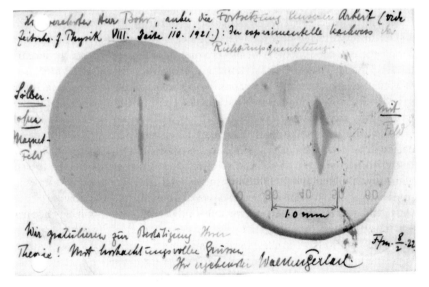

Post card written by Gerlach to Bohr on 8 February 1922. It contains photographs of glass plates with silver deposits obtained without (left) and with magnetic field (right). The same photographs appeared in [3]. From Niels Bohr Archive, Copenhagen.

In November 1921 Stern and Gerlach observed a broadening of the beam. Its size increased from 0.1 mm to about 0.3 mm if the field was turned on [4]. This result proved that silver atoms possess a magnetic moment. With a still better collimated beam in February 1922 the splitting of the beam into two was observed [3]. Spatial quantization was established. At that time, Gerlach ran the experiment by himself because Stern had just been appointed professor in Rostock and could come to Frankfurt only occasionally. The result made the Stern–Gerlach experiment famous. Expectations had varied a good deal. Gerlach later recalled [5] that Debye[4], who himself had used spatial quantization in calculations, had told him: 'you don't really believe that the orientation of atoms is something physically real; that is a prescription for calculations – time table of the electrons'. He also remembered that Bohr firmly believed in a twofold splitting and that Sommerfeld vacillated between threefold splitting and symmetric broadening. (Stern, however, quoted Sommerfeld as predicting only two spatial orientations [1].)

Attempts to understand the dynamics of the *Stern–Gerlach effect* failed, i.e., it was not possible to explain how the magnetic moments of the silver atoms, unordered before they entered the field, were oriented in or against the field direction in the very short time which they spent in the field. Only after the advent of quantum mechanics it became clear that the atoms themselves are not turned but that their quantum-mechanical wave function assumes one of its possible values in the apparatus. In quantum mechanics the concept of electrons orbiting the nucleus loses its meaning. There is still an *orbital angular momentum* but for the ground state of an electron in the outermost shell of an atom this angular momentum and its associated magnetic moment is zero. However, the electron possesses an *intrinsic angular momentum* or *spin*, the z component of which can have exactly two spatial orientations. Also the electron spin is associated with a magnetic moment and it is that magnetic moment which was observed in the Stern–Gerlach experiment. The stories of the creation of quantum mechanics and the discovery of spin are told in later episodes.

In 1923 Stern accepted a professorship in Hamburg with an Institute of Physical Chemistry erected for him. We shall meet him there in Episode 52 measuring the magnetic moment of the proton. In 1925 Gerlach succeeded his teacher Paschen as professor in Tübingen. In 1929 he was appointed professor in Munich as successor of Wien, a position he held until his retirement in 1957.

I saw Walther Gerlach at two occasions. The first time was around 1959 when I was working on my Diploma Thesis under Wolfgang Paul in Bonn. Gerlach, passing through the town, paid Paul a visit and I happened to be in Paul's secretariat when they met. On that occasion I learned that Gerlach had held the Chair in Bonn from 1946 to 1948. In 1945/46 he had been interned with other German atomic scientists in England and after the return to Germany for some time had been asked not to leave the British Occupation Zone. Only later was he allowed to return to his proper Chair in Munich. The second time was around 1970 when I was Privatdozent in Heidelberg and Gerlach gave a talk on Kepler[5] in the Physics Colloquium there. Gerlach was a tall, imposing figure. He had carefully researched the subject and gave a lucid talk in vivid language including Kepler's narration of a travel from earth to moon.

[1] Stern, O., *Zeitschrift für Physik*, **7** (1921) 249.

[2] Heinrich, R. and Bachmann, H.-R. (eds.), *Walther Gerlach: Physiker – Lehrer – Organisator, Dokumente aus seinem Nachlaß*. Deutsches Museum, München, 1989.

[3] Gerlach, W. and Stern, O., *Zeitschrift für Physik*, **9** (1922) 349.

[4] Gerlach, W. and Stern, O., *Zeitschrift für Physik*, **8** (1921) 110.

[5] Gerlach, W., *Physikalische Blätter*, **25** (1969) 472.

---

[4] Petrus (Peter) Josephus Willhelmus Debye (1884–1966), Nobel Prize 1936   [5] Johannes Kepler (1571–1630)

The Compton Effect – The Light Quantum Gains Momentum (1923)

The concept of the *light quantum* or *photon*, as it is now mostly called, is one of the most interesting of modern physics. It took nearly the whole first quarter of the twentieth century for it to become fully developed and accepted. The development was essentially completed by the work of Compton[1] [1,2].

We have seen in Episode 7 how Planck, in 1900, had to assume that electromagnetic radiation is emitted or absorbed by matter only in the form of *energy quanta* of energy $E = h\nu$, where h is Planck's constant and ν is the frequency of the radiation. In Episode 10 we saw how Einstein, in 1905, separated the idea of *energy quantization* from the processes of emission and absorption and made it a property of radiation itself. In 1909, he began to study the momentum of the light quantum, albeit in a rather indirect way. He considered a mirror suspended in a cavity containing radiation and discussed the changes in momentum of the mirror caused by light quanta being reflected by the mirror. But at that time he did not explicitly attribute momentum to the light quantum. Einstein presented this work at a meeting of the Society of German Scientists and Physicians in Salzburg in 1909. This, by the way, was the first occasion on which Einstein became known personally to a larger number of physicists. Among them was Stark[2] who, still in 1909, was the first to write down [3] the momentum of a light quantum as a vector with the absolute value $p = h\nu/c$, with c, as usual, being the speed of light. The equation is easily derived in the framework of Einstein's special theory of relativity, once one assumes that the light quantum is a *particle* of rest mass zero. As an example, Stark took the process of *bremsstrahlung* in which a photon is produced when an electron hits an atom and he wrote down the equation for momentum conservation in this process, i.e., that the sum of the momenta of the particles in the initial state (electron, atom) is equal to the momentum sum of the particles in the final state (electron, atom, photon). Although correct, this idea was not sufficient to describe the details of *bremsstrahlung* and this part of Stark's work was soon forgotten.

Einstein himself explicitly attributed momentum to light quanta in 1916 in a paper [4], which is chiefly remembered for a new and more satisfactory derivation of Planck's radiation law and for the introduction of the concept of *stimulated emission* of radiation, a concept which is the foundation for the functioning of the *laser* invented decades later. At the time all these theoretical considerations about a possible momentum of the light quantum went nearly unnoticed. It had become accepted practice to use the light-quantum concept in the discussion of certain processes of *emission* and *absorption* like the pho-

Compton

---

[1] Arthur Holly Compton (1892–1962), Nobel Prize 1927 [2] Johannes Stark (1874–1957), Nobel Prize 1919

toelectric effect and level changes in the Bohr atom but to describe the *scattering* of light (or of radiation of other wavelengths), i.e., a process in which the light changes direction, with the help of Maxwell's theory of electromagnetic waves. That certainly seemed appropriate after Laue's discovery. J. J. Thomson had formulated a theory for the scattering by the electrons in matter. The incident wave makes the electrons (which are light compared to the atomic nuclei) oscillate in the frequency of the wave and the oscillating electrons radiate secondary radiation just as a dipole antenna does. The theory of *Thomson scattering* gives a complete description of the intensity and angular distribution of the secondary radiation which has exactly the same frequency (and wavelength) as the primary one. Measurements showed that for X rays this theory did not work. It was modified and extended in several ways, also by Compton, until he realized that the scattering of X rays by electrons was best described in analogy to the collision between two billiard balls in which the two objects involved in the collision exchange both energy and momentum. This is the *Compton effect*.

Compton was born in Wooster, Ohio, and did his undergraduate studies at the College of Wooster, where his father was Professor of Philosophy. For his graduate studies he chose Princeton as his brother Karl had done before him and who was working there under Richardson[3]. Richardson, who is best known for his experimental and theoretical work on the thermionic emission of electrons, the basis of all vacuum tubes, returned to England early in 1914 and left his X-ray equipment to Compton, who then had only completed one term in Princeton. With a thesis on the intensity distribution of X rays reflected from crystals, suggested by Richardson, Compton received his Ph.D. in 1916. He then married a classmate from his undergraduate days and began to work as Instructor in Physics in the University of Minnesota at Minneapolis, where he continued to work with X rays. In 1917 he accepted a position as research engineer in the Lamp Division of the Westinghouse Company in East Pittsburgh, Pennsylvania. There he was not able to continue experimental X-ray work but, more or less in his spare time, he developed theoretical models aiming to bridge the gap between the theory of Thomson scattering and experimental results. He tried out models in which the electrons were not point-like but had about the size of an atom and were spherical or ring-like in shape. In this way he could obtain angular distributions different from Thomson's but not a difference in wavelength between primary and scattered radiation.

In 1919 Compton, who wanted to return to fundamental research, obtained a National Research Council Fellowship, established shortly before by Millikan, and used it to spend a year at the Cavendish Laboratory in Cambridge, where Rutherford had just taken over from J. J. Thomson. In the stimulating atmosphere of the Cavendish, Compton resumed his experiments on the scattering of radiation. In the spirit of the laboratory, which concentrated on radioactivity, he used γ rays rather than X rays. It became more and more apparent that there were components in the scattered radiation with wavelengths longer than those of the primary. Compton and others for some time believed in a new type of fluorescence in which atoms would absorb primary radiation and later emit ra-

---

[3] Owen Willans Richardson (1879–1959), Nobel Prize 1928

diation with a greater wavelength. However, whereas the known fluorescence was characteristic for the absorbing atom and therefore called *characteristic radiation*, the new one was not.

In 1920, Compton became Head of the Physics Department at Washington University in St. Louis, where he continued his work on X-ray scattering using a Bragg spectrometer he had brought from England. It was only in late 1922, when considering all data available to him, that Compton saw the necessity for a light quantum with energy and momentum to explain the scattering of X rays. Compton read a paper entitled *a quantum theory of the scattering of X rays by light elements* at a meeting of the American Physical Society in Chicago, which took place on 1 and 2 December 1922. Its abstract [5] begins as follows:

> The hypothesis is suggested that when an x-ray quantum is scattered it spends all of its energy and momentum upon some particular electron. The electron in turn scatters the ray in some definite direction. The change in momentum of the x-ray quantum due to the change in its direction of propagation results in a recoil of the scattering electron. The energy in the scattered quantum is thus less than the energy in the primary quantum by the kinetic energy of recoil of the scattering electron. The corresponding increase in the wave-length of the scattered beam is
> ...

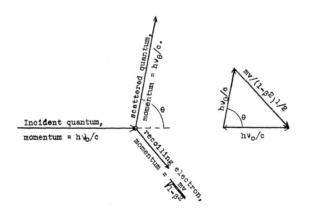

Compton's diagrams showing the momentum vectors of the incident light quantum, the scattered quantum, and the recoil electron. Figure reprinted with permission from [6]. Copyright 1923 by the American Physical Society.

The full paper was published a little later [6]. It contains diagrams showing the momentum vectors of the particles involved in the typical Compton style with the lettering done in typewriter rather than in draughtsman's ink. It also contains the calculation of the shift in wavelength between primary and secondary radiation as a function of scattering angle. Compton had data only for one scattering angle, 90°, for which he had measured the spectrum. It was obtained with a primary radiation which was the K_α radiation of molybdenum produced by an X-ray tube with a molybdenum anticathode.

Before Compton's paper [6] was published in the *Physical Review* a paper by Debye [7] appeared in the *Physikalische Zeitschrift* containing the same calculation which was even presented in a more elegant form, starting from the conservation laws of relativistic energy and momentum. This is the way in which the Compton effect is now taught to students. Debye, then at the ETH in Zurich, quoted an earlier report in which Compton had mentioned the possible importance of quantum effects in X-ray scattering but had not presented calculations. Although Debye's paper was published earlier, Compton's had been received earlier by the journal. Debye never claimed priority for himself.

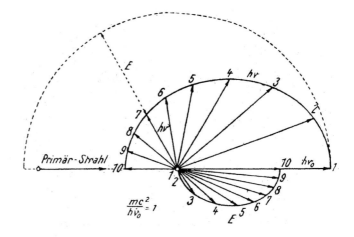

Diagram presented by Debye, showing the momentum vectors of the scattered photon (top) and the recoil electron (bottom) for various scattering angles distinguished by the numbers $1, 2, \ldots, 10$. From [7]. Reprinted with kind permission of S. Hirzel Verlag.

Having developed his theory Compton set out to test it systematically. Using his Bragg spectrometer he carefully measured the spectrum of the original K_α line of molybdenum and the spectra of the scattered radiation under three different angles, $45°$, $90°$, and $135°$, and found the shift in wavelength to be exactly as computed [8].

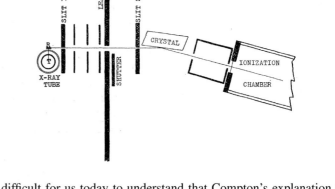

Compton's experimental set-up.   Figure reprinted with permission from [8]. Copyright 1923 by the American Physical Society.

It is difficult for us today to understand that Compton's explanation of X-ray scattering was regarded as revolutionary. Physicists until then had learned that electromagnetic radiation, depending on the process studied, had to be described as consisting of *either* waves *or* energy quanta but now had to accept that *both* the wave *and* the particle property were needed. Possible ways out were widely discussed. Bohr for some time was even willing to give up the laws of energy and momentum conservation for an individual elementary process and to assume them to be valid only on the average. But all of Compton's predictions were eventually confirmed by experiment. Individual recoil electrons were observed in the Wilson cloud chamber first by Bothe and shortly afterwards also by C. T. R. Wilson himself. The final assertion came from an experiment by Bothe and Geiger who showed that the scattered photon and the recoiling electron indeed appear simultaneously (see Episode 34).

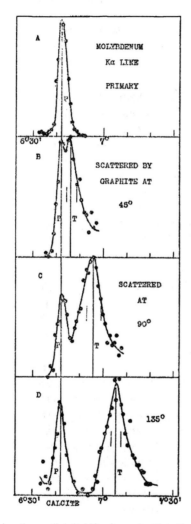

Compton's spectra showing the predicted shift of wavelength with scattering angle. Figure reprinted with permission from [8]. Copyright 1923 by the American Physical Society.

Compton moved to the University of Chicago in 1923. In the 1930s he led a world-wide study of the geographic variations of the intensity of cosmic rays. In 1941 Compton was appointed Chairman of the National Academy of Sciences Committee to Evaluate Use of Atomic Energy in War and later described this work in a book [9]. After the war, he returned to Washington University in St. Louis, where he served as Chancellor from 1945 to 1954 and subsequently became Distinguished Service Professor of Natural Philosophy.

[1] Compton, A. H., *Personal Reminiscences.* In Johnston, M. (ed.): *The Cosmos of Arthur Holly Compton.* Knopf, New York, 1967.

[2] Stuewer, R. H., *The Compton Effect – Turning Point in Physics.* Science History Publications, New York, 1975.

[3] Stark, J., *Physikalische Zeitschrift,* **10** (1909) 912.

[4] Einstein, A., *Mitteilungen der Physikalischen Gesellschaft Zürich,* **16** (1916) 47. Also published as [10].

[5] Compton, A. H., *Physical Review,* **21** (1923) 207.

[6] Compton, A. H., *Physical Review,* **21** (1923) 483.

[7] Debye, P., *Physikalische Zeitschrift,* **24** (1923) 161.

[8] Compton, A. H., *Physical Review,* **22** (1923) 409.

[9] Compton, A. H., *Atomic Quest – A Personal Narrative.* Oxford University Press, London, 1956.

[10] Einstein, A., *Physikalische Zeitschrift,* **18** (1917) 121.

32 Matter Waves Proposed by de Broglie (1923)

Louis de Broglie

We have seen that it took about two decades to fully realize that light or, generally, electromagnetic radiation has both wave and particle properties. The final proof came with the discovery of the *Compton effect* (Episode 31). Even Compton himself needed years before he saw that the particle picture of radiation was unavoidable in view of the experimental evidence. Still in the year of Compton's publications, Louis de Broglie[1] [1] on purely theoretical grounds proposed that the *particle–wave duality* not only holds for radiation but also for material particles, i.e., for particles with a rest mass such as the electron.

At that time de Broglie was practically unknown. But his family was not. Originally called Broglia and living in Piemont, it developed a French branch in the seventeenth century. Among de Broglie's ancestors were four Marshals of France, several cabinet ministers and members of the Académie Française. One fell victim to the guillotine. The family motto was *pour l'avenir* – for the future. There were two important titles in the family. The oldest son was *duc* but all were *prince*. The latter title had been awarded by the German Emperor Franz, who was the husband of the Austrian Arch Duchess Maria Theresia, to a de Broglie who won a battle in the Seven Years War when France and Russia sided with Austria against Frederick the Great of Prussia.

Louis de Broglie grew up in this extraordinary family and until the age of 15 was educated by private teachers. When he was 14, his father died and his brother began to influence his further education. He suggested that he visit a school, the Lycée Janson-de Sailly, and concentrate on mathematics and science. This brother, Maurice de Broglie[2], having been a naval officer, had studied physics, set up his own private laboratory, and had made important contributions to X-ray research. At the age of 16 de Broglie began his university studies. He obtained a degree in history. Then he studied languages and law but late in 1911 decided to devote all his future efforts to physics. And this was why: Maurice de Broglie had participated in the first Solvay Conference in Brussels. Together with Lindemann he had served as scientific secretary and now helped to edit its proceedings. Young Louis read the manuscripts by the leading scientists of the time such as Lorentz and Einstein and decided to try and join their ranks.

Louis de Broglie obtained his science degree in 1913 when he was drafted for military service and attached to a wireless regiment. Until 1919 he was stationed near the Eiffel tower which carried the antennas of his unit. Only then he could return to science. He worked with his brother's group on X-ray experiments and also published on related theoretical models. Within a few

---

[1] Prince Louis-Victor Pierre Raymond de Broglie (1892–1987), Nobel Prize 1929   [2] Duc Louis-César-Victor-Maurice de Broglie (1875–1960)

weeks in 1923 the ideas about what is now called a *matter wave* or *de Broglie wave* were published in three short notes. They were the basis for his Ph.D. thesis of 1924. In the preface to the reprint of this famous thesis, published to mark his retirement as professor at the University of Paris, de Broglie wrote about this period:

> I think back to these short years where my thoughts, nourished by un-
> countable readings on the most different subjects, always returned to the
> serious problem of the double aspect – granular and wave-like – of light
> which Einstein had laid down, almost twenty years earlier, in his inge-
> nious theory of light quanta. Having pondered for a long time in solitude
> and meditation, in the year 1923 I suddenly had the idea that there was a
> need to generalize the discovery made by Einstein in 1905 by extending
> it to all material particles and in particular to electrons.

The central idea of de Broglie's work, contained in the first note [2], was that the formula $E = h\nu$ by which Einstein had related the frequency ν of light to the energy E of light quanta should not only apply to light but also to material particles. For a particle at rest with mass m he concluded, since its energy is $E = mc^2$, that it performs an internal oscillation with frequency $\nu = mc^2/h$. He considered the motion of a particle along the x axis, carefully taking into account the effects of the special theory of relativity, and was able to construct a wave which was always *in phase* with the internal oscillation of the particle. This finding he called the *law of the harmony of phases*. He found that for a particle with velocity v, which is a fraction β of the velocity c of light, $v = \beta c$, and which is, of course, smaller than c, the wave has a *phase velocity* $v_{\mathrm{p}} = c/\beta$ larger than c. He noted that such phase velocities are not in conflict with relativity since no energy transport takes place with this velocity. He pointed out that in his opinion also light quanta possess a (very small) rest mass. Therefore, in his theory, the wave character of both light and massive particles was traced back to the same reason. At the end of the first note he gave an application of his theory by showing that he could naturally explain the discrete electron orbits in Bohr's model of the hydrogen atom. Each stable orbit should be closed in the sense that the same phase should be assumed by the matter wave after completion of an orbit.

In his second note [3] de Broglie wrote that one could consider the velocity of a particle as the *group velocity* of his wave which he now described as the *guiding wave* of the particle. The note also predicts the *diffraction* of particles:

> … any particle can in certain cases suffer diffraction. A stream of elec-
> trons traversing a very small aperture will show the phenomenon of dif-
> fraction. It is in this direction that one has probably to look for experi-
> mental confirmation of our ideas.

He went on to propose that a new dynamics should be created on the basis of his waves and that *'the new dynamics of the free material point is to the old dynamics (including that of Einstein) what wave optics is to geometric optics'*. Such a theory, now called *wave mechanics*, was indeed formulated less than three years later by Schrödinger (see Episode 39).

In the third note [4], de Broglie applied his ideas to the kinetic theory of gases and treated the atoms of a gas in a vessel as standing waves. Although here, as in the two preceding notes, the wavelength is important and was im-

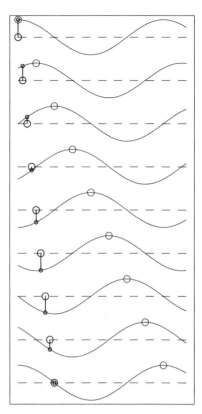

Illustration of de Broglie's original idea, drawn for nine equidistant instants in time. A particle (indicated by a circle) moves along the x axis (dashed line). Its internal oscillation is symbolized by the movement of the dot connected to it by a line. This oscillation is in phase with a harmonic wave, which travels with a phase velocity illustrated by the motion of the circle on the wave crest. In this example the particle velocity is half and the phase velocity twice the speed of light.

De Broglie Wave in Modern Description

A *harmonic* (or sinusoidal) wave is a sine-like pattern in space moving with a given *phase velocity* v_p without changing its shape. A wave travelling in x direction is described by the function $\Psi(x, t)$ of the space coordinate x and the time t,

$$\Psi(x,t) = \Psi_0 e^{-i(\omega t - kx)} \quad , \qquad k = 2\pi/\lambda \quad , \qquad \omega = 2\pi\nu \quad .$$

Here Ψ_0 is the *amplitude*, ω the *circular frequency*, k the *wave number*, ν the *frequency* and λ the wavelength. The phase velocity is $v_p = \omega/k$. In the case of light the phase velocity is equal to the velocity c of light, i.e., $c = \omega/k$. For the energy E and the momentum p of a light quantum we have (with $\hbar = h/2\pi$)

$$E = h\nu = \hbar\omega \quad , \qquad p = h/\lambda = \hbar k \quad .$$

Assuming these two relations to be valid also for material particles we can write the wave as

$$\Psi(x,t) = \Psi_0 e^{-(i/\hbar)(Et - px)} \quad .$$

For a particle of *rest mass* m moving with *velocity* v one has

$$E = mc^2\gamma \quad , \qquad p = \gamma m v \quad , \qquad \gamma = 1/\sqrt{1 - v^2/c^2} \quad .$$

The phase velocity is larger than the speed of light,

$$v_p = \frac{E}{p} = \frac{c}{\beta} \quad , \qquad \beta = \frac{v}{c} \leq 1 \quad .$$

---

plicit in his formulae de Broglie did not write it down explicitly. This was only done in the thesis [5]. In its last chapter the famous formula appears. Written down in its general form it relates the *de Broglie wavelength* λ to the momentum p via Planck's constant h,

$$\lambda = h/p \quad .$$

There was no immediate resonance in the scientific world to de Broglie's three notes. The thesis, however, was read by Einstein to whom Langevin, his supervisor, had sent a copy. Einstein was impressed by de Broglie's ideas and helped to spread them. But only after Schrödinger had used matter waves with great success in 1926 and after the diffraction of electrons had been observed in 1927 by Davisson[3] and Germer[4] [6,7] and in 1928 by G. P. Thomson[5] [8], de Broglie became a celebrity.

In 1945 Louis de Broglie officially became *immortal*. That is the attribute given to the members of the Académie Française founded in 1635 by Richelieu. (There are always forty immortals; should one die the remaining elect a successor, hence the name.) The sessions of the academy take place in the Palais de l'Institut which is crowned by a beautiful cupola. In its official biography of de Broglie the academy writes: 'The entry of Louis de Broglie under the cupola was one of the most moving, since there he was received

---

[3] Clinton Joseph Davisson (1881–1958), Nobel Prize 1937   [4] Lester Halber Germer (1896–1971)   [5] (Sir) George Paget Thomson (1892–1975), Nobel Prize 1937

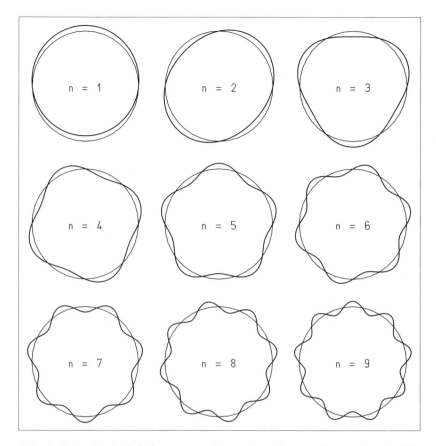

A circular Bohr orbit of principal quantum number n accommodates exactly n wavelengths of the de Broglie wave describing the orbiting electron. (The orbits are not drawn to scale but are all given the same radius in this figure.)

by his own brother, the duc Maurice de Broglie, on 31 May 1945, [a scene which] had never been seen in three hundred years [i.e., since the creation of the academy].'

What has survived of de Broglie's work is the formula $\lambda = h/p$ and the concept of matter waves. His ideas of a material point particle embedded in a guiding field and of light quanta with rest mass are now considered to have been only stepping stones towards wave mechanics. But de Broglie himself held on to them all his life. In modern textbooks, a matter wave is usually obtained by starting out from a formula describing any wave and then replacing the circular frequency and the wave number in that formula by expressions taken from special relativity and assumed to hold for both light quanta and massive particles (see Box).

[1] Lochak, G., *Louis de Broglie*. Flammarion, Paris, 1992.

[2] de Broglie, L., *Comptes Rendus Acad. Sci.*, **177** (1923) 507.

[3] de Broglie, L., *Comptes Rendus Acad. Sci.*, **177** (1923) 548.

[4] de Broglie, L., *Comptes Rendus Acad. Sci.*, **177** (1923) 630.

[5] de Broglie, L., *Recherches sur la théorie des quanta*. Masson, Paris, 1924. Reprinted 1962.

[6] Davisson, C. J. and Germer, L. H., *Nature*, **119** (1927) 558.

[7] Davisson, C. J. and Germer, L. H., *Physical Review*, **30** (1927) 705.

[8] Thomson, G. P., *Proceedings of the Royal Society*, **A117** (1928) 600.

33

Bose and Einstein – A New Way of Counting (1924)

Bose

There are two rather different phenomena that furthered the development of quantum theory: blackbody radiation and atomic spectra. In fact quantum theory began in 1900 with Planck's radiation law and his introduction of the quantum of action h (Episode 7). His derivation of the law was by no means stringent. He assumed the existence of many resonators (atoms or molecules) of different but fixed frequencies ν, which could emit or absorb radiation only in quanta of energy $E = h\nu$. Einstein, in his light-quantum hypothesis of 1905, postulated that the energy quantization is a property of the radiation field itself and not due to the material resonators. He found a similarity between a cavity filled with radiation consisting of quanta and a vessel filled with gas consisting of atoms. However, he could not obtain Planck's law but only Wien's (Episode 10).

After completion of his general theory of relativity Einstein gave a new derivation of Planck's law in 1916 [1,2] by treating the interaction between light quanta and atoms in the formalism of his A and B coefficients. They describe the probabilities for the absorption and emission of a light quantum of a given energy. The probability of absorption is, of course, proportional to the density of the radiation of that frequency. The emission is in part spontaneous, like radioactivity. But there is also *stimulated emission* which, again, is proportional to the radiation density: If the atom is 'shaken' in the right frequency, it will more probably emit radiation of that frequency. Einstein's concept of stimulated emission is the operating principle of the now ubiquitous *laser*.

If Einstein had made progress by considering the possible interactions between radiation and matter, Bose [3,4] succeeded in finding an even closer

---

Box I. Finding an Energy Distribution Using Statistics

Problem: Given a system with a large number of identical objects (e.g., light quanta or particles of the same type) with a given total energy in thermal equilibrium. What is the number N_i of objects having an energy E between E_i and $E_i + \Delta E$?
Answer: (a) Enumerate all possible *states* which the system can assume. All these states are considered to be equally probable. (b) Define energy intervals of size ΔE beginning at $E_1, E_2, \ldots$. The interval size is chosen to be small but still large enough to contain many states. (c) Of all possible energy distributions, i.e., sets of numbers N_i, that one is found, which can be realized with the available states in the maximum number of ways, each way being called a *complexion*. It is the most probable distribution and taken to be the answer to the problem. In the process of finding the maximum the total energy and, for material particles, also the number of objects has to be kept constant.

Box II. Counting in Classical and in Bose–Einstein Statistics

Consider a number of cells (in our example there are three, labelled I, II, III) and a number of objects (in our example there are two). A *macroscopic state* is defined if the number of objects in each cell is known. In a *microscopic state* the attribution of every particle to its cell is known. There are, in general, several different microscopic states corresponding to one macroscopic state. If the objects are *distinguishable*, we give them different labels (in our example the two labels A and B) and we speak of classical or *Boltzmann statistics*. If they are *indistinguishable*, there is only one label (A) and we are dealing with *Bose–Einstein statistics*. In our example there are six macroscopic states. For the first three, there is only one realization in each type of statistics. The other three are realized twice in Boltzmann statistics but only once in Bose–Einstein statistics. In fact, all macroscopic states are realized only once in Bose–Einstein statistics; there is no longer any difference between macroscopic and microscopic state.

Macroscopic State			Microscopic State					
			Boltzmann			Bose–Einstein		
I	II	III	I	II	III	I	II	III
2	0	0	AB			AA		
0	2	0		AB			AA	
0	0	2			AB			AA
1	1	0	A B	B A		A	A	
1	0	1	A B		B A	A		A
0	1	1		A B	B A		A	A

For our example in the Boltzmann case 3 out of the possible 9 microscopic states contain more than one object in the same cell, in the Bose–Einstein case 3 out of 6. This is a general tendency. In Bose–Einstein statistics the objects tend to be closer to each other in phase space compared to classical statistics.

---

connection between a photon gas and a gas of material atoms.

Before sketching Bose's work let us consider a conventional gas. Its atoms have an average kinetic energy proportional to the temperature. Their energy distribution, the *Maxwell–Boltzmann distribution*, can be computed in the framework of *statistical mechanics* (see Box I). Here the concept of *phase space* is used. It is a fictitious six-dimensional space spanned by the three space coordinates x, y, z and the three momentum components p_x, p_y, p_z of an atom. The phase space is divided into small elements, called *cells*. Each cell corresponds to a particular position and momentum of a particle within small deviations. These can be made arbitrarily small by making the cells small. Each possible state of the gas can be described in a microscopic but also in a macroscopic way (see Box II). The difference between the two makes sense if the particles are identical in kind but still considered distinguishable. This was Boltzmann's conception: He thought that two identical particles could in principle be distinguished by constantly observing them along their paths, just as two identical billiard balls can be told apart even if they collide while being

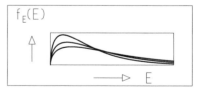

Maxwell–Boltzmann distribution. The frequency for an atom to be in the energy interval between E and $E + \mathrm{d}E$ is $f_E(E)\,\mathrm{d}E$. The distribution is shown for three different temperatures T. It shifts to higher energies and broadens as T increases.

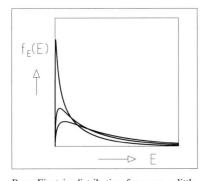

Bose–Einstein distribution for a gas a little above the critical temperature T_c (leftmost curve) and at two higher temperatures.

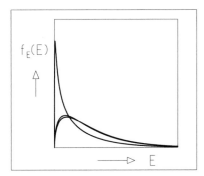

Bose–Einstein distribution for a gas a little above the critical temperature T_c (leftmost curve). The other two curves correspond to the same temperature but with the gas density reduced in two steps.

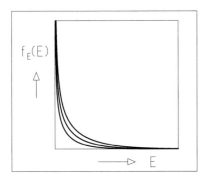

Bose–Einstein distribution for a gas at the critical temperature T_c (top curve) and at two lower temperatures. As the temperature decreases, more atoms are removed from the continuous distribution by *Bose–Einstein condensation* and accumulate in the state of lowest energy.

[1] Satyendra Nath Bose (1894–1974)

looked at.

What we have just described is *classical statistics*, now also called *Boltzmann statistics*. We mentioned that the size of the phase-space elements is arbitrary, if only small enough. But Sommerfeld (who had introduced the quantization of the action integral, Episode 27) and Planck [5] gave reasons for its size to be h^3. The dimension of the phase-space element is the dimension of action (position times momentum) raised to the third power; and action is quantized, its smallest amount being h.

This is the point where Bose comes in. Bose[1] was born in Calcutta as son of a railway employee. He took a B.Sc. degree in 1913 and an M.Sc. in 1915 and continued to study at University College, Calcutta. In the early 1920s he became reader at the newly founded University in Dacca. Already during his studies Bose had access to modern publications and textbooks in both English and German and even had translated a collection of papers on relativity into English, which was published as a book in India. In 1924 he succeeded to find a new derivation of Planck's radiation law. He wrote it up as a rather short paper in English, which he sent to Einstein, asking him to arrange for the paper to be translated into German and published in *Zeitschrift für Physik*. The letter (quoted in [4]) contains the passage: 'Though a complete stranger to you, I do not feel any hesitation to make such a request. Because we are all your pupils though profiting only by your teachings and through your writings.' Einstein answered Bose by a handwritten postcard, praising his work and announcing its publication. The postcard became a door opener for Bose. He got leave of absence and a stipend to study for two years in Europe.

Within about a week after receiving it, Einstein himself translated Bose's paper and made the following addition:

> *Translators Note:* Bose's derivation of Planck's formula in my opinion constitutes an important progress. The method used here also yields the quantum theory of the ideal gas, as I will explain elsewhere.

Bose's paper [6] is short but not easy to read, Pais [7] calls it *a confused masterpiece*. Bose takes light quanta to be particles with both energy and momentum and he takes h^3 to be the size of the phase-space element. He allows for an additional degree of freedom of light quanta, their polarization, by multiplying the number of cells by a factor of 2, which, as he wrote, 'seems to be called for' and applies the methods of statistical physics. Contrary to Boltzmann statistics he treats the quanta as indistinguishable (see Box II). This means, he introduces a new way of counting complexions. Moreover, in the process of maximizing the number of complexions, only the total energy is kept constant, not the number of quanta. The result is a different distribution of quanta over phase space. For the energy distribution he obtains Planck's law. Nowhere in his derivation is there any need to discuss the interaction between light quanta and matter.

Einstein used Bose's new method of counting to develop a statistical theory of a material gas, which he published in two papers entitled *On the Quantum Theory of the Mono-Atomic Ideal Gas* [8,9]. The essential difference between a 'gas' of light quanta and one of atoms or molecules is that the number of objects is kept constant during the maximizing process of Box I in the latter case but not in the former. In the second paper Einstein explains the fundamental difference between classical statistics and Bose's approach: Quanta (in Einstein's case particles) are treated as *indistinguishable*; the exchange of two particles does not lead to a new state. He finds that Nernst's heat law (Episode 12) is fulfilled for the Bose–Einstein gas, also called the *ideal Bose gas*, which thus shows the property of *gas degeneracy* that Nernst had expected.

Ehrenfest and others noticed that in Bose–Einstein statistics the objects are no longer *statistically independent* in the conventional sense. There is a tendency of particles to form groups in a common cell. This is acknowledged by Einstein in [9], in which he also describes a phenomenon which now goes by the name *Bose–Einstein condensation*. There is a critical temperature T_c, which depends on the mass of the atoms and their density in space. At temperatures appreciably higher than T_c the *Bose–Einstein distribution* just becomes the Maxwell–Boltzmann distribution. A little above T_c it is quite different. As T is lowered below T_c, a rising fraction of all atoms is transferred to the ground state, the state of lowest energy, and removed from the distribution. Although he predicted it, Einstein himself was not too sure about the reality of such a condensation. In a letter to Ehrenfest he wrote [10]: 'Die Theorie ist hübsch, aber ob auch was Wahres dran ist?' (The theory is pretty, but is there also some truth to it?) It is typical for Einstein's far-reaching predictions that only 12 years later Bose–Einstein condensation was first used to explain an observed phenomenon, namely, superfluidity (Episode 58) and that it took 70 years until the effect was established experimentally beyond doubt (Episode 99).

With their work Bose and Einstein established the field of *quantum statistics* one year before the appearance of quantum mechanics. They only used the 'old' quantum theory, which shows up in the size of the cells of phase space. We shall see in Episode 41 how Fermi, using the 'old' theory plus the *Pauli principle*, found yet another way of counting, and how Dirac and, independently, Heisenberg showed that quantum mechanics naturally justifies both ways.

The work described here was the last outstanding research Einstein did in Germany. In 1932 he accepted to work for half of each year at the recently founded Institute for Advanced Study in Princeton and he spent the Winter 1932/33 in the United States. On 30 January 1933, Hitler came to power in Germany. Still on board the ship, which brought him back to Europe, he wrote his letter of resignation [11] to the Academy in Berlin:

> The circumstances prevailing at present in Germany cause me to resign herewith my position at the Prussian Academy of Sciences. For 19 years the Academy has given me the possibility to devote myself to scientific work free from all professional obligation. I know to which high degree I owe it a debt of gratitude. Reluctantly I leave its circle also because of the stimulations and the beautiful human relationships which I enjoyed and highly appreciated during the long time as its member.
>
> But under the present conditions I find unbearable the dependence on the Prussian Government implied by my position.

Einstein never returned to Germany.

[1] Einstein, A., *Verh. d. Dt. Phys. Gesellschaft*, **18** (1916) 318.

[2] Einstein, A., *Mitteilungen der Physikalischen Gesellschaft Zürich*, **16** (1916) 47. Also published as [12].

[3] Mehra, J., *Biographical Memoirs of Fellows of the Royal Society*, **21** (1975) 117.

[4] Mehra, J. and Rechenberg, H., *The Historical Development of Quantum Theory, Vol. 1. – The Quantum Theory of Planck, Einstein, Bohr, and Sommerfeld. Its Foundation and the Rise of Its Difficulties 1900–1925.* Springer, New York, 1982.

[5] Planck, M., *Vorlesungen über Wärmestrahlung.* Barth, Leipzig, 4th ed., 1921.

[6] Bose, S. N., *Zeitschrift für Physik*, **26** (1924) 178.

[7] Pais, A., *Inward Bound – Of Matter and Forces in the Physical World.* Oxford University Press, Oxford, 1986.

[8] Einstein, A., *Sitzungsber. Preuss. Akad. Wiss. (Berlin)*, (1924) 261.

[9] Einstein, A., *Sitzungsber. Preuss. Akad. Wiss. (Berlin)*, (1925) 3.

[10] Ch. 23d in [13].

[11] Kirsten, C. and Treder, H.-J. (eds.), *Albert Einstein in Berlin 1913–1933, Teil I Darstellung und Dokumente.* Akademie-Verlag, Berlin, 1979.

[12] Einstein, A., *Physikalische Zeitschrift*, **18** (1917) 121.

[13] Pais, A., *'Subtle is the Lord ...' – The Science and Life of Albert Einstein.* Oxford University Press, Oxford, 1982.

34 Bothe and Geiger – Coincidence Experiments (1925)

In 1914 Bothe[1] obtained his Ph.D. in Berlin with work done under Planck. Already a little earlier, at the end of 1913, he had become assistant to Geiger at the Physikalisch-Technische Reichsanstalt. Geiger who had returned to Germany in 1912 had set up the laboratory for radium research there. At the same time as Bothe another promising young scientist joined Geiger in Berlin: James Chadwick, coming from Rutherford's group in Manchester where Geiger had worked before. When the First World War broke out, Chadwick was interned in Berlin. Bothe volunteered for the army, was taken prisoner in Russia in 1915, and sent to Siberia where he stayed till 1920. Fortunately, the two young men were able to continue doing science in their captivity although with severely restricted means. On his return to Germany, Bothe brought with him his Russian bride, to whom he was soon married, and continued to work with Geiger at the Reichsanstalt.

Bothe

The discovery of the Compton effect in 1923 stirred up the physics community. It had shown (Episode 31) that X rays, scattered off electrons under a given angle, changed their wavelength. If one assumed that photons, like material particles, possess energy and momentum, the Compton effect could

---

[1] Walther Bothe (1891–1957), Nobel Prize 1954

140

be explained easily. But this assumption at that time was hard to accept for many, in particular for Bohr. Together with Kramers[2] and Slater[3] he tried to reconcile quantum effects with Maxwell's theory of the electromagnetic field. In a paper entitled *The Quantum Theory of Radiation* [1] they tried to elaborate Planck's original idea of 1900 in which the properties of the electromagnetic field were left unchanged and all quantum effects were caused by the structure of the atoms. In what is now called the *BKS theory*, the three authors conjectured that the atom contains many virtual oscillators, one for every possible transitions between stationary states, and that these oscillators communicate with those of other atoms through virtual fields by probability laws. In that theory they had to accept, however, that energy and momentum are not conserved in each individual process but only in the statistical average. For the Compton effect that meant that there was not one scattered photon for any recoiling electron and vice versa.

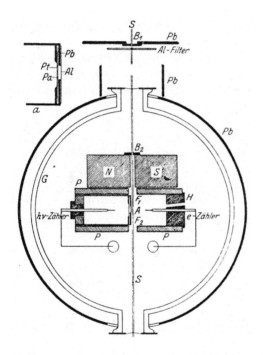

The coincidence apparatus of Bothe and Geiger. There are two needle counters, separated by a thin aluminium window. The recoil electron of the primary Compton scattering is recorded in the counter on the right. The scattered photon, if emitted in the proper direction, can traverse the window and give rise to a secondary Compton effect in the counter on the left. Reprinted with kind permission from Springer Science+Business Media: [2], Fig. 2.

Shortly after the appearance of that theory in print Bothe and Geiger published a note [3] announcing that they would test it by searching experimentally for the *simultaneous* appearance of a scattered photon and a recoiling electron. They devised the apparatus for what is believed to be the first *coincidence experiment* [2]. The apparatus was contained in a glass sphere filled with hydrogen at atmospheric pressure. The most important components were two of Geiger's *needle counters* (see Episode 13), which were able to register individual electrons. The two counters were separated by a very thin *window* of aluminium foil. A beam of carefully collimated X rays, cleaned by a strong magnet of charged particles produced in the collimator material, was passed

---

[2] Hendrik Anton Kramers (1894–1952)   [3] John Clarke Slater (1900–1976)

through one of the counters near the window and parallel to it. This counter, called the *e counter*, could register the recoil electron of a Compton-scattering process. The window between the two counters was chosen so thick that no recoil electron could traverse it. A secondary photon emitted in the direction of the window, however, could pass it, and give rise to another Compton effect in a thin platinum foil placed a short distance behind the window and the recoil electron of this secondary Compton scattering would be registered in the second counter. Since it served to register the photon, this counter was called the *hν counter*. The question to be answered by this experiment was whether the signal of the *hν* counter occurred *in coincidence*, i.e., in the same short time interval, with a signal in the *e* counter more often than expected by pure chance. If so, the two signals were connected to the same cause, a primary Compton scattering in the *e* counter.

A piece of film recording. The distance between the black vertical stripes is 1/1000 of a second. The signals from the *e* and the *hν* counters are the white lines recorded on the top and bottom parts of the film, respectively. One coincidence event is seen at the point in time where both signals change rapidly. Reprinted with kind permission from Springer Science+Business Media: [2], Fig. 7.

The determination of coincidences was made not by an electronic circuit but by registering the signals from the two counters on a fast-moving photographic film and by inspecting the developed film carefully long after the actual experiment had been performed. In this way Bothe and Geiger measured the signal times with an accuracy of 0.1 ms and defined coincidences as events in which both counters showed a signal within a time interval of 1 ms. They used a simple reasoning to determine the number of *chance coincidences* counting events with a finite time difference, e.g., with a time difference between 5 ms and 6 ms. For zero time difference they found five times as many events in a 1 ms interval than for finite time differences. Thus the simultaneous occurrence of scattered photon and recoil electron was demonstrated.

Geiger informed Bohr of the results before they were published. Bohr gracefully acknowledged and the BKS theory was abandoned. It should be mentioned that a few months later Compton and Simon published work performed with a cloud chamber [4] in which they had been able to register primary Compton scattering in the cloud-chamber gas together with secondary scattering in lead foils placed in the chamber. The line joining the positions of the primary and secondary scattering, if taken as the direction of the scattered photon, was within 20° in the direction computed from the theory of the Compton effect taking into account the angle of the scattered electron.

When Geiger left Berlin in 1925 for a professorship at the University of Kiel, Bothe took over as director of the laboratory of radium research at the Reichsanstalt. The two spoke about future projects and Geiger left the coincidence method to Bothe, who, gratefully, made good use of this generosity.

In 1929, still at the Reichsanstalt in Berlin, Bothe and Kolhörster made an important discovery with the help the coincidence method [5,6]. They used two Geiger–Müller counters, one above the other, placed a thick brick of gold between the two, surrounded the whole assembly with a shielding of iron and lead and looked for coincidences in the two counters. Signals could only be produced by cosmic radiation because the shielding prevented signals due to radioactivity. At the time cosmic radiation was still believed to consist of very penetrating high-energy γ rays. But Bothe and Kolhörster, who observed many coincidences, were unable to explain them as due to γ rays, whereas it was natural to interpret them as caused by charged particles traversing both counters. In their paper [5], to which they gave the title *The Nature of Cosmic Rays*, they concluded:

> To this whole evidence we believe we must give the interpretation that the cosmic radiation, at least in so far as it manifests itself in phenomena observed up to now, is no γ radiation but a corpuscular radiation.

In the course of time the corpuscular radiation observed by Bothe and Kolhörster became known as the *penetrating component* of the cosmic radiation. That, at least near the surface of the earth, it consists of a hitherto unknown type of particles was shown in 1937 (Episode 57).

At the end of 1929 Bothe published a method of registering coincidences between the signals of two Geiger counters by electronic means [7]. He used a special vacuum tube, fitted with two grids, which produced a signal at its anode only if the voltage at both grids (determined by the two counters) was right. Inspired by Bothe's paper, Rossi[4] devised a simpler circuit, based on single-grid tubes, which, in addition, was able to decide on the coincidence (within a short time interval) of more than two signals [8]. Realized in electronic circuits, the coincidence method became of vital importance for all experiments with several electronic detectors.

In 1930 Bothe accepted a professorship at the University of Giessen. There, with his student Becker, he observed a new type of radiation, which was eventually identified by Chadwick as consisting of neutrons, see Episode 48. In 1932 he succeeded Lenard as professor in Heidelberg. Lenard who had made important contributions to the study of cathode rays and the photoelectric effect, in his later years, took to anti-Semitism and the Nazi ideology. When Hitler came to power in 1933, he used his influence and tried to remove from office people he disliked. Bothe was one of them. But Planck, then the president of the Kaiser-Wilhelm Society (now Max-Planck Society), saw to it that a position was created for Bothe in the society's Institute for Medical Research in Heidelberg. Here Bothe set up a highly successful group of nuclear research, using the excellent workshop facilities of the institute to build two particle accelerators, a van de Graaff band generator and a cyclotron. After the war he resumed his professorship in Heidelberg while retaining his position at the Max-Planck Institute to which he returned full-time in 1954, the year in which he was awarded the Nobel Prize 'for the coincidence method and his discoveries made therewith'.

[1] Bohr, N., Kramers, H. A., and Slater, J. C., *Philosophical Magazine*, **47** (1924) 785. Also in [9], Vol. 5, p. 101.

[2] Bothe, W. and Geiger, H., *Zeitschrift für Physik*, **32** (1925) 639.

[3] Bothe, W. and Geiger, H., *Zeitschrift für Physik*, **26** (1924) 44.

[4] Compton, A. H. and Simon, A. W., *Physical Review*, **26** (1925) 289.

[5] Bothe, W. and Kolhörster, W., *Naturwissenschaften*, **17** (1929) 271.

[6] Bothe, W. and Kolhörster, W., *Zeitschrift für Physik*, **56** (1929) 751.

[7] Bothe, W., *Zeitschrift für Physik*, **59** (1929) 1.

[8] Rossi, B., *Nature*, **125** (1930) 636.

[9] Bohr, N., *Collected Works, 10 vols.* North Holland, Amsterdam, 1972–1999.

---

[4] Bruno Benedetto Rossi (1905–1993)

35 Pauli's Exclusion Principle (1925)

Sommerfeld and Pauli

The exclusion principle is one of the most far-reaching theorems in physics. In a telegram to the Nobel committee, quoted in [1], Einstein wrote: 'Nominate Wolfgang Pauli for physics prize stop his contributions to modern quantum theory consisting of so-called Pauli or exclusion principle became fundamental part of modern physics being independent from other basic axioms of that theory stop Albert Einstein.' It is interesting to note that quantum mechanics, the theory Einstein alluded to, was formulated a little after the *Pauli principle*.

Pauli[1] [1,2] was born in Vienna. His father was a medical doctor, who had turned to chemical research and later became full professor and director of the Institute for Colloid Chemistry created for him. Until 1926 Pauli signed his papers *Wolfgang Pauli jr.* because his first name was also his father's. Pauli's middle name, Ernst, was the first name of the famous physicist Mach[2], his godfather, who had been mentor and friend of his father since the latter's student days in Prague. Pauli was a prodigy child. While attending the Döblinger Gymnasium in Vienna which offered, of course, a classical education, he had private lessons in mathematics and physics and actively followed the current literature including Einstein's work on general relativity. In his class at school, also called the *class of geniuses*, there was a whole group of unusually talented boys including another future Nobel Prize winner, Richard Kuhn[3], and several future university professors, actors, and industrialists.

In 1918 Pauli also formally became a student of physics. He had chosen to study in Munich under Sommerfeld. As a first-year student he published three papers on relativity theory. In his second year Sommerfeld asked him to write a comprehensive article on relativity for the *Encyclopädie der mathematischen Wissenschaften* taking into account the extensive literature. This work [3] became a standard text and was highly praised by Einstein himself. In 1921 Pauli obtained his Doctor's degree with a dissertation on the ionized hydrogen molecule, i.e., a molecule of two atomic nuclei and one electron. In this research serious discrepancies showed up between quantum theory as it then existed and experimental facts. Pauli worked the following year as Born's assistant in Göttingen. There he met Bohr who gave a set of lectures on atomic physics which was later called the *Bohr festival* and who invited him for a year to Copenhagen. From there he went as assistant to Lenz in Hamburg.

Pauli's *exclusion principle* (so called by Dirac in 1926) was found in Hamburg early in 1925. The story has been told twice [4,5] by Pauli himself and is carefully presented in its historical and physics context in [6]. In 1924 Pauli became Privatdozent in Hamburg and presented an inaugural lecture on the Periodic Table of the elements. Later he wrote [5]:

---

[1] Wolfgang Ernst Friedrich Pauli (1900–1958), Nobel Prize 1945   [2] Ernst Mach (1838–1916)   [3] Richard Kuhn (1900–1967), Nobel Prize 1938

The contents of this lecture appeared very unsatisfactory to me, since the problem of the closing of the electronic shells had been clarified no further. The only thing that was clear was that a closer relation of this problem to the theory of multiplet structure must exist. I therefore tried to examine again critically the simplest case, the doublet structure of the alkali spectra. According to the point of view then orthodox, which was also taken over by Bohr in his already mentioned lectures in Göttingen, a non-vanishing angular momentum of the atomic core was supposed to be the cause of this doublet structure.

Born, then his boss, pulling Pauli by the ear.

It was generally accepted that the atoms of the noble gases (helium, neon, argon, ...) have closed electron shells because they are chemically inert and that therefore the atoms of the alkali metals (lithium, sodium, potassium, ...), which directly follow the noble gases in the Periodic Table, have exactly one electron in the outer shell. It was well known that the length of each period of the Periodic Table, i.e., the number of elements counting from an alkali metal to the next noble gas is $2n^2$, where n is the principal quantum number of the electron in the outer shell of the alkali atom. Thus there are periods of length $2, 8, 18, \ldots$. But it was not clear how these numbers could be explained.

We have mentioned in Episode 27 that Sommerfeld had associated one quantum number to every mechanical degree of freedom of an electron in the atom but that there were some features of the spectra which could not be accounted for by these quantum numbers alone. Such features were, in particular, the multiplet structure of the spectra and the anomalous Zeeman effect. Most spectral lines are *multiplets*, i.e., they consist of several, closely spaced, single lines. The frequency of each individual line is related to the difference between two energy values or *terms* of an electron in the corresponding atom. The multiplet structure of a spectrum is therefore related to a multiplet structure in the energy terms. The best-known spectral line is the *D line* of sodium which gives rise to the yellow light seen when a little kitchen salt is put into a flame. This line is a doublet consisting of two lines, D_1 and D_2. It is emitted when the outer electron of sodium makes a transition from a particular excited state (which

---

Angular Momentum and Magnetic Moment of an Orbiting Electron

In the Bohr–Sommerfeld model, the vector $\mathbf{L}$ of angular momentum of an orbiting electron has a magnitude L, which is an integer multiple of $\hbar$ (Planck's constant h divided by 2π), i.e., $L = n_\varphi \hbar, n_\varphi = 1, 2, \ldots$. The magnetic moment is simply proportional to the angular momentum, $\boldsymbol{\mu} = \gamma \mathbf{L}$, and the proportionality constant $\gamma = -e/2m$ is called *gyromagnetic ratio*. Because of the quantization of angular momentum the magnetic moment is also quantized. Its smallest nonvanishing magnitude is $\mu_\mathrm{B} = e\hbar/2m$, the *Bohr magneton*, so called by Pauli already in 1920.

Pauli in 1924

is a doublet) to its ground state (which is a singlet). It had been assumed, as mentioned in the quotation above, that the doubling of the excited state was caused by an interaction between the angular momenta of the outer electron and the rest of the atom.

Before continuing we want to discuss the *magnetic moment* of an orbiting electron. An orbiting electron corresponds to an electric current on the orbit and thus to a small magnet, which is characterized by its magnetic moment, proportional to the angular momentum. Since the angular momentum is quantized so is the magnetic moment (see Box). The 'orthodox view' alluded to by Pauli was the following. If the inner electrons, which form a closed shell, possess a resulting angular momentum and therefore a magnetic moment, then this gives rise to a magnetic field in which the magnetic moment of the outer orbiting electron can assume different orientations, each with a different energy. Thus there is a splitting of energy levels. In a careful analysis [7] Pauli showed that this assumption led to serious discrepancies with experiment and stressed the point that closed shells have zero angular momentum. He saw the cause for the splitting of the energy levels of the orbiting outer electron not in a magnetic moment connected to the orbit but in a 'particular two-valuedness of the quantum-theoretic properties of the electron, which cannot be described from the classical point of view'. He attributed a quantum number to the electron itself, in addition to the quantum numbers which had been given by Bohr and Sommerfeld to its different possible orbits. Already Sommerfeld had realized that his original three quantum numbers were not sufficient to explain details of the spectra and had therefore introduced an additional *inner quantum number*. However, before Pauli's analysis there were different sets of definitions of the four quantum numbers (the numbers in each set being linear combinations of the numbers in other sets) and there was no clear separation between quantum numbers due to the electron orbit and due to the intrinsic properties of the electron.

We have seen in Episode 27 that in the original Sommerfeld model with only orbital quantum numbers there was a total of n^2 different combinations of the two quantum numbers (n_φ, n_ψ), which could accompany a given principal quantum number n. In special cases the energy levels belonging to all these combinations were degenerate (i.e., identical). But that degeneracy was lifted if there was at least one inner shell of electrons and an external magnetic field. One then could account for n^2 different energy levels or terms for a given value of n. Experimentally the situation was more complicated. Spectroscopy of the

Dedication by Pauli to Sommerfeld on a copy of his Nobel Lecture. In the possession of Prof. Fritz Bopp, Siegen; reproduced with his permission.

light emitted by alkali atoms in a magnetic field, in other words, analysis of the anomalous Zeeman effect of alkali atoms, revealed more levels. Pauli had already done extensive work on the anomalous Zeeman effect when he read the following remark in a paper by Stoner[4] [8]:

> For a given value of the principal quantum number the number of energy levels of a single electron in the alkali metal spectra in an external magnetic field is the same as the number of electrons in the closed shell of the rare gases which corresponds to the principal quantum number.

As mentioned, for a given n there are $2n^2$ electrons in the closed shell of a rare gas and thus there are the same number of energy levels in the anomalous Zeeman effect of an alkali. Pauli now was able to explain both facts by the following assumptions:

1. In the conditions of the anomalous Zeeman effect the degeneracy with respect to all quantum numbers, including his own 'classically not describable two-valuedness', is lifted. Thus, not only is the number of energy levels is doubled to $2n^2$ but also the number of combinations of other quantum numbers for a given principal quantum number n.
2. In the closed shell of a noble gas all electrons have the same principal quantum number n but each electron has its own particular set of other quantum numbers. Since there are $2n^2$ different combinations of these, there is room for exactly $2n^2$ electrons in that shell. If one more electron is added, another shell with a new value of n has to be begun.

In mid-January 1925 Pauli completed the paper [9] in which he formulated his exclusion principle:

> There can never be two or more equivalent electrons in the atom for which in strong fields the values of all quantum numbers [...] coincide. If an electron is present in the atom for which the quantum numbers (in the external field) have certain values then this state is "occupied".

If Bohr had explained the general structure of atoms and Sommerfeld had been able to calculate quite a number of details, Pauli could now account for the structure of the Periodic Table. Unlike the Bohr–Sommerfeld model the Pauli principle, expressing that two electrons cannot exist in the same state, remains valid in quantum mechanics and is even true for a larger class of particles. We shall return to it in Episode 41.

---

[4] Edmund Clifton Stoner (1889–1973)

[1] Pais, A., *The Genius of Science*, chap. Wolfgang Ernst Pauli. Oxford University Press, New York, 2000.

[2] Enz, C. P., *No Time to be Brief – A Scientific Biography of Wolfgang Pauli*. Oxford University Press, Oxford, 2002.

[3] Pauli, W., *Encyclopädie der mathematischen Wissenschaften, Vol. 5, Part 2*. Teubner, Leipzig, 1921. Also in [10], Vol.1, p. 1.

[4] Pauli, W., *Science*, **103** (1946) 213. Also in [10], Vol. 2, p. 1073.

[5] Pauli, W., *Nobel Lectures, Physics 1942–1962*, p. 27. Elsevier, Amsterdam, 1964.

[6] van der Waerden, B. L., *Exclusion Principle and Spin*. In Fierz, M. and Weisskopf, V. F. (eds.): *Theoretical Physics in the Twentieth Century – A Memorial Volume to Wolfgang Pauli*, p. 199. Interscience, New York, 1960.

[7] Pauli, W., *Zeitschrift für Physik*, **31** (1925) 373. Also in [10], Vol. 2, p. 201.

[8] Stoner, E., *Philosophical Magazine*, **48** (1924) 719.

[9] Pauli, W., *Zeitschrift für Physik*, **31** (1925) 765. Also in [10], Vol. 2, p. 214.

[10] Pauli, W., *Collected Papers, 2 vols.* Interscience, New York, 1964.

36 Spin (1925)

We have seen in the preceding episode that early in 1925 Pauli attributed to the electron a 'classically not describable two-valuedness'. Within a few months this property of the electron was given a physical description which, it is true, is not satisfactory from the viewpoint of classical physics. This description is regarded as the discovery of the *intrinsic angular momentum* or *spin* of the electron. In fact the description was given twice. Already in January Kronig[1] discussed it with Pauli who was completely opposed to the idea so that Kronig did not publish it. The concept of the electron spin was independently proposed and published [1] by Uhlenbeck[2] and Goudsmit[3] who thus became the discoverers of spin.

Kronig's story has been told by himself [2]. In 1925 he was 20 years old and had won a travelling fellowship from Columbia University, which he used to spend time in different research groups in Europe. On 7 January, he arrived in Tübingen which was a centre of high-precision optical spectroscopy led by Paschen and Back[4]. The theoretical side was represented by Landé[5], who was a specialist in empirical and phenomenological analysis of complicated spectra. Kronig met with Landé and told him of his idea that the electron might have an intrinsic angular momentum due to a rotation about its own axis in addition to the orbital angular momentum, which is due to its rotation around the atomic nucleus. The idea was the same as in astronomy, where a planet possesses an angular momentum because it circles the sun and another one because it turns around its own axis. Kronig proposed to use this new degree of freedom of the electron to account for hitherto unexplained properties of the spectra. Landé suggested that Kronig should discuss this with Pauli who had announced his arrival in Tübingen for the very next day. In fact, Pauli had essentially finished his paper with the exclusion principle but wanted to check it against the latest experimental facts on the Zeeman effect before submitting it for publication. Pauli, when hearing of Kronig's thoughts, said: 'Das ist ja ein ganz witziger Einfall [a rather witty idea].' But, as Kronig put it, he did not believe that the suggestion had any connection with reality.

Here the fundamental difficulty is the following. The electron has a very small mass and a very small radius. So far it has not been possible to measure a finite radius and modern very successful theories treat it as point-like. But then there was at least the concept of the *classical electron radius* $r_e = 2.84 \times 10^{-13}$ cm. This is a length which can be formed from the fundamental constants charge e and mass m of the electron and the velocity c of light. If one considers the charge e to be smeared out uniformly on the surface of a sphere of radius r_e then the energy contents of the electric field due to the

---

[1] Ralph de Laer Kronig (1904–1995) [2] George Eugene Uhlenbeck (1900–1988) [3] Samuel Abraham Goudsmit (1902–1978) [4] Ernst Back (1881–1959) [5] Alfred Landé (1888–1975)

Kronig and Pauli in 1955.

charge in special relativity corresponds to the mass m. Now, for a sphere of that mass and radius to have an angular momentum on the order of $\hbar = h/2\pi$ the velocity of rotation on the surface of the sphere would have to be very much higher than the speed of light! Such an angular momentum cannot be described classically.

Nevertheless the electron has angular momentum and in September 1925 the idea was born a second time in discussions between Goudsmit and Uhlenbeck who at that time were graduate students working under Ehrenfest[6], the Austrian scientist who had succeeded Lorentz on the Chair of Theoretical Physics in Leiden. Uhlenbeck [3,4] was born in Batavia (now Jakarta) as son of a Dutch officer and raised in The Hague to where his father had retired early. He studied physics in Leiden and after graduation spent three years in Rome, working as private tutor to a son of the Dutch ambassador and at the same time keeping some contact with physicists there including the young Fermi. Goudsmit [5] was a native of The Hague, where both his parents ran successful businesses. He, too, studied in Leiden and became quite an expert in the analysis of optical spectra. When in the summer of 1925 Uhlenbeck returned to Holland to resume physics, Ehrenfest suggested that he first get some coaching from Goudsmit. It was during these coaching sessions, which took place mainly in the home of Uhlenbeck's parents in The Hague, that they first discussed the spinning electron. In a lecture given in 1955, quoted in [6], Uhlenbeck recalled:

> Goudsmit and I hit upon this idea by studying a paper by Pauli, in which the famous exclusion principle was formulated and in which, for the first time *four* quantum numbers were ascribed to the electron. This was done rather formally; no concrete picture was connected with it. To us, this

---

[6] Paul Ehrenfest (1880–1933)

was a mystery. We were so conversant with the proposition that every quantum number corresponds to a degree of freedom, and on the other hand with the idea of a point electron, which obviously had three degrees of freedom only, that we could not place the fourth quantum number. We could understand it only if the electron was assumed to be a small sphere which could rotate.

Uhlenbeck, Kramers, and Goudsmit (from the left) around 1928.

Landé had found that Sommerfeld's formula for the fine structure of the hydrogen spectrum (Episode 27) could also describe the alkali spectra if an additional factor, Landé's g factor, was introduced. That formula was a guide-line for Uhlenbeck and Goudsmit. It showed that for the doublet structure to be explained by the intrinsic properties of the outer electron, its spin had to have the magnitude $\hbar/2$ and the magnetic moment had to be one Bohr magneton μ_B. Thus the magnitude of spin was half the value of the orbital angular momentum of an electron on the lowest Bohr orbit. But the magnetic moment was the same in both cases. In the formula relating spin to a magnetic moment an additional factor of 2 appeared, the *g factor* of the electron. They discussed this with Ehrenfest who drew their attention to calculations by Abraham[7], published in 1903, from which they found that, for a rotating sphere with surface charge, the ratio of magnetic moment to angular momentum was indeed a factor of two larger than for an orbiting point charge. Ehrenfest also asked them to write a short note for possible publication and suggested that they should also discuss their work with Lorentz. Uhlenbeck and Goudsmit wrote the note and gave it to Ehrenfest to read. Lorentz started calculations on the spinning sphere and found that it produced a magnetic field containing such a large amount of energy that a classical electron radius, which would take that into account, would be as large as the radius of the atom itself. It was clear that he did not believe in the spinning electron. Therefore, Uhlenbeck and Goudsmit went to

---

[7] Max Abraham (1875–1922)

reclaim their manuscript from Ehrenfest. But he had already sent it off to the journal and told them: 'Sie beide sind jung. Sie können sich eine Dummheit leisten. [You both are young. You can afford a stupidity.]' The paper [1], or rather the note – it was only one page, appeared in print in November 1925.

Apart from the g factor of the electron there is another factor of two in the story of spin. When the paper was out, Goudsmit got a letter from Heisenberg with congratulations on the 'brave note' and the question 'how you got rid of the factor 2'. Heisenberg had computed quantitatively the splitting of the energy levels and got a result twice as large as the experimental value. The two young Dutchmen had not even attempted to do that calculation but, of course, they tried to do it now. Einstein himself, who happened to be in Leiden, set them on the right track. The spinning electron is orbiting the atomic core which carries electric charge. In a reference system in which the electron is at rest that charge orbits around the electron. A moving charge means an electric current which generates a magnetic field. The interaction between that field and the spin magnetic moment leads to the splitting of energy levels.

In December 1925 the golden jubilee of Lorentz' doctorate was celebrated. Among many others Bohr attended and he urged Uhlenbeck and Goudsmit to write a more detailed paper. This they completed in the same month. It is entitled *Spinning Electrons and the Structure of Spectra* [7], has two pages, and mentions the missing factor of two. Bohr was very much in favour of spin. In a letter to Ehrenfest, who had been his host in Leiden, he wrote that on his way back to Copenhagen he felt 'wie ein Profet des Elektromagnet-Evangeliums [like a prophet of the spin gospel]' when meeting with Pauli and Heisenberg. (For a facsimile of the whole letter see [3].)

The missing factor of two was found by Thomas[8] early in 1926 then working in Bohr's institute [8]. The transformation between the two reference systems (orbiting electron and electron at rest) is, of course, a Lorentz transformation. After all, in his first paper on special relativity Einstein had used it to connect the electric field of a charge at rest with the magnetic field of a moving charge. But a Lorentz transformation can be applied only if the two reference systems are in uniform (unaccelerated) motion with respect to each other. The circular motion appearing here, however, is accelerated. By taking this fact properly into account Thomas was able to explain the mysterious factor of two, later called the *Thomas factor*.

Now everything fell into place and, at last, also Pauli was convinced of the reality of spin. Although the arguments used in the discovery of spin belong to the *old quantum theory*, spin can only be properly understood in the framework of *quantum mechanics*, which was born in the same year 1925 in which spin was found (see Episode 37). The quantum-mechanical description of spin is due to Pauli (see Episode 43).

[1] Uhlenbeck, G. E. and Goudsmit, S., *Naturwissenschaften*, **13** (1925) 953.

[2] Kronig, R., *The Turning Point*. In Fierz, M. and Weisskopf, V. F. (eds.): *Theoretical Physics in the Twentieth Century – A Memorial Volume to Wolfgang Pauli*, p. 15. Interscience, New York, 1960.

[3] Uhlenbeck, G. E., *Physics Today*, **June** (1976) 43.

[4] Pais, A., *The Genius of Science*, chap. George Uhlenbeck. Oxford University Press, New York, 2000.

[5] Goudsmit, S., *Physics Today*, **June** (1976) 40.

[6] van der Waerden, B. L., *Exclusion Principle and Spin*. In Fierz, M. and Weisskopf, V. F. (eds.): *Theoretical Physics in the Twentieth Century – A Memorial Volume to Wolfgang Pauli*, p. 199. Interscience, New York, 1960.

[7] Uhlenbeck, G. E. and Goudsmit, S., *Nature*, **117** (1926) 264.

[8] Thomas, L. H., *Nature*, **117** (1926) 514.

---

[8] Llewellyn Hilleth Thomas (1903–1992)

37 Heisenberg and the Creation of Quantum Mechanics (1925)

Heisenberg

There is a certain similarity in the early lives of Planck and of Heisenberg[1] [1,2]. Both had fathers who were professors at the University of Munich, both visited the Maximilians-Gymnasium in Munich, and both went to see a professor in Munich to seek advice for their studies. We have mentioned at the beginning of the Prologue how Jolly in vain tried to dissuade Planck from studying physics. Heisenberg, in his autobiographic book [3], which is devised as a series of conversations, tells us of a similar experience. In the autumn of 1920, when he was about to enter the university, his first choice was mathematics. He wanted not only to do course work but also learn about problems of current interest. This, he thought, he could do best by participating in a professor's seminar usually reserved to advanced students and Ph.D. candidates. He asked for an interview with Lindemann[2], the most distinguished mathematician in Munich, who was famous for having proved that the number π is transcendental. But the old professor and the young student found no understanding, partly because Lindemann's dog noisily interfered. The result was that Heisenberg opted for theoretical physics and went to ask Sommerfeld to be admitted to his seminar. Sommerfeld agreed. At that time Pauli, only a year older than Heisenberg, was already actively doing research under Sommerfeld. The two young men became life-long friends.

Thus Heisenberg became a member of Sommerfeld's seminar at the very beginning of his studies. A few weeks later Sommerfeld suggested that he, like all others, should do some work on a research project and report on it in the seminar. He asked him to look for empirical rules explaining the peculiar splitting of spectral lines by a magnetic field, called the *anomalous Zeeman effect*. Only a little later Heisenberg found such rules which, however, comprised not only the usual integer quantum numbers but also half-integer ones. When, a year later, Landé published a set of similar rules Sommerfeld suggested that Heisenberg should construct a physical model of an atom obeying his rules. This model became the subject of Heisenberg's first publication. It consisted of an atomic rump (or core) containing the nucleus and the inner electrons and a single electron in an outer shell. Both rump and shell had angular momenta which were not integer but half-integer multiples of $h/2\pi$, contrary to the basic foundations of the Bohr–Sommerfeld model of the atom.

In his second semester, Heisenberg attended Sommerfeld's course on hydrodynamics where he became aware of a problem which gave rise to his second paper. He was able to calculate two constants in the *vortex street* which its originator von Kármán[3] had to take from experiment. Thus, already in his first year

---

[1] Werner Karl Heisenberg (1901–1976), Nobel Prize 1932   [2] Carl Louis Ferdinand (von) Lindemann (1852–1939)   [3] Theodore von Kármán (1881–1963)

at the University, Heisenberg did original work in both classical and atomic physics. He continued to work in both fields. *Stability and Turbulence in the Flow of Liquids*, his doctoral dissertation, became a work of lasting value. But in the oral Ph.D. exam, taken in the summer of 1923 when Heisenberg was twenty-one, he did less well. With all his research work he had neglected experimental physics and therefore annoyed Wien who was the examiner in that subject. Nevertheless, Heisenberg was now *Herr Doktor* and went to spend a year as assistant to Born in Göttingen.

Heisenberg already knew Göttingen quite well. In the summer of 1922 he had been taken along by Sommerfeld to a series of lectures which Bohr gave there and which later were referred to as the *Bohr festival*. Also, he had spent the winter semester 1922/23 in Göttingen since Sommerfeld then was in the United States and had suggested that some of his students should have contact with Born during his absence. There is an often-quoted passage in Born's memoirs [4] about the Heisenberg of that time: 'He looked like a simple peasant boy, with short, fair hair, clear bright eyes and a charming expression ... His incredible quickness and acuteness of apprehension has always enabled him to do a colossal amount of work without much effort: he finished his hydrodynamical thesis, worked on atomic problems partly alone, partly in collaboration with me, and helped me to direct my research students.' Born was convinced that a fundamentally new approach was needed in quantum theory. The Bohr–Sommerfeld model of the atom was based on *classical mechanics*, which described the electron orbiting around the nucleus, with more or less artificial *quantum conditions* added to it. He called for a new *quantum mechanics*, in fact he coined this term which he used as title of a programmatic paper in 1924, a paper in which he quotes Heisenberg's work on the Zeeman effect and work that Born and Heisenberg had done in which they adapted the venerable *perturbation theory* of the astronomers (where the motion of a planet around the sun is perturbed by the presence of another planet) to the quantum physics of an atom with more than one electron.

The scientific atmosphere in Göttingen was even more stimulating than the one in Sommerfeld's group. There were excellent relations to the experimental group of Franck, who was a personal friend of Born, and there were also close connections to the famous Göttingen school of mathematics with its towering figure Hilbert[4]. Moreover, there were visitors from many countries, among them, at that time, Blackett and Fermi. Born also stayed in close touch with Einstein in Berlin and Bohr in Copenhagen and helped to arrange for a stay of Heisenberg in Bohr's group for eight months starting in September 1924.

In Copenhagen, where Heisenberg worked closely with Bohr and his senior assistant Kramers, he found out that Bohr's attitude to physics was much less mathematically inspired than he was used to from Munich and Göttingen. Bohr let himself be guided in his thinking by rather general principles. One was his own *correspondence principle* which stated that any quantum theory, in a suitably chosen limit, should approach the theory of classical physics. Heisenberg would later say [5]: 'From Sommerfeld I learnt optimism, in Göttingen mathematics and from Bohr physics.' In work he was doing with Kramers, the idea of *virtual oscillators*, which had appeared in the abandoned *BKS theory* (Episode

---

[4] David Hilbert (1862–1943)

Box I. Heisenberg's Sharpening of the Correspondence Principle

In the transition between two stationary states with energies E_m, E_n a light quantum appears with circular frequency

$$\omega_{mn} = \omega_m - \omega_n = \frac{E_m - E_n}{\hbar} \quad , \quad \hbar = \frac{h}{2\pi} \quad .$$

For $m \to \infty$ the circular frequency $\omega_{m,m-1}$ coincides with the classical circular frequency ω of orbital motion. This motion can most easily be described by a complex position coordinate x depending on the time t,

$$x(t) = a e^{i\omega t} \quad .$$

Heisenberg wanted to extend the correspondence between classical and quantum-mechanical description also to small values of m and n. He replaced the position coordinate x by the 'ensemble of quantities'

$$x_{mn} = a_{mn} e^{i\omega_{mn} t} = a_{mn} e^{\frac{i}{\hbar}(E_m - E_n)t} \quad .$$

To define a product of two such quantities, he used the 'simplest and most natural assumption'

$$z_{mn} = \sum_{\ell} x_{m\ell} \, y_{\ell n} \quad .$$

---

34) was taken up again. There was one such oscillator for every possible transition in the atom. These were to become the starting point of Heisenberg's quantum mechanics.

Back in Göttingen in the spring of 1925, he thought about *sharpening* the correspondence principle. If the electron in Bohr's model of the hydrogen atom was in a highly excited state, corresponding to a large quantum number, and made a transition to a state with the next lower quantum number, then the frequency of the emitted light was the rotation frequency of the electron on its orbit. This satisfied the correspondence principle. The position of the electron could be written down as a time-dependent quantity containing this frequency. In general, the light frequency depends on the quantum numbers of both the initial and the final state. Heisenberg now wrote down a position function containing the quantum numbers of these two states and, in general, of all possible pairs of states (see Box I). His ingenious idea was now this: In quantum mechanics, not the equations of classical mechanics had to be changed but the quantities and operations appearing in these equations had to be *reinterpreted*. In particular, that meant that the position and also the momentum of a particle became more complicated, like the position he had written down. The equations of classical mechanics, like Newton's and Hamilton's equations (see Box II), should, however, remain valid. Heisenberg worked out this theory by himself and completed his work on the island of Heligoland in the North Sea, where he had fled because of a bad attack of hay fever. After his return he sent the draft of his paper about it to his friend Pauli, who by now worked at the University of Hamburg, for criticism and asked him to return it within a few days.

Box II. Newton's and Hamilton's Equations of Classical Mechanics

The momentum p of a particle with mass m is $p = m\dot{x}$, where $\dot{x}$ is the time derivative of the position x, i.e., the velocity. According to Newton's law of motion the time derivative of p is equal to the force F,

$$\dot{p} = F \quad , \qquad p = m\dot{x} \quad .$$

The force is the negative spatial derivative $F = -\mathrm{d}V(x)/\mathrm{d}x$ of the potential energy (or simply potential) $V(x)$. We can write for the kinetic, potential, and total energies, which we abbreviate as T, V, and H, respectively,

$$E_{\text{kin}} = T = p^2/2m \quad , \qquad E_{\text{pot}} = V(x) \quad , \qquad E_{\text{tot}} = H(p,x) = T + V \quad .$$

Writing down the derivatives of the Hamiltonian (function) H, one obtains Hamilton's equations,

$$\frac{\partial H}{\partial p} = \dot{x} \quad , \qquad \frac{\partial H}{\partial x} = -\dot{p} \quad .$$

They are equivalent to the two equations $\dot{p} = F, p = m\dot{x}$ with which we began.

Heisenberg's paper has the title *Quantum-Theoretical Reinterpretation of Kinematic and Mechanical Relations*. Its abstract reads:

The present paper seeks to establish a basis for theoretical quantum mechanics founded exclusively on relationships between quantities which in principle are observable.

In the paper [6] Heisenberg points out that on the atomic level the orbits of electrons and their period of revolution are not measurable but that theory should be based only on quantities that can, at least in principle, be experimentally observed. He went on to replace the usual position x of a point-like particle by an *ensemble of quantities* x_{mn} (we use a modernized notation) and proposed a rule for the multiplication of such *ensembles*. He postulated that Newton's law of motion remains valid if the new quantities are used and proceeded to show that this was indeed the case for the motion of an anharmonic or harmonic oscillator (i.e., a pendulum) moving in one dimension only.

After Pauli's approval Heisenberg gave the paper to Born. He asked him to study it and, in the case he agreed, to forward it to a journal for publication. He himself would have to travel to Cambridge, where he was invited to talk about his work on the Zeeman effect. (We shall see in Episode 38 how this journey triggered a quite different approach to quantum mechanics.) Only some days later Born studied the paper. He was impressed, sent the paper off for publication and began to think about more formal aspects of Heisenberg's approach. The multiplication rule Heisenberg had used for his *ensembles* seemed vaguely familiar to him and then he realized that this was the rule of matrix multiplication. The ensembles could be taken as *matrices* which were well studied by mathematicians. It was well known that, in general, matrix multiplication is not commutative, i.e., the result of a product depends on the order in which the factors are written. For the matrices x of position and p of momentum this means, that the elements of the matrix px are not necessarily equal to those of the matrix xp. Born conjectured that the *commutator* px − xp was equal to

Born in Göttingen, 1932.

Box III. Matrices of Position and Momentum. Heisenberg Equations

The position becomes a matrix x with the elements x_{mn} and its time derivative a matrix ẋ with the elements

$$\dot{x}_{mn} = \frac{i}{\hbar}(E_m - E_n)x_{mn} \quad .$$

For the time derivative of an arbitrary matrix g one obtains

$$\dot{g} = \frac{i}{\hbar}(Hg - gH)$$

for every Hamiltonian H = H(p, x), if, in addition, the *commutation relation*

$$px - xp = \frac{\hbar}{i}1$$

is satisfied. Here 1 is the unit matrix with the elements $(1)_{mn} = \delta_{mn}$. By inserting matrices for position and momentum and by using the above rule for the time derivative, the Hamilton equations turn into the *Heisenberg equations*

$$\frac{\partial H}{\partial p} = \dot{x} = \frac{i}{\hbar}(Hx - xH) \quad , \quad \frac{\partial H}{\partial x} = -\dot{p} = \frac{i}{\hbar}(pH - Hp) \quad .$$

Jordan

$(\hbar/i)$ times the unit matrix 1 although he could show that property only for the diagonal elements of the commutator. For the general proof he asked the help of Jordan who found it within two days.

Jordan[5] was born in Hanover as son of a painter. Both his parents were very interested in nature and science. Much like Pauli, already as a schoolboy, Jordan studied intensively higher mathematics and contemporary physics, including relativity. In 1921 he began his university studies at the Technische Hochschule in Hanover; a year later he transferred to Göttingen where he took his Ph.D. under Born in 1924.

Born and Jordan began to work out quantum mechanics in matrix notation on the basis of Heisenberg's ideas. Born, who felt very tired, went for summer holidays in August and left the actual writing of their paper *On Quantum Mechanics* [7] to Jordan.

Heisenberg, who had again gone to Copenhagen for a while, Jordan, and Born now began a collaboration – mostly by correspondence which is in part reproduced in [8] – in which they worked out a comprehensive exposition of quantum mechanics. Their publication, *On Quantum Mechanics II* [9], is usually referred to as the *three-men paper* (Drei-Männer-Arbeit). It contains the fundamental assumptions of the theory, i.e., the reinterpretation of physical quantities as matrices with their special multiplication laws or *commutation relations*, and Hamilton's equations written down for these quantities (now called *Heisenberg equations*), see Box III. Moreover, it presents a systematic way for the solution of these equations, and there is a discussion of perturbation theory and of several examples. The three men were together in Göttingen for only about two weeks before Born, who had written the mathematical part,

---

[5] Pascual Jordan (1902–1980)

departed for the United States and left the final editing to Heisenberg and Jordan, both twenty-three years old at the time. The paper was completed in mid-November. Its last section carries the title *Coupled Harmonics Oscillators. Statistics of Wave Fields.* It was written by Jordan alone and practically no notice was taken of it at that time. Now it is recognized as the first description of the electromagnetic field in terms of quantum mechanics and thus as the very first step towards quantum electrodynamics, a subject we shall discuss only much later in Episode 71.

Meanwhile Pauli in Hamburg, kept informed on the developments by Heisenberg through letters, joined in the effort and showed in a tedious calculation that quantum mechanics applied to the hydrogen atom, i.e., one electron in the electric field of a proton, indeed yielded the right energy levels [10]. In this important paper, Pauli did not only reproduce the energy levels which had been obtained by Bohr and Sommerfeld in the framework of the old quantum theory, that ingenious mixture of classical physics and *quantum conditions*. He also got new results differing from those of the Bohr–Sommerfeld model. We remember from Episode 27 that in addition to Bohr's principal quantum number n, which is used to enumerate the energy levels $E_1, E_2, \ldots$, Sommerfeld had introduced two quantum numbers – we called them n_φ and n_ψ – which described the angular momentum of the electron on its orbit and its component with respect to a given axis. For a given value of n the quantum number n_φ could take the values $1, 2, \ldots, n$. The value $n_\varphi = 0$ had to be excluded. Likewise, for a given value of n_φ the quantum number n_ψ could take the values $-n_\varphi, -n_\varphi + 1, \ldots, n_\varphi$. But again one value had to be excluded, $n_\psi = 0$. Pauli found that in quantum mechanics n_φ has to be replaced by the *angular-momentum quantum number* ℓ, which takes the values $\ell = 0, 1, \ldots, n-1$, and n_ψ has to be replaced by a quantum number m (sometimes called *magnetic quantum number*) with the values $m = -\ell, -\ell + 1, \ldots, \ell$. No artificial exclusion is necessary. It is particularly interesting that in the ground state the electron in the hydrogen atom has angular momentum zero: This state certainly cannot be pictured any more by an electron orbiting the nucleus. At the end of his paper Pauli mentions the spin hypothesis of Goudsmit and Uhlenbeck (Episode 36) and suggests to perform a Stern–Gerlach experiment with hydrogen atoms to measure the electron spin. In fact, that measurement had already been done in the original experiment of Stern and Gerlach (Episode 30), because the silver atom also has zero orbital angular momentum. Its angular momentum is only that of the spin of the outer electron. This, in a magnetic field, has two possible orientations. Pauli's results convinced most physicists that the new mechanics involving such strange objects as matrices had to be right.

[1] Cassidy, D. C., *Uncertainty – The Life and Science of Werner Heisenberg.* Freeman, New York, 1992.

[2] Mehra, J. and Rechenberg, H., *The Historical Development of Quantum Theory, Vol. 2. – The Discovery of Quantum Mechanics 1925.* Springer, New York, 1982.

[3] Heisenberg, W., *Der Teil und das Ganze. Gespräche im Umkreis der Atomphysik.* Piper, München, 1969. Engl. transl. [11].

[4] Born, M., *My Life: Recollections of a Nobel Laureate.* Scribner's, New York, 1978.

[5] Hermann, A., *Werner Heisenberg.* Rowohlt, Reinbek, 1976.

[6] Heisenberg, W., *Zeitschrift für Physik*, **33** (1925) 879. Also in [12], Vol. A I, p. 382, Engl. transl. in [8], p. 261.

[7] Born, M. and Jordan, P., *Zeitschrift für Physik*, **34** (1925) 858. Also in [13], Vol. 2, 124, Engl. transl. in [8], p. 321.

[8] van der Waerden, B. L., *Sources of Quantum Mechanics.* Dover, New York, 1968.

[9] Born, M., Heisenberg, W., and Jordan, P., *Zeitschrift für Physik*, **35** (1926) 557. Also in [12], Vol. A I, p. 397 and in [13], Vol. 2, p. 155, Engl. transl. in [8], p. 321.

[10] Pauli, W., *Zeitschrift für Physik*, **36** (1926) 336. Also in [14], Vol. 2, p. 252, Engl. transl. in [8], p. 387.

[11] Heisenberg, W., *Physics and Beyond. Encounters and Conversations.* Harper and Row, New York, 1971.

[12] Heisenberg, W., *Collected Works, Series A, 3 vols.* Springer, Berlin, 1985–1993.

[13] Born, M., *Ausgewählte Abhandlungen, 2 vols.* Vandenhoeck & Ruprecht, Göttingen, 1963.

[14] Pauli, W., *Collected Papers, 2 vols.* Interscience, New York, 1964.

38

Dirac's Mechanics of q Numbers (1925)

Dirac (left) with Heisenberg in Brussels, 1933.

The celebrated *three-men paper*, which gave the first extensive account of the new quantum mechanics, started by Heisenberg in the summer of 1925, appeared in print in February 1926 (see Episode 37). On its first page it contains a *note added in proof*: 'In a paper by P. Dirac [1] which has appeared in the meantime independently some of the results [in the Born–Jordan paper and in the three-men paper] and new conclusions drawn from the theory have been given.' How did it happen that in the short time available an unknown outsider could have done independent and new work?

We remember that after completion of his paper Heisenberg had left Göttingen to present a talk in Cambridge. That talk was entitled *Term Zoology and Zeeman Botany*. Only in a private discussion with the theoretician Fowler[1] did Heisenberg mention his recent work on quantum mechanics which was as yet unpublished. Fowler asked him to send him page proofs as soon as he got them from the printer's. It was customary at the time for an author to ask for more copies of the proofs than were needed for corrections in order to send them to interested colleagues to read even before the paper appeared in print. Heisenberg in due time complied and Fowler gave the proofs to his student Dirac, who thus became one of the first to learn of Heisenberg's original ideas.

Dirac[2] [2,3] was even younger than Heisenberg. He was born in Bristol as son of a Swiss citizen who was a teacher of French there. His biographers invariably mention his taciturnity. Dirac himself explained that his father would require the children to speak correct French at the dinner table and that he himself had preferred to say nothing rather than to make mistakes. He excelled at school and from 1918 to 1921 studied electrical engineering in Bristol. Unable to find a job in his profession, he continued to study mathematics and in the final examination in 1923 came out as the best student. After this he was accepted by St. John's College, Cambridge, as postgraduate student in mathematics and Fowler became his research supervisor.

Dirac, at first, could not make much of Heisenberg's paper and put it aside for some time. When he came back to it he found the multiplication rule which Heisenberg had postulated for his quantities the most interesting part. The rule implies that a product of two quantities x and y depends on the order of the factors and Dirac began to study the *commutator* $(xy - yx)$ which, in general, is not equal to zero. He vaguely recalled that in a particular formulation of classical mechanics a difference of two products, called a *Poisson bracket*, appears, but could not remember exactly what is was. Since it was a Sunday and the libraries in Cambridge were closed, he had to await the next morning.

---

[1] (Sir) Ralph Howard Fowler (1889–1944)   [2] Paul Adrien Maurice Dirac (1902–1984), Nobel Prize 1933

Box I. Poisson Brackets and Classical Mechanics

For a single particle with position x and momentum p the *Poisson bracket* of two variables u and v is defined as

$$\{u, v\} = \frac{\partial u}{\partial x}\frac{\partial v}{\partial p} - \frac{\partial v}{\partial x}\frac{\partial u}{\partial p} \quad .$$

For particular arguments one has

$$\{x, x\} = 0 = \{p, p\} \quad , \qquad \{x, p\} = 1 \quad .$$

Hamilton's equations (see Box II in Episode 37) take the form

$$\dot{x} = \{x, H\} \quad , \qquad \dot{p} = \{p, H\} \quad .$$

In general, for any function $F = F(x, p)$ one gets the equation of motion

$$\dot{F} = \{F, H\} \quad .$$

Indeed, around 1840 Jacobi had given a very general equation of motion containing a difference of two products of derivatives (see Box I). When he found out that this difference had been used already in 1809 by Poisson[3], he called it a Poisson bracket. Dirac now saw his way clearly before him. He had to establish the *algebra*, i.e., the set of rules to use in calculations, for the quantum-mechanical variables. Moreover, he had to find out which quantum-mechanical expression corresponds to the Poisson bracket. Then he would be able to make full use of Jacobi's theory. Dirac's program was in some respect similar to what Born, Jordan, and Heisenberg did in Göttingen. But it was more general. He did not limit himself to considering matrices but his quantum-mechanical variables were any objects which satisfied his algebra. In his second paper on the subject [4] he called these variables *q numbers* as opposed to classical variables or *c numbers*.

In the introduction to his first paper, entitled *The Fundamental Equations of Quantum Mechanics* [1], Dirac writes:

> In a recent paper [5] Heisenberg puts forward a new theory, which suggests that it is not the equations of classical mechanics that are in any way at fault but that the mathematical operations by which physical results are deduced from them require modification. All the information supplied by the classical theory can thus be made use of in the new theory.

He goes on to develop the algebra of his quantum variables and the rules to differentiate them. Independently of Born, Jordan, and Heisenberg he presents the *commutation relations* of position x and momentum p, which he calls his *quantum conditions*. Relating the quantum-mechanical commutator to the classical Poisson bracket he then derives fundamental equations (see Box II), which are the *Heisenberg equations* in a somewhat more general form. Dirac completed his paper early in November 1925 and gave it to Fowler, who, realizing its importance, arranged for shortcuts in the publishing process.

Dirac in 1938

---

[3] Siméon Denis Poisson (1781–1840)

Box II. Dirac's Fundamental Equations of Quantum Mechanics

We denote Dirac's quantum-mechanical variables or q numbers by a special type, e.g., v, and the corresponding classical variable by normal type (v). The *commutator* or *Dirac bracket* of two q numbers is

$$[\mathsf{u},\mathsf{v}] = \mathsf{uv} - \mathsf{vu} \quad .$$

Dirac postulated that in quantum mechanics a Poisson bracket is replaced by the corresponding commutator multiplied by $ih/2\pi$,

$$\{u,v\} \to (ih/2\pi)[\mathsf{u},\mathsf{v}] \quad .$$

Thus, in quantum mechanics the equation of motion for a quantity F becomes

$$(ih/2\pi)\dot{\mathsf{F}} = [\mathsf{F},\mathsf{H}] \quad .$$

---

[1] Dirac, P. A. M., *Proceedings of the Royal Society*, **A109** (1925) 642.

[2] Kragh, H., *Dirac – A Scientific Biography*. Cambridge University Press, Cambridge, 1990.

[3] Mehra, J. and Rechenberg, H., *The Historical Development of Quantum Theory, Vol. 4, Part 1 – The Fundamental Equations of Quantum Mechanics 1925–1926*. Springer, New York, 1982.

[4] Dirac, P. A. M., *Proceedings of the Royal Society*, **A110** (1926) 561. Reprinted in [9], p. 307.

[5] Heisenberg, W., *Zeitschrift für Physik*, **33** (1925) 879. Also in [10], Vol. A I, p. 382, Engl. transl. in [9], p. 261.

[6] Dirac, P. A. M., *Proceedings of the Royal Society*, **A113** (1926) 611.

[7] Dirac, P. A. M., *The Principles of Quantum Mechanics*. Oxford University Press, Oxford, 1st ed., 1930.

[8] Polkinghorne, J. C., *Dirac and the Interpretation of Quantum Mechanics*. In Taylor, J. G. (ed.): *Tributes to Paul Dirac*. Adam Hilger, Bristol, 1987.

[9] van der Waerden, B. L., *Sources of Quantum Mechanics*. Dover, New York, 1968.

[10] Heisenberg, W., *Collected Works, Series A, 3 vols*. Springer, Berlin, 1985–1993.

At the end of January 1926, Dirac completed his second paper [4] containing the theory in a more formalized way but also its application to the hydrogen atom yielding the Balmer spectrum. The same computation for the framework of matrix mechanics had been completed by Pauli only days earlier. In May Dirac obtained his Ph.D. with the first thesis on quantum mechanics.

In December 1926, Dirac published a paper on what he called *transformation theory* [6]. The work on it was mainly done while Dirac was in Bohr's institute in Copenhagen. At that time Schrödinger's wave mechanics (Episode 39) was not yet a year old. Transformation theory is a presentation of quantum mechanics in its most general form, from which Heisenberg's matrix mechanics and Schrödinger's wave mechanics are obtained as particular representations. The theory is greatly simplified by the introduction of *Dirac's δ function*, a curious function (now called a *distribution* by mathematicians) which is zero everywhere except at one point, where it becomes infinite in a well-defined way. It allows to deal with discrete and continuous variables on the same footing.

In 1927 Dirac was elected Fellow of St. John's College, Cambridge. He went to live in College and also to do most of his work there. In 1930 he published his beautiful book *The Principles of Quantum Mechanics* [7], still highly readable, written in typical Dirac style with not a word too much. For the third edition, published in 1947, he introduced his notation of *bra vectors* $\langle\alpha|$ and *ket vectors* $|\beta\rangle$ which are both q numbers and which, if multiplied in the order $\langle\alpha|\beta\rangle$, forming a *bracket*, yield a c number. But if they are multiplied the other way round, the product $|\beta\rangle\langle\alpha|$ is something completely different, an operator. One of Dirac's former students wrote about the lectures based on the book [8]: 'Dirac was scrupulous in not underlining his own formidable contributions to quantum theory. However, one did gain the impression from a slight smile that played around his features when he introduced bras and kets that this invention (and the harmless joke enshrined in the nomenclature) had given him great satisfaction.'

Schrödinger Creates Wave Mechanics (1926)

The breakthrough to quantum mechanics was, as we have seen in Episode 37, achieved by Heisenberg in the summer of 1925 and the theory, in the form now known as *matrix mechanics*, was completed a little later by Born, Heisenberg, and Jordan. From his first days as a student Heisenberg had belonged to groups which worked on quantum theory. He was, so to speak, totally immersed in the field and reached his solution at the age of twenty-three. At the turn of the year 1925/26 quantum mechanics was created again, now clad as *wave mechanics* [1], by Schrödinger, who did not belong to an active research group in the field and who was then thirty-eight. As soon as it appeared wave mechanics was popular with physicists, because it was based on familiar differential equations and not on matrix formalism which was practically unknown.

Schrödinger[1] [2,3] grew up in Vienna. His grandfather Alexander Bauer was professor of chemistry and married to an Englishwoman. His father Rudolf Schrödinger also was a chemist. He married one of Bauer's daughters and later ran a small inherited factory. Schrödinger's mother and his aunt made sure that he spoke English as well as German. He attended the *Akademisches Gymnasium*, where only classical languages were taught, and always was top of his class, excelling particularly in mathematics and physics. In 1906 he began to study physics at the University of Vienna, where his teachers were Exner in experimental and Hasenöhrl[2] in theoretical physics. He was attracted to theory by Hasenöhrl's course extending over four years with five lectures each week. This was probably the most extensive course on the subject offered anywhere at that time. Schrödinger would always speak of Hasenöhrl, who fell in 1915, with affection and admiration. Nevertheless, he obtained his doctor's degree with experimental work and, after a year of military service, became assistant to Exner. In 1914 he became Privatdozent. For the whole of the First World War he served as officer in the Austrian Army, three years in the artillery and at the end in the meteorological service. The dismantlement of the Austro-Hungarian Empire cost him a professorship he had been promised in Czernowitz because that town fell to Romania (it later became part of the Soviet Union and now belongs to Ukraine). Therefore he returned to the Physics Institute in Vienna and did research in both experiment and theory. In 1920 he married, and the young couple left for Germany. Schrödinger was first assistant at the University of Jena, and later in the same year associate professor in Stuttgart and then in Breslau.

In 1921 Schrödinger followed a call to the University of Zurich as full professor of theoretical physics. Here he soon established good contacts with De-

Schrödinger

---

[1] Erwin Schrödinger (1887–1961), Nobel Prize 1933   [2] Fritz Hasenöhrl (1874–1915)

Box I. Wave Equations in Classical Physics

A classical harmonic wave can be described as a function of the space coordinate x and the time t in the form

$$\Psi(x,t) = \Psi_0 e^{-i(\omega t - kx)} \quad , \qquad k = 2\pi/\lambda \quad , \qquad \omega = 2\pi\nu \quad ,$$

see Box I in Episode 32. Here Ψ_0 is the *amplitude*, ω the *circular frequency*, k the *wave number*, ν the *frequency*, and λ the wavelength. The second derivatives with respect to x and t are

$$\frac{\partial^2}{\partial x^2}\Psi = -k^2\Psi \quad , \qquad \frac{\partial^2}{\partial t^2}\Psi = -\omega^2\Psi \quad .$$

By comparing the two expressions one obtains the *time-dependent wave equation*

$$\frac{\partial^2}{\partial x^2}\Psi - \frac{1}{v_{\mathrm{p}}^2}\frac{\partial^2}{\partial t^2}\Psi = 0 \quad , \qquad v_{\mathrm{p}}^2 = \frac{\omega^2}{k^2} \quad ,$$

where v_{p} is the phase velocity. Eliminating the time dependence by replacing $\partial^2\Psi/\partial t^2$ with $-\omega^2\Psi$ one gets the *stationary wave equation*

$$\frac{\partial^2}{\partial x^2}\Psi + k^2\Psi = 0 \quad .$$

Box II. Schrödinger Equation

Using, as de Broglie had done, the relations

$$E = h\nu = \hbar\omega \quad , \qquad p = h/\lambda = \hbar k \quad ,$$

the energy E and the momentum p of a particle can be introduced in the expression for a wave and in the wave equations. For a free particle one has relativistically

$$E_{\mathrm{rel}} = mc^2\gamma \quad , \qquad p = \gamma m v \quad , \qquad \gamma = 1/\sqrt{1 - v^2/c^2}$$

and non-relativistically

$$E = E_{\mathrm{kin}} = p^2/2m \quad , \qquad p = mv \quad .$$

If the particle moves non-relativistically in a potential V, the kinetic energy is

$$E_{\mathrm{kin}} = E - V = p^2/2m \quad \text{and therefore} \quad k^2 = 2m(E-V)/\hbar^2 \quad .$$

By inserting the latter expression in the classical stationary wave equation one obtains the *stationary Schrödinger equation*

$$\frac{\hbar^2}{2m}\frac{\partial^2}{\partial x^2}\Psi + (E-V)\Psi = 0 \quad ,$$

where the *wave function* Ψ depends on x only. The *time-dependent Schrödinger equation* reads

$$i\hbar\frac{\partial}{\partial t}\Psi = \left[-\frac{\hbar^2}{2m}\frac{\partial^2}{\partial x^2} + V\right]\Psi \quad .$$

In this equation Ψ depends on both x and t.

bye, his colleague in experimental physics and with the mathematician Weyl[3]. In 1925 he became aware of de Broglie's work (Episode 32) and, following a suggestion of Debye, in November gave a talk about it in the joint physics colloquium of the University and the ETH. He mentioned de Broglie's idea of closed matter waves for the electron on a Bohr orbit. Half a century later Felix Bloch, who was present as a student, wrote about Debye's reaction [4]: "Debye casually remarked that he thought this way of talking was rather childish. As a student of Sommerfeld he had learned that, to deal properly with waves, one had to have a wave equation ... Just a few weeks later he [Schrödinger] gave another talk in the colloquium which he started by saying: 'My colleague Debye suggested that one should have a wave equation, well I have found one!'"

Schrödinger found his equation during a holiday over Christmas and the New Year in the Swiss mountain resort Arosa. He started out from the conventional wave equation of a classical wave as, for instance, a sound wave (see Box I). Essential quantities in such an equation are the *angular frequency* and the *wave number*. Through de Broglie's work, based on Einstein's special relativity, they are related to the *energy* and *momentum* of a particle, respectively. By inserting these relations Schrödinger obtained a *relativistic wave equation* for matter waves. He tried immediately to use it to compute the spectrum of the hydrogen atom. Since his equation was based on relativity theory he had good reasons to expect that he would find the spectrum including its fine structure as computed by Sommerfeld (Episode 27). He wrote down his relativistic equation in three space dimensions. For the potential energy he inserted the potential of the electric attraction which the atomic nucleus exerts on the electron in the hydrogen atom. Because of the symmetry of that potential the wave equation is best written in polar coordinates and can then be separated into three differential equations, each depending on one coordinate only. These have the mathematical form of *eigenvalue equations*, i.e., there are solutions only for certain characteristic patterns of the wave, each pattern corresponding to a particular energy of the electron. There is a similarity to the patterns of standing waves on a string of a musical instrument corresponding to different tones.

After his return to Zurich, Schrödinger asked his colleague Weyl for advice concerning the solutions of his equations. These are now known to most physics students but then were hidden away in the mathematical literature where Weyl found them. To Schrödinger's disappointment he could not reproduce Sommerfeld's result. His relativistic formula failed. But Schrödinger did not give up. He used the non-relativistic expressions for energy and momentum to obtain a new wave equation, now called the *stationary Schrödinger equation* (see Box II). Applying it to the hydrogen atom he got the same energy levels for the electron as in the Bohr model. Still in January 1926 he submitted his first paper on the subject, entitled *Quantization as Eigenvalue Problem (First Communication)* for publication [6]. In the same year he completed another five papers which essentially contain *wave mechanics* as we now know it. In his first paper he did not indicate the way in which he arrived at the equa-

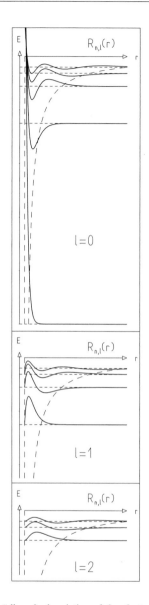

Schrödinger's description of the electron in the hydrogen atom. Shown are the Coulomb potential of the nucleus (long-dashed line), the five lowest energy levels or eigenvalues (short-dashed lines), and the radial part $R_{n,\ell}$ of the wave function for three values of the angular-momentum quantum number ℓ. The eigenvalues are independent of ℓ, except that for $\ell = 1$ the lowest eigenvalue is missing, for $\ell = 2$ the lowest two, etc. The wave functions, however, are different for different values of ℓ. From [5], Fig. 13.15. © 2001 by Springer, New York. With kind permission from Springer Science+Business Media.

---

[3] Hermann Weyl (1885–1955)

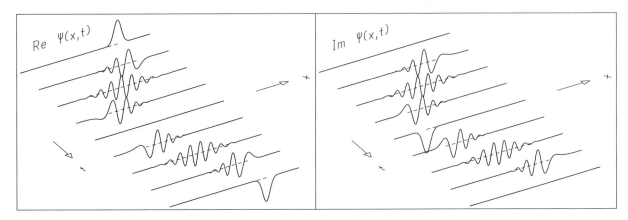

Wave function of a particle in a harmonic-oscillator potential for several consecutive times. Schrödinger noted that the 'wave packet stays together'. Shown are the real part Re Ψ and the imaginary part Im Ψ of the complex wave function Ψ.

tion but gave other grounds to motivate it. In the *Second Communication* [7] he used a much more elegant way showing that it is indeed possible, as de Broglie had suggested, to construct a relation between wave mechanics and Hamilton's mechanics of a mass point, which is similar to the relation between wave optics and geometrical optics. In the *Third Communication* [8] he incorporated perturbation theory into his formalism and in the *Fourth Communication* [9] he presented the *time-dependent Schrödinger equation*.

Box III. Operators and Commutation; Eigenvalue Equation

In Heisenberg's formulation of quantum mechanics physical quantities are expressed by matrices. Schrödinger, in general, uses differential operators which act on the wave function. In particular, position x and momentum p are replaced as follows:

$$x \rightarrow x \quad , \qquad p \rightarrow \frac{\hbar}{i} \frac{\partial}{\partial x} \quad .$$

With this replacement one easily arrives at the commutation relations of Heisenberg,

$$px - xp \rightarrow \frac{\hbar}{i} \left(\frac{\partial}{\partial x} x - x \frac{\partial}{\partial x} \right) = \frac{\hbar}{i} \left(1 + x \frac{\partial}{\partial x} - x \frac{\partial}{\partial x} \right) = \frac{\hbar}{i} \quad ,$$

simply by using the rules of differential calculus.

The operator of total energy $p^2/2m + V$, called *Hamiltonian operator* or simply *Hamiltonian*, is

$$H = -\frac{\hbar^2}{2m} \frac{\partial^2}{\partial x^2} + V \quad .$$

With it the *stationary Schrödinger equation* takes the simple form

$$H\Psi = E\Psi \quad .$$

It is an *eigenvalue equation*: Application of the operator H to a solution (or *eigenfunction*) Ψ of the equation simply reproduces Ψ, multiplied by a constant, the *eigenvalue E*. With H the *time-dependent Schrödinger equation* becomes

$$i\hbar \frac{\partial}{\partial t} \Psi = H\Psi \quad .$$

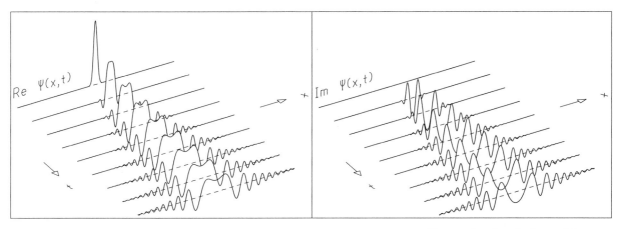

Re $\psi(x,t)$ Im $\psi(x,t)$

The wave function of a free particle at rest, well localized at some initial time, disperses indefinitely as time proceeds.

Schrödinger published two other important papers in 1926. One of them, written already in March, he called *On the Relation of the Heisenberg–Born–Jordan Quantum Mechanics to Mine* [10]. In a footnote he stresses that his own approach was in no way influenced by the Göttingen work and he continues: 'Of course, I knew of his [Heisenberg's] theory but by the methods of transcendental algebra, which seemed very difficult to me and by the lack of graphicness [Anschaulichkeit] I felt deterred, not to say repelled.' But then he goes on to show that the two seemingly different theories are mathematically equivalent to each other. The key to the equivalence is that Schrödinger used differential operators where Heisenberg used matrices and that, of course, the order matters in which operators are written in front of the function on which they act (see Box III).

Indeed, wave mechanics was easier to picture than matrix mechanics. In particular, Schrödinger's *wave function* is a function of space and time and not a matrix with infinitely many elements. But what exactly was its meaning? Schrödinger himself tried to answer this question in a paper entitled *The Continuous Transition from Micro- to Macro-Mechanics* [11]. He considered a particle oscillating about an equilibrium point along a fixed line. The corresponding wave function is a wave group or *wave packet* oscillating along that line. The wave function is essentially zero except in a small region. Schrödinger thought that in quantum mechanics the particle is no longer point-like

Entries for the winter semester 1931/32 by my mother, Maria Brandt, then a student at the University of Berlin, in her *Studienbuch*. Students had to list their courses and professors had to testify the attendance by their signatures at the beginning and the end of the semester. Schrödinger signed for his course *Feldphysik II* and accompanying exercises.

[1]   Mehra, J. and Rechenberg, H.,
      *The Historical Development of
      Quantum Theory, Vol. 5, Part 1 –
      Erwin Schrödinger and the Rise of
      Wave Mechanics*. Springer, New
      York, 1987.

[2]   Schrödinger, E., *Meine Weltan-
      sicht*, chap. Mein Leben. Zsolnay,
      Vienna, 1961.

[3]   Moore, W., *Schrödinger – Life
      and Thought*. University Press,
      Cambridge, 1989.

[4]   Bloch, F., *Physics Today*, **Decem-
      ber** (1976) 23.

[5]   Brandt, S. and Dahmen, H. D.,
      *The Picture Book of Qauntum
      Mechanics, 3rd ed.* Springer, New
      York, 2001.

[6]   Schrödinger, E., *Annalen der
      Physik*, **79** (1926) 361. Also
      in [12], Vol. 3, p. 82. Engl.
      transl. [13], p. 1.

[7]   Schrödinger, E., *Annalen der
      Physik*, **79** (1926) 489. Also
      in [12], Vol. 3, p. 98. Engl.
      transl. [13], p. 13.

[8]   Schrödinger, E., *Annalen der
      Physik*, **80** (1926) 437. Also
      in [12], Vol. 3, p. 166. Engl.
      transl. [13], p. 62.

[9]   Schrödinger, E., *Annalen der
      Physik*, **81** (1926) 109. Also
      in [12], Vol. 3, p. 220. Engl.
      transl. [13], p. 102.

[10] Schrödinger, E., *Annalen der
      Physik*, **79** (1926) 734. Also
      in [12], Vol. 3, p. 143. Engl.
      transl. [13], p. 45.

[11] Schrödinger, E., *Naturwissen-
      schaften*, **14** (1926) 664. Also
      in [12], Vol. 3, p. 137. Engl.
      transl. [13], p. 41.

[12] Schrödinger, E., *Gesammelte
      Abhandlungen / Collected Papers,
      4 vols.* Vieweg, Braunschweig,
      1984.

[13] Schrödinger, E., *Collected Papers
      on Wave Mechanics*. Blackie,
      London, 1928.

but somehow smeared out over that region. He stressed that, at least in his example, the wave packet 'stays together', i.e., does not indefinitely widen with time. Alas, this is not so in other cases as, for example, for a free particle. The interpretation of the wave function in terms of probabilities, which became generally accepted, is due to Born (see Episode 40). But Schrödinger himself and also Einstein and de Broglie, while seeing the advantages of this interpretation, disagreed with it.

Wave mechanics, almost overnight, made Schrödinger famous. In 1927 he succeeded Planck in Berlin on Germany's most prestigious chair of theoretical physics. In 1933, although not a victim of the Nazi purges himself, he left Germany and with the help of Lindemann got a position in Oxford. In 1936 he accepted a professorship at the University of Graz in Austria, a decision which he would later call a 'stupidity without example', explaining that it was without example because it ended so luckily. When Austria was annexed by Germany he had already an offer from the Irish President de Valera who wanted to found in Dublin an Institute of Advanced Studies modelled after the one in Princeton. De Valera, a mathematician, greatly appreciated that Schrödinger had 'derived' wave mechanics from the mechanics of Hamilton, the great Irish scientist. From 1940 to 1956 Schrödinger worked at the *Dublin Institute for Advanced Studies*. He then returned to Austria, where a special chair had been created for him at the University of Vienna.

Schrödinger in Alpbach, Tyrol, 1956.

Born's Probability Interpretation of Quantum Mechanics (1926)

Schrödinger's wave mechanics was using the mathematics familiar to all physicists and was therefore preferred by most over matrix mechanics. Moreover, it contained a function Ψ, the *Schrödinger function* or *wave function*, which was expected to be connected in some way to the position of a particle in space. For some time this connection was unclear. Bloch quotes a verse (original and translation) by Hückel[1], recited in the summer of 1926 in Zurich, which describes the situation in a humorous way [1]:

Gar Manches rechnet Erwin schon	Erwin with his psi can do
Mit seiner Wellenfunktion	Calculations quite a few.
Nur wissen möcht' man gerne wohl	But one thing has not been seen:
Was man dabei sich vorstell'n soll.	Just what does psi really mean?

The answer was given by Born and is now considered to be one of the fundamental statements of modern physics. But it was not accepted by all, in particular, not by Schrödinger himself.

We already met Born[2] in several episodes, as director of the Physics Institute in Frankfurt, when the Stern–Gerlach experiment was done there, and as director of the Institute of Theoretical Physics in Göttingen with Pauli and later Heisenberg as his assistants. Born [2,3] was a native of Breslau (now Wrocław, Poland), where his father was Professor of Anatomy. He studied in Breslau, Heidelberg, and Zurich and then went to Göttingen, having heard that it was the 'mecca of German mathematics and that three prophets lived there: Felix Klein, David Hilbert, and Hermann Minkowski'. He enjoyed Hilbert's and Minkowski's lectures but not Klein's, which he found 'too perfect for my taste.' But, having aroused Klein's interest because of a good seminar talk, he could not refuse to work on a prize problem concerning a question from elasticity theory suggested by 'the Great Felix'. This work was accepted as his doctoral dissertation in 1906.

After a short spell of military service with the cavalry in Berlin, he was discharged because of asthma attacks which plagued him since childhood. Born now was a mathematician with interest in physics. Wanting to learn more about that subject, he went to Cambridge for half a year, where he particularly enjoyed J. J. Thomson's lectures with experimental demonstrations. On his return to Breslau, he planned to work experimentally on optics with Lummer and Pringsheim but found he was more interested to do some calculations in relativity theory, recently founded by Einstein. He wrote about his work to Minkowski who promptly invited him to return to Göttingen as his assistant. Alas, Minkowski, who had just given relativity an elegant mathematical

---

[1] Erich Hückel (1896–1980) [2] Max Born (1882–1970), Nobel Prize 1954

form (Episode 11), died within only weeks after Born's arrival. Born for some years lived in the same house as von Kármán and from him learned about Einstein's application of quantum theory to the specific heat problem (Episode 12). Together they improved Einstein's results by taking into account a whole spectrum of vibrations of atoms within a crystal. This work made Born begin an extended research program on the physics of crystals.

In 1915 Born moved as associate professor to Berlin. He and his wife became close friends of Einstein [4]. He was soon drafted and for most of the First World War worked in an artillery research group, but in the spare time with colleagues like Landé could do also some basic research. He became full professor in Frankfurt in 1919 and in 1921 was offered to lead both experimental and theoretical physics in Göttingen. However, he was able to convince the authorities to divide the institute and offer the experimental chair to Franck whom he knew from common student times in Heidelberg.

In Episode 37, we have told how quantum mechanics was created by Heisenberg, how it was given the form now called *matrix mechanics* in Göttingen, and how Born travelled to the United States, leaving most of the actual writing of the *three-men paper* to Jordan and Heisenberg. Born gave lectures, which appeared as the first book on the new theory, at the Massachusetts Institute of Technology (MIT) and at various other places. At MIT, together with Wiener[3], he extended the original matrix mechanics, which could only cope with states of discrete energies such as those of an electron in the hydrogen atom, to states taking a continuum of energy values. As example they took the motion of a force-free particle.

When Born returned to Göttingen in April 1926, Schrödinger had already begun to publish his *wave mechanics* and Born realized that it was well suited to describe *scattering*, i.e., a process in which a particle interacts with some object, like the α particle with the atomic nucleus in *Rutherford scattering* (Episode 16) or the electron with mercury atoms in the *Franck–Hertz experiment* (Episode 25). Born was interested in the subject because in Franck's group, of course, scattering experiments were performed. He described the incoming particles by a wave function in the spirit of a de Broglie wave. This wave function is called a *plane wave* because its wave fronts are planes which are perpendicular to the direction of motion. After the scattering the particle is moving away from the scattering centre. It has to be described by a *spherical wave*. The wave fronts form spheres around the scattering centre which move away from it. It was Born's original idea that this spherical wave or scattered wave, when inspected at a point far away from the scattering centre, carries information about the *probability* of scattering the particle in the direction towards this point. He hastily wrote a short note, entitled *On the Quantum Mechanics of Scattering Processes (Preliminary Communication)* and submitted it to the journal *Die Naturwissenschaften*, where he hoped it would be published fast. But in fact it was rejected 'because of lack of space'. He therefore submitted it at the end of June 1926 to a regular research journal. Born denoted the scattered wave function by $\Phi_{mn}(\alpha, \beta, \gamma)$. His paper contains the sentence [5]: 'If one wants to interpret this result [the scattering

---

[3] Norbert Wiener (1894–1964)

Born (in front) as host of the 1922 'Bohr festival', a very influential series of lectures given by Bohr in Göttingen. Standing are Oseen[4], Bohr, Franck, and Oscar Klein.

wave function] in terms of corpuscles only one way is possible: $\Phi_{mn}(\alpha, \beta, \gamma)$ determines the probability for the scattering [. . .] in the direction determined by α, β, γ.' There is a footnote attached to the word probability which reads: 'Note added in proof: A more precise consideration shows that the probability is proportional to the square of Φ_{mn}.' Neither statement was quite correct. The footnote should have ended 'to the absolute square of Φ_{mn}' because the wave function is a complex function and only its absolute square $|\Phi_{mn}|^2$ will never be negative, as required of a probability. But Born had made his point and he followed it up with a statement pointing out the decisive difference between classical and quantum mechanics:

> Schrödinger's quantum mechanics therefore gives a well-defined answer to the question about the effect of a collision but that is not a causal relation. One gets *no* answer to the question "how is the state after the collision" but only to the question "how probable is a given effect of the collision" . . .

Born is precise and very detailed in his final paper on the subject, completed

---

[4] Carl Wilhelm Oseen (1879–1944)

less than a month later [6]. In the introduction he states that matrix mechanics started out from the assumption that it is impossible in principle to exactly describe a process in space and time. He mentions that, on the other hand, in wave mechanics Schrödinger had tried to construct 'wave groups [now called wave packets] which have small dimensions in all directions and obviously are supposed to represent the moving particle' (see end of Episode 39). Instead, he proposes to try a 'third interpretation' and to test its usefulness on scattering processes. He summarizes his interpretation in the words:

> The motion of a particle follows probability laws but the probability propagates according to the laws of causality.

Let us consider Schrödinger's wave function $\Psi(\mathbf{x})$ describing a particle as function of the position $\mathbf{x}$. Born required it to be normalized, i.e., $\int |\Psi(\mathbf{x})|^2 \, dV = 1$, where the integral is extended over all space. (That can be easily done for bound states. In the example we give for scattering we use a wave packet for which normalization is also simple. Born himself treated scattering in a more formal way.) Born's statement is: The particle is in the small volume dV, which contains the point $\mathbf{x}$, with the probability $dP = |\Psi(\mathbf{x})|^2 \, dV$, i.e., $\rho(\mathbf{x}) = |\Psi(\mathbf{x})|^2$ is a *probability density*, i.e., a probability per cubic centimetre.

We now apply this statement to the wave function of the electron in the hydrogen atom. The probability density is a function of all three space coordinates, but is has rotational symmetry with respect to the z axis. (If we introduce an arbitrarily small magnetic field we can choose its direction to orient the z axis, thought to pass through the nucleus, parallel to it.) It is convenient to introduce spherical coordinates r, ϑ, φ. The probability density ρ then has the same form in any half-plane bounded by the z axis. We show graphics of ρ over such half-planes for states with principal quantum numbers $n = 1, 2, 3$. Bearing in mind the rotational symmetry about the z axis, one sees that the spatial probability density in three dimensions is spherically symmetric for the ground $n = 1$ state and for all exited states, provided that $\ell = 0$. For excited states with the highest possible angular momentum, $\ell = n - 1$, oriented in z direction, i.e., $m = \ell$, the probability is concentrated near $\vartheta = 90°$ and at a distance r from the axis which increases quadratically with n. In three dimensions this region is a torus including a circle of radius r. This torus for very large n, in the spirit of the correspondence principle, becomes one of Bohr's circular orbits. But for the ground state there is no hint of a circle. In its ground state the atom is not flat but round, as it should be!

Born did not consider bound states, because he saw no direct connection to experiments, but studied scattering in one and in three dimensions. If a particle can move only along one direction in can still be scattered if it encounters an obstacle. There are only two possible results: transmission (continuation in the original direction) or reflection. Born studied this problem in a rather general form; we shall give a specific example.

Let us take as obstacle a region where the particle obtains a higher potential energy than elsewhere. Expressed in classical terms, we study a particle on a rail which runs straight in flat country except for some region, where it passes a hill. In classical physics the particle is transmitted if its energy is larger than the potential energy it would have on the hilltop. If the energy is smaller, it is reflected. Born showed that in quantum mechanics, for any given energy,

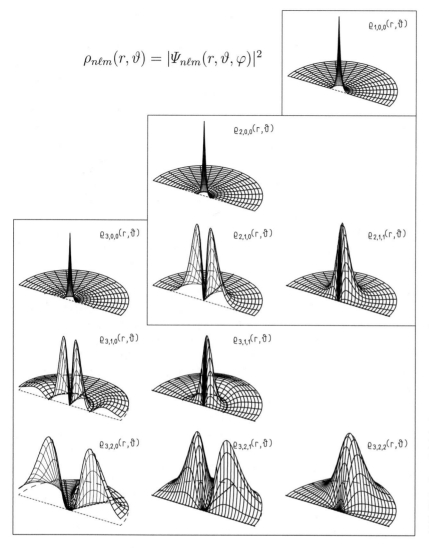

$$\rho_{n\ell m}(r, \vartheta) = |\Psi_{n\ell m}(r, \vartheta, \varphi)|^2$$

The absolute squares $\rho_{n\ell m}(r, \vartheta) = |\Psi_{n\ell m}(r, \vartheta, \phi)|^2$ of the full three-dimensional wave functions for the electron in the hydrogen atom. They are functions only of r and ϑ. All eigenstates having the same principal quantum n number have the same energy eigenvalue E_n. The possible angular-momentum quantum numbers are $\ell = 0, 1, \ldots, n - 1$. Each figure gives the probability density for observing the electron at any point in a half-plane containing the z axis. All pictures have the same scale in r and ϑ. They do, however, have different scale factors in ρ. From [7], Fig. 13.16. © 2001 by Springer, New York. With kind permission from Springer Science+Business Media.

there can be reflection and transmission. But one cannot say which of the two happens in a single scattering process. Only the probabilities for reflection and for transmission (their sum is 100%) can be computed. By the way, the quantum-mechanical phenomenon that a particle is transmitted, which, classically, could not pass the hill, is called the *tunnel effect*. In our illustration a wave packet (shown is the absolute square of the wave function, properly normalized) moves towards the region of high potential energy, called a potential barrier, where it is partly transmitted and partly reflected. It is obvious that the packet does not stay together, as Schrödinger had thought, and can therefore not be taken to indicate the position of the particle. The integrals over its two parts are the probabilities of reflection and of transmission.

Since Newton, physicists had believed in strict determinism. If the state of a system was completely known at a given time, then for any later time it could

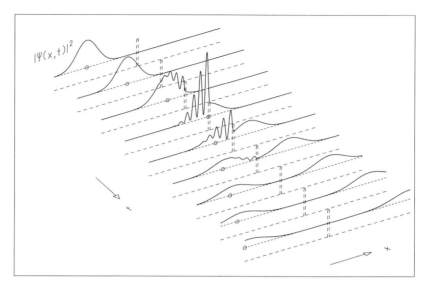

$|\psi(x,t)|^2$

Illustration of the tunnel effect. Shown is the time development of the probability density for a wave packet incident from the left onto a potential barrier. The potential is shown as a long-stroked dashed line; it is zero except for a small region in the middle of the range in x. The mean energy of the wave packet, indicated by the short-stroked horizontal line, is only half the height of the barrier. In classical physics a particle would be reflected at the front of the barrier. Its behaviour would be that of the little circle in the figure. From [7], Fig. 5.6b. © 2001 by Springer, New York. With kind permission from Springer Science+Business Media.

[1]   Bloch, F., *Physics Today*, **December** (1976) 23.

[2]   Born, M., *My Life: Recollections of a Nobel Laureate.* Scribner's, New York, 1978.

[3]   Born, M., *My Life and Views.* Scribner's, New York, 1968.

[4]   Einstein, A., Born, H., and Born, M., *Briefwechsel 1916–1955.* Nymphenburger, München, 1969. Engl. transl. [9].

[5]   Born, M., *Zeitschrift für Physik*, **37** (1926) 863. Also in [10], Vol. 2, p. 228.

[6]   Born, M., *Zeitschrift für Physik*, **38** (1926) 803. Also in [10], Vol. 2, p. 233.

[7]   Brandt, S. and Dahmen, H. D., *The Picture Book of Qauntum Mechanics, 3rd ed.* Springer, New York, 2001.

[8]   Einstein, A., *Verh. d. Dt. Phys. Gesellschaft*, **18** (1916) 318.

[9]   Einstein, A., Born, H., and Born, M., *The Born–Einstein Letters.* Macmillan, London, 1971.

[10]   Born, M., *Ausgewählte Abhandlungen, 2 vols.* Vandenhoeck & Ruprecht, Göttingen, 1963.

be computed with certainty. This certainty was now replaced by a mere probability. It was not the first time for probability to enter modern physics. In his statistical mechanics Boltzmann had used it successfully to derive the laws of heat theory. But in this case it had to be used only because the measurement of positions and velocities of all atoms in a system was considered to be impossible in practice, not in principle. More serious was Rutherford's discovery of the law of radioactive decay, which is a probability law, Episode 8. There was no cause for an atom to decay at a specific time. Causality itself was endangered. But then there was still hope to eventually find a cause within the as yet unknown structure of the atom. Einstein had to introduce probability for the derivation of Planck's law he gave in 1916 [8] and was troubled by it. But now probability appeared as a central concept in quantum mechanics which was to replace Newtonian physics. It is surprising that this concept was immediately accepted by Bohr and the younger physicists. But Planck, Einstein, de Broglie, and Schrödinger at best saw it as a provisional help to get results, hoping for some later theoretical development which would bring back determinism.

Together with Heisenberg's uncertainty principle (Episode 42), the probability interpretation now goes by the name *Copenhagen interpretation* of quantum mechanics. In many textbooks Born's name is not mentioned in this context. However, it appears invariably when it comes to scattering, because in paper [6] Born also introduced an iterative method of solving problems of scattering, which is taught to every physics student as the *Born approximation*.

Born, who led a flourishing and outstandingly successful school of theoretical physics, had to leave Germany in 1933. He first went to Cambridge and was appointed Tait Professor of Natural Philosophy in Edinburgh in 1936. After his retirement in 1953, he gracefully accepted the status of Professor Emeritus in Göttingen, to which he was entitled, and lived in the little spa Bad Pyrmont not far from it.

Fermi–Dirac Statistics – Yet Another Way of Counting (1926)

Fermi[1] [1,2] is one of the towering figures in twentieth-century physics. He made important contributions to both theoretical and experimental physics. In fact, being a theoretician essentially by self-training, he was awarded the Nobel Prize for experimental work. His first outstanding achievement was the discovery of another way of counting in statistical physics which, like the way of Bose and Einstein, was able to explain the third law of thermodynamics.

Fermi was born in Rome. His father was a railway employee from the north of Italy, his mother a school teacher from the south. Both Enrico Fermi and his brother Giulio, older by one year, were exceptionally bright children, very close to each other. Together they built electrical motors and other complicated toys. Completely unexpectedly, when only fourteen, Guilio died in a seemingly trivial operation on an abscess in his throat. This tragedy forever changed the atmosphere in the Fermi household to deep depression. Fermi began to study a lot by himself using second-hand books, some of them in Latin, which he bought from his meagre pocket money. In school, he found a lifelong friend with a similar inclination towards science, Persico[2]. Later, they became the first two professors of theoretical physics in Italy. A colleague of Fermi's father, Adolfo Amidei, was an engineer with a university degree. He lent Fermi one of his own mathematics books and when he found that Fermi had studied it within a few weeks and solved all the problems posed there he decided that Fermi was a prodigy. He set up a rather careful plan for Fermi's studies in mathematics as prerequisite for physics, which already then was Fermi's primary interest, and advised him to learn German (not taught in school) so that he could read more scientific literature. Following another advice by Amidei, Fermi took part in a competition for a fellowship at the Scuola Normale Superiore in Pisa, which had been founded by Napoleon and had become something like an elite college attached to the university. He passed first and, beginning in 1918, he studied in Pisa where he obtained his Ph.D. with an experimental dissertation on X-ray spectroscopy in 1922.

At that time Fermi 'was almost entirely self-taught; all that he knew he had learned from books or rediscovered himself. He had found no mature scientists who could guide him, as he would have found at the time in Germany, Holland, or England' [2]. He returned to Rome where Corbino[3], the director of the physics institute who was also a senator of the Kingdom of Italy and a minister in several Italian cabinets, began to foster him. He won a scholarship for postdoctoral studies and used it to join Born's group in Göttingen. A little later he was invited to Leiden by Ehrenfest who was interested in work done

---

[1] Enrico Fermi (1901–1954), Nobel Prize 1938   [2] Enrico Persico (1900–1969)   [3] Orso Mario Corbino (1876–1937)

Counting in Bose–Einstein Statistics and in Fermi–Dirac Statistics

Consider a number of cells (in our example there are three, labelled I, II, III) and a number of identical indistinguishable objects (in our example there are two, both labelled A). Shown are the allowed states in the case of Bose–Einstein statistics, where any number of particles may be in one cell and in Fermi–Dirac statistics, which allows at most one particle to occupy one cell.

States in Case of					
Bose–Einstein			Fermi–Dirac		
I	II	III	I	II	III
AA					
	AA				
		AA			
A	A		A	A	
A		A	A		A
	A	A		A	A

Obviously there are much less allowed states in Fermi–Dirac statistics compared to Bose–Einstein statistics. No states are possible with more than one particle in the same cell. For our example in Bose–Einstein statistics 3 out of 6 possible states contain more than one particle and in classical Boltzmann statistics 3 out of 9. These numbers reflect the general picture: In Bose–Einstein statistics particles tend to be nearer to each other than in classical statistics. In Fermi–Dirac statistics they are further apart.

---

by Fermi in Göttingen, again essentially all by himself. On his return from Leiden, Fermi got a position as lecturer at the university in Florence. It is here that he conceived *Fermi statistics*.

Pauli, Heisenberg, and Fermi taking a boat trip on lake Como in 1927.

The essential idea is to combine the Bose–Einstein concept of indistinguishability of identical particles with Pauli's exclusion principle (see Box). The result is that the number of states available for a gas of particles obeying these rules is drastically reduced, since never more than a single particle can occupy one state. To make explicit use of Pauli's prescription, that no pair of two particles shares all quantum numbers, Fermi considered all particles to be attracted by a spring-like force towards the coordinate origin. In this way, he could treat them as harmonic oscillators and assign them quantum numbers in the framework of Sommerfeld's quantum theory. He might just as well have used phase-space cells to distinguish quantum states. Fermi worked out the complete theory for the new type of gas, now called the *ideal Fermi gas*. He showed that it obeys Nernst's theorem, i.e., that it shows *gas degeneracy* just as the ideal Bose gas does, but that its other properties are quite different. For high temperature and low particle density, the differences between the three types of ideal gas, classical, Bose, and Fermi, are negligible. But for low temperature and high density they can become important. Under such circumstances the Fermi gas differs appreciably from the classical one. Fermi considered a classical gas, thought to consist of helium atoms, at a temperature of 5 Kelvin and a pressure of 10 atmospheres and showed that to keep a Fermi gas at the same temperature with the same density the pressure had to be increased by 15%. Because of the Pauli principle particles tend to repel each other. It is interesting to note that for the Bose gas the pressure is lower than in the classical case.

While he lived in Italy, Fermi submitted all his results for publications to Italian journals. In addition, he published the papers which he considered particularly important in German to make them internationally more accessible. (After 1933 he used English for that purpose.) Therefore the usual reference to his work on Fermi statistics is in German [3], the original Italian paper [4] is somewhat shorter.

Dirac, only a little later, attacked the problem from a different point of view, using quantum mechanics [5]. (Decades later he told [6] that he had read Fermi's paper but forgotten about it until Fermi, in a letter triggered by Dirac's paper, reminded him.) Dirac studied the possible quantum-mechanical wave functions of a system of many identical particles in a rectangular box and found a fundamental symmetry property. The stationary wave functions of a many-particle system are functions of the positions of all particles. As an example, we consider the simplest case, the influence of one space coordinate x in a two-particle system. We denote by x_1 and x_2 the values of x for the particles 1 and 2 and the wave function by $\Psi(x_1, x_2)$. Dirac observed that the latter has either one or the other of the two properties

$$\Psi(x_1, x_2) = \Psi(x_2, x_1) \qquad \text{or} \qquad \Psi(x_1, x_2) = -\Psi(x_2, x_1) \quad ,$$

i.e., *symmetry* or *antisymmetry* under the exchange of the two particles, respectively.

It is clear that in the antisymmetric case the wave function must vanish for $x_1 = x_2$ when the x coordinate for both particles is the same. That suggested to Dirac that only the antisymmetric set of wave functions should be used to describe particles which obey the Pauli principle.

At this point in his paper Dirac wrote a footnote: 'Prof. Born has informed me that Heisenberg has independently obtained results equivalent to these.' And when he received the proofs he added the reference to Heisenberg's paper [7]. Indeed, Heisenberg had found this symmetry property about two months

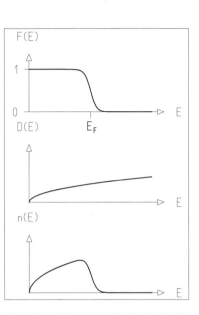

Illustration of the energy distribution for the hypothetical case of a Fermi gas of free electrons in a box. The Fermi function $F(E)$ is the probability for a state of energy E to be occupied. At zero absolute temperature one has $F(E) = 1$ for $E < E_F$ and $F(E) = 0$ for $E > E_F$, where E_F is the Fermi energy. With increasing energy the abrupt fall-off at E_F softens. The density $D(E)$ of available quantum-mechanical states rises like the square root of the energy E. The energy distribution $n(E)$ is equal to $2D(E)F(E)$; the factor 2 is due to the two possible spin states of the electron.

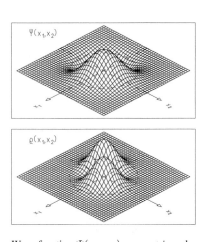

Wave function $\Psi(x_1, x_2)$, *symmetric* under exchange of x_1 and x_2 and the corresponding probability density $\rho(x_1, x_2)$.

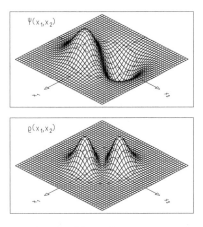

Wave function $\Psi(x_1, x_2)$, *antisymmetric* under exchange of x_1 and x_2 and the corresponding probability density $\rho(x_1, x_2)$.

earlier. He used it in one of his next papers to solve the riddle of the helium spectrum which had withstood all attempts to describe it in the old quantum theory.

Dirac, in his paper, applied symmetry properties in his quantum theory of the ideal gas. He showed, that by using only symmetric wave functions, he obtained the results of Einstein about the ideal Bose gas, including Bose–Einstein condensation. If, on the other hand, he used only antisymmetric wave functions, he got an ideal gas with new properties, obeying a new statistics and showing *gas degeneracy* but not condensation. This was, of course, the ideal Fermi gas, described by Fermi five months earlier but which Dirac did not know. Fermi's way of counting is now called *Fermi–Dirac statistics*.

About the two gases Dirac writes [5]:

> The solution with symmetric eigenfunctions must be the correct one when applied to light quanta, since it is known that the Einstein–Bose statistical mechanics leads to Planck's law of black-body radiation. The solution with antisymmetric eigenfunctions, though, is probably the correct one for gas molecules, since it is known to be the correct one for electrons in an atom, and one would expect molecules to resemble electrons more closely than light quanta.

This conjecture turned out to be too simple. One soon learned to distinguish two families of particles:

- Particles with *integer spin*, $0, \hbar, 2\hbar, \ldots$ obey Bose–Einstein statistics. They are now called *bosons*.
- Particles with *half-integer spin*, $\frac{1}{2}\hbar, \frac{3}{2}\hbar, \frac{5}{2}\hbar, \ldots$ obey Fermi–Dirac statistics. They were given the name *fermions*.

The connection between spin and statistics was rigorously proved by Pauli in 1940 [8]. About this *spin–statistics theorem* Feynman writes [9]: 'It appears to be one of the few places in physics where there is a rule which can be stated very simply, but for which no one has found a simple and easy explanation. The explanation is deep down in relativistic quantum mechanics. This probably means that we do not have a complete understanding of the fundamental principle involved.'

As a simple example, we consider a system of two particles which move in one direction only. Its wave function depends on the particle coordinates x_1, x_2. As illustrations we show the wave function for the cases of symmetry and antisymmetry with respect to the exchange of the two coordinates. The absolute square of the wave function is also shown. It is a measure of the probability to find the particles at the positions x_1, x_2. This probability is large if the two particles are near one another, i.e., $x_1 \approx x_2$ in the case of symmetry which is the case of two identical bosons. For two identical fermions, the antisymmetric case, it is small and even exactly zero for $x_1 = x_2$. This behaviour can be described as an attractive force between identical bosons and a repelling force between identical fermions. Such a force, appearing in quantum mechanics only, is often called an *exchange force*. The name is misleading, because nothing is exchanged, except the arguments (in our case x_1 and x_2) in the wave function.

Atoms and other composite objects can be either bosons or fermions. Their state is determined by total spin found by combining the spins of their constituents. Let us consider the helium atom. The most common isotope is ^{4}He. Its nucleus consist of two protons and two neutrons. All are fermions of spin $\frac{1}{2}\hbar$ which combine to spin zero, as do the two electrons in the atomic shell. Therefore the ^{4}He atom is a boson. The atom of the rare isotope ^{3}He is, however, a fermion; its spin is that of the unpaired neutron in its nucleus.

Fermi–Dirac statistics is of great technical importance. In a metallic conductor electrons form a Fermi gas of high density with properties completely different from those of a Boltzmann gas. Practically all states up to a particular energy, the *Fermi energy*, are occupied; nearly all states beyond are empty. At very low temperature, the energy distribution has a sharp edge at the Fermi energy, the *Fermi edge*. At higher temperatures the edge gets smoothed out a little but still dominates the form of the distribution. An electron theory of metals was developed by Sommerfeld [10–12], following, as he wrote, the discovery by 'the young Italian physicist, Fermi'.

Fermi, Corbino, and Sommerfeld.

[1] Fermi, L., *Atoms in the Family – My Life with Enrico Fermi.* University of Chicago Press, Chicago, 1954.

[2] Segrè, E., *Enrico Fermi, Physicist.* University of Chicago Press, Chicago, 1970.

[3] Fermi, E., *Zeitschrift für Physik,* **36** (1926) 902. Also in [13], Vol. I, p. 186.

[4] Fermi, E., *Rend. Lincei,* **3** (1926) 145. Also in [13], Vol. I, p. 178.

[5] Dirac, P. A. M., *Proceedings of the Royal Society,* **A112** (1926) 661.

[6] Mehra, J. and Rechenberg, H., *The Historical Development of Quantum Theory, Vol. 5, Part 2 – The Creation of Wave Mechanics: Early Response and Applications 1925–1926.* Springer, New York, 1987.

[7] Heisenberg, W., *Zeitschrift für Physik,* **38** (1926) 411. Also in [14], Vol. A I, p 411.

[8] Pauli, W., *Physical Review,* **58** (1940) 716. Also in [15], Vol. 2, p. 911.

[9] Feynman, R. P., Leighton, R. B., and Sands, M., *The Feynman Lectures in Physics,* vol. III. Addison-Wesley, New York, 1965.

[10] Sommerfeld, A., *Naturwissenschaften,* **15** (1927) 825. Also in [16], Vol. II, p. 385.

[11] Sommerfeld, A., *Naturwissenschaften,* **16** (1928) 374. Also in [16], Vol. II, p. 393.

[12] Sommerfeld, A., *Zeitschrift für Physik,* **47** (1928) 1, 43. Also in [16], Vol. II, pp. 426, 458.

[13] Fermi, E., *Collected Papers, 2 vols.* University of Chicago Press, Chicago, 1962.

[14] Heisenberg, W., *Collected Works, Series A, 3 vols.* Springer, Berlin, 1985–1993.

[15] Pauli, W., *Collected Papers, 2 vols.* Interscience, New York, 1964.

[16] Sommerfeld, A., *Gesammelte Schriften, 4 vols.* Vieweg, Braunschweig, 1968.

42 Heisenberg's Uncertainty Principle and Bohr's Complementarity (1927)

In his memoirs Heisenberg reports on his first encounter with Einstein in April 1926. He had been invited to present his work on quantum mechanics in the prestigious Berlin colloquium and after the talk they walked together to Einstein's apartment. Einstein began the conversation by criticizing Heisenberg's postulate that quantum mechanics must be based exclusively on quantities which are in principle observable (Episode 37). Astonished, Heisenberg remarked that he thought, just that had been Einstein's approach in special relativity when giving up the concept of absolute time, Einstein replied [1]: 'Probably I did use this kind of reasoning, but it is nonsense all the same [...] In reality the opposite happens. It is the theory which decides what we can observe.'

The *uncertainty principle* describes a fundamental property of quantum mechanics and was found in a period of intense and heated discussions in 1926 and early 1927 among physicists about the meaning of this new theory [2]. Although the theory had been developed in Göttingen, in Cambridge, and in Zurich (Episodes 37, 38, and 39), most of these discussions took place in Bohr's institute in Copenhagen. There Heisenberg had become lecturer in suc-

Heisenberg and Bohr in 1936.

178

Box I. The Gamma-Ray Microscope

In order to measure the position of an electron, a light quantum of short wavelength (gamma quantum or gamma ray) is scattered off the electron. Its wavelength λ is related to its momentum p_γ by

$$\lambda = h/p_\gamma \quad ,$$

where h is Planck's constant. The accuracy Δx, with which the electron's position x can be determined, is, essentially, the wavelength λ, i.e., $\Delta x \approx \lambda$. In the scattering process the momentum p of the electron is changed. If it was exactly known before the position measurement, its momentum uncertainty is now on the order of the light-quantum momentum, $\Delta p \approx p_\gamma$. Therefore, as result of the measurement, for the product of the uncertainties Δx in position and Δp in momentum we obtain the *uncertainty relation*

$$\Delta x \, \Delta p \approx h \quad . \tag{1}$$

---

cession to Kramers who had accepted a chair in Utrecht. There Dirac had gone for a first postdoctoral stay abroad, and there Schrödinger was invited by Bohr for a visit in the summer of 1926.

Schrödinger had shown that his wave mechanics and the more abstract matrix mechanics of Göttingen were mathematically equivalent. Dirac, in Copenhagen, developed his *transformation theory* [3], which allowed to express quantum-mechanical problems in a large variety of forms, including those of matrix and wave mechanics. Parts of this theory were independently created by Jordan. This was an important progress, but mainly of technical nature, and, although Dirac had entitled his paper *The Physical Interpretation of Quantum Dynamics*, questions concerning just that interpretation remained.

Like most physicists, Bohr had welcomed wave mechanics and was convinced that *particle–wave duality*, the necessity to describe the electron both as a point-like particle and as an extended wave, was fundamental to the difficulties. Heisenberg, on the other hand, as he freely admitted in his narrations about this period [1,4], strove to tackle the problem using matrix mechanics only or, at most, Dirac's theories that had been inspired by it. He describes that his discussions with Bohr were aimed to 'express the connection between mathematics and experiment in a way free of contradictions' [4], but that they could not agree and 'occasionally parted somewhat discontented'. At the end of February 1927, Bohr left for a skiing holiday in Norway and Heisenberg had more time to work by himself. As he wrote later the 'obvious idea' occurred to him 'that one should postulate that nature only allowed experimental situations described within the framework of the formalism of quantum mechanics' [4] and that this idea was inspired by the words of Einstein quoted above [1].

Heisenberg now conceived his imaginary gamma-ray microscope and wrote down his ideas in a long letter to Pauli in Hamburg. When Pauli signalled approval, he completed the paper and submitted it for publication a few days after Bohr's return. It is probable but not certain that Bohr read it before submission. At any rate, he objected to details, in particular, to the fact that Heisenberg had not taken into account the resolving power of the microscope. Heisenberg

Bohr and Heisenberg in Tyrol, 1932.

duly included this criticism, which is not of fundamental importance, as a note, added in proof, to his paper.

The paper on uncertainty [5] carries the title *On the Intuitive Content of the Quantum-Theoretical Kinematics and Mechanics*. It begins with a criticism of the notion of a particle trajectory which implies knowledge of position and velocity along the particle's path. To obtain the velocity, the position has to be repeatedly measured in small time intervals. For such measurements, Heisenberg writes, in principle a gamma-ray microscope could be used which would allow to determine the position with arbitrarily high precision. (The precision of a microscope is limited by the wavelength of light and gamma rays have a much shorter wavelength than visible light.) But the scattering process of a gamma quantum off the electron, which would reveal the electron's position, is the *Compton effect* in which momentum is transferred from the quantum to the electron. Thus, the measurement itself necessarily changes the momentum. Heisenberg obtains a relation between the measurement errors in position x and in momentum p: Their product is essentially equal to Planck's constant (see Box I). If one of the quantities is to be measured with very high precision, that can be achieved only by sacrificing accuracy of the other. He relates this finding to the fact that position and momentum do not commute (Episode 37), and he states that a similar relation holds for the product of the uncertainties connected with the measurements of energy E and time t. Thus he has the two *uncertainty relations*,

$$\Delta x \, \Delta p \approx h \quad , \qquad \Delta E \, \Delta t \approx h \quad .$$

Heisenberg then goes on to derive these results in the framework of quantum mechanics using the methods by Dirac and Jordan mentioned above.

In a section on the *Transition from Micro- to Macromechanics*, Heisenberg gives an application of his findings. He shows that it is utterly impossible to observe the orbit of an electron in the ground state of the hydrogen atom. Obviously, for such an observation light quanta with wavelengths smaller than the Bohr radius, i.e., the radius of the ground-state orbit in the Bohr model, are needed. Such quanta have an energy of more than $4000 \, \mathrm{eV}$. Since the ground-state electron is bound to the nucleus with an energy of only $13.6 \, \mathrm{eV}$, by the measurement it would be separated from the nucleus and would not orbit it any more. Heisenberg points out that the situation is quite different for a stationary state with a large radial quantum number, say $n = 1000$. Here the radius is so large that much softer quanta can be used, still disturbing the orbit but changing the radial quantum number only by $\Delta n = \pm 50$ or thereabouts.

In the concluding paragraph of his paper Heisenberg writes as follows about *causality* and quantum mechanics:

> What is wrong in the sharp formulation of causality, "When we know the present precisely, we can predict the future," is not the conclusion but the assumption. Even in principle we cannot know the present in all detail. For that reason everything observed is a selection from a plenitude of possibilities and a limitation on what is possible in the future. As the statistical character of quantum theory is so closely linked to the inexactness of all perceptions, one might be led to the presumption that behind the perceived statistical world there still hides a "real" world in which causality

holds. But such speculations seem to us, to say it explicitly, fruitless and senseless. Physics ought to describe only the correlation of observations. One can express the true state of affairs better in this way: Because all experiments are subject to the laws of quantum mechanics, and therefore to equation (1), it follows that quantum mechanics establishes the final fall of causality.

Whereas Heisenberg had obtained his results without reverting to wave mechanics, Bohr, on his return from Norway, computed them quite easily for a particle represented by a *wave packet* as proposed by Schrödinger. A wave function with fixed wavelength offers no way to localize it in some region of space. But if wave functions of different frequencies are superimposed, i.e., simply added, by interference a pattern can be created which is different from zero only in some finite region of space. The range of wavelengths used in the superposition, by virtue of de Broglie's relation between momentum and wavelength, corresponds to a range in momentum. The larger that range Δp is, the better is the localization, i.e., the smaller is the spatial width Δx of the packet (see Box II). Since the absolute square $|\Psi(x)|^2$ of the wave function is the probability density for the position x, and since a similar density can be defined for the momentum p, the quantities Δx and Δp are *uncertainties* just as they were in Heisenberg's reasoning.

Box II. Wave Packet

Let us denote by Δx the spatial width of a wave packet, by Δt the time it needs to move along by the distance Δx, by $\Delta \nu$ the interval of frequencies ν contained in the wave packet, and by $\Delta \lambda$ the corresponding interval in wavelengths λ. Then it is easy to show that the relations

$$\Delta t \, \Delta \nu \approx 1 \quad , \qquad \Delta x \, \Delta \frac{1}{\lambda} \approx 1$$

hold, quite independently of the nature of the wave packet. It may be a packet of acoustical waves, electromagnetic waves, or matter waves. If we now insert the Einstein–de Broglie relations which connect energy E and momentum p with frequency and wavelength, respectively, namely

$$E = h\nu \quad , \qquad p = \frac{h}{\lambda} \quad ,$$

we immediately arrive at the *uncertainty relations*

$$\Delta E \, \Delta t \approx h \quad , \qquad \Delta x \, \Delta p \approx h \quad .$$

Three wave packets: The width in position decreases (from top to bottom) while the width in frequency and wavelength increases.

It can be shown, that for a particular form of wave packet, called a *Gaussian wave packet* or a *minimum-uncertainty wave packet*, the product of uncertainties can reach its smallest value, which is $\hbar/2$. Thus, in its accurate form, the uncertainty relation (1) reads

$$\Delta x \, \Delta p \geq \frac{1}{2}\hbar \quad .$$

Since in Bohr's argument both a wave property (through the use of a wave function) and a particle property (implied by the probability density) appeared,

[1] Heisenberg, W., *Der Teil und das Ganze. Gespräche im Umkreis der Atomphysik.* Piper, München, 1969. Engl. transl. [8].

[2] Mehra, J. and Rechenberg, H., *The Historical Development of Quantum Theory, Vol. 6. – The Completion of Quantum Mechanics 1926–1941.* Springer, New York, 2000.

[3] Dirac, P. A. M., *Proceedings of the Royal Society,* **A113** (1927) 621.

[4] Heisenberg, W., *Quantum Theory and its Interpretation.* In Rozental, S. (ed.): *Niels Bohr – His life and work as seen by his friends and colleagues.* North Holland, Amsterdam, 1967.

[5] Heisenberg, W., *Zeitschrift für Physik,* **43** (1927) 172. Also in [9], Vol. A I, p. 478. Engl. transl. in [10], p. 62.

[6] Bohr, N., *Nature,* **121** (1928) 580. Also in [11], Vol. 6, p. 147.

[7] Bohr, N., *Discussion with Einstein on epistemological problems in atomic physics.* In Schilpp, P. A. (ed.): *Albert Einstein: Philosopher–Scientist.* Open Court, La Salle, Ill., 1949.

[8] Heisenberg, W., *Physics and Beyond. Encounters and Conversations.* Harper and Row, New York, 1971.

[9] Heisenberg, W., *Collected Works, Series A, 3 vols.* Springer, Berlin, 1985–1993.

[10] Wheeler, J. A. and Zurek, W. H. (eds.), *Quantum Theory and Measurement.* Princeton University Press, Princeton, 1983.

[11] Bohr, N., *Collected Works, 10 vols.* North Holland, Amsterdam, 1972–1999.

he referred to it by the name *complementarity*: The concepts of wave and of particle complement each other. Bohr first reported about this idea in September 1927 at an international conference in Como and published several papers on the subject. An early reference is [6].

Born's probability interpretation, Heisenberg's uncertainty principle, and Bohr's complementarity essentially form what is now called the *Copenhagen interpretation* of quantum mechanics. This is now generally adopted, but was at that time contested by many.

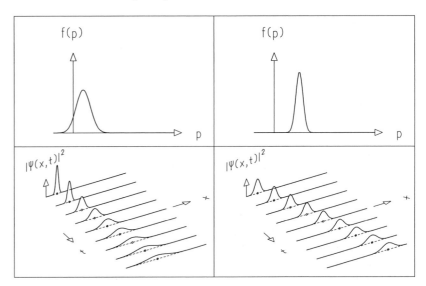

Illustration of the uncertainty principle. The two columns represent two different Gaussian wave packets. In the top row the momentum distributions of the two packets are shown, in the bottom row the time development of the probability density in the space coordinate x. For the initial time the packet with the larger width in momentum p (left column) has the narrower distribution in position x. However, since it contains a broad range in momenta, it widens rapidly as time increases, since the different momentum components travel at different speeds. At the initial time we have $\Delta x \, \Delta p = \hbar/2$ and at later times $\Delta x \, \Delta p > \hbar/2$.

In particular, de Broglie, Schrödinger, and Einstein never really accepted it. While recognizing the successes of quantum mechanics, they hoped for another theory beyond it which would respect causality in the conventional sense and would have no need for probabilities rather than certainties. Famous are the arguments which Bohr and Einstein, who deeply respected one another, had on the subject of uncertainty during the *Solvay Conference* in 1927 and still at the conference in 1930. These mental battles, in which Einstein constructed thought experiments that should contradict the uncertainty relations and in which Bohr was able to point out flaws in Einstein's arguments, did not take place in the conference proceedings but at the breakfast and dinner tables. They were retold by several participants and written down in detail, two decades later, by Bohr in a volume edited in Einstein's honour [7].

Quantum Mechanics and Relativity – The Dirac Equation (1928)

In 1955 Dirac gave lectures in Moscow. When asked about his philosophy of physics, he wrote on the blackboard:

PHYSICAL LAWS SHOULD HAVE MATHEMATICAL BEAUTY.

Dalitz, who relates this story, adds that 'this has been preserved to this day' [1]. It was this philosophy that Dirac used when he found the equation that now bears his name. Unlike many results in theoretical physics it was neither inspired by unexplained measurements nor by physical insight but only by considerations of mathematical 'beauty' or, in other words, simplicity. In the *Dirac equation* not only quantum mechanics and the special theory of relativity were married, but also the spin of the electron is contained in it without any ad hoc assumption. So far, so good. But the equation not just beautifully described known phenomena, it did more. It predicted the existence of electrons with negative energy. This was at first held to be a severe problem of the theory but was finally understood as great progress, because negative-energy electrons could be interpreted as hitherto unknown particles. Thus, the existence of new particles was predicted which had all properties of the electron except for the electric charge, which must be positive rather than negative. These particles were indeed found four years after the equation. Dirac is often quoted to have said that his equation 'contains most of physics and all of chemistry'. This, however, is not the case, although in a paper on the (non-relativistic) *Quantum Mechanics of Many-Electron Systems*, quite unrelated to the Dirac equation, similar words appear [2]: 'The underlying physical laws necessary for the mathematical theory of a large part of physics and the whole of chemistry are thus completely known, and the difficulty is only that the exact application of these laws leads to equations much to complicated to be soluble.' We begin this episode by mentioning briefly the work, previous to Dirac's, on the reconciliation of quantum mechanics with special relativity and with spin, respectively.

We have seen in Episode 39 how Schrödinger, starting out from de Broglie's ideas, first wrote down a relativistic wave equation, but gave it up, because, contrary to his expectations, it did not yield the fine structure of the hydrogen spectrum. In fact he did not publish the relativistic equation. But that was soon done by others, in particular, by Klein[1] and also by Gordon[2] after whom the equation (see Box I) is now usually named. At that time the equation did not meet with much interest, because quantum mechanics was used to describe the electron. An essential property of the electron, spin, was accounted for neither in the non-relativistic nor the relativistic Schrödinger equation.

Dirac in Göttingen, 1928.

---

[1] Oscar Benjamin Klein (1894–1977)   [2] Walter Gordon (1893–1939)

Box I. Relativistic Schrödinger Equation

Einstein's relation between energy E, momentum p, and rest mass m for a free particle (a particle not influenced by a force) reads

$$E^2 = p^2 c^2 + m^2 c^4 \quad .$$ (1)

By replacing E and p by the operators $-(\hbar/i)(\partial/\partial t)$ and $(\hbar/i)(\partial/\partial x)$, respectively, acting on a wave function Ψ, one obtains the equation

$$\left[-\frac{\hbar^2}{c^2}\frac{\partial^2}{\partial t^2} + \hbar^2 \frac{\partial^2}{\partial x^2} \right] \Psi = m^2 c^2 \Psi \quad .$$

This relation, usually referred to as *Klein–Gordon equation*, is of *second order* in both x and t, because the second derivatives of these variables appear in it. If we introduce convenient units in which the numerical values of the universal constants $\hbar$ and c are equal to one,

$$\hbar = c = 1 \quad ,$$

the equation simplifies to

$$\left[-\frac{\partial^2}{\partial t^2} + \frac{\partial^2}{\partial x^2} \right] \Psi = m^2 \Psi \quad .$$

If we now use the notations of special relativity, i.e., write $t = x^0$, $x = x^1$ and employ the Einstein summing convention (see Box I in Episode 11), if, furthermore, we abbreviate the derivatives by writing $\partial_0 = \partial/\partial x^0, \ldots$, and if, finally, we extend our equation from one to three spatial dimensions, we obtain the equation in the very compact form

$$(\partial_\mu \partial^\mu + m^2)\Psi = 0 \quad .$$ (2)

The first to successfully incorporate spin into quantum mechanics and to compute the anomalous Zeeman effect were Heisenberg and Jordan [3]. However, they treated spin as a vector similar to the vector of orbital angular momentum. This was not easily acceptable, especially not by Pauli who, after all, had called the property of the electron which was later called spin 'a two-valuedness which cannot be described from a classical point of view'. Pauli developed a genuinely quantum-mechanical description of spin by constructing a vector with three components, which are themselves matrices in the sense of matrix mechanics [4]. Using these *Pauli matrices*, which have two rows and two columns, he was able to fulfil all requirements posed by quantum mechanics on spin, including the mysterious two-valuedness. He described the electron by a *spinor* (see Box II), in which both elements are ordinary wave functions, and in this way accounted for both spin and position. His *Pauli equations* are a system of two coupled Schrödinger equations.

The Pauli equations, apart from being non-relativistic, had one essential drawback. An electron orbiting an atomic nucleus with a certain angular momentum behaves like a little magnet, it has a magnetic moment. So has the electron by virtue of its intrinsic angular momentum or spin. But the ratio of magnetic moment and angular momentum for spin is twice as large as for the

---

Box II. Pauli's Description of Spin. Pauli Matrices

In Episode 39, Box III, the stationary Schrödinger equation was written as eigen-value equation, $H\Psi = E\Psi$, where the operator H of total energy is a differential operator, E an energy eigenvalue, and Ψ an eigenfunction. In the language of matrix mechanics H is a matrix and Ψ an eigenvector. Similar equations hold for other physical quantities, as, for instance, orbital angular momentum and spin.

The operator **L** of angular momentum is a vector with 3 components which obey certain commutation relations. A state with quantum numbers ℓ (angular momentum) and m (its z component) fulfills the two eigenvalue equations

$$\mathbf{L}^2 Y_{\ell m} = \ell(\ell+1)\hbar^2 Y_{\ell m} \quad , \qquad L_z Y_{\ell m} = m\hbar Y_{\ell m} \quad ,$$

which again can be interpreted either as differential or as matrix equations. The angular-momentum eigenfunctions $Y_{\ell m}$ are well known in mathematics under the name *spherical harmonics*.

For spin, however, as found by Pauli, only a matrix representation is possible. He defined the spin operator **S** with the 3 components $S_x = (\hbar/2)\sigma_x, S_y = (\hbar/2)\sigma_y, S_z = (\hbar/2)\sigma_z$ containing the *Pauli matrices*

$$\sigma_x = \sigma_1 = \begin{pmatrix} 0 & 1 \\ 1 & 0 \end{pmatrix} , \quad \sigma_y = \sigma_2 = \begin{pmatrix} 0 & -i \\ i & 0 \end{pmatrix} , \quad \sigma_z = \sigma_3 = \begin{pmatrix} 1 & 0 \\ 0 & -1 \end{pmatrix} .$$

Then indeed eigenvalue equations hold, analogous to those for orbital angular momentum,

$$\mathbf{S}^2 \chi_\pm = s(s+1)\hbar^2 \chi_\pm \quad , \qquad S_z \chi_\pm = s_z \hbar \chi_\pm \quad ,$$

where $s = 1/2$ is the spin quantum number and $s_z = \pm 1/2$ the quantum number of its z component. The corresponding eigenvectors,

$$\chi_+ = \begin{pmatrix} 1 \\ 0 \end{pmatrix} , \qquad \chi_- = \begin{pmatrix} 0 \\ 1 \end{pmatrix} ,$$

are two-component objects called *spinors*.

---

orbital momentum. This factor of two in the *gyromagnetic ratio* of the spinning electron had to be introduced 'by hand' or 'ad hoc' in the Pauli equations. It is, essentially, the *Thomas factor* mentioned in Episode 36.

When Dirac began to work on the problem he, at first, did not consider spin but compared the non-relativistic and the relativistic Schrödinger equations. These are differential equations in space and time. If a differential has only a first derivative in a variable it is said to be of *first order* in that variable. If there is a second derivative, it is said to be of *second order*. The non-relativistic Schrödinger equation is of first order with respect to time but of second order with respect to the space coordinates. In special relativity, of course, time and space are treated on equal footing. The relativistic Schrödinger equation is of second order in time and in space. Following his sense of beauty or simplicity, Dirac searched for an equation that was only of first order in these variables. He made an *ansatz* of the form (3) (see Box III). It was clear that the equation made no sense if the quantities γ^μ were mere numbers. Dirac said about his work on the equation [5]:

Gamma Matrices

$$\gamma^0 = \begin{pmatrix} 1 & 0 & 0 & 0 \\ 0 & 1 & 0 & 0 \\ 0 & 0 & -1 & 0 \\ 0 & 0 & 0 & -1 \end{pmatrix} = \begin{pmatrix} I & 0 \\ 0 & -I \end{pmatrix}$$

$$\gamma^1 = \begin{pmatrix} 0 & 0 & 0 & 1 \\ 0 & 0 & 1 & 0 \\ 0 & -1 & 0 & 0 \\ -1 & 0 & 0 & 0 \end{pmatrix} = \begin{pmatrix} 0 & \sigma_1 \\ -\sigma_1 & 0 \end{pmatrix}$$

$$\gamma^2 = \begin{pmatrix} 0 & 0 & 0 & -i \\ 0 & 0 & i & 0 \\ 0 & i & 0 & 0 \\ -i & 0 & 0 & 0 \end{pmatrix} = \begin{pmatrix} 0 & \sigma_2 \\ -\sigma_2 & 0 \end{pmatrix}$$

$$\gamma^3 = \begin{pmatrix} 0 & 0 & 1 & 0 \\ 0 & 0 & 0 & -1 \\ -1 & 0 & 0 & 0 \\ 0 & 1 & 0 & 0 \end{pmatrix} = \begin{pmatrix} 0 & \sigma_3 \\ -\sigma_3 & 0 \end{pmatrix}$$

This was a problem for some months, and the solution came rather, I would say, out of the blue, one of my undeserved successes. It came from playing about with mathematics. [...] It took me quite a while, studying over this dilemma, that there was no need to stick to quantities σ, which can be represented by matrices with just two rows and two columns. Why not go to four rows and four columns?

The equation was indeed fulfilled if the quantities γ^{μ} are taken to be four-by-four matrices with two-by-two submatrices, which are either the matrix 0 (containing zeroes only), or the unit matrix I, or the Pauli matrices. In his first paper on the subject [6] Dirac gives credit to Pauli for having introduced the σ matrices but, in reminiscences, half a century later [5], he said: 'I believe I got these variables independently of Pauli, and possibly Pauli also got them independently of me.'

Pauli and Dirac in 1938.

The effect of the new matrices was that the wave function now had four components and not just two as in the case of the Pauli equations. They had to be interpreted as to correspond to two states of positive energy, each with another spin orientation, and two states of negative energy. This situation is not changed when a force is introduced into the equation, for instance, the force between the electron and an atomic nucleus, the simplest and most obvious case which Dirac studied immediately. In his first papers [6,7] he chose to just ignore the negative energy states. Doing this, he could reap a triple harvest from his equation:

- It yielded the correct *fine structure* of the hydrogen spectrum, which had been computed as a relativistic effect by Sommerfeld in the old quantum theory and which Schrödinger had been unable to reproduce.
- It gave the correct results for the anomalous Zeeman effect.
- It automatically gave the right gyromagnetic ratio $g = 2$ for the electron.

Box III. Dirac Equation

Instead of equation (2) which is of second order in the derivatives ∂_μ Dirac tried an *ansatz* of first order only,

$$(i\gamma^\mu \partial_\mu - m)\Psi = 0 \quad , \tag{3}$$

and he could show that this equation is fulfilled if the four quantities $\gamma^0, \gamma^1, \gamma^2, \gamma^3$ are matrices with four rows and four columns and if the wave function Ψ is an object with four components, now called a *Dirac spinor*. The *Dirac equation* is really a system of four coupled equations. That becomes apparent, once the indices of the matrix elements and of the wave function and the necessary summations are explicitly written down,

$$\sum_{k=1}^{4}\left[\sum_{\mu=0}^{3} i\,(\gamma^\mu)_{jk}\,\partial_\mu - m\delta_{jk}\right]\Psi_k = 0 \quad ; \qquad j = 1, 2, 3, 4 \quad .$$

Here δ_{jk} is the usual Kronecker symbol (1 for $j = k$ and 0 otherwise).
The four components of Ψ allow for the description of electron and positron, each in the two possible orientations of spin.

In particular, the last point indicated that spin is a truly relativistic effect, since it was described correctly in an equation based on the special theory of relativity.

In spite of these successes, which made it clear that Dirac's equation must be of fundamental importance, the difficulty of negative-energy states remained. The reason for these states is, of course, the relativistic equation (1). Here the energy E comprises the kinetic energy and the rest energy mc^2 of the electron. If we take the square root of that equation, then, because of the ambiguity of the sign of the square root (it can be $+$ or $-$), energies are possible in two regions E greater than or equal to mc^2 and E less than or equal to $-mc^2$,

$$E \leq -mc^2 \quad , \qquad E \geq mc^2 \quad .$$

Obviously, also the Klein–Gordon equation shows this behaviour. In classical mechanics quantities show a continuous (or smooth) variation with time. A sudden jump over the gap of size $2mc^2$ from positive to negative energy cannot occur and therefore negative-energy states could simply be excluded as *unphysical*. Discontinuous transitions are, however, typical of quantum mechanics. It seemed that Dirac's equation predicted for all electrons their transition to states of lower and lower negative energy and the production of radiation in the process, i.e., the instability of all atoms.

There was no progress in this question for nearly two years when Dirac came up with a bold assumption [8]. He simply postulated that all states of negative energy are occupied. Since electrons obey Pauli's *exclusion principle*, electrons with positive energy find no free negative-energy state to which they could transit. Stability is restored but at a price: One has to assume the existence of infinitely many negative-energy electrons. All these electrons are called the *Dirac sea*. It carries an infinite electric charge and yet does not affect the world, except for one aspect. An electron can be 'lifted' out of the

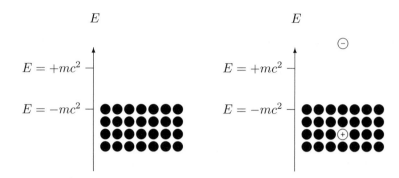

Illustration of the Dirac sea. Left: All states of negative energy are occupied by electrons. Right: One electron is lifted out of the sea, acquiring a positve energy. It leaves a 'hole' in the sea which behaves like a positively charged particle.

[1]   Dalitz, R. H., *A Biographical Scetch of the Life of Professor P A M Dirac, O M, F R S.* In Taylor, J. G. (ed.): *Tributes to Paul Dirac.* Adam Hilger, Bristol, 1987.

[2]   Dirac, P. A. M., *Proceedings of the Royal Society,* **A123** (1929) 714.

[3]   Heisenberg, W. and Jordan, P., *Zeitschrift für Physik,* 37 (1926) 263. Also in [10], Vol. A I, p. 516.

[4]   Pauli, W., *Zeitschrift für Physik,* **43** (1927) 601. Also in [11], Vol. 2, p. 306.

[5]   Dirac, P. A. M., *Recollections of an exciting era.* In Weiner, C. (ed.): *History of Twentieth Century Physics.* Academic Press, New York, 1977.

[6]   Dirac, P. A. M., *Proceedings of the Royal Society,* **A117** (1928) 610.

[7]   Dirac, P. A. M., *Proceedings of the Royal Society,* **A118** (1928) 351.

[8]   Dirac, P. A. M., *Proceedings of the Royal Society,* **A126** (1930) 360.

[9]   Dirac, P. A. M., *Proceedings of the Royal Society,* **A 133** (1931) 60.

[10] Heisenberg, W., *Collected Works, Series A, 3 vols.* Springer, Berlin, 1985–1993.

[11] Pauli, W., *Collected Papers, 2 vols.* Interscience, New York, 1964.

sea by a photon with an energy of more than the size of the energy gap, i.e., more than $2mc^2$, into the positive-energy region. It leaves behind an unoccupied state, called a *hole*, which behaves like a positively charged electron with positive energy. The quantum theory of similar *holes* had already been developed for atoms whose outer shell was nearly completely occupied with only a single electron missing, and good use had been made of the concept to explain chemical bonds. This postulate provided a way out but it also predicted the existence of electrons with positive charge. At that time when Dirac wrote it down, in December 1929, only two particles were known, the (ordinary) electron and the proton. (In addition, the light quantum had gradually become a particle.) Dirac simply made the assumption that the particle with positive charge was the well-known proton and that the large difference in mass (the proton is nearly 2000 times as heavy as the electron) might possibly be explained by the fact that the electric forces within the sea were not taken into account.

Dirac sea and holes were much discussed in the literature and Dirac became convinced that he could not uphold the asymmetry between electrons and holes. Therefore, in mid-1931, he wrote [9]:

A hole, if there were one, would be a new kind of particle, unknown to experimental physics, having the same mass and opposite charge to an electron. We may call such a particle an anti-electron. We should not expect to find any of them in nature, on account of their rapid rate of recombination with electrons, but if they could be produced experimentally in high vacuum they would be quite stable and amenable to observation.

This was the first theoretical prediction of a new particle, or was it? Pauli, a year earlier, had publicly advertised his *neutrino hypothesis* (Episode 46) but not published it. And, of course, Einstein, in 1905, had published his *light-quantum hypothesis*. But then, as we have seen, the light quantum took a long time to develop into a particle and was helped along in this process by experiment, notably the Compton effect. Thus, one can say that Dirac was the first to predict a new particle in a publication; certainly he was the first to predict an *anti-particle*. We shall see later, that anti-particles were to stay but not the Dirac sea.

The Band Model of Conductors and Semiconductors (1928–31)

In 1927, when he was twenty-five, Heisenberg was appointed full professor of theoretical physics at the University of Leipzig and soon established a very productive group. His first doctoral student was Bloch[1], whom we met briefly already as a student in Zurich (Episode 39). Following an advice by Debye, who himself had moved to Leipzig, Bloch [1] had arrived for the winter semester 1927/28. On Heisenberg's suggestion he thought about a subject for his thesis over the Christmas break and concluded that he would like to reformulate an important theorem, obtained by Ehrenfest in the 'old quantum theory', in the framework of the new quantum mechanics. But he was told by Heisenberg [2]: 'Yes, one might do that, but I think you had better leave such things to the learned gentlemen in Göttingen', alluding to the group of Born, who, as we have seen, was originally a mathematician. Instead, Heisenberg suggested something 'more down to earth, such as, for example, ferromagnetism or the conductivity of metals'.

Heisenberg, who, in 1925, had laid the foundations of quantum mechanics (Episode 37), had also been the first to apply the new theory to a system with more than one electron, the helium atom which had been intractable in the old theory. In Leipzig he turned his attention to extended systems containing a very large number of electrons. He himself had already begun to develop the theory of *ferromagnetism*, based on the electron spin. Bloch, rightly, got the impression that Heisenberg had already mastered the essentials of that field and that 'there was no point of my going into it'. Therefore he chose to look into the conductivity of metals. His thesis, as well as a paper summarizing it [3], was completed within half a year. In the introduction Bloch mentions the two major steps taken until then towards an understanding of the conduction of electricity in metals. The first was the postulate of the existence of a gas of free electrons inside the metal, pioneered by Drude in 1900 and extended by Lorentz a little later. The second step, taken recently by Pauli and, in particular, by Sommerfeld was the application of Fermi–Dirac statistics to this gas of free electrons. The very existence of such a gas, however, had to be accepted as an empirical fact. More than half a century later Bloch described the situation [4]:

> I had never understood how anything like free motion could even approximately be true. After all, a metal wire with all its densely packed ions is far from being a hollow tube and as I started to think about it, I felt that the first thing to be done in my thesis was to face this striking paradox. From the beginning I was convinced that the answer, if at all, could be found only in the wave nature of the electron . . .

A year earlier Hund[2] had shown [5] that, in a molecule, an electron is not

Bloch in 1938

---

[1] Felix Bloch (1905–1983), Nobel Prize 1952   [2] Friedrich Hund (1896–1997)

Heisenberg and some members of his group in Leipzig. From the left: Placzek[3], Gentile[4], Peierls, Wick[5], Bloch, Heisenberg, Weisskopf, and Sauter[6].

confined to a particular atom. Instead of a single molecule Bloch considered an extended piece of solid material, taken to be a crystal, i.e., a regular arrangement or *lattice* of atoms. In a simple model of a metal these atoms were considered ionized with one of the electrons allowed to move within the lattice. For his quantum-mechanical study Bloch described the influence of the lattice of ions as an electric potential which was periodic in all three dimensions (just as the lattice) and of infinite extension. He wrote down the Schrödinger equation for the electron in such a potential and, by the same evening [4], he found that the solutions were free waves, modulated by the potential but extending through the whole lattice. These are now called *Bloch waves*. He also constructed wave packets travelling through the lattice by superimposing solutions of slightly different energies. The phenomenon of free electrons was now explained. In a perfectly periodic potential, i.e., in a perfectly regular crystal with each atom in its nominal lattice position, there would be no electric resistance if an external electric field was applied. The observed resistance was explained by Bloch as due to imperfections and, in particular, to thermal vibrations of the lattice. With later refinements his theory gave a good description of the temperature dependence of resistivity in metals.

Early in his paper Bloch stresses the fact that, although the electrons can travel without hindrance through the lattice, the relation between momentum and energy of these electrons is, in general, different from that of a free electron in vacuum. He mentions that Heisenberg had pointed out to him that the changed energy–momentum relation leads to particular properties of the quantum-mechanical electron gas (see Box).

In the spring of 1928, when Bloch had worked out the essentials of his theory, Peierls[7] [6] arrived in Leipzig. He was a Berliner and had begun to study at the university of his home town and then moved to Sommerfeld in Munich.

---

[3] Georg Placzek (1905–1955)   [4] Giovanni Gentile (1906–1942)   [5] Gian Carlo Wick (1909–1992)   [6] Fritz Sauter (1906–1983)   [7] (Sir) Rudolf Ernst Peierls (1919–2000)

Effective Mass of an Electron in a Lattice

For a free classical particle of mass m, momentum p, and energy E we write down the energy–momentum relation and its first and second derivative,

$$E = \frac{p^2}{2m} \quad , \qquad \frac{dE}{dp} = \frac{p}{m} \quad , \qquad \frac{d^2E}{dp^2} = \frac{1}{m} \quad .$$

For the quantum-mechanical description of an electron in a lattice these relations no longer hold. Instead of the first one there is some more general relation $E = E(p)$. In analogy to the third one the *effective mass* m^* can be defined by

$$\frac{d^2E}{dp^2} = \frac{1}{m^*} \quad .$$

In Newton's equation $F = ma$, the mass m connects between the force F and the acceleration a for a classical particle. Likewise the effective mass m^* can be taken to connect an external force, acting on an electron in a lattice, with the acceleration experienced by the electron. In the particular case of a negative effective mass, force and acceleration act in opposite directions.

When Sommerfeld left for a sabbatical year, Peierls joined Heisenberg's group. After some initial work and the summer break Heisenberg suggested that he look into the so-called *anomalous Hall effect*. If a current is passed through a conducting plate, situated in a magnetic field perpendicular to the current, then a voltage, the *Hall voltage*, appears across the plate, perpendicular to the current and to the magnetic field. That had first been observed in 1879 by Hall[8], while working for his Ph.D. at Johns Hopkins University in Baltimore. The 'normal' Hall effect is easily explained: The electrons, making up the current, experience the Lorentz force which is perpendicular to the current and to the magnetic field. This leads to a charge depletion on one side of the plate and to a surplus on the other side – hence the Hall voltage. The sign of the of the Hall voltage shows that the current, as expected, is due negatively charged carriers, namely electrons. For some materials, however, the sign of that voltage was 'wrong'; to explain it, one had to assume that the current was carried by positive charges. That was the anomaly. (A completely different, unexpected feature of the Hall effect is discussed in Episode 95.)

Peierls studied the connections between the momentum p and the energy E of the quantum-mechanical solutions in more detail. For a free particle in classical physics this relation is simply $E = p^2/2m$. In an infinite lattice in one dimension, however, this simple relation, graphically a parabola, holds only for low energies and momenta. With rising energy the curve flattens off and reaches a point at which the energy jumps abruptly to a higher value. No states exists in the energy gap between the two values.

To illustrate the nature of that gap we look at the spectrum of energy values and at the corresponding wave functions, obtained by computer calculations, for tiny model crystals consisting of a single potential well (the model of one atom), two adjacent wells, and, finally, ten wells. In our simple example there

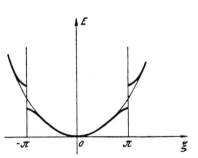

The energy–momentum relation as found by Peierls. The variable ξ is proportional to the momentum p. (Its exact definition is $\xi = p/\hbar G$; $\hbar$ is Planck's constant and G the number of atoms in the lattice per unit length.) From [7]. Copyright Wiley-VCH Verlag GmbH & Co. KGaA. Reproduced with permission.

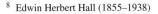

[8] Edwin Herbert Hall (1855–1938)

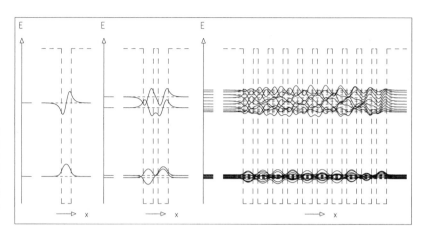

Solutions of the Schrödinger equation for systems of (from the left) one, two, and ten potential wells. The horizontal bars at the energy scale are the energy eigenvalues. The long-dashed broken line indicates the potential as function of the position x. The wave functions for the different eigenvalues are shown as continuous lines.

are two eigenstates in the single well. Each of them splits into two in the double well and into ten in the tenfold well. In the general case each eigenvalue of a single atom splits into a very large number, equal to the number of atoms in the crystal, of neighbouring energy values. In essence, the eigenvalues of the single atom are broadened to form continuous *energy bands* with well-defined upper and lower boundaries. In a one-dimensional model, as used in our illustration, the bands do not overlap; there is always a *band gap* of forbidden energies between two bands.

Peierls looked at the case in which nearly all energy states of the lowest band are occupied. For states with an energy near the upper boundary of the band, the effective mass becomes negative, since the curvature of his graph for $E = E(p)$ is negative there (see Box). As a consequence, particles in these states are not accelerated in the direction of the Lorentz force, acting in the presence of a magnetic field, but against that direction. They behave as if they carry positive, rather than negative charge. In contrast, for a band with many unoccupied states, the Hall effect comes out normal.

Like Bloch, Peierls was a fast worker. He completed his paper [8] on the Hall effect just before Christmas 1928 and then still wrote a short conference report [9]. In the spring of 1929, when Heisenberg left for a while for the United States, Peierls moved to Zurich. On Pauli's suggestion he worked on the quantum-mechanical treatment of lattice vibrations and thermal conductivity. He submitted this work, completed within a few months, as thesis in Leipzig, since he had been in residence in Zurich for only one semester but in Leipzig for two, the minimum requirement for a doctoral student.

Both Bloch and Peierls had only considered electrons in the band of lowest energy. In fact, the term *band* was not yet in use. Peierls had pointed out [8] that there could be no electric conduction if all the states (in that band) were occupied. In that case the motion of an electron with a certain momentum in one direction was exactly balanced by that of another one with the opposite momentum. He also mentioned that, obviously, there would also be no conductivity if none of the states was occupied. Finally, he stated that 'a priori little can be said about the number of occupied states'. It took another newcomer to Leipzig to achieve further progress.

Wilson[9] [10], like Dirac, was a student of Fowler in Cambridge and had become interested in conductivity by experiments performed there by Kapitza (Episode 58) on metals under extreme conditions, namely, very high magnetic fields and very low temperature. When he was awarded a Rockefeller Travelling Fellowship he decided to spend most of the nine months granted in Leipzig. Later he remembered [11]:

> I arrived in Leipzig in the first week in January 1931, and Heisenberg immediately pressed me into giving a colloquium on magnetic effects in metals, remarking 'Peierls' work is undoubtedly important, but the mathematics is complex and the physical ideas not easy to disentangle. We ought to have a thorough discussion of the whole subject, and, as you have more free time than the rest of us, would you agree to give the first talk?' ... [While preparing that talk] just at the end of January it suddenly occurred to me that the Bloch–Peierls theories could be enormously simplified and made intuitively more plausible if one assumed that quasi-free electrons, like valency electrons in single atoms, could form either open or closed shells.

Wilson had realized that, for a given energy band, there were exactly as many quantum-mechanical states as there were atoms in the crystal. Because of the Pauli principle each state could, at most, be occupied by two electrons (with opposite spin orientations). In the simplest crystals all atoms are alike and have the same number Z of electrons. At first Wilson concluded that, if Z is even, then the lowest $Z/2$ energy bands are completely occupied and all others are empty. There could be no conduction; such a crystal would be an insulator. For Z odd, on the other hand, there was one half-filled band; the crystal would be a good conductor.

The next day Wilson went to see Heisenberg who, on hearing these ideas, immediately asked Bloch to join in the discussion. Bloch, recently returned to Leipzig, objected. He pointed out, in particular, that the alkaline earths, for whom Z is even, are known to be conductors, not insulators. Wilson was able to refute that argument a day later. He demonstrated that in a three-dimensional lattice two or more energy bands can overlap and thus form a single band with a multiple of $2Z$ states, while such an overlap does not occur for the simplified model of a one-dimensional lattice which was often used in calculations.

Wilson now gave a first colloquium and also worked out his theory, which provided a clear separation of conductors and insulators, in a publication [12]. More difficult was the situation for a class of crystals, classified as *semiconductors*. It was not even clear if the conductivity, in between that of conductors and insulators, measured for semiconductors, was due to the material itself or to an oxide layer which was possibly formed on their surface.

In parallel to Wilson's work, Heisenberg published a paper [13] in which he introduces the elegant concept of a *hole* in atomic and solid state physics. In the introduction he summarizes it as follows:

> If N denotes the number of electrons in a closed shell, then it shall be shown in the following, that one can also replace a Schrödinger equation for n electrons by an equivalent Schrödinger equation of similar build for $N - n$ holes.

---

[9] (Sir) Allan Harries Wilson (1906–1995)

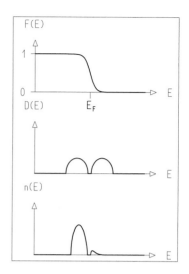

Fermi function $F(E)$ at non-zero temperature, density of states $D(E)$ in the valence band (left hump) and the conduction band (right hump), and the electron density $n(E) = 2D(E)F(E)$ for the case of an intrinsic (undoped) semiconductor. The Fermi energy E_F is located in the middle of the band gap between valence and conduction band. Because of the band structure the situation is different from the ideal case of a free electron gas illustrated on page 175.

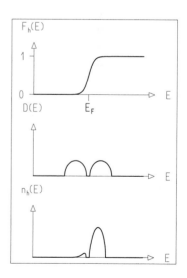

By defining $F_h = 1 - F(E)$ as the Fermi function for holes, i.e., unoccupied states, one immediately gets the energy distribution n_h of holes. Only the few electrons in the otherwise empty conduction band (top figure) and the few holes in the valence band can be considered free particles.

[1]   Hofstadter, R., *Biographical Memoirs (Nat. Acad. Sci.)*, **64** (1994) 34.

[2]   Bloch, F., *Physics Today*, **December** (1976) 23.

[3]   Bloch, F., *Zeitschrift für Physik*, **52** (1928) 555.

[4]   Bloch, F., *Proceedings of the Royal Society*, **A371** (1980) 24.

[5]   Hund, F., *Zeitschrift für Physik*, **40** (1927) 742.

[6]   Peierls, R., *Bird of Passage – Recollections of a Physicist.* University Press, Princeton, 1985.

[7]   Peierls, R., *Annalen der Physik*, **396** (1930) 121.

[8]   Peierls, R., *Zeitschrift für Physik*, **53** (1929) 255.

[9]   Peierls, R., *Physikalische Zeitschrift*, **30** (1929) 273.

[10]   Sondheimer, E. H., *Biographical Memoirs of Fellows of the Royal Society*, **45** (1999) 547.

[11]   Wilson, A. H., *Proceedings of the Royal Society*, **A371** (1980) 39.

[12]   Wilson, A. H., *Proceedings of the Royal Society*, **A133** (1931) 458.

[13]   Heisenberg, W., *Annalen der Physik*, **10** (1931) 888. Also in [16], Vol. A I, p. 610.

[14]   Hoddeson, L., Braun, E., Teichmann, J., and Weart, S., *Out of the Crystal Maze – Chapters from the History of Solid-State Physics.* Oxford University Press, Oxford, 1992.

[15]   Wilson, A. H., *Proceedings of the Royal Society*, **A134** (1931) 277.

[16]   Heisenberg, W., *Collected Works, Series A, 3 vols.* Springer, Berlin, 1985–1993.

Obviously, many problems can be much simplified by such a treatment in cases in which there are many (negative) electrons and only few (positive) holes. After the formal derivation of his statement, Heisenberg applies it to atomic spectra and to the Hall effect. The complicated behaviour of the very many electrons in a nearly full band now is described quite simply as the free motion of a few holes. Peierls' recalls a conversation in which Heisenberg pointed out the analogy between nearly empty and nearly closed shells [6]: 'I wish I could remember when he said this. If it was when I presented my results, it simply meant that he appreciated their significance immediately. If it was when he introduced me to the problem, then he knew the answer from the beginning (which is quite believable, considering his powerful intuition) and left me just to work out the mathematical details.'

What concerned semiconductors, Wilson was told by Heisenberg that Gudden[10] from the Technical University in Erlangen had expressed the view that a pure substance was never a semiconductor and that the observed properties of semiconductors were caused by impurities. Based on this assumption Wilson now essentially described a semiconductor as an insulator crystal with an admixture of a certain kind of atoms, later called *donors*. In the semiconductor crystal there is a highest band, completely filled with electrons, and, at still higher energy, an empty band. The donor atoms are easily ionized by thermal energy, yielding electrons that can only go into the upper band where they lead to a little conductivity. Wilson gave a second colloquium, attended by Gudden and other experimentalists and turning into a discussion of several days [14]. His paper [15] on this concept contains the first figure of energy bands.

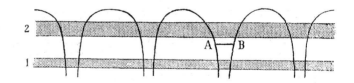

Wilson's picture of energy bands. In the absence of donors band 1 is full and band 2 is empty. Donors sit at fixed positions and have energy levels only a little below band 2. One such level is shown as the line AB. Thermal energies suffice to lift electrons from such levels to band 2 where they can move freely and transport current. From [15], Fig. 1. Reprinted with permission.

The highest filled band is now called the *valence band* and the lowest empty one the *conduction band*. Wilson's concept of *donors* was soon complemented by that of *acceptors* with energy levels just above the valence band. Some electrons from that band can be trapped in these levels and leave holes in the valence band, which then display the particular properties described above.

The band model is the basis of all modern semiconductor technology. It was developed long before applications were possible. Indeed, it took years to find that germanium and silicon are *intrinsic semiconductors*, i.e., that for them thermal energies suffice to lift electrons from the valence band to the conduction band. With donor or acceptor atoms they conduct electricity with negative or positive charge carriers, respectively. The combination of such materials led to very useful devices, such as the transistor (see Episode 69).

---

[10]   Bernhard Friedrich Adolf Gudden (1892–1945)

Hubble Finds that the Universe is Expanding (1929)

45

In 1929, through the work of Hubble[1] [1,2], our present picture of the *expanding universe*, consisting of galaxies moving away from one another, was established. Hubble is considered by many to be the greatest astronomer of the twentieth century. Certainly he made good use of the fact that, nearly from the outset of his career, he worked with the world's largest telescope and concentrated on work that could be done with this instrument only.

The *Mount Wilson Observatory* in California, Hubble's base, owes its existence to Hale[2] [3] who was an extraordinary person as astronomer and astrophysicist and also as fund-raiser and organizer. In astronomy, Hale is best known for his discovery of magnetic fields in sunspots. Deeply interested in science, he was successful in convincing wealthy people and organizations to provide funds, beginning with his own father who was an engineer and producer of hydraulic lifts for high-rise buildings. Successively he founded three important observatories which for nearly a century took turns in boasting the most powerful telescopes. The first was the *Yerkes Observatory* in Wisconsin with its 40-inch-refractor (lens) telescope which opened in 1897. It was connected to the University of Chicago where Hale then was professor of astronomy. The Mount Wilson Observatory was founded in 1904. Its big instrument, a 100-inch-reflector (mirror) telescope, was completed in 1917 and remained the world's largest until 1948. Only then the 200-inch reflector took over, which belongs to the *Mount Palomar Observatory*, initiated by Hale in 1928. The 200-inch, named the *Hale Telescope*, remained the world's largest until 1993.

Hubble was born in Marshfield, Missouri, as son of a lawyer who worked as an insurance agent. Having won a scholarship, he studied at the University of Chicago and in 1910 obtained a B.Sc. in mathematics and astronomy. For some time he was laboratory assistant to Millikan. He also was a fine athlete and excelled particularly in boxing. The years from 1910 to 1913 Hubble spent as Rhodes scholar at Queen's College, Oxford, studying law and taking a B.A. in that subject. He became fond of everything English and even developed British mannerisms which were not to everybody's liking when he returned to America. Back in the United States, Hubble did not practise law but worked for some time as high-school teacher before he decided in 1914 to return to Chicago to work for a Ph.D. in astronomy.

For his dissertation Hubble decided to study *nebulae*, diffuse objects which, if observed with a telescope of sufficient power, usually turn out to consist of a very large number of stars. Known nebulae are listed in catalogues named M (for *Messiers*) or NGC (for *New General Catalogue*). In 1915 and 1916, he

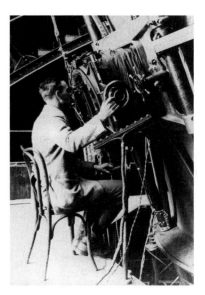

Hubble at the Mt. Wilson 100-inch telescope.

---

[1] Edwin Powell Hubble (1889–1953)   [2] George Ellery Hale (1868–1938)

took photographic plates of the nebula NGC 2261 and compared them carefully with plates taken by other astronomers in 1913, 1908, and 1900. To his surprise he found that in the short period of time the nebula had measurably changed in shape [4]. Impressed by this work, Hale, looking out for young scientists for the new Mount Wilson Observatory, offered Hubble a position to assume after completion of his Ph.D. However, Hubble joined the staff of Mount Wilson only in the autumn of 1919. When the United States entered the First World War he rushed the completion of his thesis and his exam and joined the army. After training he arrived in France shortly before the armistice, ending the war as Major Hubble, a title which he liked to be addressed by even as a scientist *on the mountain.*

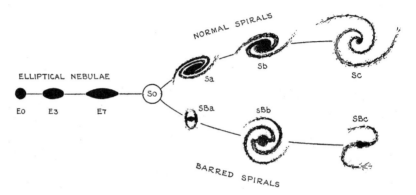

Hubble's 'tuning fork' diagram, a classification of nebulae. From [5]. © 1936 Yale University Press. Reprinted with permission.

Hubble made the observation of nebulae the scientific subject of his life. He devised a classification scheme for nebulae, still in use, which he described in technical terms in papers published between 1922 and 1926 and which is summarized best by the *tuning fork diagram* he included in his popular book *The Realm of the Nebulae* [5]. In 1924 he married the daughter of a wealthy Los Angeles banker. The couple lived in a splendid house built for them and was able to travel extensively abroad and to cultivate friendships with millionaires, fashionable writers, and Hollywood actors.

In 1925 Hubble gained wide recognition for observations putting an end to what was later called the *great debate.* At that time there was still no clear answer to the old question whether all objects, observed in the sky, are contained within the Milky Way or not. Already in the eighteenth century the German philosopher Kant[3] had pronounced the opinion that nebulae are milky ways or, to use the Greek name, *galaxies* in their own right. He based it on the (correct) assumption that the Milky Way, appearing as a band spanning the sky, has the form of a round flat disk, and that nebulae, which have partly circular but mostly elliptical forms, are just such disks viewed from different directions with respect to their axes of symmetry. Although many astronomers shared this opinion, it could not be put to a final test, because no way was known to measure the distance to the nebulae. The measurement of distances is the weak point of astronomy, whereas angles are measured with the proverbial astronomic precision. The distance to relatively near objects is determined by

Measures of Distance in Astronomy

One *light-year,*

$$1\,\mathrm{ly} = 9.463 \times 10^{12}\,\mathrm{km} \quad,$$

is the distance light in empty space travels in one year.
One *parsec,*

$$1\,\mathrm{parsec} = 1\,\mathrm{pc} = 3.259\,\mathrm{ly} \quad,$$

is the distance of a star from which the average radius of the earth's orbit about the sun would appear under an angle of $1''$, i.e., one second of an arc.

---

[3] Immanuel Kant (1724–1804)

Hubble's Law. Hubble Constant. Age of the Universe

Hubble's Law: A nebula with distance r from the solar system moves away from us with a velocity v, which is proportional to r,

$$v = H_0 r \quad .$$

Hubble Constant: The proportionality constant H_0 is called the *Hubble constant*. It is now customary to write it in the form

$$H_0 = h \times 100\,\mathrm{km\,s}^{-1}\,\mathrm{Mpc}^{-1} = h \times (9.778 \times 10^9\,\mathrm{years})^{-1} \quad .$$

The numerical factor $h = 0.71 \pm 0.04$ contains the remaining uncertainty in the measurement of extragalactic distances.

Age of the Universe: Under the assumption that the Hubble constant always had its present-day value H_0, the age of the universe is obtained as

$$T = \frac{1}{H_0} = \frac{1}{h} \times 9.778 \times 10^9\,\mathrm{years} = 13.7 \times 10^9\,\mathrm{years} \quad .$$

---

conventional triangulation with the diameter of the earth's orbit about the sun serving as base. For distances very much larger than the base indirect methods are used. There is a class of stars, called *Cepheids* or *Cepheid variables*, whose brightness changes periodically with time in a very characteristic fashion. That was first observed by Leavitt[4]. The absolute brightness is connected to the period which, typically, is on the order of one month. Therefore the former can be computed from the latter. The distance of near Cepheids was roughly known when Hubble began his work. (It would take more episodes to relate how that was achieved.) That of Cepheids farther away could be inferred from their apparent brightness observed on the earth which, essentially, is the absolute brightness divided by the square of the distance. In 1925 Hubble published an extensive paper, in which he reports that he observed over a longer period the nebula NGC 6822 and found 11 Cepheids in it [6]. From these he determined the distance of that nebula to be 700 000 light-years and concluded:

> N.G.C. 6822 lies far outside the limits of the galactic system [...] and hence may serve as a stepping-stone for speculations concerning the habitants of space beyond.

A little later Hubble corroborated his results by observations on M33 and M31. The great debate was over: The world consists of very many galaxies. It has therefore also been called the *island universe*, the galaxies being the islands. Our own island is the Milky Way; the others, if visible, appear as nebulae. Hubble himself, by the way, stuck with the term *nebulae* and did not use the now common word *galaxies*.

The velocity of celestial objects with respect to an observer on the earth can be measured by observing the *Doppler effect* of light reaching us from that object. Light with a given wavelength in the rest system of the object appears of larger wavelength, i.e., shifted in the direction of the red end of the visual spec-

---

[4] Henrietta Swan Leavitt (1868–1921)

trum, if the object moves away from us. One speaks of *redshift*. It is observed by comparing known spectral lines from celestial objects with the corresponding lines from laboratory sources. It was known, in particular, through the work of Slipher[5], that nebulae could have large redshifts. In his six-page paper of 1929 [7] Hubble presents measurements of velocity v and distance r for 24 nebulae and finds that, within errors, there is a direct proportionality between the two quantities. The relation is now written in the form $v = H_0 r$ and is called *Hubble's law* with H_0 being the *Hubble constant*.

Five years earlier evidence for the rise of redshift with the distance of galaxies had been published by Wirtz[6] in Kiel [8]. He had estimated the relative distances of nebulae from their apparent sizes, assuming that their real sizes were about equal. From the data he used Wirtz concluded that the velocity increase was less pronounced than later found by Hubble (logarithmic in distance, not linear). His important observation was largely ignored.

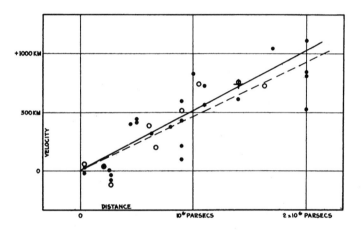

Hubble's results of 1929. Velocity (in kilometres per second) versus distance (in parsecs). From [7]. Reprinted with permission of Huntington Library.

The fastest nebula in Hubble's paper had a radial velocity of $1000 \, \mathrm{km \, s^{-1}}$. In the journal, in which it was published, it was preceded by a paper by Humason[7], reporting on a nebula nearly four times as fast [9]. The two men had been working together since Hubble's arrival on Mount Wilson. At that time Humason was *night assistant*, taking care of technical questions, developing plates, and the like. His first position on the mountain had been one of helper and janitor. His abilities and his keen interest in astronomy eventually gained him a position as *assistant astronomer*, a position like the one Hubble had when he came to Mount Wilson. In 1931, Hubble and Humason published a comprehensive paper [10] containing data on 40 new objects and extending by a factor of nearly 20 the range of distances and velocities with respect to Hubble's 1929 paper. Hubble's law still stood. It holds to this day. The numerical value of the *Hubble constant*, however, has changed with respect to the one Hubble himself had obtained. This is due to a better understanding of the scale of extragalactic distance measurements.

Hubble's law has a number of important implications. As we mentioned at the beginning, it shows that the universe is expanding. It singles out a partic-

---

[5] Vesto Melvin Slipher (1875–1969)   [6] Carl Wirtz (1876–1939)   [7] Milton La Salle Humason (1891–1972)

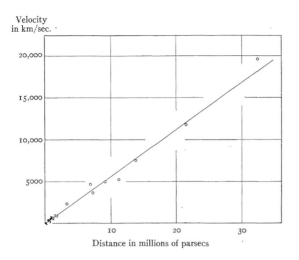

Velocity in km/sec.

Distance in millions of parsecs

Hubble's and Humason's results of 1931. From [10]. Reprinted with permission of the American Astronomical Society.

ular class of solutions of Einstein's field equations of general relativity. Other allowed solutions would describe a static universe or a contracting universe. Einstein himself, in 1915, informed by astronomers that the main structures observed in the sky gave no indications of change, had written down a static solution. The fact that all galaxies are moving away from us does not mean that we are the centre of the universe. Hubble himself used the analogy with a balloon which is being inflated. Points painted on the balloon's surface then all move away from one another. Hubble's law suggests that at some time all matter was concentrated in a small region and has been expanding into space ever since. This is the theory of the *Big Bang*, a term coined by Hoyle[8] (see also Episode 89). In the very simplest Big Bang model the age T of the universe is the inverse of the Hubble constant (see Box). Its numerical value agrees, within errors, with those found in other ways. Hubble himself, who, through his observations and analyses, provided so much information about the cosmos, carefully stayed away from theoretical cosmology, let alone speculations. Not even a comment on the importance of the simple proportionality form of his law can be found in his writings.

Galaxies as seen through the Hubble Space Telescope. Photo: NASA.

In 1990, a satellite consisting of a powerful telescope with associate equipment was launched, the *Hubble Space Telescope*. In the first 15 years of operation it took 700 000 photographs. It added considerably to our knowledge of the cosmos and also, since the most spectacular photographs were often reproduced in the press, drew attention to Hubble's name.

[1] Sandage, A., *Journal of the Royal Astronomical Society of Canada*, **83** (1989) 351.

[2] Christianson, G. E., *Edwin Hubble – Mariner of the Nebulae*. Ferrar, Straus & Giroux, New York, 1995.

[3] Wright, H., *Explorer of the Universe – A Biography of George Ellery Hale*. Dutton, New York, 1966.

[4] Hubble, E., *Astrophysical Journal*, **44** (1916) 190.

[5] Hubble, E., *The Realm of the Nebulae*. Yale University Press, New Haven, 1936.

[6] Hubble, E., *Astrophysical Journal*, **62** (1925) 409.

[7] Hubble, E., *Proceedings of the National Academy of Sciences*, **15** (1929) 168.

[8] Wirtz, C., *Astronomische Nachrichten*, **222** (1924) 21.

[9] Humason, M. L., *Proceedings of the National Academy of Sciences*, **15** (1929) 167.

[10] Hubble, E. and Humason, M. L., *Astrophysical Journal*, **74** (1931) 43.

---

[8] (Sir) Fred Hoyle (1915–2001)

46 Pauli Presents His Neutrino Hypothesis (1930)

All through the 1920s, the greatest discoveries took place in quantum theory, culminating in the development of quantum mechanics and its applications. Radioactivity or, as it gradually was called, *nuclear physics* stood in the background. But not quite. Experimental facts, concerning the β decay, were accumulated that were difficult to explain until Pauli resorted to a 'desperate remedy' by postulating the existence of a new particle, later called the *neutrino*. A beautiful account of the history of the neutrino was written by Pauli himself in his last year [1].

Information about nuclei is carried by α, β, and γ rays which are produced in nuclear transitions. If a nucleus emits an α particle or a β particle (which is an electron), it changes its electric charge and becomes a nucleus of another element; one speaks of α decay and β decay, respectively. If a nucleus is in an excited state, i.e., if it has an energy larger than the lowest possible energy of the same nucleus, it can emit a γ ray (which is a light quantum of short wavelength) and assume a state of lower energy. This process in the nucleus is quite similar to what happens in the atomic shell when a light quantum is emitted or absorbed. The α and γ rays from nuclear transitions have a *discrete* energy spectrum, i.e., only certain fixed energy values are observed. The situation is different for β rays. Their spectrum is *continuous*. This had been discovered, in contrast to expectations and also to previous experiments, by Chadwick who had introduced a new technique, using counters instead of photographic plates. That was in 1914 when Chadwick worked in Geiger's group in Berlin. With the outbreak of war, Chadwick was interned in a camp in Ruhleben at the outskirts of Berlin. One of his comrades in captivity was Ellis[1], whom Chadwick recruited to assist him in the experiments he was able to continue in the camp under difficult conditions. Ellis had planned a career as army officer but now opted for physics. After the war, with Chadwick, he joined Rutherford's group in Cambridge and continued to work on the β spectrum.

It was clear that the electron, observed in a β decay, did not carry away all the energy lost by the nucleus. One could, however, assume that the maximum β energy observed in the decay of a particular nucleus was the transition energy. Meitner in Berlin was convinced that, after leaving the nucleus, still in the shell of the atom, it lost energy by scattering on other electrons and that this energy, in the form of γ rays was not detected. Ellis disagreed. To prove his point without doubt, together with Wooster[2], he designed and performed an unusual experiment: He placed a β source in a calorimeter which would absorb β and γ rays and measure the total energy in the form of heat. Knowing the number of decays, from the heat he could compute the average energy per β

---

[1] Charles Drummond Ellis (1895–1980)   [2] William Alfred Wooster (1903–1984)

particle, which turned out to be much less than the maximum [2]. Lise Meitner was not convinced of the result but of the beauty of the method. She repeated the experiment with the assistance of Orthmann[3], a co-worker of Nernst, the world expert in calorimetry. They found, with higher precision, that Ellis was right [3]. Energy was missing.

Bohr was willing to give up the law of energy conservation for the special case of β decay. This seems very implausible to us now. But one should remember that, for what happened in the atomic shell, Newtonian physics had to be given up and replaced by quantum mechanics. Although energy was conserved in quantum mechanics, it was expected by many that quantum mechanics would not describe the nucleus and that further sacrifices had to be made.

But there was a connection between the nucleus and quantum mechanics, provided by the nuclear spin, which showed itself in the optical spectra of atoms and molecules. Just as the spin of the electron had led to a *fine structure* of the spectra, the spin of the nucleus, coupling magnetically to the spins of the shell, gave rise to a *hyperfine structure*. Experiments and computations were difficult; but it was possible to extract the spins of some nuclei from the measured spectra. The results were puzzling. In his celebrated *Bakerian Lecture* of 1920 [4] Rutherford had described the atomic nucleus as composed of protons and electrons. A nucleus, characterized by two integer numbers, the *mass number* A and the charge number Z, had to contain A protons to account for its mass. Since it did not possess A but only Z elementary charges, it had also to contain $A - Z$ electrons compensating the charge of $A - Z$ protons. The spin of the proton, the nucleus of the hydrogen atom, was found to be $\frac{1}{2}\hbar$; the proton, like the electron, is a *fermion*. The number of protons and electrons in the nucleus was therefore $2A - Z$ and the spin of the nucleus would be half integer if that number was odd and integer if it was even. Accordingly, the nucleus would be a fermion in the former case and a boson in the latter. This 'alternation law', however, was not verified in the spectra: The nitrogen nucleus $^{14}_{7}\text{N}$ with $A = 14, Z = 7$ behaved like a boson rather than a fermion and so did the lithium nucleus $^{6}_{3}\text{Li}$. They displayed the 'wrong statistics', Bose rather than Fermi.

Pauli solved both the energy problem and the statistics puzzle with his new particle. He first wrote about it in an open letter [5] to be read to the participants of a small physics conference which was held in Tübingen in December 1930. The lady addressed in the letter was Lise Meitner. She preserved it for posterity. We quote from the English translation by Schlapp[4], published in [1].

Pauli in 1933

> Dear Radioactive Ladies and Gentlemen,
> As the bearer of these lines, for whom I pray the favour of a hearing will explain in more detail, I have, in connection with the "wrong" statistics of the N and Li6-nuclei as well as the continuous β-spectrum, hit upon a desperate remedy for rescuing the "alternation law" of statistics and the energy law. This is the possibility that there may exist in the nuclei electrically neutral particles, which I shall call neutrons, which have spin 1/2, obey the exclusion principle and moreover differ from light quanta in not travelling with the velocity of light. The mass of the neutrons

[3] Wilhelm Orthmann (1901–1945) [4] Robert Schlapp (1899–1991)

would have to be of the same order as the electronic mass and in any case not greater than 0.01 proton masses. – The continuous β-spectrum would then be understandable on the assumption that in β-decay, along with the electron a neutron is emitted as well, in such a way that the sum of the energies of neutron and electron is constant. [...]

I do not in the meantime trust myself to publish anything about this idea, and in the first place turn confidently to you, dear radioactive folk, with the question – how would things stand with regard to the experimental detection of such a neutron if it possessed an equal or perhaps ten times greater penetrating power than a γ ray? [...]

Unfortunately I cannot appear personally in Tübingen, since on account of a dance which takes place in Zurich on the night of 6 to 7 December I cannot get away from here. [...]

Your most humble servant,   W. Pauli

The particle, which Pauli called *neutron*, was given properties to resolve the difficulties: An additional spin of $\frac{1}{2}\hbar$ would turn a boson into a fermion; the high penetrating power would assure that the energy carried by the particle would go unobserved; its low mass assured that the nuclei were not made appreciably heavier by its presence.

Half a year later, in June 1931, Pauli gave a lecture at a meeting of the American Physical Society in Pasadena, where he talked about the new particle but still he did not have the talk printed. On the way back, he went to another conference in Rome where he met Fermi, who, as Pauli writes, 'showed a lively interest in my idea'. In fact, in 1934 Fermi published a theory of the β decay in which Pauli's particle plays a crucial role (see Episode 53). Then, in 1932, the *neutron* was discovered by Chadwick (Episode 48). This was not Pauli's particle but a fermion of about the mass of the proton. The nucleus was now understood to be composed of Z protons and $A - Z$ neutrons and to be a boson or a fermion for A even or odd, respectively. Chadwick's neutron solved the statistics puzzle but not the energy problem. For that Pauli's particle was still needed and Fermi, in discussions in his group in Rome, called it the *neutrino*, the little neutron. The name stuck.

Meanwhile also the *positron* had been discovered so that the existence of a new particle was no longer such an outlandish idea. Pauli talked again on the neutrino hypothesis at the Solvay Conference in 1933 and this time he had his contribution printed [6]. He criticizes Bohr's idea of energy non-conservation and states that he sees no reason why, in conservation laws, electric charge, which was obviously conserved, should take preference over energy. The neutrinos are described as follows:

Concerning the properties of these neutral particles, the atomic weights of the radioactive elements teach us first of all that their mass cannot surpass much that of the electron. To distinguish them from the heavy neutron, Mr Fermi has proposed the name "neutrino". It is possible that the proper mass of the neutrinos be equal to zero, so that they would propagate with the speed of light, like photons. In any case their penetration power would exceed much that of photons of the same energy. It seems admissible to me that the neutrinos possess a spin of $1/2$ and that they obey the statistics of Fermi, although the experiments provide us with no direct proof of this hypothesis.

From its conception, the neutrino has been important for the theory of nuclear and particle physics and it has turned out to be a fundamental constituent of the world, on equal footing with the electron. We will therefore meet with it again in future episodes. Concerning its mass, for many decades most physicists believed that is was zero. Only as late as 2001 it was shown that neutrinos have tiny, as yet unknown, masses (see Episode 100).

Pauli was an exceptional man in many respects. Not only did he enter physics as a prodigy and contribute outstanding discoveries under his own name but he contributed also to the success of others through personal discussions and through an extensive correspondence. The preserved letters [7] allow us still now to witness how ideas grew before they were published in polished form. His criticism was sought but also dreaded. He has therefore been called the 'Conscience of Physics' (Gewissen der Physik). Ehrenfest, who was very fond of him, conferred on him the title 'God's Whip' (Geissel Gottes), because he would not relent in a discussion until all the weak points had been made perfectly clear. Pauli himself enjoyed this title and signed his letters to Ehrenfest with it. He maintained that it recognized his divine mission.

Pauli and Ehrenfest on a boat to Copenhagen.

There are innumerable Pauli anecdotes. The best one is like a running gag, it was called the *Pauli effect*: Experiments invariably would go wrong when Pauli was near; be he in the room, in the institute, or just riding through town in a train. Pauli was proud of his fame and a colleague wanted to please him. He devised a mechanism that would make the lamp drop from the ceiling when the laboratory door was opened. It worked perfectly well when rehearsed. But when Pauli himself entered through that door it failed [8]. I also like the remarks he made about his young co-workers in Zurich who later became celebrities [9]. About Peierls, a future knight of the British Empire, he said: 'He talks so fast that by the time you understand what he is saying, he is already asserting the opposite.' And when Landau was discouraged in a discussion and asked if Pauli thought that he had been talking nonsense, the future Nobel laureate got the answer: 'Oh no, far from it, far from it. What you said was so confused, one could not tell if it was nonsense!'

In 1940, a German invasion of Switzerland seemed possible and Pauli moved to Princeton where he worked at the Institute for Advanced Studies. He did not travel to Stockholm when he was awarded the Nobel Prize in 1945 but there was a banquet in Princeton. A number of talks were given and then, unexpectedly, Einstein rose to say a few improvized very personal words. There is no written record, only a letter by Pauli to Born, written shortly after Einstein's death. It shows how much Pauli was touched. We quote from a translation in [10]:

> Never will I forget the discourse he has pronounced about me and for me in 1945 in Princeton after I had received the Nobel Prize. It was like a king who abdicates and who installs me as sort of a 'son of choice' [Wahl-Sohn] as the successor.

[1]   Pauli, W., *Writings on Physics and Philosophy*, chap. 20. Springer, Berlin, 1994.

[2]   Ellis, C. D. and Wooster, W. A., *Proceedings of the Royal Society*, **A117** (1927) 109.

[3]   Meitner, L. and Orthmann, W., *Zeitschrift für Physik*, **60** (1930) 143.

[4]   Rutherford, E., *Proceedings of the Royal Society*, **A97** (1920) 374.

[5]   Pauli, W. In [7], Vol. II, p. 39.

[6]   Pauli, W., *Septième Conseil de Physique Solvay 1933*. Gauthier–Villars, Paris 1934. Also contained in [1].

[7]   Pauli, W., *Wissenschaftlicher Briefwechsel, 4 vols.* Springer, Berlin, 1985–2005.

[8]   Telegdi, V. In [11], p. 115.

[9]   Peierls, R., *Bird of Passage – Recollections of a Physicist*. University Press, Princeton, 1985.

[10] Enz, C. P., *No Time to be Brief – A Scientific Biography of Wolfgang Pauli*. Oxford University Press, Oxford, 2002.

[11] Enz, C. P. and von Meyenn, K. (eds.), *Wolfgang Pauli – Das Gewissen der Physik*. Vieweg, Braunschweig, 1988.

Einstein and Pauli in 1926.

Although Pauli was a permanent member of the Institute in Princeton and had become an American citizen, in 1946 he returned to Europe and resumed his chair at the ETH. It is conjectured that he disapproved of the influence of the military on research which had grown with the development of the atomic bomb. He died in Zurich in 1958 at the age of only fifty-eight.

Lawrence and the Cyclotron (1931)

47

One night early in 1929 in Berkeley, California, Ernest Lawrence[1] [1,2], in his habitual effort to keep abreast of the literature, came across an article in a German journal on electrical engineering. Lawrence had recently arrived from Yale University although his friends there had warned him of the 'unscientific climate of the West' [2]. In 1930 he would become the youngest full professor on the Berkeley faculty. The article that caught his eye was based on the doctoral dissertation which Wideröe[2] [3], a young Norwegian engineer, had submitted to the Technische Hochschule in Aachen [4]; it was entitled *On a New Principle for the Production of High Voltages*. Lawrence later described how he extracted the relevant information from that paper [5]: 'Not being able to read German easily, I merely looked at the diagrams and photographs of Wideröe's apparatus and from the various figures in the article was able to determine his general approach to the problem ...' Lawrence was interested because the paper demonstrated a new way to accelerate charged particles.

Lawrence

A charged particle can be accelerated by an electric field and in this way gain energy (see Box in Episode 4). If it carries one elementary electric charge e and travels a distance corresponding to the potential difference (or voltage) U, the energy gained is $E = eU$. The energy unit used in this context is one *electron volt* or $1\,\text{eV}$. It is the energy gained in a potential difference of $1\,\text{V}$. Radioactivity provided particles with energies of a few million electron volts (MeV). If one wanted projectiles with even higher energies to induce nuclear reactions, one would have to accelerate them in the laboratory and that seemed to imply voltages of many million volts. This was awkward because such voltages would be very difficult to produce and even more difficult to maintain without breakdown due to sparks in the apparatus. Wideröe was the first to demonstrate that no excessive voltage was needed because the particles could be repeatedly accelerated by one moderate voltage. He was inspired by a proposal made by the Swedish physicist Gustav Ising in 1924 who wanted to accelerate particles by making them 'surf' on an electric 'travelling wave'. Wideröe realized that the necessary travelling waves could not be achieved. Instead, he designed a system of metal tubes, which the particles would have to pass through. The tubes were placed one after the other and, in alternation, connected to ground and to a radio-frequency voltage. When the particles passed a gap between two tubes they experienced an accelerating voltage. Within the tubes there was no field, and while the particles drifted through them, the voltage went through a complete cycle of the radio frequency so that the particles experienced the same acceleration at the next gap. Of course, the tubes had to grow in length along the apparatus to take into account the increasing speed of the particles.

---

[1] Ernest Orlando Lawrence (1901–1958), Nobel Prize 1939   [2] Rolf Wideröe (1902–1996)

Wideröe built an apparatus with two gaps and showed that he could accelerate particles with twice the peak voltage of his radio-frequency source.

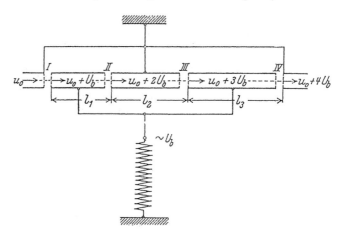

Illustration of the drift-tube principle in Wideröe's thesis. Reprinted with kind permission from Springer Science+Business Media: [4], Fig. 1.

Livingston

Lawrence was fascinated with the idea of using the same voltage over and over again but realized that an accelerator for nuclear physics would be inconveniently long. He thought of 'winding up' the path of the particles using a magnetic field. A simple calculation showed him that a homogeneous magnetic field would do the trick (see Box). The revolution frequency, now called the *cyclotron frequency*, of a charged particle in a plane perpendicular to a magnetic field does not depend on the particle's energy. The idea of the *cyclotron* was born: A charged particle moves on a circular path in a plane perpendicular to a magnetic field. If it is accelerated by an electric field, the revolution frequency stays the same, only the radius of the trajectory increases. The acceleration can be performed at discrete places in each revolution and the field at these places can be provided by a radio-frequency source with a frequency directly related to the cyclotron frequency. Lawrence set to work immediately and already in 1930 there was the first preliminary report [6].

In April 1931 quantitative results were reported by Lawrence and Livingston[3], then a graduate student to whom Lawrence had assigned the construction of the accelerator as subject of his thesis. The report was given at the Washington Meeting of the American Physical Society. We quote the short printed version [7] almost in its entirety because it is still a very neat description of the cyclotron:

A method for producing high speed hydrogen ions without the use of high voltages was described at the September Meeting of the National Academy of Sciences [6]. The hydrogen ions are set in resonance with a high frequency oscillating voltage between two hollow semicircular plates[4] in vacuum, and are made to spiral around in semicircular paths inside these plates by a magnetic field. Each time the ions pass from the interior of one plate to that of the other they gain energy corresponding to the voltage across the [gap between the] plates. This method has now been tried out with the following results: Using a magnet with pole faces

---

[3] Milton Stanley Livingston (1905–1986)   [4] These structures, because of their similarity with the capital letter 'D', were later simply called 'dees'.

Cyclotron Basics

Cyclotron Radius: We consider a particle with charge q, mass m, and velocity v in a homogeneous magnetic field of strength B for the case in which the motion is in a plane perpendicular to the magnetic field. The particle trajectory is a circle of radius

$$R = \frac{m}{q}\frac{v}{B} \quad,$$

which increases with the velocity and therefore also with the energy of the particle.
Cyclotron Frequency: The revolution frequency ν on the circles is the inverse $1/T$ of the time needed for one revolution. The corresponding angular or circular frequency is

$$\omega = 2\pi\nu = 2\pi\frac{1}{T} = 2\pi\frac{v}{2\pi R} = \frac{q}{m}B \quad.$$

This *cyclotron frequency* is independent of the particle energy. It becomes energy dependent, however, once the particle velocity v is no longer small compared to the velocity c of light. Then, because of special relativity, the particle mass m depends on the velocity.

---

10 cm in diameter and giving a field of 12,700 gauss, 80,000 volt hydrogen molecule ions have been produced using 2000 volt high frequency oscillations on the plates. [...] These preliminary experiments indicate clearly that there are no difficulties in the way of producing one million volt ions in this manner. A larger magnet is under construction for this purpose.

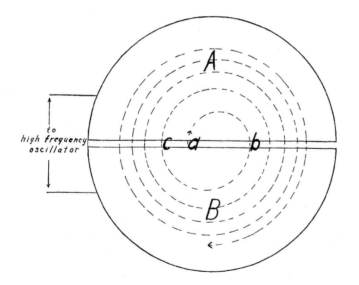

Operating principle of the cyclotron. The two D-shaped electrodes are A and B. Three consecutive positions of an accelerated ion are denoted by a, b, c. Figure reprinted with permission from [8]. Copyright 1932 by the American Physical Society.

The first cyclotron had a diameter of 4 inch. The next one, which indeed yielded protons of more than $1\,\text{MeV}$, had an 11-inch diameter and was described by Lawrence and Livingston in a comprehensive paper [8]. They were astonished and pleased that with relative ease they got protons of such a high energy and rather large intensity. Two decades later Lawrence would write [5]: 'We soon recognized that the focussing actions of the electric and magnetic fields were responsible for the relatively large currents of protons that reached the periphery of the apparatus; but we must acknowledge that here experiment preceded theory!' The new discipline of *accelerator theory* soon was born and helped to create new generations of particle accelerators (Episode 64), which now reach energies of more than 1 TeV, i.e., more than a million MeV.

A few months before the second cyclotron was completed, the disintegration of nuclei by artificially accelerated particles was reported from the Cavendish Laboratory (Episode 50). Rather than disappointed by having missed a discovery the Berkeley group was 'overjoyed with this news, for it constituted definite assurance that the acceleration of charged particles to high speeds was a worth-while endeavour' [5] and promptly verified and extended the Cavendish results [9].

The next cyclotron [10], 27 inch in diameter, was completed in 1934 and yielded protons of up to 5 MeV. It was built with 'incredible speed'. Alvarez writes about these times [2]:

> The great enthusiasm for physics with which Ernest Lawrence charged the atmosphere of the Laboratory will always live in the memory of those who experienced it. The Laboratory operated around the clock, seven days a week, and those who worked a mere seventy hours a week were considered by their friends "not to be very interested in physics".

The 27-inch cyclotron was later extended to 37 inch [11] and would also accelerate deuterons, the nuclei of a heavy isotope of hydrogen consisting of a proton and a neutron, and helium nuclei. The latter, because they carry two elementary charges, were given energies up to 11 MeV. Only in 1936 the name *cyclotron* appears in publications. Before, it had been called *apparatus for multiple acceleration*. Later even more powerful cyclotrons were built in Berkeley and in many other laboratories. The cyclotron became the workhorse of nuclear physics.

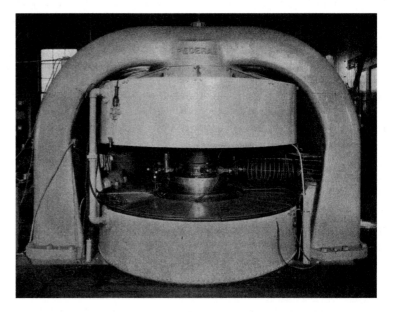

The 37-inch cyclotron of 1936. Figure reprinted with permission from [11]. Copyright 1936 by the American Physical Society.

From the outset Lawrence was striving not only for high energy but also for large intensity of the accelerated particles, which would allow the study of rare reactions. The publication [11] contains a spectacular photograph: A beam of 5.8-MeV deuterons leaves the vacuum chamber of the cyclotron through a thin platinum window and begins to propagate in the air of the laboratory. The beam corresponds to an electric current of $5\,\mu$A, i.e., to about 3×10^{13} deuterons per second. Air molecules hit are excited and emit light until the deuterons have lost their energy after travelling 25 cm in air.

A deuteron beam, in air, produced by the 37-inch cyclotron. Figure reprinted with permission from [11]. Copyright 1936 by the American Physical Society.

The Berkeley group missed another great discovery, that of *artificial radioactivity*, made in Paris in 1934 (see Episode 54). Due to the high beam intensity a lot of artificial radioactivity was produced in the laboratory and also observed to give pulses in Geiger–Müller counters. But these were thought to misbehave and less sensitive detectors were used. Lawrence writes [5] that after the discovery was announced his group again 'was overjoyed at the richness of the domain in the nucleus accessible to particles of several million electron volts'.

From 1931 to 1950 and again for a few years in the mid-1950s Berkeley accelerators provided the highest-energy particles in the world. Important discoveries were made, especially in the production of new isotopes and even new elements. In 1937, in a scrap of molybdenum, irradiated in a cyclotron and given by Lawrence to Segrè, the element 43 in the Periodic Table was observed by Segrè and Perrier[5] in Palermo. This element, although placed in the middle of the Periodic Table, is radioactive and no stable isotope exists. It can only be artificially produced and therefore was given the name *technetium* by its discoverers. In 1940 Kamen[6] and Ruben[7] produced the carbon isotope ^{14}C by bombarding graphite with a beam of deuterons. From 1940 onwards the Berkeley group, in particular, McMillan and Seaborg, produced *transuranic elements*, i.e., elements which occupy positions in the Periodic Table beyond uranium (number 92). The first two were called *neptunium* (number 93) and *plutonium* (number 94), see Episode 61. Many more transuranic elements were produced in Berkeley and later also elsewhere. An element found in 1949 (number 97) received the name *berkelium*. Element number 103, first produced and identified in 1961, not long after Lawrence's premature death was named *lawrencium* in his honour.

Ernest Lawrence not only created a thriving centre of fundamental research but he also transformed the way research was done. His name is often connected to what is now sometimes called 'Big Science'. Accelerators resembled large machines in a factory more than laboratory equipment. Work around them had to be organized with different teams doing different experiments according to an agreed time schedule. It is remarkable that the scientific spirit of individual researchers was not seriously hampered by Big Science. Nowadays large groups successfully cooperate in experiments at accelerator centres that could in no way be performed by a few individuals.

[1] Childs, H., *An American Genius: The Life of Ernest Orlando Lawrence*. Dutton, New York, 1968.

[2] Alvarez, L. W., *Biographical Memoirs*, **XLI** (1970) 251.

[3] Waloschek, P., *The Infancy of Particle Accelerators – Life and Work of Rolf Wideröe*. Vieweg, Braunschweig, 1994.

[4] Wideröe, R., *Archiv für Elektrotechnik*, **21** (1928) 387.

[5] Lawrence, E. O., *Nobel Lectures, Physics 1922–1941*, p. 430. Elsevier, Amsterdam, 1965.

[6] Lawrence, E. O. and Edelfsen, N. E., *Science*, **72** (1930) 376.

[7] Lawrence, E. O. and Livingston, M. S., *Physical Review*, **37** (1931) 1707.

[8] Lawrence, E. O. and Livingston, M. S., *Physical Review*, **40** (1932) 19.

[9] Lawrence, E. O., Livingston, M. S., and White, M. G., *Physical Review*, **47** (1932) 150.

[10] Lawrence, E. O. and Livingston, M. S., *Physical Review*, **45** (1934) 608.

[11] Lawrence, E. O. and Cooksey, D., *Physical Review*, **50** (1936) 1131.

---

[5] Carlo Perrier (1886–1948) [6] Martin Daniel Kamen (1913–2002) [7] Samuel Ruben (1913–1943)

48 Chadwick Discovers the Neutron (1932)

In June 1920, Rutherford presented his *Bakerian Lecture* on the *Nuclear Constitution of Atoms*. Among experimental facts and the assertion, that the atomic nucleus is composed of protons and electrons, it contains the following speculation [1]:

> Under some circumstances, however, it may be possible for an electron to combine much more closely with the H nucleus [the hydrogen nucleus, i.e., the proton], forming a kind of neutral doublet. Such an atom would have novel properties. Its external [electric] field would be practically zero, and in consequence it should be able to move freely through matter. Its presence would probably be difficult to detect by the spectroscope, and it may be impossible to contain it in a sealed vessel. On the other hand it should enter readily the structure of atoms, and may either unite with the nucleus or be disintegrated by its intense field, resulting possibly in the escape of a charged H atom or an electron or both.

In 1921, the name *neutron* appears in reports of the Cavendish group which made several attempts to find it. A decade later Rutherford's conjecture seems to have been forgotten everywhere else but not in Cambridge.

In June 1930, Bothe and his co-worker Herbert Becker reported unexpected results they had obtained at the Physikalisch-Technische Reichsanstalt in Berlin when bombarding light elements with α particles from a polonium source [2]. With a Geiger counter they observed a radiation which was absorbed by lead to a much lesser degree than ordinary γ radiation, which of all known radiations was the most penetrating. The intensity was by far the largest when they irradiated beryllium. A smaller effect was found with lithium and boron. Bothe and Becker concluded that they had discovered a new kind of very *hard*, i.e., very penetrating γ rays. They continued to study their new radiation but, although it showed some odd features, did not change their interpretation.

While Bothe and Becker had used a Geiger counter to study the radiation, more subtle methods were employed by Curie and Joliot. Irène Curie was the daughter of Marie Curie and worked in her mother's institute in Paris. Her husband, Frédéric Joliot, had introduced new experimental techniques in the laboratory. At first they used an ionization chamber connected to a very sensitive electrometer. Irradiating the chamber with Bothe–Becker radiation they registered a small current as expected from γ rays, but when they placed material containing hydrogen atoms, like paraffin, in front of the ionization chambers they observed a much higher current. This current, they argued, was caused by the high ionization of protons, which were liberated from the paraffin by the radiation. Protons could not be produced in the ordinary Compton effect by incoming γ rays. But Curie and Joliot thought that the new hard γ radiation might give rise to a Compton effect not on electrons in the atomic shell but on protons in the nucleus, in particular, the hydrogen nucleus, which consists of just one proton. Using the formulae governing the Compton effect, they showed that γ rays with an energy of $50\,\mathrm{MeV}$ were needed to explain their measurements, much higher than were known to appear in radioactivity. Their findings are reported in a paper, read at a session of the French Academy of Sciences on 18 January 1932 [3].

A month later, on 22 February 1932, they reported on the observation of proton tracks in a cloud chamber which was irradiated by Bothe–Becker radiation and which contained a plate of paraffin [4]. The tracks came from that material, and, in some cases, originated directly in the water vapour which filled the cloud chamber. Cloud-chamber photographs with beautiful proton tracks were presented in the session of 7 March 1932 [5]. By that time, however, their correct interpretation had been given by Chadwick in Cambridge.

Chadwick

Cloud-chamber photograph by Curie and Joliot. It shows the track of a proton, knocked out of a paraffin plate near the bottom of the picture, by a neutron. From [4]. Reprinted with permission.

We met Chadwick[1] [6] already in Episodes 34 and 46 when he discovered the continuous β spectrum in Berlin where, against his will, he had to stay through the war years 1914–18. He had gone to school in Manchester and studied physics there under Rutherford. After the war he resumed the work with his teacher who had moved to Cambridge, studying nuclear disintegration though α particles. The two scientists, without success, also searched for Rutherford's neutrons as late as 1929. Chadwick was startled by the news from Paris about the appearance of protons. He discussed the matter with Rutherford and became convinced that the most natural explanation was the assumption that protons were hit by particles of about their own mass and were recoiling under the impact. These particles would have to be electrically neutral, for otherwise they would themselves leave tracks. They could be Rutherford's neutrons. To test this idea he irradiated an ionization chamber, equipped with a signal amplifier, and studied the signal height obtained when different materials were exposed to the Bothe–Becker radiation. The results were as expected

---

[1] (Sir) James Chadwick (1891–1974), Nobel Prize 1935

from atoms recoiling under the influence of incoming particles with a mass near that of the proton.

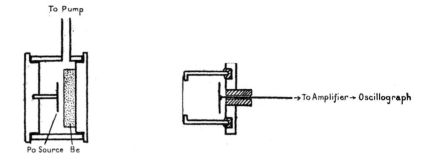

Chadwick's apparatus. Neutrons are produced in the polonium–beryllium source on the left. Recoil protons are detected in the ionization chamber on the right. Different materials, e.g., paraffin, can be inserted between the two instruments. From [7], Fig. 1. Reprinted with permission.

Chadwick used a powerful energy argument to exclude the idea of γ rays. Such γ rays, he argued, might be produced in the reaction $\alpha + {}^9\text{Be} \rightarrow {}^{13}\text{C} + \gamma$ and from the known masses of the nuclei involved, the γ energy could be computed to be about 14 MeV. A photon of that energy could not make protons recoil in the way they were observed to, because they are not *matched* to protons just as a tiny marble is not matched to a full-size billiard ball. But neutrons, having about the same mass as protons, are matched to protons as one billard ball is to another. Chadwick's first paper [8], entitled *Possible Existence of a Neutron* and only a little more than half a page long, was received by *Nature* on 17 February 1932, only a month after the Curie–Joliot report of 18 January.

The Bothe–Becker radiation indeed consists of neutrons which are produced, as Chadwick proposed, in the reaction of an α particle, which is a helium nucleus ${}^4\text{He}$, with a beryllium nucleus,

$$\alpha + {}^9\text{Be} \rightarrow {}^{12}\text{C} + n \quad .$$

More details were presented by Chadwick in a conference session, chaired by Rutherford, on 28 April. His complete paper, confidently entitled *The Existence of a Neutron* [7], was received on 10 May. The arguments are now also based on a cloud-chamber experiment by Feather, who, also in the Cavendish Laboratory, had studied the tracks of protons and nitrogen nuclei recoiling from collisions with neutrons. Using the simple laws of elastic collisions of bodies with different mass, the mass of the neutron could be estimated from the ratio of the maximum velocities of the two types of recoil particles. It turned out to be the same as the proton mass within a measurement error of 10%. A better value was obtained from the observation, already made by Bothe and Becker, that also boron emitted the new radiation if it was exposed to α particles. The reaction would be $\alpha + {}^{11}\text{B} \rightarrow {}^{14}\text{N} + n$. From the known masses of the nuclei involved and the measured kinetic energies a mass value for the neutron was found which was equal to the proton mass within 0.3%.

There is an important section in Chadwick's paper about *The Passage of the Neutron through Matter*. He argued, as Rutherford had done in 1920, that neutrons would travel freely through matter, colliding only with atomic nuclei. From the observed frequency of such collisions the nuclear radii of some elements were estimated to be on the order of 5×10^{-13} cm.

Chadwick still advances Rutherford's idea of the neutron as a complex particle consisting of a proton and an electron as the most plausible but does not exclude a more radical point of view:

> It is, of course, possible to suppose that the neutron may be an elementary particle. This view has little to recommend at present, except the possibility of explaining the statistics of such nuclei as ^{14}N.

This was in fact the interpretation soon generally accepted. The neutron was a particle in its own right; it was a *fermion* with spin $\frac{1}{2}\hbar$, which it could not be if composed by a proton and an electron, which are both fermions themselves. The 'statistics problem' to which Chadwick alluded and which we discussed in Episode 46 was thus solved.

At a conference on the history of science, thirty years later, Chadwick presented some *personal notes* [10] on the search of the neutron in Cambridge which started right after Rutherford's Bakerian Lecture. Rutherford had convinced him that such a particle was needed in the nucleus and Chadwick in different experiments, some of them 'so far-fetched as to belong to the days of alchemy', tried to produce and detect it. He even considered beryllium as a good source of neutrons; since it was known that the mineral beryl contains helium he thought the beryllium nucleus might split into two α particles and a neutron under cosmic radiation. Thus, he irradiated beryllium with α, β, and γ rays and, without success, used the scintillation method and later counters trying to detect an effect. When, later, an effect was found by the Joliots in the form of recoiling protons, he was the first to interpret it correctly. In a *postscript* to this story Chadwick writes:

> The decisive clue had indeed been supplied by others. This after all is not unusual; advances in knowledge are generally the result of the work of many minds and hands. But I could not help but feel that I ought to have arrived sooner. I could offer myself many excuses: lack of facilities, and so on. But beyond all excuses I had to admit, if only to myself, that I failed to think deeply enough about the properties of the neutron, especially those properties which would most clearly furnish evidence on its existence.

The discovery of the neutron completely revolutionized the physics of the atomic nucleus, both experimentally and theoretically. Since they are not electrically repelled they provide an ideal probe to study the nucleus. The first nuclear reaction induced by a neutron was reported by Feather [9]. In a cloud chamber filled with nitrogen he observed the process

$$n + {}^{14}\text{N} \to {}^{11}\text{B} + \alpha \quad ;$$

after absorption of a neutron a nitrogen nucleus splits up into a nucleus of boron and one of helium, i.e., an α particle.

We shall see in Episode 51 how Heisenberg in Leipzig used the neutron to develop the concept of *isospin* and to devise the first quantum theory of nuclear structure. Fermi and his group in Rome were to make the most of the neutron as an experimental tool (Episode 55) before uranium fission by neutrons was discovered by Hahn and Strassmann in Berlin (Episode 60).

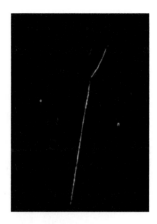

The nuclear reaction $n + {}^{14}\text{N} \to {}^{11}\text{B} + \alpha$ as observed by Feather in a cloud chamber. The neutron (leaving no track) enters from the left and reacts with a nitrogen nucleus of the chamber gas. The short track is due to the boron nucleus, the long one to the α particle. From [9], Fig. 18. Reprinted with permission.

[1] Rutherford, E., *Proceedings of the Royal Society*, **A97** (1920) 374.

[2] Bothe, W. and Becker, H., *Naturwissenschaften*, **18** (1930) 705.

[3] Curie, I. and Joliot, F., *Comptes Rendus Acad. Sci.*, **194** (1932) 273.

[4] Curie, I. and Joliot, F., *Comptes Rendus Acad. Sci.*, **194** (1932) 708.

[5] Curie, I. and Joliot, F., *Comptes Rendus Acad. Sci.*, **194** (1932) 876.

[6] Brown, A., *The Neutron and the Bomb – A biography of Sir James Chadwick*. Oxford University Press, Oxford, 1997.

[7] Chadwick, J., *Proceedings of the Royal Society*, **A136** (1932) 692.

[8] Chadwick, J., *Nature*, **129** (1932) 312.

[9] Feather, N., *Proceedings of the Royal Society*, **A136** (1932) 709.

[10] Chadwick, J., *Some Personal Notes on the Search of the Neutron*. In Guerlac, H. (ed.): *Proceedings of the Tenth International Congress of the History of Science*, p. 159. Hermann, Paris, 1964.

<table>
<tr><td>**49**</td><td># Anderson Discovers the Positron (1932)</td></tr>
</table>

Anderson

Anderson[1] was born in New York but grew up in Los Angeles, where he attended a technically oriented high school. From his physics teacher he heard of the recently founded California Institute of Technology (Caltech) and, although discouraged by other teachers, he applied and was admitted as student of physics. The domineering figure there was Millikan. Besides leading Caltech as a whole he was also Chairman of the Division of Physics, Mathematics, and Astronomy. He chose the subject of Anderson's Ph.D. thesis, the spatial distribution of electrons produced by X rays, but as Anderson recollected, the busy 'Chief' never came to see him in his lab.

Before the completion of his thesis Anderson saw Millikan to inquire if he could stay on at Caltech for post-doctoral studies but was told that, having done both his undergraduate and his graduate work there, he should go elsewhere with a National Research Council fellowship. He applied for a fellowship and got an invitation by Compton to work in Chicago in case he would get one. But some time later, to his surprise, Millikan asked him to stay on for a year and set up equipment for cosmic-ray measurements. His chances of getting a fellowship would be much better with an additional year of experience. Since Millikan was a member of the fellowship selection committee, Anderson, who by then was looking forward to move to Chicago, felt that he had no choice in the matter. He stayed and shifted his work to cosmic rays.

Until the end of the 1920s cosmic rays were thought to be high-energy gamma rays. But in 1929 Bothe and Kolhörster had registered very penetrating charged particles (see Episode 34) and also cloud-chamber photographs [2] had been obtained by Skobelzyn[2]. Having obtained his Ph.D. in June 1930, Anderson designed and built a cloud chamber which could be placed in a high magnetic field. If a charged particle traverses such a 'magnetic cloud chamber', it is deflected by the field and leaves a curved track. The momentum of the particle can be obtained by measuring the curvature. The tracks of a positive and a negative particle are curved in opposite directions. The magnetic field, which could be as high as 2.5 Tesla or 25 000 Gauss, was provided by an electromagnet with water-cooled copper windings powered by a large generator, originally used for a wind tunnel at Caltech.

Anderson produced his first photographs with tracks from cosmic rays in 1931. Millikan was convinced that the primary cosmic rays were photons of high energy and he expected the tracks to be all electrons, resulting from Compton scattering of the primary photons off electrons in the matter somewhere above the cloud chamber. Anderson, therefore, was surprised when he

---

[1] Carl David Anderson (1905–1991), Nobel Prize 1936   [2] Dimitry Skobelzyn (1892–1980)

found tracks from positive and negative particles to be about equally frequent. Millikan was in Europe and showed the photographs he got from Anderson in Paris and in Cambridge. He took the positive particles to be protons. In April 1932 Millikan and Anderson submitted a paper [3] in which they explained the findings by stating that a primary photon can also react with an atomic nucleus. Their reasoning: 'Positive particles obviously come only from the nucleus.' In other words, the positive tracks were taken to be caused by protons or even heavier fragments of the nucleus.

More detailed studies by Anderson showed, however, that at least some of the positive tracks could not be caused by protons. One way to classify tracks in a cloud chamber is to look at the density of droplets along their path. It is proportional to the number of ion pairs per unit length of path formed by the particle traversing the chamber gas. For a particle with one elementary charge, i.e., a proton or an electron, it is small if the particle moves with a velocity not far from the speed of light. One speaks of *minimum ionization*. Since the proton is so much heavier than the electron, for not too large momenta, less than, say, $1\,\mathrm{GeV}/c$, the speed of a proton is much smaller than that of an electron with the same momentum, i.e., the same curvature in the magnetic field. In this momentum range the tracks of protons and electrons look quite different: Proton tracks have many droplets per inch; they are thick. Electrons tracks have few and are thin. Anderson found that his tracks of positive particles in that momentum range had the track density of electrons rather than of protons.

There was still one way out: A negative particle entering the chamber from below and moving *upwards* would leave the same kind of track as a positive particle entering the chamber from above and moving *downwards*. (If the negative particle is deflected to the right along its path, then the positive particle is deflected to the left.) Anderson was inclined to take the tracks in question to be caused by upward-moving electrons. But Millikan, pointing out that it was common knowledge that cosmic-ray particles travel downwards, insisted that they were produced by protons, in spite of the anomalous droplet density.

Anderson found a way to measure the direction taken by a particle. He inserted a lead plate in the cloud chamber. A particle traversing the plate would lose part of its energy and momentum. The radius of curvature of the track would be smaller after the traversal. In 1961 Anderson recalled the following about this stage of the experiment [4]:

> It was not long after the insertion of the plate that a fine example was obtained in which a low-energy light-weight particle of positive charge was observed to traverse the plate from below and moving upward through the lead plate. Ionization and curvature measurements clearly showed this particle to have a mass much smaller than that of a proton and, indeed, a mass entirely consistent with an electron mass. Curiously enough, despite the strong admonitions of Dr. Millikan that upward-moving cosmic-ray particles were rare, this indeed was an example of one of those very rare upward-moving cosmic-ray particles.

More photographs with similar features were obtained, but the one described above was the main illustration in Anderson's paper entitled *The Positive Electron* and received by the *Physical Review* on 28 February 1933 [5]. It is also reproduced in many textbooks. In this paper Anderson uses the name *positron* for the new particle. Half a year earlier he had already published a short note [6], not accompanied by a photograph, about his findings. Here the wording was more careful but he concluded that the most likely interpretation of the positive tracks was that they were caused by 'a positively-charged particle comparable in mass and magnitude of charge with an electron'. It was this note which established Anderson's priority as discoverer of the positron.

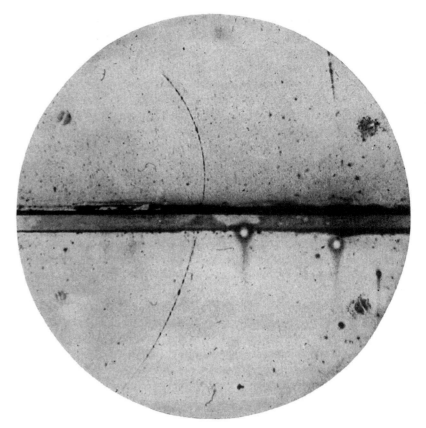

Anderson's cloud-chamber photograph showing the track of a positron flying upwards and losing about two thirds of its energy in the lead plate, which is mounted horizontally in the middle of the chamber. Figure reprinted with permission from [5]. Copyright 1933 by the American Physical Society.

Two physicists in the Cavendish Laboratory in Cambridge were close at his heels. One of them, Blackett, was already very experienced with cloud chambers, with which he had verified Rutherford's experiment on the transformation of the nucleus of nitrogen (see Episode 28). He was assisted by Occhialini, a young Italian physicist from the University of Florence. They, too, had designed a magnetic cloud chamber to study tracks of the cosmic radiation and they were the first to use a new experimental technique. After expansion a cloud chamber is sensitive for a short time. If photographed upon expansion, most of the photographs are empty, since tracks of high-energy charged particles are rare. Blackett and Occhialini found a way by which the particles 'can be made to take their own cloud photographs', as they wrote. Their method was not unlike that used in today's radar trap, in which a driver triggers a camera by the very fact that he or she is speeding. One Geiger counter was placed below the chamber and one above it and, when both counters, in coincidence, gave a signal, it could be assumed that the chamber as well as the two counters had been traversed by a charged particle. The chamber was then automatically expanded and a photograph was taken. This *self-triggering* mechanism increased the efficiency of the chamber by a large factor.

Blackett and Occhialini submitted an extensive paper [7] with many photographs three weeks before the *Physical Review* received Anderson's full paper. They confirm Anderson's early report; there are positive electrons. Moreover, they develop the hypothesis of *pair production*: '... one may imagine that negative and positive electrons may be born in pairs during the disintegration of light nuclei.' (We know now that pair production takes place in the collision of a photon of sufficiently high energy with an electric charge which can be carried by a nucleus or an electron.) Also the fate of positrons is discussed, in particular, that they are likely to 'disappear by reacting with a negative electron to form two or more quanta.' The two authors report about a discussion they had with Dirac about this *annihilation* process and report numbers that Dirac computed. The mean life of positrons with energies below 100 keV in water is given as 3.6×10^{-10} s.

In their paper Blackett and Occhialini not only confirm the existence of the positron and propose mechanisms for its production and annihilation. They are also the first to report a phenomenon they called a cosmic-ray *shower*, an avalanche of reactions in which more and more charged particles are produced.

Although the Cavendish scientists give a reference to Dirac's prediction of an anti-electron, neither their nor Anderson's experiment were motivated by it. They wanted to study cosmic rays. Later, Anderson wrote [4] about the discovery:

> The discovery of the positron was wholly accidental. Despite the fact that Dirac's relativistic theory of the electron was an adequate theory of the positron, and despite the fact that the existence of this theory was well known to nearly all physicists, it played no part whatsoever in the discovery of the positron. [...] Its highly esoteric character was apparently not in tune with most of the scientific thinking of the day. Furthermore, positive electrons were not needed to explain any other observations.

Anderson also speculated on what a 'sagacious person' might have done and writes [4]: 'Had he been working in a well equipped laboratory and had he taken the Dirac theory at face value he would have discovered the positron in a single afternoon.' Indeed, a high-energy γ-ray source and a cloud chamber with a modest magnetic field would have sufficed. There was no need for cosmic rays. Anderson himself performed this experiment [8] after the positron was discovered. One wonders if he did it in a single afternoon. In the papers [5,8] he acknowledges the assistance of Mr. Seth H. Neddermeyer. Only three years later, the two men together discovered another unknown particle (see Episode 57).

Occhialini (left) and Blackett.

[1] Anderson, C. D., *The Discovery of Anti-Matter*. World Scientific, Singapore, 1999.

[2] Skobelzyn, D., *Zeitschrift für Physik*, **59** (1929) 686.

[3] Millikan, R. A. and Anderson, C. D., *Physical Review*, **40** (1932) 325.

[4] Anderson, C. D., *American Journal of Physics*, **29** (1961) 825.

[5] Anderson, C. D., *Physical Review*, **43** (1933) 491.

[6] Anderson, C. D., *Science*, **76** (1932) 238.

[7] Blackett, P. M. S. and Occhialini, G. P. S., *Proceedings of the Royal Society*, **A139** (1933) 699.

[8] Anderson, C. D., *Science*, **77** (1933) 432.

50

Nuclear Reaction Brought About by Machine (1932)

In Episode 47, we mentioned that Lawrence and Livingston failed to be the first to observe nuclear reactions by artificially accelerated particles although they had built an excellent accelerator. Instead, the discovery was made [1] in the Cavendish Laboratory, which was the centre of nuclear physics. Since 1919 the laboratory was directed by Rutherford who had reaped great successes by bombarding nuclei with radiation, particularly α particles from radioactive decays. But, like others, he saw that the limited energy (a few MeV) and also the limited intensity made further progress difficult. At a meeting of the Royal Society in November 1927, addressing the Society as its president, Rutherford spoke about the importance to produce high electric and magnetic fields. He pointed out the advances in the production of high voltages, which had been developed to transport electric energy over large distances [2]. He reported that so far a vacuum tube suitable for the acceleration of particles withstood voltages only up to 300 000 V, i.e., 300 kV. That was only about one-twentieth of the voltage needed to reach the energy provided by natural radioactivity. Still, Rutherford stressed the importance of trying to obtain particles with energies beyond those provided by radioactivity: 'This would open up an extraordinarily interesting field of investigation which could not fail to give us information of great value, not only on the constitution and stability of atomic nuclei but in many other directions.'

Only a few days later, a research student who had recently joined the Cavendish asked Rutherford to let him try out a scheme to accelerate electrons as his Ph.D. work. The student was Walton[1], an Irishmen, who had obtained his master's degree at Trinity College, Dublin, with a thesis on hydrodynamics and who had then won an *Exhibition of 1851* scholarship allowing him to go to Cambridge. Walton's idea was to build a machine, which is now called the *betatron* and which, in some sense, is similar to an electric transformer. A transformer is an iron core carrying two coils. If the ends of one coil, the primary, are connected to an alternating voltage, then an alternating magnetic field is created in the core. The field, in turn, gives rise to a circular voltage in the secondary coil and another alternating voltage appears between the ends of that coil. It was Walton's idea to replace the secondary coil with a circular vacuum tube and have electrons accelerated in it. But only in 1941 was a betatron operated successfully after it had been analysed theoretically and designed to always 'focus' the electrons on a circular path. After a year of fruitless work Walton came across the paper by Wideröe [3], which had induced Lawrence to develop the cyclotron. But for Walton it was rather frustrating to read: Wideröe, a gifted engineer, had in fact designed and built the machine

---
[1] Ernest Thomas Sinton Walton (1903–1995), Nobel Prize 1951

218

Walton had in mind. He had called it a *ray transformer* and only because he could not make it work, had he built his small linear accelerator. Walton might have followed him along that path, had there not been new developments in nuclear physics and in the Cavendish Laboratory.

Gamow[2] [4], a young theoretician, who was born in Odessa and had studied in Leningrad, was on his first trip abroad in Göttingen, when he applied quantum mechanics to a question of nuclear physics and came up with a theory of α decay [5]. The theory explained an empirical rule, the *Geiger–Nuttal law*, connecting the energy of an α particle to the half-life of the nucleus emitting it. Gamow's theory is based on a plausible assumption for the electric potential which the α particle experiences inside and outside the nucleus. Once outside the nucleus, the positively charged α particle is repelled by the positively charged nucleus; the potential is simply given by Coulomb's law. It rises toward the nucleus. When inside the nucleus, the α particle is generally confined there; there has to be a repelling force near the surface of the nucleus. The potential has to rise near the surface. Combining these two considerations one is led to assume the existence of a *potential barrier* near the nuclear surface. Obviously, if the energy of every α particle in the nucleus is smaller than the potential far outside, the nucleus is stable. It is unstable, i.e., radioactive, if there is an α particle with an energy which is higher, even if that energy is smaller than the height of the barrier. That is due to the quantum-mechanical tunnel effect (see Episode 40). Because of its wave nature, the α particle can with a certain probability 'tunnel' through the barrier. The probability increases for decreasing barrier width and decreasing barrier height, the latter being measured from the α particle energy in the nucleus. The result of this chain of arguments is that α particles with high energy have a high probability to leave the nucleus. The half-life of nuclei emitting high-energy α particles is small, just as described by the Geiger–Nuttal law.

What is interesting for the present episode is that Gamow, meanwhile in Copenhagen, also reversed the argument and computed the probability for an α particle to enter a given nucleus if it had a certain energy [6,7]. Because of the tunnel effect, reasonable probabilities were obtained for much lower energies than previously considered necessary. Gamow gave his German paper the rather dramatic title *On the Quantum Theory of Atom Smashing* (Quantentheorie der Atomzertrümmerung). The title of the publication in English is more sober: *The Quantum Theory of Nuclear Disintegration*. Gamow's ideas reached Cambridge before they appeared in print. Cockcroft realized that protons would be even more useful as projectiles to provoke nuclear reactions; they carry only a single elementary charge and accordingly for them the potential barrier is lower than for α particles. He wrote a short memorandum for Rutherford, pointing out that one could expect to observe reactions with light nuclei if protons of $300\,\mathrm{keV}$ were used as projectiles. This was just the voltage Rutherford had described as manageable in his talk given a year earlier.

Like Walton, Cockcroft[3] was not originally a Cambridge man. He had begun to study mathematics at the University of Manchester in 1914 and a year later had joined the army. After the war he studied electrical engineering

Gamow in 1938

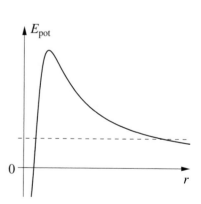

Potential energy of an α particle as a function of the distance r from the nuclear centre according to Gamow. An α particle with positive energy, as indicated by the broken line, may leave (or enter) the nucleus by tunnel effect.

---

[2] George Gamow (1904–1968)   [3] (Sir) John Douglas Cockcroft (1897–1966), Nobel Prize 1951

in Manchester, for two years worked for the Metropolitan–Vickers Electrical Company, resumed his mathematics studies in Cambridge, and finally joined the Cavendish Laboratory. Here his qualities as an engineer and his connections to industry had already been useful to several experimental groups. He had been working, in particular, with Kapitza, a Russian physicist who had come to England in 1922 with a Soviet Mission but who had stayed in Cambridge and was very successful in his work with high magnetic fields and low temperatures (see Episode 58).

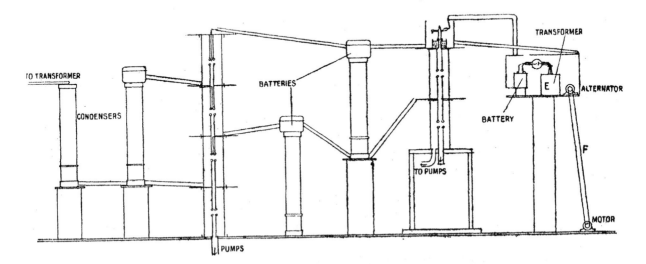

The cascade generator of Cockcroft and Walton. The rectifiers are in the third column from the left. The tube in which protons are accelerated is in the sixth column. It ends in a little hut used for experiments. From [8], Fig. 6. Reprinted with permission.

Early in 1929 Gamow visited the Cavendish. It was about the same time when Walton found that his original acceleration scheme would not work. The overall result was that Rutherford asked Cockcroft and Walton to join forces to provide protons of 300 keV. This goal was already achieved in the next year [9]. However, no nuclear reactions were observed, because Cockcroft and Walton looked for γ rays and the energy was not high enough to excite nuclei to emit them. Consequently a new accelerator aimed at 800 keV was designed and set up, based on what had been learned from the first. The high voltage, practically constant in time, was produced by what is now called a *cascade generator*, a system of capacitors and rectifiers. Each of the components had to sustain only a quarter of the total voltage. The rectifiers, high-voltage vacuum diodes, were made in-house. The protons were provided as *canal rays*. The evacuated glass tube in which they travelled was mechanically similar in construction to the rectifiers. It was segmented so that the acceleration took place in two stages of half the total voltage. The whole apparatus filled a large room from floor to ceiling. At the lower end of the vertical accelerating tube there was a tiny wooden hut with barely the room for an observer to sit. Of course, the hut was covered with lead sheets to protect it from radiation. The lead, in turn, was grounded to keep out electric fields. In February 1932, the acceleration of protons to 800 keV was reported [10].

This time, Cockcroft and Walton used another indicator for nuclear reactions: Rutherford's trusted α particles, easily identified by eye because of the scintillations they cause on a zinc-sulphide screen. They fitted a small chamber to the end of the accelerating tube. It contained a sheet of lithium metal *A* which was exposed directly to the proton beam and formed an angle of 45° with the beam direction. There was a glass window *B* in the chamber, the inside of which was covered with zinc sulphide and which could be viewed by an observer in the hut through a microscope. A mica foil *C* in front of the zinc sulphide prevented all stray protons to reach the screen. Already with the modest beam energy of 125 keV, α particles were observed. In the first short note on their findings [12] Cockcroft and Walton cautiously write that

> it seems not unlikely that the lithium isotope of mass 7 occasionally captures a proton and the resulting [beryllium] nucleus of mass 8 breaks up into two α-particles, each of mass four and each with an energy of about eight million electron volts. The evolution of energy on this view is about sixteen million electron volts per disintegration, agreeing approximately with that to be expected from the decrease of atomic mass involved in such a disintegration.

This was indeed the correct interpretation. Denoting the proton by the symbol of the hydrogen isotope ^1_1H and the α particle by that of the helium isotope ^4_2He, the reaction can be written in the form

$$^1_1\text{H} + {}^7_3\text{Li} \rightarrow {}^8_4\text{Be} \rightarrow {}^4_2\text{He} + {}^4_2\text{He} \quad .$$

The reaction is *exothermic*, i.e., it yields more energy than needed to bring it about. The released energy corresponds to the mass difference between the beryllium nucleus and the two α particles, it is *nuclear energy*.

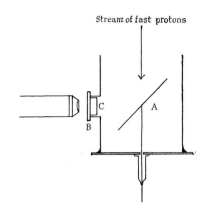

Target *A* and zinc-sulphide screen *B*, used by Cockcroft and Walton to observe α particles from their first nuclear reactions. From [11], Fig. 1. Reprinted with permission.

Walton (left), Rutherford, and Cockcroft.

Cockcroft and Walton verified their interpretation by measuring the energy of the α particles and also by making sure that the α particles appeared in pairs. They performed a *coincidence experiment* in which two observers with microscopes and scintillation screens looked for α particles emitted back to

[1]   Cathcart, B., *The Fly in the Cathedral – How a small group of Cambridge scientists won the race to split the atom.* Viking, London, 2004.

[2]   Rutherford, E., *Proceedings of the Royal Society*, **A117** (1928) 300.

[3]   Wideröe, R., *Archiv für Elektrotechnik*, **21** (1928) 387.

[4]   Gamow, G., *My World Line: An Informal Autobiography.* Viking, New York, 1970.

[5]   Gamow, G., *Zeitschrift für Physik*, **51** (1928) 204.

[6]   Gamow, G., *Zeitschrift für Physik*, **52** (1928) 51.

[7]   Gamow, G., *Nature*, **122** (1928) 805.

[8]   Cockcroft, J. D. and Walton, E. T. S., *Proceedings of the Royal Society*, **A136** (1932) 619.

[9]   Cockcroft, J. D. and Walton, E. T. S., *Proceedings of the Royal Society*, **A129** (1930) 477.

[10] Cockcroft, J. D. and Walton, E. T. S., *Nature*, **129** (1932) 242.

[11] Cockcroft, J. D. and Walton, E. T. S., *Proceedings of the Royal Society*, **A137** (1932) 229.

[12] Cockcroft, J. D. and Walton, E. T. S., *Nature*, **129** (1932) 649.

[13] Dee, P. and Walton, E. T. S., *Proceedings of the Royal Society*, **A141** (1933) 733.

[14] Gentner, W., Maier-Leibnitz, H., and Bothe, W., *Atlas typischer Nebelkammerbilder.* Springer, Berlin, 1940. Engl. transl. [15].

[15] Gentner, W., Maier-Leibnitz, H., and Bothe, W., *An atlas of typical expansion chamber photographs.* Pergamon, London, 1954.

back from a thin lithium target and indeed found more pairs of α particles to appear within a short time interval than could be explained by pure chance, although the time resolution of their method was only about one second. Comprehensive descriptions of the apparatus [8] and the physics results [11] were written in the first half of 1932 and other reactions reported, in particular, the disintegration of boron nuclei. A year later, Dee[4] and Walton presented cloud-chamber photographs [13], which clearly show pairs of α particles leaving the lithium target in opposite directions. They had managed to direct the proton beam onto a small target located in the chamber.

Two pairs of α particles (tracks marked by arrows) leave a lithium target in opposite directions. Cloud-chamber photograph by Dee and Walton. Published in [14], Fig. 3, © 1940 by Julius Springer in Berlin. With kind permission from Springer Science+Business Media.

The Cockcroft–Walton machine was the first *atom smasher*, later so called by the press. The importance of artificially accelerated particles for fundamental science had been impressively demonstrated. Walton returned to Trinity College, Dublin, in 1934 to pursue an academic career, becoming professor and eventually senior fellow. When, also in 1934, Kapitza was not allowed by the Soviet authorities to return to England from a visit to his home country, Rutherford asked Cockcroft to take over the *Royal Society Mond Laboratory*, recently created for Kapitza. Cockcroft also led various research projects connected to the British war effort and in 1946 became director of the *British Atomic Energy Research Establishment* at Harwell.

---

[4] Philip Ivor Dee (1904–1983)

Heisenberg on Nuclear Forces: Isospin (1932)

51

Until 1932, Heisenberg was convinced that quantum mechanics was insufficient to describe the atomic nucleus just as classical physics had been insufficient to describe the atom. Indeed, experimental results on nuclei seemed to contradict quantum mechanics. In Episode 46 we mentioned the problem of 'wrong statistics' and its cure through Pauli's neutrino hypothesis. But Pauli himself was not too sure and published about the neutrino only in 1933. Before the discovery of the neutron the nucleus was considered to consist of protons and electrons. If one computes the momentum uncertainty Δp of an electron which is confined to the small spatial region Δx of a nuclear diameter, the uncertainty relation yields $\Delta p \geq \hbar/(2\,\Delta x)$. Although 'on the average' at rest, such an electron could easily attain a momentum on the order of Δp. The nuclear binding energy known from mass spectroscopy was insufficient to keep such electrons inside the nucleus. Now, most of the known nuclei were stable; but their stability was incompatible with the uncertainty relation and thus incompatible with quantum mechanics.

The situation changed when, early in 1932, Chadwick announced his discovery of the neutron. Although, originally, the neutron was thought of as being composed of a proton and an electron, it became clear that it might also be considered as a particle in its own right. The first to do so in writing was Iwanenko[1], who, in a note of a few lines only [1], asked the question 'how far neutrons can be considered as elementary particles'.

At the end of June 1934, Heisenberg completed a paper entitled *On the Structure of the Atomic Nuclei I* [2]. In the introductory paragraph he points out that the structure of nuclei can be explained in the framework of quantum mechanics if proton and neutron are taken as elementary but that fundamental difficulties remain if the decay of a neutron into a proton and an electron is considered. He thus clearly separated the question of nuclear structure from that of β decay. In a letter to Bohr, quoted in [3], he explained:

> The basic idea is to shift all difficulties of principle to the neutron and to deal with the nucleus by [ordinary] quantum mechanics.

In the paper and its two sequels [4,5] he tried to answer both questions. We shall discuss only his work on nuclear structure, because Heisenberg, like Bohr, suspected that not only quantum mechanics but even energy conservation was no longer valid in β decay. A year later, Fermi was able to formulate a quantum theory of β decay as we shall see in Episode 53.

Once it was accepted that the nucleus was composed of protons and neutrons and that quantum mechanics could be applied to it, the question remained which force acted between its constituents. Heisenberg assumed it to be an

---

[1] Dmitrij D. Iwanenko (1904–1994)

exchange force, i.e., a force based on the symmetry properties of a quantum-mechanical wave function. We introduced the concept in Episode 41 and mentioned that Heisenberg had been the first to exploit it. Except for a single proton, the simplest nucleus is the *deuteron*, which consists of one proton and one neutron. It is the nucleus of deuterium, a heavy isotope of hydrogen. Heisenberg described the deuteron in analogy to the binding of two protons in H_2^+, an ionized hydrogen molecule. A hydrogen molecule, H_2, is composed of two hydrogen atoms, each of which has a proton as nucleus and a shell containing a single electron. In the molecule the two protons stay at a large distance from one another but they share the electrons between them. If one of the electrons is removed, the molecule becomes an ion H_2^+; the single remaining electron can bind the two protons. That is due to an *exchange force* similar to the one discussed in Episode 41. The quantum-mechanical wave function for the position of the electron is a function of that position x but also of the positions x_1 and x_2 of the two protons. If we place the coordinate origin in the middle between the two protons, we have $x_1 = -x_2$. Symmetry or antisymmetry with respect to the exchange $x \rightarrow -x$ is equivalent to symmetry or antisymmetry under the exchange of x_1 and x_2. In the symmetric case there is a higher probability to find the electron in the region between the two protons than in the antisymmetric case. The probability density between the protons corresponds to a region of negative charge pulling the positively charged protons towards it. Thus, in the symmetric case there is an attractive force pulling the protons towards the origin. Let us call the potential energy corresponding to that force $J(r)$, where r is the distance from the origin. It can be computed as a certain integral in quantum mechanics, called exchange (*Platzwechsel*) integral. The repulsive electrostatic force between the two positively charged protons acts against the attractive force. As a result, the protons stay at some well-defined equilibrium distance from each other. So far we described the H_2^+ molecule in a static way. Dynamically, one might say the electron oscillates between the two protons, alternately neutralizing one of them. The oscillation frequency is on the order of $\nu = J(a)/h$, where a is the equilibrium distance of the protons and h Planck's constant.

The deuteron, composed of a proton and a neutron, is quite different from the H_2^+ ion, which is composed of two protons and one electron. But still Heisenberg saw an analogy between the two systems. His principal assumption was that proton and neutron are two states of one and the same particle, differing only in their electric charge (and also slightly in mass). This particle was given the name *nucleon* by Møller[2] in 1941. Heisenberg mentioned as a possible model the exchange of 'electrons that have no spin' between the two nucleons of the deuteron, which would change the proton into a neutron and at the same time the neutron into a proton. No spin was allowed to such 'electrons' because the two nucleons had always to stay fermions, carrying spin $\hbar/2$. But he did not develop this model in detail, writing that is was

> probably more correct (*wohl richtiger*) to look at the exchange integral $J(r)$ as a fundamental property of the pair proton and neutron, without wanting to reduce it to electron motions.

---

[2] Christian Møller (1904–1980)

Thus Heisenberg discarded the idea of particle exchange. Instead, he expressed the symmetry of the proton–neutron system in a way independent of space coordinates. He decided to describe the two possible states (proton and neutron) of the nucleon with the same formalism that Pauli had developed for the two states of the electron spin (see Episode 43, Box II). He introduced matrices $\rho^\xi, \rho^\eta, \rho^\zeta$ which were identical to the Pauli matrices, except that the space ξ, η, ζ, as he writes, 'of course has nothing to do with the real $[x, y, z]$ space'. The two states proton and neutron correspond to Pauli's states $\chi_\pm$ with the quantum numbers $s_z = \pm\frac{1}{2}$ in the case of spin. They are eigenstates of the matrix $\frac{1}{2}\rho^\zeta$ with eigenvalues $I_3 = \pm\frac{1}{2}$. The matrix $\frac{1}{2}\rho^\zeta$ is constructed in analogy to the spin matrix $S_z = (\hbar/2)\sigma_z$. Heisenberg's description of different charge states is now called the formalism of *isospin*; hence the letter I in the quantum number I_3 (the 3 refers to the third coordinate ζ). It is now customary to attribute the $I_3 = \frac{1}{2}$ to the proton and $I_3 = -\frac{1}{2}$ to the neutron. Then the charge Q of a nucleon is given by

$$Q = \tfrac{1}{2} + I_3$$

in units of the elementary charge e.

In quantum mechanics the Hamiltonian operator of total energy determines the state of a system, including its symmetry properties. With the help of his isospin matrices Heisenberg was able to include a term in the Hamiltonian, describing the attraction between nucleons, which yielded the desired symmetry under the exchange of nucleons. He could then not only explain the stability of the deuteron but also other qualitative features of stable nuclei:

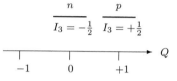

Neutron (n) and proton (p) as two states of different isospin quantum number I_3 and different charge Q.

- The nuclear binding energy is proportional to the number A of nucleons in the nucleus.
- For light nuclei the number Z of protons and the number $N = A - Z$ of neutrons are approximately equal, $N \approx Z$ for $Z \lesssim 20$.
- For heavier nuclei the electrostatic repulsion of the protons becomes important. It is then necessary that the nucleus contains more neutrons than protons to provide sufficient attraction between the nucleons.

These results were obtained from symmetry arguments only. No quantitative formula of a binding potential $J(r)$ could be given, because the distance r between two nucleons did not appear in the theory. We did not write down Heisenberg's Hamiltonian. Heisenberg thought that there was attraction only for a pair proton–neutron and a pair neutron–neutron but that for a pair proton–proton there was only the electrostatic repulsion. This made his Hamiltonian complicated. It was later revealed by experiment that also in the latter case for very small distances there is an attraction due to nuclear forces. With this information the relevant term in the Hamiltonian for the interaction between two nucleons k and ℓ is proportional to $\rho_k^\xi \rho_\ell^\xi + \rho_k^\eta \rho_\ell^\eta + \rho_k^\zeta \rho_\ell^\zeta = \boldsymbol{\rho}_k \cdot \boldsymbol{\rho}_\ell$, where the symbol $\boldsymbol{\rho}$ stands for a vector with the components $\rho^\xi, \rho^\eta, \rho^\zeta$. That type of term, in the form $\boldsymbol{\sigma}_k \cdot \boldsymbol{\sigma}_\ell$, was first introduced [6] for the interaction between the ordinary spins of two nuclei by the brilliant young Italian physicist Majorana[3] in 1933 when he was a visitor in Heisenberg's group.

---

[3] Ettore Majorana (1906–1938)

Heisenberg's theory was quantitatively insufficient to explain nuclear forces. In fact, it did not introduce a genuinely new force but used symmetry aspects of the electromagnetic force which became apparent in quantum mechanics. The riddle of nuclear forces stayed a subject of research for decades to come. The lasting value of Heisenberg's approach lies in the revelation of *inner symmetries* of elementary particles and of quantum numbers associated with these symmetries. The discovery of further symmetries of this type would lead first to a classification of particles and then to an understanding of the forces between them.

Heisenberg (left) and Landau in the 1950s.

[1] Iwanenko, D. D., *Nature*, **129** (1932) 798.
[2] Heisenberg, W., *Zeitschrift für Physik*, **77** (1932) 1. Also in [9], Vol. A II, p. 197.
[3] Mehra, J. and Rechenberg, H., *The Historical Development of Quantum Theory, Vol. 6. – The Completion of Quantum Mechanics 1926–1941*. Springer, New York, 2000.
[4] Heisenberg, W., *Zeitschrift für Physik*, **78** (1932) 156. Also in [9], Vol. A II, p. 208.
[5] Heisenberg, W., *Zeitschrift für Physik*, **78** (1932) 156. Also in [9], Vol. A II, p. 217.
[6] Majorana, E., *Zeitschrift für Physik*, **82** (1933) 137.
[7] Tomonaga, S. I., *The Story of Spin*. University of Chicago Press, Chicago, 1997.
[8] Heisenberg, W. In Ferretti, B. (ed.): *1958 Annual International Conference on High Energy Physics at CERN*, p. 119. CERN, Geneva, 1958.
[9] Heisenberg, W., *Collected Works, Series A, 3 vols.* Springer, Berlin, 1985–1993.

We conclude this episode by a few sentences on Heisenberg's later life. He became particularly interested in collisions of particles at very high energy, e.g., collisions of cosmic rays with nuclei, because there he expected quantum mechanics to fail at some point. No such failure has been observed yet. After 1933 many of his students and collaborators had to leave Germany; he himself decided to stay. In 1937, when Heisenberg was about to succeed Sommerfeld as professor in Munich, he came under attack of the Nazi press which called him a 'white jew' and was advised to stay in Leipzig. A visitor in the late 1930s was Tomonaga[4], who, incidentally, in his book *The Story of Spin* [7] includes a chapter on isospin full of historical and technical detail. During the Second World War, Heisenberg played an important role in the (unsuccessful) German attempt to build a nuclear reactor. This led to his internment in England at the end of the war together with other German scientists. In 1946 he became director of the Kaiser-Wilhelm Institute (soon named Max-Planck Institute) of Physics in Göttingen which later was moved to Munich. In the 1950s together with Pauli he worked on a unified field theory, dubbed *world formula* by the press. Pauli, as critical as ever, finally withdrew. For a public discussion of the two old friends on the subject, following a talk by Heisenberg in 1958, see [8]. Besides his scientific work, Heisenberg decisively helped to rebuild and shape science in Germany and to reintegrate it internationally.

---

[4] Sin-Itiro Tomonaga (1906–1979), Nobel Prize 1965

The Proton Displays an 'Anomalous' Magnetic Moment (1933)

We told in Episode 30 that Stern, at the end of 1921, had moved to Rostock, a small town on the Baltic Sea, while the Stern–Gerlach experiment was completed. There he had been appointed associate professor. He was accompanied by Estermann[1], who worked closely with Stern until the latter's retirement. Already the following year Stern was offered a chair of physical chemistry and the directorship of an institute still to be erected at the recently founded university in Hamburg.

Stern [1], who moved to Hamburg in 1923, decided to concentrate his work on experimental studies with molecular beams. While the institute was being built he wrote a paper containing his research program in this field. It was published as No. 1 of a series of 30 publications called *Untersuchungen zur Molekularstrahlmethode* and contained eight distinct subjects [2]. In two of these Stern would be particularly successful, the determination of nuclear magnetic moments and the demonstration of the wave nature of matter. Here we can mention only in passing that Estermann and Stern demonstrated the existence of de Broglie waves for atoms and molecules [3] by observing the interference occurring when a beam of helium atoms or hydrogen molecules was 'reflected' by a crystal surface; they performed experiments analogous to Bragg scattering (Episode 21) with atoms and molecules instead of X rays, i.e., electromagnetic waves. During his ten years in Hamburg, Stern established a flourishing research group which also attracted very capable visitors from abroad [4].

In 1927 Rabi came to Hamburg to work with Pauli on theoretical physics. Impressed by Stern's work he soon changed to the subject of molecular beams and after his return to Columbia University set up his own research group in this field. He writes [1]:

Stern in 1937

> Hamburg University at the time was one of the leading centres of physics in Germany, therefore in the world. What characterized physics in Hamburg was the extremely close collaboration between Stern and Pauli, experiment and theory. Some of Pauli's great theoretical contributions came from Stern's suggestions, or rather questions; for example, the theory of magnetism of free electrons in metals. Conversely, Pauli's researches, which were at the time devoted to the construction of quantum electrodynamics, were important in directing Stern's thinking. The frequent visits of Niels Bohr, Paul Ehrenfest and others helped to maintain a high level of interest and achievement in all fields.

Although Stern is famous for the original and sophisticated experiments he designed and performed, he did not like to touch his own apparatus. Frisch

---

[1] Immanuel Estermann (1900–1973)

remembers [5]:

> Stern was rather clumsy, and moreover one of his hands invariably held a
> cigar (except when it was in his mouth); so he was disinclined to handle
> any breakable equipment and always left that to his assistants. I still
> remember what he did when anything appeared to topple. He would never
> try to catch it; he lifted both hands in a gesture of surrender and waited.
> As he explained to me 'You do less damage if you let the thing fall than
> if you try to catch it.'

Stern's first great success was his and Gerlach's measurement of the magnetic moment of the silver atom and the quantization of its direction. When the
experiment was performed, it was still believed that this magnetic moment was
due to the orbiting of a single electron in the outer shell of the atom. Only with
the advent of quantum mechanics and the discovery of spin it became clear that
there is no orbital angular momentum in silver and that the magnetic moment
is connected only to the spin of this one outer electron. Dirac, in his theory of
the electron (Episode 43) had shown that the magnetic moment of the electron,
caused by its spin, was 1 *Bohr magneton*, $1\,\mu_B = 1\,\hbar e/2m_e$. In this formula
$\hbar$ is Planck's constant divided by 2π, e the elementary electric charge and m_e
the electron mass. It seemed therefore completely clear that the proton, known
to have the same spin as the electron, would have a magnetic moment of one
nuclear magneton, $1\,\mu_N = 1\,\hbar e/2m_p$, where m_p is the proton mass. Since the
mass ratio of the two particles is $m_e : m_p \approx 1 : 1836$; the magnetic moment
expected for the proton was nearly 2000 times smaller than that of the electron.

In the Stern–Gerlach experiment the electron's magnetic moment was measured by observing the deviation of a beam of silver atoms as it traversed an
inhomogeneous magnetic field, which exerted a force proportional to the moment. It was Stern's aim to determine the proton's magnetic moment using the
same principle. Because of the much smaller magnetic moment a field with the
highest possible inhomogeneity and a sensitivity for very small deviations had
to be reached. The measurement would be not an easy undertaking. Moreover
there seemed to be no doubt about the result. Estermann recalls that, when
Stern together with Otto Frisch and himself began the experiment, 'they were
told more than once by eminent theoreticians that they were wasting their time
and effort'. But when the result was finally obtained and published in 1933,
the magnetic moment of the proton turned out not to be 1 but about 2.5 nuclear
magnetons. It was 'anomalous'.

Let us now turn to the measurement itself. A magnetic moment could only
be measured in an electrically neutral object. If a charged object was used, the
effect would be completely masked by its deflection due to the Lorentz force in
the magnetic field. The obvious choice was the hydrogen molecule H_2, composed of two hydrogen atoms with one proton and one electron each. In its
ground state the two electron spins are antiparallel to each other, there is no effective magnetic moment due to the electrons. The proton spins can be antiparallel (para-hydrogen) or parallel to each other (ortho-hydrogen). The rules of
spin combination tell that the frequency of para- to ortho-hydrogen molecules
is 1 : 3. Para-hydrogen has zero magnetic moment, ortho-hydrogen twice the
magnetic moment of the proton. It also has twice its spin, i.e., $2 \times \hbar/2 = \hbar$. In a
magnetic field the spin, and with it the magnetic moment, can take three orien-

tations: parallel, perpendicular, and antiparallel to the field. In a Stern–Gerlach magnet a deviating force acts only on ortho-hydrogen molecules in the parallel or antiparallel orientation. Thus a beam of hydrogen atoms passing through the field is expected to split into three parts. One part is unaffected, the other two are deflected along or against the field direction. The experiment had to be designed in such a way that the three parts could be clearly distinguished and the deflection measured.

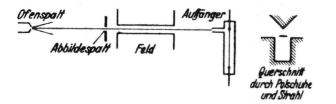

Principle of the apparatus. A beam of hydrogen molecules leaves the 'oven' (on the left) through a narrow slit, is further collimated by a second slit in front of the magnetic field, passes the field, and finally reaches a pressure gauge. A cut through the pole shoes of the magnet is shown on the right. The beam is indicated as a narrow band. Reprinted with kind permission from Springer Science+Business Media: [6], Fig. 1.

The original Stern–Gerlach apparatus had four essential elements mounted in an evacuated vessel: An oven in which silver vapour was produced, which could leave through a small opening, a system of collimators defining a fine beam of atoms, the inhomogeneous magnetic field provided by an electromagnet with pole shoes in the form of an edge and a groove, and a detector for the atoms placed behind the field. The design by Stern, Estermann, and Frisch contained the same four elements. The 'oven' was simply a container for hydrogen with a very small orifice. It was not heated but could be cooled down to liquid-air temperature, i.e., $T = 90\,\mathrm{K}$, to provide slow hydrogen molecules. The collimators were designed to define a beam with widths as small as 0.03 mm. The field between the pole shoes (groove width 1 mm, distance edge to groove plane 0.5 mm, length 10 cm) was designed to yield, for 1 nuclear magneton, a deviation of 0.044 mm at $T = 90\,\mathrm{K}$, a little less than the smallest beam width. The most delicate part of the design was the detector, since hydrogen molecules, unlike silver atoms, do not conveniently stick to a glass plate. The opening of a small cylindrical container was moved through the beam of molecules and the tiny increase of pressure caused by the molecules falling on the opening was measured electrically. As already mentioned, the results, first presented by Frisch and Stern [6] and a little later with increased precision by Estermann and Stern [7], were a real surprise: The registered deviation corresponded to 5 nuclear magnetons, i.e., the proton possesses a magnetic moment of about $2.5\,\mu_\mathrm{N}$ instead of the predicted value of $1\,\mu_\mathrm{N}$. The experimental error was estimated to be at most 10% [7].

By using a beam of heavy hydrogen molecules, Estermann and Stern were also able to measure the magnetic moment of the deuteron, the nucleus containing one proton and one neutron. They found a surprisingly small value for the magnetic moment of the deuteron, located between $0.5\,\mu_\mathrm{N}$ and $1.0\,\mu_\mathrm{N}$ [8]. The difference from the proton value, obviously, was due to a magnetic moment of the neutron. This was the next surprise. In Dirac's theory the neutron, a particle without electric charge, was not expected to have any magnetic moment at all. Since it was known from optical spectroscopy that the deuteron has a spin of $1\,\hbar$, i.e., that the spins of the proton and the neutron are oriented in the same direction, it had to be concluded that the neutron possesses a mag-

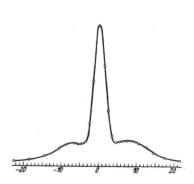

The beam splitting into three parts observed by Frisch and Stern. The scale is in $1/100$ of a millimetre. Reprinted with kind permission from Springer Science+Business Media: [6], Fig. 12.

Present Values of Proton and Neutron Magnetic Moments

$$\mu_\mathrm{p} = 2.792\,847\,351\,(28)\,\mu_\mathrm{N}$$
$$\mu_\mathrm{n} = -1.913\,042\,7\,(5)\,\mu_\mathrm{N}$$

The numbers in parentheses are the experimental errors in the last digits.

netic moment, oriented opposite to the spin direction and accounting for the difference between deuteron and proton magnetic moment. Introducing a minus sign to indicate the orientation relative to the spin direction, Estermann and Stern concluded that the neutron magnetic moment was between $-1.5\,\mu_N$ and $-2.0\,\mu_N$. The unexpected results led to the development of a peculiar language. The theoretically expected values for the proton, $\mu_p = 1\,\mu_N$, and for the neutron, $\mu_n = 0\,\mu_N$, were called the *normal magnetic moments* of these particles. The differences between the measured and the 'normal' values were given the name *anomalous magnetic moments*. We now know that the 'anomaly' is caused by the fact that the proton and the neutron, unlike the electron, are not truly elementary particles but have an inner structure. They are composed of more fundamental particles. Although we know these particles, the proton and neutron magnetic moments cannot, at present, be computed in a satisfactory way. Experimentally, on the other hand, these moments belong to physical quantities known with the highest precision.

Even before the papers [6,7] were completed, Hitler took power in Germany and soon after most Jews were dismissed from public office. Stern himself, who, as a veteran of the First World War, was not in immediate danger of losing his professorship, resigned. He and Estermann accepted an offer from the Carnegie Institute of Technology in Pittsburgh to set up a new laboratory there. However, there were few resources in the depression era and the group did not regain the importance it had until 1933. It was Rabi, one of Stern's visitors in Hamburg, who, at Columbia University, would lead atomic beam physics and, in particular, the study of nuclear magnetic moments, to new heights. In November 1944, the Royal Academy of Sciences in Stockholm announced that the Noble Prize in Physics for 1943 was awarded to Stern and that for 1944 to Rabi.

Stern in the Swiss Alps, winter 1953/54.

Stern retired soon thereafter and in 1946 settled in Berkeley, California, but also spent much time in Zurich. At the age of 81, while attending a performance in a cinema in Berkeley, he suffered a heart attack from which he did not recover.

[1] Rabi, I. I., *Physics Today*, **October** (1969) 103.

[2] Stern, O., *Zeitschrift für Physik*, **39** (1926) 751.

[3] Estermann, I. and Stern, O., *Zeitschrift für Physik*, **61** (1930) 95.

[4] Estermann, I., *Molecular Beam Research in Hamburg 1922–1933*. In Estermann, I. (ed.): *Recent Research in Molecular Beams – A Collection of Papers Dedicated to Otto Stern On the Occasion of his Seventieth Birthday*, p. 1. Academic Press, New York, 1959.

[5] Frisch, O. R., *What little I remember*. Cambridge University Press, Cambridge, 1979.

[6] Frisch, R. and Stern, O., *Zeitschrift für Physik*, **85** (1933) 4.

[7] Estermann, I. and Stern, O., *Zeitschrift für Physik*, **85** (1933) 17.

[8] Estermann, I. and Stern, O., *Physical Review*, **45** (1934) 761.

Fermi's Theory of Beta Rays (1933)

In 1926, not long after Fermi had published the work on his new statistics (Episode 41), Corbino, his mentor, succeeded to have a chair for theoretical physics established at the University of Rome. The customary competition was won by Fermi who thus, in 1927, at the age of only 26 years, became professor in Rome. As a result of the same competition Persico, his friend from school days in Rome, became professor in Florence. Fermi's chair was attached to the Institute of Physics headed by Corbino and the two men decided to thoroughly modernize physics in Rome in both theory and experiment. As a first step Rasetti[1], a young experimentalist with wide interests whom Fermi knew from Florence, was hired as Corbino's assistant. He began with important work on optical spectroscopy, some together with Fermi. There were only very few young people studying physics. But additional students were won over from electrical engineering, the first two being Segrè[2] and Amaldi[3].

Fermi in Rome, mid-1930s.

International contacts were established to acquire experimental know-how in modern fields: Rasetti for some time worked with Millikan in Pasadena, then with Meitner in Berlin. Segrè went to Zeeman in Amsterdam and to Stern in Hamburg, and Amaldi visited Debye in Leipzig. In 1928 Fermi married; his wife Laura later wrote the charming book 'Atoms in the Family' [1] about their common life. When the new Accademia d'Italia was founded in 1929, Fermi was the only physicist to be a member. The honour came with a handsome extra salary, a fancy uniform, and a title. Of the latter Fermi jokingly complained that, although he was to be addressed as 'Your Excellency', he was not expected to speak of himself as 'My Excellency'.

Fermi travelled to the United States for the first time in 1930. At a summer school in Ann Arbor, Michigan, he gave a series of lectures entitled *Quantum Theory of Radiation* [2]. In this course Fermi presented in a lucid way the new field of *quantum electrodynamics* (*QED*), started by Jordan, Heisenberg, Pauli, Dirac, Klein, Wigner, and others. In this field, which at the time was still developing, *emission* and *absorption* or – as one came to say – the *creation* and *destruction* of light quanta was described. Fermi had not participated in the development of quantum mechanics and had not used it for formulating his statistics. But now he became thoroughly acquainted with its current status.

In October 1933, Fermi participated in the 7th Solvay Conference in Brussels, which was devoted to *The Structure and Properties of Atomic Nuclei*. Only the year before the neutron and the positron had been discovered. Dirac presented his theory of the positron and Heisenberg reported about his theory of nuclear structure in which the neutron plays an essential role. In Episode 46, we mentioned already that Pauli spoke about his neutrino at the conference

---

[1] Franco Rasetti (1901–2001)   [2] Emilio Gino Segrè (1905–1989), Nobel Prize 1959   [3] Eduardo Amaldi (1908–1989)

Segrè, Persico, and Fermi on the beach, 1927.

and for the first time had his remarks on that hypothetical particle printed.

After returning to Rome, Fermi developed his theory of β decay which he based on results by Dirac, Pauli, and Heisenberg. A nucleus can undergo β decay, he argued, if one of the neutrons in the nucleus transforms into a proton and if the energy, set free in this transformation, is sufficient to create an electron and a neutrino. He postulated that not only photons could be created (or annihilated) but also electrons and neutrinos. The case was different because these particles are fermions (not bosons like the photon) and because they are created (or annihilated) in pairs. Fermi completed his theory within a few weeks and submitted a note on it to *Nature*, which was promptly refused as 'containing abstract speculations too remote from physical reality to be of interest to the readers' [3]. He sent a slightly extended paper to an Italian journal [4] and published a full account of his theory in German [5] early in 1934. Both papers are entitled *Attempt of a Theory of β-Rays*.

Fermi devised a Hamiltonian for the β decay (see Box) containing a constant g, which had to be determined from experiment. It was defined in analogy to the elementary charge e determining the strength of the electromagnetic interaction between charged particles. It is now called the *Fermi coupling constant* and usually denoted by G_F. While the electron and the neutrino were described by relativistic quantum mechanics, the neutron and the proton in the nucleus were treated in conventional non-relativistic quantum mechanics because the total energy of these heavy particles was still dominated by their rest mass. With the complete theory Fermi was able to compute the lifetime of β-active nuclei as a function of the maximum energy of the decay electrons. The only unknown parameter of the theory was the coupling constant g, which then could be obtained by analysing the experimental data. The introduction of this coupling constant by Fermi was in fact the introduction of a new force of nature. Until then only Newton's gravitational force and Coulomb's electromagnetic force were known. Fermi's force is now called the *weak force* or the force of the *weak interaction*.

Segrè reports that he heard Fermi's first account of this work during a skiing vacation of several Roman physicists in the Christmas holidays of 1933 [6]:

Fermi's Theory of Weak Interaction

Fermi considered a very general Hamiltonian describing the creation of an electron–neutrino pair (or its annihilation). With ψ and φ being Dirac spinor wave functions of the electron and the neutrino, respectively, and Q an isospin matrix taking care of the transition neutron $\to$ proton (or proton $\to$ neutron), it reads

$$H = QL(\psi\varphi) + Q^*L^*(\psi^*\varphi^*) \quad .$$

Here $L(\psi\varphi)$ stands for any bilinear function, i.e., a function linear in the elements of ψ and of φ and where the symbol * stands for complex conjugation. He then restricted the generality by assuming a similarity of H to the potential energy $E_{\text{pot}} = eV(\mathbf{x})$ of a particle with elementary charge e in an electrostatic potential $V(\mathbf{x})$. Under Lorentz transformations $V(x)$ behaves like A^0, the component with the index 0 of a four-vector, the four-potential A of the electromagnetic field. Requiring the same transformation properties of H, Fermi determined the function L. The place of the charge e was taken by a constant g, which is a measure of the strength of the interaction and which is now called the *Fermi coupling constant*.

'Fermi was fully aware of the importance of his accomplishment and said that he thought he would be remembered for this paper, his best so far.' Only a few months after this remarkable theoretical progress Fermi would start a new line of experimentation (Episode 55). His theory of weak interactions was significantly extended only after a quarter of a century (Episode 79).

Fermi's theory was developed to explain the β decay of nuclei, not the binding of protons and neutrons in a stable nucleus. But soon efforts were made to make the new force responsible also for nuclear binding. We mentioned in Episode 51 that Heisenberg discussed the possibility of a 'boson electron' as exchange particle between proton and neutron but discarded it in favour of his isospin formalism. An electron–neutrino pair, if considered as a single particle, had the properties of a charged boson and was therefore a candidate exchange particle. It did not, however, yield the required strong binding [7,8].

Fermi's theory not only described the β *decay* proper as, for instance, that of the neutron n into a proton p, an electron e^-, and a neutrino (since the electron and the neutrino form a particle–antiparticle pair it is now called an antineutrino $\bar{\nu}$),

$$n \to p + e^- + \bar{\nu} \quad .$$

It could also be applied to *K capture*, a process occurring in nuclei with an excess of protons over neutrons, in which an electron from the innermost atomic shell, the K shell, interacts with a proton of the nucleus,

$$e^- + p \to n + \nu \quad .$$

Even the cross section for the *inverse β decay*,

$$\bar{\nu} + p \to n + e^+ \quad ,$$

was soon computed [9] and turned out to be ridiculously small: It seemed to be hopeless to observe this reaction induced by a neutrino.

[1] Fermi, L., *Atoms in the Family – My Life with Enrico Fermi*. University of Chicago Press, Chicago, 1954.

[2] Fermi, E., *Reviews of Modern Physics*, **4** (1932) 87.

[3] Rasetti, F. In [10], Vol. I, p. 538.

[4] Fermi, E., *Ricerca Scientifica*, **4(2)** (1933) 491. Also in [10], Vol. I, 540.

[5] Fermi, E., *Zeitschrift für Physik*, **88** (1934) 161. Also in [10], Vol. I, 575.

[6] Segrè, E., *Enrico Fermi, Physicist*. University of Chicago Press, Chicago, 1970.

[7] Tamm, I., *Nature*, **133** (1934) 981.

[8] Iwanenko, D., *Nature*, **133** (1934) 981.

[9] Bethe, H. A. and Peierls, R. E., *Nature*, **133** (1934) 532.

[10] Fermi, E., *Collected Papers, 2 vols*. University of Chicago Press, Chicago, 1962.

54 Irène and Frédéric Joliot–Curie – Artificial Radioactivity (1934)

In Episode 48, we told the story how, early in 1932, the Joliot–Curies barely missed the discovery of the neutron but made observations which helped Chadwick find it. In 1935, the year in which Chadwick was awarded the Nobel Prize in physics, the young couple received the same honour in chemistry. They had produced radioactive elements not normally existing on earth.

Irène and Marie Curie in 1921.

In the Curie family [1] there was for a second time a team of husband and wife, highly gifted and outstandingly successful. Irène Curie[1] was eight years old when her father died. Her grandfather Eugène Curie, a medical doctor and widower, did his best to help Marie Curie to care for Irène and her younger sister Ève. Irène also attended lessons given by her mother and other university professors like Perrin and Langevin for their children. In 1914 she entered the University of Paris and obtained degrees both in mathematics and physics in 1920. At the beginning of the First World War, Marie Curie recognized the lack of X-ray equipment for the wounded and organized a system of mobile

---

[1] Irène Curie (1897–1956), Nobel Prize 1935

X-ray stations, ably assisted by Irène who was still in her teens. In 1919 she began to work without pay in her mother's *Institut du Radium*; in 1920 she got a position as assistant. For her doctoral dissertation, Irène Curie made a detailed study of the energy loss of α particles along their path through the gas of a cloud chamber. As source of α particles she used polonium, the first element discovered by her mother. The radium institute in Paris could produce the strongest polonium sources in the world. As a token of respect, many doctors sent ampoules, which had contained radon gas that had decayed, to the institute. Among the decay products was polonium which could be separated and provided α particles nearly without background radiation.

Frédéric Joliot[2] [2,3] was the son of a merchant. After his father's death in 1917, only shortly before he would have got his *baccalaureat* required for university studies, he decided to aim at a more practical profession. He passed the entrance exam for the *Ecole de Physique et de Chimie* where Pierre and Marie Curie had found polonium and radium, obtained his degree as physics engineer, and then had to do his military service. A few weeks before that would end, Joliot, now aged 25, saw Marie Curie, because he wanted to do research in her institute. At the end of the interview she asked him: 'Could you begin work tomorrow?' When Joliot explained, she said: 'I shall write to your colonel'. Indeed, from the next day he worked as her assistant. Irène is described as rather shy and reclusive, Frédéric as outgoing, brilliant in conversations and particularly good looking. The two seemingly different young scientists were married in 1926, the year after Frédéric had come to the institute. In 1928 the first joint publication appeared, signed Irène Curie and Frédéric Joliot. This was the way they would sign all their papers. But since in the press they were called *Joliot–Curie*, Frédéric more and more used that name. He was a gifted experimenter who invented and introduced new techniques in Marie Curie's institute. In 1930 he got his Ph.D. with a thesis on the electro-chemistry of radioactive elements, not without having to fulfil before the requirements for the baccalaureat he had missed years earlier.

The young couple was not discouraged by the events in early 1932. They could prepare very strong sources of α particles and they continued to investigate the Bothe–Becker radiation, which α particles provoke beryllium to emit and in which the neutron had been observed. In a cloud chamber, in which they placed a polonium source covered with beryllium or other material containing light atoms, they observed the tracks of recoil protons caused by neutrons. (Indeed, as we have seen, they were the first to note them.) In addition, they saw also other tracks, caused by positrons. There is an interesting little story here. The Joliot–Curies actually observed these tracks before the positron was known and mistook them to be electrons travelling the other way, not coming out of the source but, possibly having their origin in the chamber, moving into the source.

After the positron had been identified by Anderson (Episode 49), they studied the new particle in detail, in particular, the production of electron–positron pairs and the annihilation of a positron with an electron into two γ quanta. When aluminium was irradiated with α particles, both neutrons and positrons

2 Frédéric Joliot–Curie (1900–1958), Nobel Prize 1935

appeared. The Joliot–Curies conjectured that this was due to a reaction

$$^{27}\text{Al} + \alpha \rightarrow {}^{30}\text{Si} + n + e^+ \quad ,$$

i.e., that the neutron n and the positron e^+ appeared together when an aluminium nucleus was transformed by an α particle into one of silicon. They decided to check this idea by measuring energy *thresholds*. If their conjecture was right, the smallest energy or threshold energy, needed for an α particle, hitting an aluminium target, to produce a neutron, had to be the same as the threshold energy to produce a positron. Polonium radiates α particles of 5.3 MeV energy. That can be reduced by letting them pass a certain amount of gas. For their experiment the Joliot–Curies used the following set-up. On the inside of one wall of a gas vessel a polonium source was attached. The opposite side had an opening covered with a sheet of aluminium. The vessel could be completely evacuated or filled with carbon dioxide of a desired pressure. In this way the energy with which the α particles hit the aluminium was under control. Outside the vessel, next to the aluminium, a detector for neutrons or positrons was placed. Neutrons were detected in an ionization chamber, its inside walls being clad with material containing hydrogen to allow the production of recoil protons which caused signals in the chamber. Protons give high signals. The threshold was easily determined. Neutrons appeared above the threshold (below a certain gas pressure) but not below it (above that pressure).

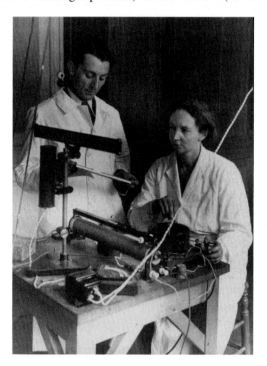

Irène and Frédéric Joliot–Curie in 1932.

Electrons and positrons give only very low signals. To detect them, a relatively new instrument was used, a *Geiger–Müller counter*, built by Gentner[3],

---
[3] Wolfgang Gentner (1906–1980)

a young postdoctoral visitor who had recently arrived from Frankfurt and became a life-long friend of Joliot. The counter was connected to an amplifier which, in turn, drove an electromechanical device actually counting the pulses due to individual particles. In the afternoon of Thursday, 11 January 1934, Joliot tried to determine the threshold for the production of positrons. With the chamber filled above the pressure corresponding to the neutron threshold there were no signals in the Geiger counter. They appeared when the pressure was lowered. But when it was raised again, even to a value at which no α particles could reach the aluminium, the counter kept going. Something must have happened to the aluminium or to the counter. Gentner, who was called by Joliot, assured him that the counter was working properly. (We remember that Lawrence and Livingston in a similar situation did not believe what their Geiger counters indicated, Episode 47.) Joliot concluded: the aluminium had become radioactive.

The next two days, Friday and Saturday, Irène and Frédéric discovered two more elements which could be made radioactive in this way. They measured their half-lives and made sure that the particles produced in the decay were really positrons. The decay of the new radioactive elements was thus a β decay in which positrons instead of electrons were emitted. Like in the ordinary β decay the energy spectrum was continuous and Irène and Frédéric determined its maximum for their three elements. They wrote a paper [4] which was presented on Monday, 15 January, by Perrin in a session of the Academy of Sciences. We quote from this paper entitled *A New Type of Radioactivity*:

> The emission of positive electrons by certain light elements irradiated by the α-rays of polonium subsists for times more or less long, which can attain more than half an hour in the case of boron, after the removal of the source of α rays.

> We place a leaf of aluminium at 1 mm from the source of polonium. After the aluminium is irradiated for ten minutes we place it above a Geiger–Müller counter with an opening closed by a screen of 7/100 of a millimetre of aluminium. We observe that the sheet emits radiation, the intensity of which decreases exponentially as a function of time with a period of 3 minutes and 15 seconds. One obtains an analogous result with boron and magnesium but the periods of decrease are *different*, 14 minutes for boron and 2 minutes and 30 seconds for magnesium.

In this paper also the correct explanation is given for the appearance of a neutron and a positron. Rather than in their first conjecture, in which it was assumed that these particles were produced together, they now appear in a two-stage process. In the first stage aluminium is transformed into phosphorus and a neutron is emitted,

$$^{27}\text{Al} + \alpha \rightarrow {}^{30}\text{P} + n \quad .$$

The new phosphorus isotope ^{30}P is radioactive and decays into silicon and a positron (and a neutrino),

$$^{30}\text{P} \rightarrow {}^{30}\text{Si} + e^+ (+\nu) \quad .$$

The neutrino does not appear in the paper because that particle and the theory of β decay, based on it, was not yet generally accepted.

Gentner when he was working
under Joliot in Paris . . .

. . . and in later years.

[1] Radvanyi, P., *Les Curies*. Belin,
Paris, 2005.

[2] Biquard, P., *Frédéric Joliot–Curie
et l'énergie atomique*. Seghers,
Paris, 1961.

[3] Pinault, M., *Frédéric Joliot–Curie*.
Éditions Odile Jacob, Paris, 2000.

[4] Curie, I. and Joliot, F., *Comptes
Rendus Acad. Sci.*, **197** (1934)
1622.

[5] Radvanyi, P. and Bordry, M., *La
radioactivité artificielle et son
histoire*. Seuil/CNRS, Paris, 1984.

Before the discovery of 'artificial' radioactivity one had thought that elements with radioactive isotopes can exist only at the end of the Periodic Table. Here nuclei could be unstable and decay spontaneously. Transmutation of lighter nuclei had been observed if those were hit by α particles or protons. But it was common belief that these nuclear reactions were so violent that the final products appeared at once. Now it turned out that also unstable, radioactive nuclei were produced. We saw in Episode 22 how Thomson found that the element neon, as it exists on earth, consists of more than one isotope and how Aston extended his findings to other elements. It was now shown that many more isotopes than those observed in nature can be produced. If they ever existed naturally, they had long since decayed. Very soon it was understood that radioactive isotopes have important applications in biology and medicine. They serve as *tracers*, which, by their radioactivity, indicate the transport of chemical compounds containing them through a living organism.

It was one of the last great pleasures of Marie Curie to witness the success of her daughter and her son-in-law. In his Nobel Lecture Joliot said:

> At the present time we know how to synthesize [. . .] more than fifty new radio-elements, a number already greater than that of the natural radio-elements found in the earth's crust. It was indeed a great source of satisfaction for our lamented teacher Marie Curie to have witnessed this lengthening of the list of radio-elements which she had the glory, in company with Pierre Curie, of beginning.

In his letter of congratulations (reproduced in [5]) Gentner, who at the time worked with Bothe in Heidelberg, wrote: 'Recalling this famous afternoon I still admire the great swiftness with which you immediately recognized the fact and all the importance of that discovery. But it was me who straight away made another discovery because I told you at once that this work is worth the Nobel Prize and I am happy that my presentiment has now come true.'

The Joliot–Curies not only became leaders of large and important research groups, Irène at the University of Paris and Frédéric at the Collège de France, where he began to build a cyclotron. Both also, to some extent, entered politics. In the time when fascism became a threat in Europe they ostentatiously sided with the political left. In 1936 Irène for some time was Secretary for Science and Research in the cabinet of prime minister Léon Blum.

During the occupation of France, Gentner, with the secret agreement of Joliot, became supervisor of the laboratory at the Collège de France. The cyclotron was completed. Gentner managed to protect the scientists and closed his eyes to Joliot's connections to the underground resistance. After the war, Gentner, among other activities, founded the important Max-Planck Institute for Nuclear Physics in Heidelberg. In these happier times he was made an officer of the French Legion of Honour.

At the beginning of 1946, Joliot was appointed High Commissioner of the French Atomic Energy Commission and led the construction of the first French nuclear reactor, which began to operate in 1948. In 1950 he lost his office as High Commissioner because of his membership in the French Communist Party. After the death of his wife in 1958, in addition to his professorship at the Collège de France, he was appointed to the Chair she had held at the Sorbonne. He survived her for only two years.

Fermi Produces Radioactivity with Neutrons (1934)

55

In Episode 53, we described the origins of Fermi's group in Rome. After the completion of quantum mechanics it had become more and more clear that research had to be redirected from the atomic shell to the nucleus. Accordingly, a cloud chamber and also Geiger–Müller counters were built by the group and put in use.

When news reached Rome of the Joliot–Curies' discovery of new radioactive isotopes, which were produced by the bombardment of light elements with α particles, it occurred to Fermi that neutrons would be useful projectiles because they are not repelled by the electric charge of the nucleus and should be able to enter even the heaviest nuclei. He suggested to Rasetti to look for artificial radioactivity. Rasetti used a polonium–beryllium source which, however, was much weaker than the one the Joliot–Curies had at their disposal. As a consequence, no effect was observed and Rasetti left for a holiday in Morocco. Now, Fermi realized that he could make much stronger sources if he did not require them, as the Paris group had done, to be virtually free of background radiation. Then there was no need to use polonium. The chief physicist of the Public Health Department, Trabacchi, had his laboratories on the same premises as the Physics Department. He had more than one gramme of radium and generously allowed Fermi's group to use the radon produced in its decay. Fermi made neutron sources in the form of small glass tubes, which were filled with beryllium powder and radon gas and then sealed off. With neutrons from such sources he systematically bombarded elements in the order of their appearance in the Periodic Table. He met with success first with fluorine, then with aluminium and submitted the first paper, entitled *Radioactivity Induced by Neutron Bombardment – I* on 25 March 1934. We quote from this short paper (only about one page) [1]:

> I used the following apparatus: The source of neutrons was a small tube of beryllium and emanation. Using about 50 millicurie of emanation [...], I could obtain more than 100,000 neutrons per second, mixed, of course with a very intense γ-radiation; however the latter does not influence experiments of this kind. Small cylindrical containers filled with the substances tested were subjected to the action of the radiation from this source during intervals of time varying from several minutes to several hours.

> Immediately after being irradiated, the targets were placed in the vicinity of a Geiger–Müller counter, whose wall was formed of aluminium sheet about 0.2 mm thick, allowing β-rays to enter the counter. Positive results have been obtained, so far, with the two following elements:

> *Aluminium.* – A small cylinder, irradiated by neutrons for about two hours, gives rise, in the first few minutes after the end of the irradiations,

to a considerable increase in the rate of pulses from the counter, the rate increasing by about 30–40 pulses per minute. A decrease follows, the rate reducing to half its initial value in about 12 minutes.

Fluorine. – Calcium fluoride, irradiated for a few minutes and rapidly brought into the vicinity of the counter, causes in the first few moments an increase of pulses; the effect decreases rapidly, reaching the half-value in about 10 seconds.

Fermi's group in Rome: D'Agostino, Segrè, Amaldi, Rasetti, Fermi.

Fermi wanted to ensure that Rome maintained the lead in the new field of neutron research. He wrote to Rasetti suggesting that he cut short his vacation and recruited Segrè and Amaldi to help. Fermi did most of the measurements himself, Amaldi took care of the delicate counter, and Segrè procured specimens of practically every element, paying directly in cash. Soon Rasetti returned. Moreover, D'Agostino[1], a chemist who had worked in the Health Department and at that time held a fellowship at the Institut de Radium in Paris, was convinced to join in. The spirit in the group of bright young men was excellent. They were fond of nicknames: Fermi was called the 'Pope', Rasetti the 'Vicar Cardinal', and Trabacchi, who provided the radon and helped out with instruments, 'The Divine Providence'.

The group wrote a total of ten papers, all bearing the same title, differing only in the Roman numeral at the end. They were typeset immediately and copies sent to important research groups before actual publication. Only about

---

[1] Oscar D'Agostino (1901–1975)

four weeks after completing the first paper, Fermi got a letter with thanks from Rutherford. In this letter (reproduced in [2]) Rutherford writes: 'Congratulations on your successful escape from the sphere of theoretical physics! You seem to have struck a good line to start with.' Also, papers in English appeared, in particular, a comprehensive report, communicated on 25 July by Rutherford to the Proceedings of the Royal Society [3]. Two-thirds of all elements had been irradiated with neutrons and in more than half of these radioactivity had been induced.

So far the measurements had been mainly qualitative. When the group tried to obtain precise measures for the amount of activity induced by irradiation under fixed conditions, it met first with difficulties, then with surprise. The induced activity was, of course, expected to depend on the strength of the source, on the distance to the target, on the size of the target, and on the duration of irradiation. Once these factors were well under control, quantitative measurements should be possible. Amaldi and Pontecorvo, a student who had recently joined the group and who was therefore nicknamed the 'cub', were asked to develop a standardized technique with silver as a target. But the measurements were not reproducible. This was a dilemma and possible sources of error were looked into. It turned out that the induced activity depended on the surroundings in which the experiment was done: If it was performed on a wooden table, the silver became much more radioactive than if it was placed on a marble shelf in the same room. For a few days the influence of lead, the usual shielding material, was studied but without conclusive results. Then, in the morning of 22 October 1934, Fermi decided to place a piece of paraffin between source and target. The activity of the silver was greatly increased. Segrè [2] writes, he at first thought that the counter had gone wrong. But the result was real and soon also obtained with other targets. Somehow, rather than to weaken the influence of the neutrons, the paraffin inserted between source and target had considerably strengthened it.

In spite of the excitement, the group adjourned for lunch and siesta and when it reassembled in the afternoon Fermi presented his explanation of the seemingly strange phenomenon: Paraffin (and also the wooden table mentioned previously) contained hydrogen atoms with protons as nuclei. As was first recognized by Chadwick, because of their similarity in mass, neutrons easily transfer energy to protons and are thus slowed down. The experiment now showed that slow neutrons are much more effective in activating silver and other material than fast ones. The same evening the group met in Amaldi's apartment. Fermi dictated a paper to Segrè while Amaldi, Rasetti, and Pontecorvo interrupted with comments. The paper [4] also contains Fermi's concept of *thermal neutrons*:

> ... one may expect that after some collisions the neutrons move in a manner similar to that of molecules of a diffusing gas, eventually reaching the energy corresponding to thermal agitation. One would form, in this way, something like a solution of neutrons in water or paraffin surrounding the neutron source. The concentration of this solution at each point depends on the intensity of the source, on the geometrical conditions of the diffusion process, and on possible neutron-capture processes due to hydrogen or to other nuclei present.

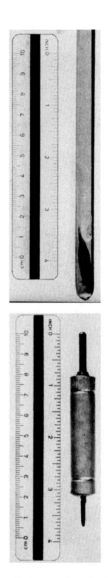

Radon–beryllium source in a glass tube (top) and Geiger counter (bottom) used in the work of Fermi and his group. Photographs: Amaldi Archives, Physics Department, Sapienza – University of Rome.

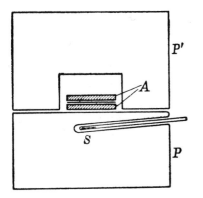

Arrangement for absorption measurements with thermal neutrons. From [5], Fig. 1. Reprinted with permission.

Thus neutrons leaving the source with an energy of several MeV, in collision with light nuclei, can lose practically all their energy until they reach thermal equilibrium with these nuclei. Such thermal neutrons, at room temperature, have an average energy of only about $0.04\,\text{eV}$.

Fermi's group measured the properties of most elements under the influence of slow neutrons, not only the production of radioactivity but also the capability to simply absorb neutrons. Absorption measurements were performed by placing a thin sheet of rhodium or silver, covered on both sides with a layer A of the absorbing material, in a cavity of a paraffin block which was made of two parts P, P'. The neutron source S was placed in another cavity of the same block. The activity which the rhodium (or silver) displayed after some time of irradiation was compared with the activity measured after irradiation without absorbers. We mention here, in particular, that cadmium was found to have a larger absorption for slow neutrons than all other elements. The extensive paper [5] contains many results and techniques which shaped the emerging field of neutron physics.

A fast as well as a slow neutron can cause different reactions in the nucleus it hits. For instance, it may enter the nucleus and knock out an α particle. In a shorthand notation, invented by Bothe, this is an (n, α) reaction (neutron in, α out). Also (n, p) reactions are possible in which a proton leaves the nucleus. Particularly simple is an (n, γ) reaction in which a neutron is absorbed and the binding energy, set free in the process, is radiated off as a γ ray. In this case, the new nucleus is an isotope of the original one. Since the balance between protons and neutrons was changed, this nucleus is often unstable, i.e., radioactive. Via a β decay, one neutron is changed into a proton, and after the decay the nucleus finds itself shifted up by one place in the Periodic Table. Slow neutrons often lead to (n, γ) reactions.

Fermi and his group found artificial radioactivities with different half-lives when they irradiated uranium, the last element in the Periodic Table carrying the largest nuclear charge number, $Z = 92$. Using chemical methods, they searched for traces of elements with somewhat smaller values of Z in the irradiated target which would have indicated (n, α) or (n, p) reactions. When none was found, reluctantly at first, they assumed to have produced one or even two *transuranic elements*, i.e., new elements with nuclear charge numbers larger than that of uranium [6,7].

In several laboratories, methods were developed to produce neutrons of well-defined energy. This allowed to study the energy dependence of neutron absorption which, in several cases, showed pronounced maxima, especially at low energies. A theory was developed by Breit and Wigner [8] in which the process was described as *resonant absorption*. It was assumed that the incoming neutron at certain energies gives rise to *metastable states* or *resonances* of the target nucleus, which subsequently decay under the emission of a photon. But even then the very large absorption cross sections could not be explained.

Bohr showed that it could be understood, if the ensemble of all nucleons in the nucleus was considered as the target of the incoming neutron, not the nucleus as a whole or one single nucleon. In his theory of the intermediate or *compound nucleus* [9], he compares the average potential of a nucleus to a flat bowl. The nucleons are confined to the region of this potential like billiard balls lying in the bowl. If an additional ball, representing the neutron, is pushed into the empty bowl, it will slip out again on the opposite side. If the bowl is filled, however, the incoming ball, in multiple collisions, will share its energy with the other balls. None of the billiard balls will be able to leave the bowl. Bohr said that the energy of the incoming neutron was distributed on the many degrees of freedom of the nucleus. As a result, several different events could happen: After some time the energy might concentrate again on a single nucleon or an α particle near the surface which could then leave the nucleus. The energy might also be radiated off in the form of one or several photons. Resonances as discussed by Breit and Wigner were possible, but only for very low energies. At high energies so many resonances existed that they were essentially washed out. In the course of time the model of the compound nucleus was developed further, leading to the *liquid-drop model* of the nucleus.

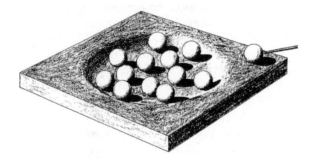

Illustration used by Bohr to visualize his compound model of the nucleus. Reprinted by permission from MacMillan Publishers Ltd: Nature [9], copyright 1936.

The research with slow neutrons and, in particular, the search for transuranic elements became a new and active field in several laboratories. Fermi's group, however, dissolved. Mainly because of the growing Fascist influence, Rasetti went to New York and Pontecorvo to France. Segrè became professor in Palermo and in 1938 went to Berkeley. In the autumn of 1935, less than two years after it all began, Fermi and Amaldi alone continued the work in Rome. Fermi had declined several offers from the United States. In 1938 Bohr let him know that he might expect the Nobel Prize that year and Fermi confidentially wrote to Columbia University in New York, asking if their offer was still valid. He was assured it was. When the good news from Stockholm arrived, the Fermis with their two children and their maid travelled to Sweden and from there proceeded to the United States.

In his Nobel Lecture, delivered on 10 December 1938, as an aside Fermi mentioned the transuranic elements. Less than two weeks later, Hahn and Strassmann finished their paper in which they show that the radioactivity produced by bombarding uranium is due to fission (see Episode 60). For the published version of his lecture [10] Fermi added a footnote drawing attention to their work.

[1]   Fermi, E., *Ricerca Scientifica*, **5** (1934) 283. Also in [11], Vol. I, 645, 674.

[2]   Segrè, E., *Enrico Fermi, Physicist*. University of Chicago Press, Chicago, 1970.

[3]   Fermi, E., Amaldi, E., D'Agostino, O., Rasetti, F., and Segrè, E., *Proceedings of the Royal Society*, **A146** (1934) 483. Also in [11], Vol. I, 732.

[4]   Fermi, E., Amaldi, E., Pontecorvo, B., Rasetti, F., and Segrè, E., *Ricerca Scientifica*, **5** (1934) 282. Also in [11], Vol. I, 757, 761.

[5]   Amaldi, E., D'Agostino, O., Fermi, E., Pontecorvo, B., Rasetti, F., and Segrè, E., *Proceedings of the Royal Society*, **A149** (1935) 522. Also in [11], Vol. I, 765.

[6]   Fermi, E., Rasetti, F., and D'Agostino, O., *Ricerca Scientifica*, **5** (1934) 536. Also in [11], Vol. I, 704.

[7]   Fermi, E., *Nature*, **133** (1934) 898. Also in [11], Vol. I, 748.

[8]   Breit, G. and Wigner, E., *Physical Review*, **49** (1936) 519.

[9]   Bohr, N., *Nature*, **137** (1936) 344. Also in [12], Vol. 9, p. 152.

[10] Fermi, E., *Nobel Lectures, Physics 1922–1941*, p. 414. Elsevier, Amsterdam, 1965.

[11] Fermi, E., *Collected Papers, 2 vols*. University of Chicago Press, Chicago, 1962.

[12] Bohr, N., *Collected Works, 10 vols*. North Holland, Amsterdam, 1972–1999.

56 Cherenkov Radiation Discovered (1934) and Explained (1937)

Vavilov

Cherenkov

At the end 1917, a young Russian soldier was taken prisoner by a German unit. Sergei Vavilov[1] [1–3] had been a physicist before the war and had worked under Lebedev[2] in Moscow. As fate had it, the officer interrogating him was a physicist, too. They discussed physics for most of the night and, as morning drew near, Vavilov was helped by his German colleague to make his escape. On returning to Moscow, he found that his father, a textile merchant, had lost his fortune and emigrated. But he and his brother Nikolai, a biologist, decided to stay. Vavilov worked in the Physics and Biophysics Institute in Moscow, mainly on luminescence, and taught at Moscow University. In 1932 he was made the youngest full member of the Soviet Academy of Sciences. By the same year, while retaining his chair in Moscow, he became head of research of the Optical Institute and of a department of the Physico-Mathematical Institute of the Academy, both in Leningrad.

Cherenkov[3] [4] was the only boy in a family of six children. His father was a peasant in the district of Voronezh who also worked as carpenter and coachman. Cherenkov graduated from Voronezh University in 1928. He had worked for some time as a mathematics and physics teacher when he became aware of vacancies for *aspirants*, i.e., graduate students, at the academy institute in Leningrad. He was accepted, took up his duties in 1930, and, beginning in 1933, did research for his thesis under Vavilov's direction.

More than 50 years later, Cherenkov wrote [5] that Vavilov was impressed by the scientific work which Rutherford had been able to do using the scintillation method which allowed the detection of individual α particles. His first suggestion to Cherenkov was to look for scintillations from β rays. When, using a weak source, Cherenkov found it impossible to observe by eye scintillations from single electrons, Vavilov told him to look for global effects in luminescent substances, especially uranium salts, with the help of a strong source of β and γ rays. Quantitative intensity measurements were done with a method, pioneered by Vavilov, based on the rather well-defined threshold of the human eye adapted in complete darkness for at least one hour.

In the course of his tedious observations, Cherenkov found that light was not only produced in luminescent solutions but also in pure solvents. A fellow Ph.D student tells that Cherenkov came to a group seminar, where he reported that he observed light in a glass container with pure sulphuric acid placed near the radium source and that he wanted to discontinue the work because of that 'background' effect [6]. Vavilov, however, convinced him to go on and study this effect in detail. The result of these observations in which Vavilov would

---

[1] Sergei Ivanovich Vavilov (1891–1951)   [2] Pyotr Nikolaevich Lebedev (1866–1912)   [3] Pavel Alekseyevich Cherenkov (1904–1990), Nobel Prize 1958

participate on occasions was published, in May 1934, in two separate papers by Cherenkov and by Vavilov. Cherenkov [7] reports that γ rays from a rather strong radium source (100 mg) lead to a blue glow in many different liquids, including distilled water. It was made sure that this glow was not due to luminescent admixtures. Moreover, the polarization of the observed light was not as expected from luminescence. Vavilov [8] tries a first explanation of the effect. He sees no way how γ radiation directly can cause the visible light and (correctly) attributes its emission to the electrons liberated in the liquid by Compton effect. These electrons, he argues, then suffer *bremsstrahlung*, i.e., emit electromagnetic radiation while travelling through the liquid and losing energy. Although the bulk of this kind of radiation has frequencies much shorter than that of visible light, he suggests (erroneously) that it contributes to the visible spectrum to explain the blue glow.

Vavilov saw to it that Cherenkov could go on studying his bluish light. This was by no means self-evident. The effect, hardly visible as it was, was considered spurious and even an 'idee fixe' by many. When it was established beyond doubt in 1937 it became known as *Cherenkov radiation* in the West but was called *Vavilov–Cherenkov radiation* in the Soviet Union.

In 1934 the department headed by Vavilov was made into the Physical Institute of the Academy of Sciences (FIAN), also called the *Lebedev Institute*. It was transferred to Moscow, rapidly grew in size, and new fields of research, in particular, nuclear physics were added to its scope. This was in line with the policy of the Soviet Union to concentrate much of its research activities in institutes of the Academy [9]. Vavilov's brother Nikolai was president of the Academy of Agricultural Sciences but was forced to resign in 1935, apparently because he disagreed with the official Lysenko ideology of genetics. In 1940 he was arrested and died in prison in 1943. Sergei Vavilov became president of the Soviet Academy of Sciences in 1945. In 1949 he wrote a letter to Stalin requesting the exoneration of his brother in which he described the accusations as slanderous. The letter was marked by Beria, Stalin's security chief, 'To be rejected' [10].

Among the first scientists added to the staff of FIAN were Tamm[4] and Frank[5]. Tamm was the son of an engineer. He was born in Vladivostok, in the far east of Russia, and had graduated from Moscow university in 1918. He had held positions at various institutes and universities when he was appointed as leader of the theoretical division of the new Lebedev Institute. Frank, son of a mathematics professor, was a pupil of Vavilov. After his graduation in 1930, he had worked at the optical institute in Leningrad. It seems [6] that Frank soon became interested in the mysterious radiation and suggested to place the liquid in a magnetic field and that in this way the connection was first observed between the direction of the Compton electrons (influenced by the magnetic field) and the direction of the emitted radiation [11]. The directional effect is, to some extent, washed out because the Compton electrons are not emitted directly along the direction of the incident γ rays and, moreover, because they are scattered in the liquid and change their direction along their path. But enough of the effect remained to allow Cherenkov to observe that essentially all radia-

Tamm

Frank

[4] Igor Yevgenyevich Tamm (1895–1971), Nobel Prize 1958   [5] Il'ja Mikhailovich Frank (1908–1990), Nobel Prize 1958

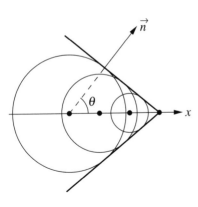

Mach's cone: A projectile travels along the x direction with a speed larger than that of the spherical waves it emits. These waves, by interference, give rise to a conical wave front with a normal **n**, which forms an angle θ with the projectile's direction of motion.

tion was emitted into the forward hemisphere, the forward direction being the γ-ray direction. The maximum intensity did, however, not occur exactly in the forward direction but under a certain angle to it. While Cherenkov performed quantitative measurements of the light intensity as a function of the emission angle, Frank and Tamm developed the theory of the *Cherenkov effect*. It was based on the analogy to the acoustic shock wave, which appears if a projectile moves through air with a velocity v_{proj} larger than the speed of sound v_{sound} in air. Sound waves are excited along the projectile's path, spreading with the velocity v_{sound}. By virtue of Huygens' principle they interfere to form a wave front in the form of a cone with the projectile at its tip. This cone was first observed and photographed by Mach and in aerodynamics is known as Mach's cone. The shock wave travels in the direction perpendicular to the cone, forming an angle θ with the projectile's line of flight. One sees immediately that $\cos\theta = v_{\text{sound}}/v_{\text{proj}}$.

The velocity of light in a medium of refractive index n is, in general, smaller than the speed c of light in vacuum; it is c/n. In such a medium a particle may therefore travel with a velocity greater than c/n but, of course, smaller than c. Denoting the particle velocity by $v = \beta c$, with $\beta < 1$, the analogy to Mach's case yields

$$\cos\theta = \frac{c/n}{v} = \frac{c/n}{\beta c} = \frac{1}{n\beta}$$

for the angle θ under which the Cherenkov light is emitted.

Having had this idea, it remained to be shown by Frank and Tamm that indeed light waves are excited by the passage of a charged particle under the conditions described. Surprisingly enough, that could be directly deduced from Maxwell's equations on the electromagnetic field in a dielectric medium. The charged particle polarizes the medium, i.e., it induces dipole moments in the molecules in its vicinity. After the particle's passage the moments disappear again and the change in dipole strength makes the molecules act like little antennas emitting radiation. The theory yields the energy loss due to Cherenkov radiation, its spectrum (increasing strongly towards small wavelengths) and also its polarization (which lies in the plane defined by the direction of the incident particle and that of the emitted radiation). The theory was published early in 1937 [12] at the same time as detailed measurements by Cherenkov [13].

Now that his effect was experimentally well established and theoretically understood, Cherenkov also wanted to publicize it in an international journal and sent a short note to *Nature* in London which was promptly returned. (The letter of refusal is reproduced in facsimile in [5]). Cherenkov was not discouraged. He simply 'transferred [the manuscript] to another envelope and sent it to the editors of the *Physical Review*' who had it printed in August 1937 [14]. The paper contains photographs, depicting the angular distribution of Cherenkov radiation, which were taken in the following way. A transparent cylindrical vessel containing liquid was surrounded by a conical mirror. The assembly was exposed to γ rays entering the vessel in the horizontal plane. Radiation emitted in that plane was reflected upwards by the mirror and photographed in an exposure taking 72 hours. If Cherenkov would have had a beam of high-energy electrons at his disposal, he would have observed radiation only in the neighbourhood of a well-defined angle θ. However, he had to rely on electrons from Compton effect due to the incident γ rays, which have a less well-defined direction. Moreover, these electrons suffer energy loss and directional changes in the liquid.

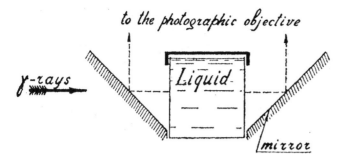

The set-up used by Cherenkov to record his effect photographically. Figure reprinted with permission from [14]. Copyright 1937 by the American Physical Society.

Still, the photographs show that emission occurs in the forward hemisphere only, with maxima around angles $+\theta$ and $-\theta$. The value of θ is seen to increase with the refractive index of the liquid, as expected.

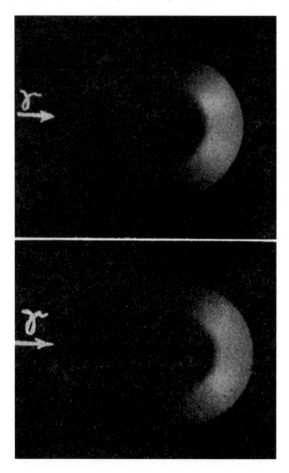

Photographs by Cherenkov showing the angular distribution of radiation for water (refractive index $n = 1.337$, top) and benzene ($n = 1.515$, bottom). Figure reprinted with permission from [14]. Copyright 1937 by the American Physical Society.

Frank and Tamm mention in their paper that they presented their theory

[1]   Ginzburg, V. L., *Physics –
      Uspekhi*, **44** (2001) 1026.

[2]   Bolotovski, B. N., Vavilov, Y. N.,
      and Shmeleva, A. P., *CERN
      Courier*, **44(9)** (2004) 37.

[3]   Ch. 7 in [9].

[4]   Cherenkov, P. A., *Celovek i
      otkrytie*. Nauka, Moscow, 1999.

[5]   Cherenkov, P. A., *Nuclear
      Instruments and Methods*, **A248**
      (1986) 1.

[6]   Dobrotin, N. A. In [4].

[7]   Čerenkow, P., *Comptes Rendus
      Acad. Sci. URSS*, **3** (1934) 455.

[8]   Wawilow, S., *Comptes Rendus
      Acad. Sci. URSS*, **13** (1934) 450.

[9]   Kojevnikov, A. B., *Stalins
      Great Science – The Times and
      Adventures of Soviet Physicists*.
      Imperial College Press, London,
      2004.

[10] Krokhin, O. N., *Physics – Uspekhi*,
      44 (2001) 1026.

[11] Čerenkow, P., *Comptes Rendus
      Acad. Sci. URSS*, **12** (1936) 413.

[12] Frank, I. and Tamm, I., *Comptes
      Rendus Acad. Sci. URSS*, **14**
      (1937) 109.

[13] Čerenkow, P., *Comptes Rendus
      Acad. Sci. URSS*, **14** (1937) 101,
      105.

[14] Čerenkow, P. A., *Physical Review*,
      52 (1937) 378.

[15] Mather, R. L., *Physical Review*, **84**
      (1951) 181.

[16] http://www.icecube.
      wisc.edu.

to Abraham Joffe from the Optical Institute in Leningrad, who told them that Sommerfeld, in 1904, had studied Maxwell's equations in vacuum for a charge moving faster than light and had shown that it loses energy by radiation. The computations became meaningless with the advent of Special Relativity in 1905, as did the even earlier work on the radiation by a charge in superluminal motion by Heaviside[6], and are of historical interest only. We also know now that the radiation had been observed before Cherenkov but had not been studied in a systematic way.

Already in 1937 Cherenkov mentions the possibility to use the threshold effect, i.e., the fact that radiation occurs only for particles with a velocity greater than c/n to determine the velocity of fast particles. In 1951 Cherenkov radiation was for the first time studied for particles other than electrons. A beam of 340 MeV protons was extracted from the Berkeley 184-inch cyclotron and passed through material of high refractive index. There was, however, no need to use the threshold effect. Since protons of this energy are scattered much less than electrons from radioactive sources, it was possible to measure the Cherenkov angle with high precision and thus to verify the theory by Frank and Tamm [15].

As mentioned, Cherenkov radiation is weak; only about 100 photons in the visible spectrum are produced by an electron travelling 1 cm in water. With the advent of *photomultipliers*, which can generate an electric signal from a single photon, it became possible to construct detectors for individual particles on the basis of the Cherenkov effect. One advantage of a *Cherenkov counter* is that no special scintillating material is needed; any transparent liquid or gas will do. Since the effect is caused only by particles surpassing the threshold velocity $v_{thr} = c/n$, a radiator with a given refractive index n allows to discriminate between particle velocities below and above threshold. A particular type of detectors, called *ring-imaging Cherenkov counters*, allows the measurement of the Cherenkov angle θ for individual particles and thus the determination of their velocity. With readily available Cherenkov media, such as water or ice, very large detectors can be realized to register rare events (see, for instance, Episode 100). The most spectacular example is the experiment *IceCube* [16], consisting of a cubic kilometre of natural ice at the South Pole into which a large number of photomultipliers is at present inserted. It is designed to serve as a neutrino telescope, detecting Cherenkov light from muons which themselves are created from cosmic neutrinos interacting in the ice.

---

[6] Oliver Heaviside (1850–1925)

Prediction of the Meson (1934) – Discovery of the Muon (1937)

Soon after the discovery of the neutron and its recognition as constituent of the nucleus Heisenberg (Episode 51) in 1932 and Fermi (Episode 53) in 1933 presented the first quantum theories of nuclear binding and of nuclear β decay, respectively. The next important step towards a theory of the nucleus was done in 1934 by Yukawa, then an unknown young scientist in Japan. His work would go unnoticed until, more than two years later, a particle that his theory predicted was apparently found.

Yukawa[1] was born in Tokyo as the fifth of seven children. For the first twenty-five years his name was Hideki Ogawa. His father, Takuji Ogawa, under the impression of a disastrous earth quake, had studied geology. In 1901 he had given a presentation at an international conference in Paris and stayed on in Europe for a year. When young Hideki was only a year old, the family moved to Kyoto, where his father became professor of geography at the university. In the small volume *Tabibito (The Traveler)* [1], he recalls his youth in the old Imperial Town. He was educated in a traditional way and taught to read Chinese classics even before entering primary school. His father, whom he remembers as severe and distant, had a large library, and reading became Hideki's favourite pastime. He depicts himself as a shy and introvert child with few friends. In high school, he learned English and German and developed an affection for mathematics and physics. One of his classmates was Sin-Itiro Tomonaga about whom we shall hear in Episode 71. Still in high school, Hideki bought and studied university text books on physics. He began in 1924 with an English translation of *Quantum Theory* by Reiche[2]. The original text had been written in 1921 and at its end the author expressed the hope that the many open questions of what is now called the old quantum theory would be answered on a 'day not far distant'. Yukawa writes about it: 'Never, in my life, have I received greater stimulation or greater encouragement from a single book than I did from that one.' Attracted by the author's name, he also bought the volume on classical mechanics, in German, by Planck and, no wonder, found it easier to understand than quantum theory.

Yukawa and Tomonaga in 1926 entered Kyoto Imperial University to study physics. They graduated three years later and stayed on as unpaid assistants, studying quantum mechanics mainly on their own from the literature; but in 1929 they were able to attend a series of lectures given by two of the theory's originators, Dirac and Heisenberg, in Tokyo and Kyoto. In 1932 Hideki Ogawa married Sumi Yukawa, the daughter of a medical doctor who had studied in Germany and later built his own hospital. He was adopted by his father-in-law and became Hideki Yukawa. This was by no means unusual. Both his

Yukawa,
self-portrait of the scientist as a young man.

---

[1] Hideki Yukawa (1907–1981), Nobel Prize 1949   [2] Fritz Reiche (1883–1969)

Coulomb Potential and Yukawa Potential

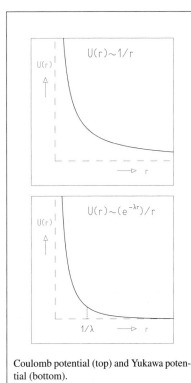

Coulomb potential (top) and Yukawa potential (bottom).

Coulomb Potential: The potential $U(\mathbf{r})$ of an electrostatic field at a point $\mathbf{r}$ in space in a region without electric charges is determined by the *Laplace equation*

$$\left(\frac{\partial^2}{\partial x^2} + \frac{\partial^2}{\partial y^2} + \frac{\partial^2}{\partial z^2} \right) U(\mathbf{r}) = 0 \quad .$$

One solution is the well-known Coulomb potential, inversely proportional to the modulus r of $\mathbf{r}$,

$$U(\mathbf{r}) \sim \frac{1}{r} \quad ,$$

corresponding to the case of a point charge at the origin $\mathbf{r} = 0$.

Yukawa Potential: Yukawa modified the Laplace equation to take the form

$$\left(\frac{\partial^2}{\partial x^2} + \frac{\partial^2}{\partial y^2} + \frac{\partial^2}{\partial z^2} - \lambda^2 \right) U(\mathbf{r}) = 0 \quad .$$

This equation yields the *Yukawa potential*

$$U(\mathbf{r}) \sim \frac{e^{-\lambda r}}{r}$$

with a range essentially limited to the region $r < 1/\lambda$. With the method of *second quantization* the field U could be written as a sum of matter waves of a particle with mass

$$m_U = \lambda \frac{h}{c} \quad ,$$

h being Planck's constant and c the speed of light in vacuum.

father and his father-in-law had been adopted in this way. That same year he became lecturer in Kyoto and the following year at the new university in Osaka. Yukawa writes that he had been searching for a quantum theory of nuclear forces, mediated by an exchange particle, for quite some time but that he had the decisive idea only in October 1934, a few weeks after the birth of his second son. On 17 November 1934, he presented his theory at a meeting of the Physico-Mathematical Society in Sendai. Early in 1935 it appeared as a paper, in English, in the proceedings of the society [2].

Yukawa's theory again is an attempt to describe the force between nucleons by the exchange of a charged bosonic particle. Heisenberg had considered the possibility that this particle was a 'boson electron' but had discarded it. In the wake of Fermi's theory of weak interaction, an electron–neutrino pair was tried out in this role without success. The central idea of Yukawa was to connect the short range $1/\lambda$ of the force between nucleons to the mass m_U of the exchanged boson (see Box). Inserting for $1/\lambda$ the radius of the proton, as estimated from experiment, $1/\lambda = 0.2 \times 10^{-12}$ cm, he obtained for m_U a value of 200 times the electron mass. In analogy to the potential energy of a negative charge $-e$ at a point $\mathbf{r}$, which is attracted by a positive charge at the origin, $U_{\text{el}}(\mathbf{r}) = -e^2/r$, Yukawa wrote his potential, if attractive, in the form

$$U(\mathbf{r}) = -g^2 \frac{e^{-\lambda r}}{r} \quad .$$

Here g is the *coupling* constant of Yukawa's field, which he estimated to be 'a few times' larger than the elementary charge e. Therefore the nuclear force is now called the *strong force* or *force of strong interaction*. Yukawa called his particle the *heavy quantum* (as opposed to the light quantum (!)) and he had a good explanation for the fact that it had not yet been observed. The mass difference between a neutron and a proton was much smaller than the mass of a heavy quantum. Therefore it could be exchanged between nucleons inside the same nucleus but there was not enough energy to separate it from a nucleus. That could well be different, Yukawa conjectured, in cosmic-ray processes with their high energies.

Yukawa expected his new particle also to be responsible for β decay. With small probability, i.e., very rarely, it should decay into an electron–neutrino pair which could leave the nucleus. The probability was kept small because Fermi's coupling constant, denoted g' by Yukawa, was very much smaller than g.

In his first paper Yukawa considered only two charge or isospin states of his particle, carrying one unit of positive or negative electric charge. But it was found later that also an electrically neutral particle was needed to account for the strong force between two protons or two neutrons. The three particles form an isospin triplet with isospin quantum numbers $I_3 = -1, 0, +1$.

Although Yukawa's paper was written in English and although complimentary copies of the *Proceedings of the Physico-Mathematical Society* were sent to research institutes abroad, no one noticed it. This sad situation changed only when a particle with properties predicted by Yukawa was indeed observed in the cosmic radiation.

In 1935 Anderson, the discoverer of the positron (Episode 49), and Neddermeyer[3], who had just obtained his Ph.D. from Caltech, had mounted their magnetic cloud chamber on a trailer. With a veteran truck they pulled it to the summit of Pike's Peak, Colorado, to collect photographs of cosmic rays at the high altitude of $4300\,\mathrm{m}$ [3]. In these photographs, they observed many more showers than at sea level. An *electromagnetic shower* is a cascade of reactions. It can be initiated by a high-energy photon which, in the electric field of a nucleus, produces an electron–positron pair. Both electrons can hit nuclei and produce photons by *bremsstrahlung* which, in turn, produce pairs etc. Of course, only a small section of a shower is seen on a cloud-chamber photograph. But for a shower event such a photograph usually shows many tracks.

After this first survey of cosmic-ray events, Anderson and Neddermeyer decided to perform a systematic measurement of the energy loss of particles. They placed a bar of platinum of 1 cm thickness in the middle of their chamber. The energy and also the energy loss in the bar was obtained by measuring the curvature of the particle's track before entering and after leaving the platinum bar. The energy range for the analysis was limited to the region below $400\,\mathrm{MeV}$, in which the curvature could be well measured. In this energy range protons have a much higher specific ionization compared to electrons: They produce cloud-chamber tracks with many more droplets per unit length, i.e.,

Yukawa in the 1950s.

---
[3] Seth Henry Neddermeyer (1907–1988)

fatter tracks. Such tracks were excluded from the analysis. Neddermeyer and Anderson discovered that the remaining tracks formed two groups:

- *Non-penetrating particles:* One group showed the behaviour expected of electrons and positrons; their energy loss was proportional to the original energy. A particle in this group was part of a shower, i.e., its track was one of several in the same photograph.
- *Penetrating particles:* The other group displayed a small energy loss, essentially independent of the energy. (Since the measurement errors were appreciable, in quite a few cases the energy loss appeared to be negative.) Particles in this group usually were not associated to showers.

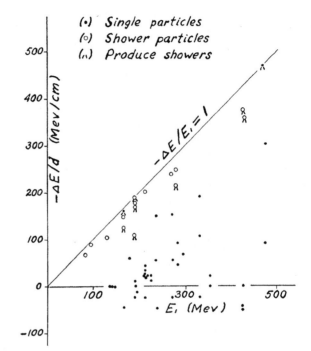

Energy loss in 1 cm of platinum for different types of cosmic-ray particles, measured as a function of their energy by Neddermeyer and Anderson. Figure reprinted with permission from [4]. Copyright 1937 by the American Physical Society.

In their paper [4], which appeared in May 1937, Neddermeyer and Anderson declare that the second group consists of particles with a mass intermediate between the masses of the electron and the proton. They could not be electrons because they do not show the high energy loss of these particles due to *bremsstrahlung* in platinum and because they do not occur in shower events. Proton tracks were excluded to begin with. In a note added in proof to the paper [4], the authors report that their discovery was confirmed by Street and Stevenson who had been able to observe for some isolated tracks a drop density somewhat higher than that of electrons but much lower than that of protons, indicating a mass in between the mass of these two particles. These findings were published later in 1937 [5]. (Incidentally, the first cloud-chamber photograph of such a particle had already been published in 1933 by Kunze[4] from the University of Rostock [6], who wrote about it, 'the nature of this particle is

---

[4] Paul Kunze (1897–1986)

unknown. For a proton, apparently, it ionizes too little, for a positive electron too much'.) Thus that part of the cosmic radiation, first observed by Bothe and Kolhörster (Episode 34), in the form of charged corpuscles traversing thick layers of matter and therefore called the *penetrating component* was identified to consist of a new type of particles.

Anderson and Neddermeyer thought about a name for their particle. Because of its intermediate mass they constructed the term 'mesoton' and wrote a short note for *Nature* suggesting that name [7]. Their 'Chief', Millikan, had been absent; but when he returned he objected, claiming it had to be *mesotron* like in 'electron'. He would not listen to arguments about particle names without *r* in the final syllable like 'proton'. Since he was the Chief, Anderson and Neddermeyer cabled the *r* to England, where it still arrived in time. Eventually the *r*, together with two more letters, was dropped again and the term *meson* was adopted for the Yukawa particle; for a while some authors also called it '*U* particle' or 'Yukon'.

Only after the observation of the new particle, Yukawa's paper was noticed by scientists abroad. Two letters appeared in *The Physical Review* [8,9] suggesting a connection of Yukawa's theory to the new particle. The theory and Yukawa suddenly became well known. He was invited to participate in the Solvay Conference planned for 1939 and was in Germany when the Second World War broke out. The conference was cancelled. Yukawa and also Tomonaga, who at the time was working in Heisenberg's group in Leipzig, were advised to leave Europe in a hurry. In Japan Yukawa's capabilities were recognized soon after the presentation of his first paper. He became Associate Professor in Osaka and with some of his very first students founded a flourishing school of theoretical physics. In 1939 he assumed a professorship in Kyoto which he would hold all his life, although after the war he spent several years in the United States. His *meson theory*, for decades, became the standard theory of strong interactions.

Over the years, the nature of the 'meson', found in 1937, was better understood. In 1939/40 its decay was observed, in which an electron appeared [10,11]. That seemed to strengthen its interpretation as the Yukawa particle. But then it was realized that the observed particle, which traverses thick layers of matter, did not display the essential feature of Yukawa's particle, strong interaction with nucleons and nuclei. In 1947, finally, another particle was found (Episode 66), which is a little more massive and which produces the particle of Anderson and Neddermeyer when it decays. This, indeed, showed strong interaction. It was called the π meson and its daughter particle the μ meson. Later the term *meson* was reserved for strongly interacting particles. Thus the μ meson became the *muon*. In many respects, it is merely a heavy electron. This, in itself, is most remarkable: There is more than one *generation* of electrons. In fact, at present, it seems certain that there are exactly three (see Episode 98).

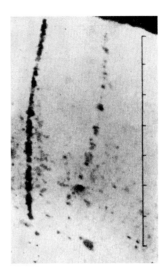

The track of unknown nature (left) observed by Kunze and that of an electron (right). Reprinted with kind permission from Springer Science+Business Media: [6], Fig. 5.

[1] Yukawa, H., *"Tabibito" (The Traveler)*. World Scientific, Singapore, 1982.

[2] Yukawa, H., *Proceedings of the Physico-Mathematical Society (Japan)*, **17** (1935) 48. Reprinted in [1].

[3] Anderson, C. D., *The Discovery of Anti-Matter*. World Scientific, Singapore, 1999.

[4] Neddermeyer, S. H. and Anderson, C. D., *Physical Review*, **51** (1937) 884.

[5] Street, J. G. and Stevenson, E. G., *Physical Review*, **52** (1937) 1003.

[6] Kunze, P., *Zeitschrift für Physik*, **83** (1933) 1.

[7] Neddermeyer, S. H. and Anderson, C. D., *Nature*, **142** (1938) 878.

[8] Oppenheimer, J. R. and Serber, R., *Physical Review*, **51** (1937) 1113.

[9] Stückelberg, E. C. G., *Physical Review*, **52** (1937) 41.

[10] Rossi, B., Hilbery, H. V. N., and Hoag, J. B., *Physical Review*, **56** (1939) 837.

[11] Williams, E. J. and Roberts, G. E., *Nature*, **145** (1940) 102.

58

A New Kind of Liquid: Superfluid Helium (1937)

Early in January 1938, two letters to the editor appeared on adjacent pages of *Nature*, describing a hitherto unknown state of matter, superfluidity of helium. Author of the first letter [1], dated 3 December 1937, was Kapitza[1] who worked in an institute in Moscow which had specially been erected for him. The second letter [2], dated 22 December, was signed by Allen[2] and Misener[3] who did their research in Cambridge, also in an institute planned and built for Kapitza.

Kapitza [3,4] was born in Kronstadt, an island fortress off St. Petersburg in a bay of the Baltic Sea, as son of a military engineer. Ever since Czar Peter the Great founded St. Petersburg in the age of enlightenment and made it the Russian capital, it had also been a city of science and learning. (The city's original name was in Dutch, *Sankt-Pieterburch*, but soon changed to its German equivalent. In 1914 it became *Petrograd* and in 1924 *Leningrad*. The traditional name was reintroduced in 1991.) The Academy of Sciences was located there as well as a university and a Polytechnical Institute, now St. Petersburg Technical University. It is this institute which Kapitza entered in 1912. (As a schoolboy, being more gifted for science than for the classics, he had been transferred from the *Gymnasium* to the *Realschule*, which gave access only to engineering schools.) From 1915 onwards the leading physicist there was Joffe[4] who had graduated from the institute and later earned his Ph.D. under Röntgen in Munich. In 1918, Joffe became department head in one of the first research institutes started after the revolution. Three years later the department was made into the independent Physico-Technical Institute, led by Joffe until 1950 and now also called the Joffe Institute. There Joffe founded an important school of which Kapitza was an early member.

After his graduation in 1919, Kapitza got a teaching post at the Polytechnical Institute. Only a little later, in the state of need and famine after the revolution, he suffered a family tragedy. In the winter of 1919/20, within a few weeks his father, his two-year old son, his wife, and her just born daughter died in an epidemic. Joffe, in order to help in this crisis, asked Kapitza to accompany him abroad on a journey in 1921 aimed at re-establishing contacts with Western European science severed by war and revolution. Kapitza was refused a German visa but met Joffe in England. Their visit culminated at the Cavendish Laboratory in Cambridge and Kapitza managed to be accepted as research student.

Rutherford suggested that he measure the energy loss of α particles near the end of their range. For his first paper on the subject, Kapitza registered the

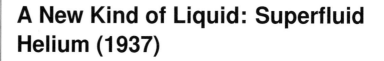

Kapitza

---

[1] Pyotr Leonidovich Kapitza (1894–1984), Nobel Prize 1978   [2] John Frank Allen (1908–2001)   [3] Austin Donald Misener (1911–1996)
[4] Abraham Feodorovich Joffe (1880–1960)

average effect for many particles. In order to measure the change of energy or, to be more precise, momentum of individual particles, he decided to use a cloud chamber placed in a magnetic field. The necessary precision, however, could be not be achieved with a conventional electromagnet. He therefore built a pulsed magnet, operated by short-circuiting a large battery through a coil for the short time in which the cloud chamber was sensitive.

This work led Kapitza away from nuclear physics to the study of materials in very high magnetic fields and, somewhat later, also at very low temperatures. To achieve these conditions, new large and costly machines had to be constructed. Astonishingly, Rutherford, whose hallmark was the ingenious table-top experiment of simple design, strongly supported him. He had become fond of Kapitza, who, besides being an excellent physicist, was gifted in mathematics and engineering and was very deft with his hands. Moreover, Kapitza had a colourful personality, full of charm and wit. Being initially a little awed by Rutherford he had nicknamed him *the crocodile*. Missing the informal discussions in Joffe's seminar he created the *Kapitza club*, which would meet in his rooms in Trinity College. He was not only quick-witted but also a man of fast decisions. In April 1927, he informed Rutherford by letter from Paris that the next day he was going to marry a young Russian lady to whom he had become engaged the week before.

Rutherford used his influence and funds, left by Ludwig Mond, the founder of Imperial Chemical Industries, to the Royal Society, to have a modern laboratory built in the courtyard of the Cavendish which was designed to meet the needs of Kapitza. In February 1934, the *Royal Society Mond Laboratory* was officially opened in style. Next to the entrance there is a life-size crocodile carved into the brick wall. The first liquid helium was produced in the Mond Laboratory in April with a machine of novel design by Kapitza. (A little to Kapitza's dismay, the first liquid helium in England was made in Oxford by Lindemann with a more conventional liquefier or, rather, by Mendelssohn, a refugee from Nazi Germany who had worked with Nernst in Berlin.) Kapitza had become a member of the scientific establishment in Britain, Fellow of Trinity College, and Fellow of the Royal Society. But he had also been made a Corresponding Member of the Soviet Academy of Sciences and never had renounced his Soviet citizenship. Once a year he travelled to Russia, visiting his mother and holidaying on the Black Sea.

In October 1934, the Soviet authorities did not permit his return from such a journey. Only his wife was allowed to travel to England to look after their two sons. Attempts from various sides to get permission for his return came to nothing. Eventually, a laboratory similar to the Mond, the Institute for Physical Problems (IFP), was built for him in Moscow and completed in 1936. The high-field equipment and copies of the liquid-helium machinery were purchased by the Soviet government from Britain. Cockcroft, who had already worked on his Ph.D. under Kapitza, succeeded him as director of the Mond Laboratory.

Liquid helium was long known to display a peculiarity. Its specific heat at normal pressure passes discontinuously through a maximum at the temperature $2.18\,\mathrm{K}$, called the λ point. The liquid was called helium I at temperatures above the λ point and helium II below it. Kapitza was intrigued by recent

The Crocodile.
Engraving by Eric Gill.

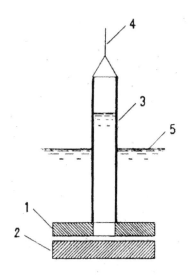

Kapitza's experimental set-up. Reprinted by permission from MacMillan Publishers Ltd: Nature [1], copyright 1938.

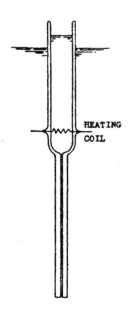

The thermomechanical effect as observed by Allen and Jones. Reprinted by permission from MacMillan Publishers Ltd: Nature [8], copyright 1938.

findings in Leiden [5] and in Cambridge [6] about an abnormally high heat conductivity of helium II. He suspected that the primary cause was not conduction but convection, i.e., the actual transport of liquid. Then helium II would have to possess an abnormally low viscosity to allow large amounts of liquid to carry along heat, because the specific heat was not particularly large. Kapitza tested his assumption in an experiment he described as follows [1]:

> The viscosity was measured by the pressure drop when the liquid flows through the gaps between disks 1 and 2; these discs were of glass and were optically flat, the gap between them being adjustable by mica distance pieces. The upper disk, 1, was 3 cm in diameter with a central hole of 1.5 cm diameter, over which a glass tube (3) was fixed. Lowering and raising this plunger in the liquid helium by means of the thread (4), the level of the liquid in the tube (3) could be set above or below the level (5) of the liquid in the surrounding Dewar flask. The amount of flow and the pressure were deduced from the difference of the two levels, which was measured by a cathetometer.

Kapitza describes the results obtained as 'rather striking'. When the glass plates were half a micrometre apart for a temperature above the λ point a change in levels could only just be detected in a time of several minutes, 'while below the λ point the liquid helium flowed quite easily, and the level in the tube 3 settled down in a few seconds'. He further reports that he can only give an upper limit for the viscosity, a value more than four orders of magnitude smaller than that of gaseous hydrogen, the least viscous fluid known up to then, and he concludes:

> The present limit is perhaps sufficient to suggest, by analogy with superconductors, that helium below the λ point enters a special state which might be called a 'superfluid'.

Kapitza gave a copy of his letter to *Nature* to a visitor from Cambridge, where similar experiments were conducted by Allen and Misener, who used the more conventional capillaries rather than a flat gap. Having seen the letter, Cockcroft wrote to Kapitza about the Cambridge results and shortly thereafter also submitted them for publication. The Cambridge work had been done independently; but publication was speeded by Kapitza's manuscript. Allen [7] was a Canadian who had earned his doctorate on superconductivity from the University of Toronto. He had written to Rutherford for permission to work at the new Mond Laboratory. But when he arrived, Kapitza already was 'captured [in the USSR] like a proton in a carbon nucleus', as Gamow described it in a letter to Bohr [4]. It is all the more remarkable that he and his graduate student Misener, also a Canadian, found superfluidity on their own.

But Allen did more. Together with Jones[5] within less than a month he found another astonishing phenomenon, now called the *thermomechanical effect* or, popularly, the *fountain effect* [8]. The original aim of the experiment was to study heat conduction in narrow capillaries. A tube, ending in a capillary, was placed in a Dewar vessel, which was partly filled with liquid helium below the λ point. Of course, the helium surface was at the same level inside and outside the tube. But when the temperature in the tube was slightly increased by running an electric current through a wire, the surface in the tube rose. The

[5] Harry Jones (1905–1986)

opposite behaviour was expected of a conventional liquid because of the rise in vapour pressure over the warmer liquid in the tube. The effect was made rather spectacular by replacing the capillary with a U-shaped container, open at both ends and filled with emery powder. When light was shone on that container a fountain of liquid helium was seen to spurt out of the tube.

First steps towards a theory of superfluidity were done by the German Fritz London[6] and the Hungarian Tisza[7], who, at that time both worked in Paris after leaving Nazi Germany. On 5 March 1938, London suggested [9] that superfluidity and the fountain effect are connected with the formation of a *Bose–Einstein condensate* from ^{4}He atoms (which are bosons, see Episode 33) at very low temperatures. Six weeks later Tisza, whom London had told of this idea, constructed his *two-fluid model* of helium II [10]. One part of the atoms are in the quantum-mechanical ground state, the state with the lowest possible energy. This part carries no entropy and therefore no heat; it forms the superfluid. Atoms in the other part are in excited states and form a normal fluid. The fraction of superfluid in the liquid varies between one (at zero temperature) and zero (at the lambda point). Within this model Tisza was able to explain the fountain effect: If two volumes filled with helium II at different temperatures are connected by a 'superleak', e.g., a capillary or a tube filled with emery powder, the concentration of superliquid is higher in the volume of lower temperature. To ease the difference in concentration, superfluid flows through the superleak from the colder to the warmer volume and raises the pressure there. Tisza also predicted the inverse effect which later was indeed observed: If the two volumes initially are at the same temperature and pressure is applied to the first, forcing superliquid into the second, a temperature difference between the two volumes appears, the temperature being higher in the first volume.

We shall see in Episode 62 that the definite quantum-mechanical description of the superfluid was given in 1941 by Landau, another Russian physicist with a life just as colourful as that of Kapitza. In the same year, when Germany invaded the Soviet Union, Kapitza broke off his work on liquid helium. Instead, he developed a new and more efficient method to liquefy air and thus to produce oxygen, needed in the steel industry, and was made a minister responsible for the production of the equipment he had invented. Having become a full Academician in 1939 and having received two Stalin Prizes and two Orders of Lenin, Kapitza was ousted in 1946 not only as a minister but also as director of his own Institute of Physical Problems. But he retained his position as member of the Academy. For years he lived in his country house, which he gradually transformed into a laboratory for microwave and plasma physics until, in 1954, he was reinstated as leader of his Moscow institute. Only in 1965 was he allowed to follow an invitation to Denmark and from then on to many countries, including England. He arranged that his old house in Cambridge, which he had made over to the Soviet Academy, was made available to house visiting scientists. *Kapitza House* still serves that purpose.

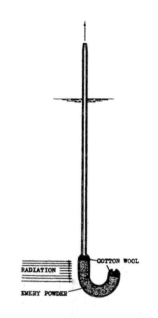

The fountain effect realized by Allen and Jones. Reprinted by permission from MacMillan Publishers Ltd: Nature [8], copyright 1938.

[1] Kapitza, P., *Nature*, **141** (1938) 74.
[2] Allan, J. F. and Misener, A. D., *Nature*, **141** (1938) 75.
[3] Boag, J. W., Rubinin, P. E., and Shoenberg, D. (eds.), *Kapitza in Cambridge and Moscow – Life and Letters of a Russian Physicist*. North Holland, Amsterdam, 1990.
[4] Ch. 5 in [11].
[5] Keesom, W. H. and Keesom, A. P., *Physica*, **3** (1936) 359.
[6] Allen, J. F., Peierls, R., and Uddin, M. Z., *Nature*, **140** (1937) 62.
[7] Donelly, R. J., *Physics Today*, **July** (2002) 76.
[8] Allan, J. F. and Jones, H., *Nature*, **141** (1938) 243.
[9] London, F., *Nature*, **141** (1938) 643.
[10] Tisza, L., *Nature*, **141** (1938) 913.
[11] Kojevnikov, A. B., *Stalins Great Science – The Times and Adventures of Soviet Physicists*. Imperial College Press, London, 2004.

---

[6] Fritz London (1900–1954)   [7] Laszlo Tisza (born 1907)

59   Why the Stars Shine (1938)

What is the source of energy that makes the sun shine? The first who tried to give a scientific answer to this question was Helmholtz. He proposed that the sun has been slowly contracting under its own gravitation and that energy was being released because matter was falling towards the centre of the sun. This mechanism could provide for the sun's energy output since around 30 million years, a long time but much too short on a geological scale.

In 1920, the year after he found astronomical proof for Einstein's general theory of relativity (Episode 29), Eddington addressed the British Association for the Advancement of Science on the subject [1]:

> A star is drawing on some vast reservoir of energy by means unknown to us. This reservoir can scarcely be other than the subatomic energy which, it is known, exists abundantly in all matter; we sometimes dream that man will one day learn how to release it and use it for his service. The store is well-nigh inexhaustible, if only it could be tapped. There is sufficient in the sun to maintain its output of heat for 15 billion years. ...

In his talk Eddington points out that Aston (with his mass spectrograph, Episode 22) found the mass of the helium nucleus to be smaller than the mass of four protons, i.e., hydrogen nuclei, by 1 part in 120. In the transformation of hydrogen into helium this mass difference, according to Einstein's formula $E = mc^2$, would appear, he thought, as electromagnetic energy. This transformation, taking place inside stars, could therefore provide for the stellar radiation. Eddington concluded his address with the dramatic sentence:

> If, indeed, the subatomic energy in stars is being freely used to maintain their great furnaces, it seems to bring a little nearer to fulfilment our dream of controlling this latent power for the well-being of the human race – or for its suicide.

Eddington's conjecture was indeed correct. The formation of helium from hydrogen is the principal source of stellar energy. In 1920, however, there was no way to verify this hypothesis. Nuclear physics was still in its infancy and quantum mechanics, needed to calculate nuclear reactions, did not exist. Nevertheless, Eddington made important advance in *The Internal Constitution of the Stars*, as he called his influential book of 1926 [2]. He devised a stationary model of a star consisting of a gaseous mixture of elements in fixed proportion. It was described by a set of two differential equations, the *Eddington equations* later improved by Chandrasekhar[1] and Strömgren[2]. These equations, at every distance from the centre of the star, ensure equilibrium between the gravitational pressure acting inwards and the pressure due to thermal motion and radiation which acts outwards. They allow, for a star of given mass

---

[1] Subramanyan Chandrasekhar (1910–1995), Nobel Prize 1983   [2] Bengt Strömgren (1908–1987)

The CNO Cycle or Bethe–Weizsäcker Cycle

The complete cycle comprises six steps. In steps $(1, 3, 4, 6)$ a nucleus of carbon (C) or nitrogen (N) absorbs a proton, i.e., a nucleus of hydrogen (H), and is transformed into a nucleus of another element. In steps $(2, 5)$ a nucleus of nitrogen or oxygen (O) undergoes β decay under emission of a positron (e^+) and a neutrino (ν). The net result is the vanishing of four protons and the appearance of an α particle, i.e., a helium nucleus (He), the carbon nucleus only serves as a catalyst:

$$^{12}\text{C} + {}^1\text{H} \rightarrow {}^{13}\text{N} + \gamma \quad , \tag{1}$$

$$^{13}\text{N} \rightarrow {}^{13}\text{C} + e^+ + \nu \quad , \tag{2}$$

$$^{13}\text{C} + {}^1\text{H} \rightarrow {}^{14}\text{N} + \gamma \quad , \tag{3}$$

$$^{14}\text{N} + {}^1\text{H} \rightarrow {}^{15}\text{O} + \gamma \quad , \tag{4}$$

$$^{15}\text{O} \rightarrow {}^{15}\text{N} + e^+ + \nu \quad , \tag{5}$$

$$^{15}\text{N} + {}^1\text{H} \rightarrow {}^{12}\text{C} + {}^4\text{He} \quad . \tag{6}$$

Although the atoms are ionized, there are, on the average, one electron per hydrogen nucleus and two electrons per helium nucleus in the solar material. The loss of four protons in the creation of one helium nucleus implies that two electrons become superfluous. They are annihilated by the two positrons from β decay. A fraction of the total energy set free is carried away by the neutrinos, which leave the sun with very little chance to interact. The rest, kinetic energy of nuclei and energy of photons (γ), is quickly turned into heat.

---

and given surface temperature, which both are measurable, to compute the density and temperature anywhere inside the star, once a particular composition of elements is assumed.

Stars are classified by their position in a diagram spanned by two variables, the absolute brightness of the star in units of the sun's brightness and its colour, which is directly related to its surface temperature. The majority of stars lies on a rather well-defined line, called the *main sequence*, in this *Hertzsprung–Russell diagram*. Eddington could show that for main-sequence stars there exists a well-defined relationship between brightness and mass. The distribution of elements, at least on the star's surface, is obtained by optical spectroscopy. Thus for the most frequent type of stars all the information to compute density and temperature in their interior is available.

The temperature is, of course, highest in the centre of the star. Because of the high temperature the atoms are ionized; the nuclei have a velocity distribution determined by the temperature according to the laws of statistical mechanics. In particular, some nuclei can have energies much exceeding the average energy. After Gamow had presented his theory (Episode 50), in which two nuclei can interact in spite of their Coulomb repulsion by virtue of the tunnel effect, it was realized by Atkinson[3] and Houtermans[4] [3] that the thermal energy of protons in the centre of the sun was sufficient to provoke nuclear reactions. The physical possibility of a *thermonuclear reaction* in the interior of stars was established.

---

[3] Robert d'Escourt Atkinson (1898–1982)   [4] Friedrich Georg Houtermans (1903–1966)

With the advent of accelerators a large number of nuclear reactions was studied in detail in the laboratory and the knowledge gained could be used to select those which could be important in stars. Since nuclei are transformed into others, the question of energy production became related to that of *nucleosynthesis*, i.e., the formation of heavier elements from hydrogen. In 1937 and 1938 Weizsäcker published two comprehensive articles on the subject, entitled *On Element Conversion in the Interior of Stars* [4,5].

Weizsäcker[5] had met Heisenberg when he was still at school and at that time had considered to study philosophy. But Heisenberg let him know: 'There is already a lot of 'beautiful' philosophy, but 'good' physics we still can use.' Later he told him [6]: 'In our century you can only make good philosophy, if you understood what is the most important philosophical event of this century, and that is modern physics.' Weizsäcker took the advice. After studying physics and earning his Ph.D. under Heisenberg in Leipzig, he continued to work in this field until, in 1957, he accepted a Chair of Philosophy at the University of Hamburg, where in his lectures he emphasized the writings of Plato and Kant.

In his paper [5], completed in July 1938, Weizsäcker considers, in particular, two processes of energy generation in stars, both involving the fusion of hydrogen nuclei:

Direct fusion Two protons combine to yield a deuteron, i.e., a nucleus of a heavy hydrogen isotope, as well as a positron and a neutrino,

$$^1\mathrm{H} + {}^1\mathrm{H} \to {}^2\mathrm{H} + e^+ + \nu \quad .$$

(The neutrino is not explicitly mentioned.)

Fusion with a catalyst in a cycle of reactions (CNO cycle) This is a series of four proton-capture processes and two β decays (see Box). Its net result is the transformation of hydrogen into helium. The carbon nucleus needed in step (1) is reproduced in step (6) and thus acts as a catalyst.

In footnotes Weizsäcker lets the reader know that, as he learned from Gamow, both processes were being studied quantitatively by Bethe.

Bethe[6] [7,8] had begun to study physics at the University of Frankfurt, where his father was professor of physiology. After his first two years, in 1926, he moved to Munich to work on his thesis on electron diffraction in crystals under Sommerfeld. He obtained his Ph.D. in 1928 and in 1930 became Privatdozent in Munich. Early that year he had completed his pioneering work on the energy loss of fast particles in matter. Although many different processes contribute, Bethe could summarize their effect in a concise formula which, including later refinements, is now called the *Bethe–Bloch formula*. It is of great importance to experimentalists since the energy loss is directly related to the signature a particle leaves in a detector, as, for instance, the pulse height in a counter or the droplet density in a cloud chamber.

With a travel fellowship Bethe was able to go to Cambridge for some time in 1930 and to Rome in the spring of 1931 and again in 1932, where he worked

von Weizsäcker

Bethe

[5] Carl Friedrich von Weizsäcker (1912–2007)   [6] Hans Albrecht Bethe (1906–2005), Nobel Prize 1967

in Fermi's group. In 1933 he lost his position as acting professor at the University of Tübingen when the Nazis came to power in Germany. After spells in Manchester and Bristol he became professor at Cornell University in 1935. Together with Livingston, who had recently arrived from Berkeley, where he had built cyclotrons with Lawrence, and with Bacher[7], he made Cornell a major centre of nuclear physics. Bethe, Livingston, and Bacher wrote three monumental reviews [9–11] in which the wealth of experimental results in nuclear physics was ordered, summarized, and complemented by new theoretical insight. These papers were repeatedly reprinted and became known as the *Bethe bible*.

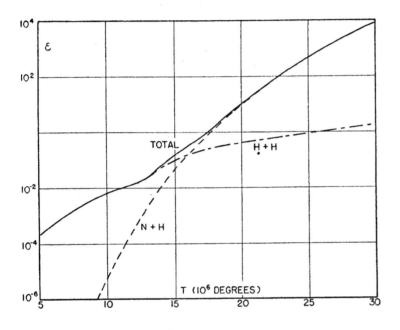

Bethe's plot, presenting the result of his computations. It shows the energy produced (in units erg/g s) as a function of temperature (in millions of Kelvin) for the process of direct fusion (dash-dotted line marked H+H) and for the carbon–nitrogen cycle (dashed line marked N+H). The full line shows the total energy production. Figure reprinted with permission from [12]. Copyright 1939 by the American Physical Society.

With the detailed knowledge, gained while assembling the 'bible', Bethe was in an ideal position to quantitatively tackle the problem of stellar energy. He had become interested in the problem while attending a small conference on the subject, which was organized in Washington in March 1938, by Gamow and Teller[8]. These two had recently improved [13] Atkinson's and Houtermans' theory of thermonuclear reactions. As a first step Bethe computed the probability and temperature dependence of the direct fusion process with Charles L. Critchfield, a Ph.D. student of Gamow [14]. They found that this could 'almost but not quite' account for the observed energy production in the sun. (By the way, the deuteron, through further processes, is transformed into a helium nucleus; the main reactions are $^1H + {}^2H \rightarrow {}^3He + \gamma$ and $^3He + {}^3He \rightarrow {}^4He + {}^1H + {}^1H$.) Moreover, they computed that the energy output increases with temperature T approximately as $T^{3.5}$. But Bethe realized that a more rapid increase was needed to explain the vastly larger energy production in hotter stars.

---

[7] Robert Fox Bacher (1905–2004)   [8] Edward Teller (1908–2003)

In his paper *The Energy Production in Stars*, which he submitted to *The Physical Review* in September 1938, Bethe independently discovered the CNO cycle and computed its properties [12]. As he told himself [15], to obtain a steeper energy dependence he had to find a reaction involving a higher potential barrier, i.e., a nucleus with a higher charge:

> So I went systematically through the Periodic Table but everything gave nonsense because whatever atom I used, lithium, beryllium, etc., would be destroyed in the reaction, and there was very little of these substances anyway, as we know from the abundances both on earth and in the stars. So these elements could not possibly give the energy production for the length of time that the universe had been functioning. Finally I got to carbon, and as you all know, in the case of carbon the reaction works out beautifully.

Bethe found that for this process the energy production rises very steeply with temperature, like T^{17}, for temperatures around 20×10^6 K, and he concluded that, while direct fusion was important in the sun, the CNO cycle provided the energy in the hotter stars. Together the two processes were sufficient to explain the observed energy yield of the stars on the main sequence.

In the Second World War, Bethe was the leader of the theory group in Los Alamos, where the atomic bomb was developed. Shortly after the war, in 1947, Sommerfeld asked Bethe if he would be willing to succeed him on the Chair in Munich. In a moving letter (reproduced in English translation in [7]) Bethe declined. He stayed at Cornell and remained active there for the rest of the century. Through his research, his teaching, and his students he was as important for theoretical physics in the Unites States as Sommerfeld had been in Germany.

[1]   Eddington, A. S., *The Observatory*, **43** (1920) 353.

[2]   Eddington, A. S., *The Internal Constitution of Stars*. Cambridge University Press, Cambridge, 1926.

[3]   Atkinson, R. and Houtermans, F. G., *Zeitschrift für Physik*, **54** (1929) 656.

[4]   v. Weizsäcker, C. F., *Physikalische Zeitschrift*, **38** (1937) 633.

[5]   v. Weizsäcker, C. F., *Physikalische Zeitschrift*, **39** (1938) 633.

[6]   von Weizsaecker, C. F., *Große Physiker*, chap. 'Heisenberg als Physiker und Philosoph'. Hanser, Munich, 1999.

[7]   Gottfried, K. (ed.), *Special Issue: Hans Bethe*. Physics Today, October 2005.

[8]   Brown, G. E. and Lee, C. H. (eds.), *Hans Bethe and his Physics*. World Scientific, Hackensack, N.J, 2006.

[9]   Bethe, H. A. and Bacher, R. F., *Reviews of Modern Physics*, **8** (1936) 82.

[10] Bethe, H. A., *Reviews of Modern Physics*, **9** (1937) 69.

[11] Livingston, M. S. and Bethe, H. A., *Reviews of Modern Physics*, **9** (1937) 245.

[12] Bethe, H. A., *Physical Review*, **55** (1939) 434.

[13] Gamow, G., Teller, E., and Critchfield, C. L., *Physical Review*, **53** (1938) 6088.

[14] Bethe, H. A. and Critchfield, C. L., *Physical Review*, **54** (1938) 248.

[15] Bethe, H. A. et al., *From a Life of Physics*. World Scientific, Singapore, 1989.

Bethe lecturing at CERN, 1964.

Nuclear Fission (1938)

Otto Hahn's father had come to the old city of Frankfurt in 1866. A few years later he bought the small glazier's shop in which he worked and gradually transformed it into a flourishing glazing business which still exists under his name. Otto Hahn[1] [1], the youngest of four brothers, studied organic chemistry in Marburg, and after a year of military service, stayed there for another two years as assistant to Zincke[2], his Ph.D. adviser. Near the end of this time he was given the prospect of a position in industry for which he would need fluency in a foreign language. His parents agreed to fund a stay abroad and Zincke wrote to Ramsay in London who, in 1904, accepted Hahn as an unpaid visitor in his institute at University College.

Ramsay gave Hahn a sample of barium chloride, telling him that the sample contained a very small amount of radium and suggesting to isolate the radium and to attempt a determination of its atomic mass number. Thus Hahn's first work on radioactivity was connected to barium and radium as was later his greatest discovery. In the course of this work, Hahn found a new radioactive substance, which he called radio thorium. In those times all such substances were called elements. Radio thorium is now classified as the isotope ^{228}Th of thorium. After this success, Ramsay convinced Hahn to give up his plans for a career in industry and to stay in fundamental research. Moreover, he recommended him to Emil Fischer[3] who invited Hahn to Berlin. Hahn felt, however, that he should learn more about radioactivity. Again at his parents' expense, he spent nearly a full year in 1905/06 with Rutherford in Montreal, certainly the best place for that purpose. He discovered another two radioactive 'elements', thorium C' (^{212}Po) and radio actinium (^{227}Th) and became very fond of Rutherford, his way to work, and the spirit of his group.

Hahn

Hahn began to work in 1906 in Fischer's Institute of Chemistry at the University of Berlin, where he established the new field of radiochemistry. In the overcrowded institute he was given a room in the basement which until then had been the wood workshop. He soon discovered more radioactive isotopes. In 1907 Lise Meitner[4] [2] arrived in Berlin. She had been a student of Boltzmann and also performed some work on radioactivity in Vienna. After Boltzmann's death she had decided to study under Planck. She also began to work with Hahn. Fischer, who did not accept female students in his institute, agreed under the condition that she would not show up in the large students' laboratories on the upper floors. Thus began the collaboration between Hahn and Meitner which lasted for more than thirty years. Hahn became Privatdozent in 1910 and associate professor in 1912. In 1911 the *Kaiser-Wilhelm Society* was founded, which was to run research institutes independent of universi-

---

[1] Otto Hahn (1879–1968), Nobel Prize 1944   [2] Ernst Carl Theodor Zincke (1843–1928)   [3] Hermann Emil Fischer (1852–1919), Nobel Prize 1902   [4] Lise Meitner (1878–1968)

Hahn, Meitner, and Rutherford in front of Hahn's institute in Berlin-Dahlem, 1929.

ties, and in 1912 the Kaiser-Wilhelm Institute of Chemistry in Berlin-Dahlem was opened in which Hahn led the small department of radioactivity. (He became director of the institute in 1928.) Here he and Meitner had much better working conditions. In 1918 they discovered the long-lived mother substance of actinium and called it *protactinium*. Although a short-lived isotope of the same element had been found before and had been given the name *brevium*, the name protactinium (Pa) was officially adopted for the element. When, after the First World War, limitations for women in the academic world were lifted, Meitner, too, became professor. Hahn remembered [1]:

> With all her modesty Lise Meitner grew more and more into the dignity of a professor, including the then still proverbial absentmindedness. On a congress, a colleague greeted her by saying: 'We have met before.' Lise Meitner, who did not remember, answered: 'You probably take me for Professor Hahn!' Since we had published so much work together, apparently she thought that such mistaking was possible.

In 1934 Fermi roused the world of radioactivity with his method of neutron bombardment (Episode 55) and still that same year reported on the possible production of transuranic elements by irradiating uranium with neutrons [3,4]. The irradiation had led to radioactive substances with different half-lives such as 10 s, 40 s, 13 min, and 90 min. Fermi's group had separated the 13-min and 90-min 'bodies' chemically from uranium and had shown that they were not isotopes of elements, which are located only a few places below uranium in the Periodic Table. As briefly discussed in Episode 55, they assumed that the uranium nucleus with the extra neutron transformed, via β decay, into a nucleus of an element with the number 93 in the Periodic Table. That could still be unstable and transit, by another β decay, into a nucleus of element 94.

The idea of more than 92 elements was, of course, contested. Ida Nod-

Meitner and Hahn in their Berlin laboratory in the 1920s.

dack[5], a renowned chemist and co-discoverer of the element rhenium, pointed out that all known elements had to be excluded before new ones were proposed [5]. This very sound advice was not taken. Nuclear physicists saw no possibility for a nucleus to fragment into large pieces. Nothing more drastic than the α decay had ever been observed. Another way out was also proposed: In spite of Fermi's interpretation, his 13-min body might be an isotope of protactinium, element number 91 [6]. Here Hahn and Meitner came in. After all, they were discoverers of protactinium and knew the properties of this element. They were able to show that the activity in question was not due to protactinium and became convinced that transuranic elements had been produced. They began intensive work in this new field, from 1935 onwards together with Strassmann[6]. Strassmann, like Hahn, was a chemist who had come to the Kaiser-Wilhelm Institute in 1929 on a fellowship. When it expired in the depression era, he stayed on without pay and got an assistantship only in 1935. Quite a number of substances with different half-lives and different chemical properties were found in uranium irradiated with neutrons. A detailed scheme for their production was proposed, which implied the creation and subsequent decay of four, possibly five, transuranic elements [7]. It was not seriously challenged by other groups working in the field.

When Austria was annexed in 1938, Lise Meitner became a German citizen and, because of her Jewish descent, was in acute danger. Helped by Hahn and other colleagues, she fled via Holland and Denmark to Sweden, where she could work in the Physical Institute of the Academy of Sciences in Stockholm.

Lise Meitner in the 1920s. The photograph carries a dedication to Max von Laue.

Meitner with Hahn in 1938 on the day before she left Germany.

Hahn and Strassmann continued alone. The decay products of the apparent transuranic elements seemed to contain three substances, which underwent β decays of different half-lives and were chemically very similar to barium. They were taken to be isomeric nuclei of the isotope ^{231}Ra of radium. (A nucleus of a radioactive isotope can exist in different energy states, called isomeric, which decay differently; the first case of isomeric states had been found by Hahn himself in 1922.) Radium is an alkaline-earth metal as is barium and

---

[5] Ida Eva Noddack (1896–1978)   [6] Fritz Strassmann (1902–1980)

Strassmann

Frisch in 1934

is located below barium in the second column of the Periodic Table, hence the similarity. Hahn and Strassmann tried to isolate the radium. Since only minute quantities could have been produced, a precipitation with barium as *carrier* from a solution was performed; the barium was to carry along the chemically similar radium. The precipitate then only contained barium and radium, which were to be separated in the next step. As mentioned above, Hahn was well versed in the method of separation, fractional crystallization, originally introduced by Marie Curie. But although they tried hard and checked and rechecked their method, Hahn and Strassmann were unable to separate any radium by chemical means. In their first paper they still conclude rather cautiously [8]:

> We come to the conclusion: Our 'radium isotopes' have the properties of barium; as chemists we should rather say the new bodies are not radium but barium. [...] As 'nuclear chemists', in a certain sense close to nuclear physics, we cannot yet decide ourselves to perform this step contradicting all previous experience of nuclear physics. A series of strange coincidences might still have faked our results.

Hahn had kept Meitner informed by letter about the work and he mailed her a copy of the manuscript of the paper on 21 December 1938, the same day it was submitted to *Die Naturwissenschaften*. The manuscript reached her in a small town near Gothenburg, where she had gone to visit Swedish friends over Christmas and where she had also invited her nephew Otto Frisch. We have already met him as collaborator of Stern in Hamburg (Episode 52). He, too, had been forced to leave Germany and at that time was working in Bohr's institute in Copenhagen. Meitner showed him Hahn's letter and the manuscript and dismissed the possibility of mistake [9]: 'Hahn was too good a chemist for that.' The two began to look for an explanation. The nucleus could not just been cracked like a nut. In fact there was evidence that it behaved rather like a droplet. Now, a droplet might divide into two smaller ones by contracting in one direction, then elongating, and so on, until it finally would split up. The uranium nucleus might need little extra energy to do so, because its many protons provided a repulsive electrical force counteracting the attractive nuclear force between all nucleons, protons as well as neutrons. This extra energy could be provided by a single neutron. Meitner calculated that the energy of 200 MeV would be released in a single process, an energy equivalent to one-fifth of the rest mass of a proton.

After a few days, Frisch returned to Copenhagen and told Bohr, who was enthusiastic [9]: 'Oh what idiots we have all been! Oh but this is wonderful! Have you and Lise Meitner written a paper about it?' Frisch told him that there would soon be a paper and asked Bohr, who was about to travel to the United States to participate in a conference, not to discuss the matter before it would appear in print. We shall see in Episode 63 that this promise could not be kept and how Bohr's reports triggered intense activities which led to the first nuclear reactor in less than four years.

Frisch was looking for a good word describing the way a droplet divides and found out that biologists call the division of cells *fission*. In the famous paper by Meitner and Frisch [10], accordingly, the term *nuclear fission* is introduced. The paper was composed during several long-distance calls between Copenhagen and Stockholm, then still quite a costly matter. Following a suggestion from a colleague (he was later astonished that it needed such a suggestion), Frisch set up a simple experiment to detect the products of fission, which carried away the energy Meitner had computed: An ionization chamber, its inside lined with uranium, was connected to an electronic amplifier. When irradiated with neutrons high ionization pulses produced by the energetic fission products were registered, much higher than the pulses of α particles. This was the first physics experiment in which fission was demonstrated [11]. It convincingly complemented Hahn's and Strassmann's excellent chemical analyses.

In their common paper, Meitner and Frisch had predicted the appearance of krypton along with barium in the fission of uranium, because the nuclear charge numbers of krypton (36) and of barium (56) add up to that of uranium (92). In a second paper [12], submitted on 28 January 1939, Hahn and Strassmann write that they received the two manuscripts from Meitner and Frisch, and they report on several new fission products, including a noble gas, either krypton or xenon. In this paper they also write: 'In our opinion, the 'transuranic elements' maintain the position ascribed to them.' This was indeed the case. Once the many products of fission were removed, true transuranic elements could be isolated. That work was done mainly in Berkeley, see Episode 61. At present transuranic elements, artificially produced in different ways, up to the number 111 in the Periodic Table have been given names. Examples are *fermium* ($_{100}$Fm) and *meitnerium* ($_{109}$Mt).

For the next few years, Hahn and Strassmann continued to identify radioactive isotopes produced in uranium fission. By spring of 1945, they had found a total of about 100 isotopes from 25 different elements. After the war Hahn and other German scientists were interned in England and kept there until the beginning of 1946. In 1945 Hahn was awarded the Nobel Prize in Chemistry for the year 1944. On his return to Germany, he succeeded Planck as president of the Kaiser-Wilhelm Society whose name was changed to *Max-Planck Society*. Lise Meitner stayed in Stockholm and in 1947 became professor at the Technical University. She retired in 1954 and in 1960 moved to England to live with her nephew Otto Frisch, who had become professor at the University of Cambridge. Strassmann became head of the radiochemical department of the Max-Planck Institute of Chemistry. Its building in Berlin had been destroyed; the institute was moved to Mainz. From 1953 onwards he concentrated on his professorship at the newly founded University of Mainz. The town later made him an honorary citizen. In 1966 Meitner, Hahn, and Strassmann were the first non-US citizens to be given the Enrico-Fermi Award by the American president.

In the summer of 1955, when I had just completed the first semester of my physics studies, I had the good fortune to have lunch with Otto Hahn. Our family and Professor and Mrs Hahn were having holidays in the same part of Bavaria. My parents went for a visit and took me along. Hahn affably asked if I enjoyed my studies and I, taking him for a physicist, complained about the chemical laboratory in which we were asked to follow certain procedures without understanding what we were doing. He answered, obviously tongue-in-cheek: 'One sees, Herr Brandt, that you are a physicist. As a chemist, I, in my time, just happily cooked along (... habe einfach drauflos gekocht)!'

[1] Hahn, O., *Vom Radiothor zur Uranspaltung: eine wissenschaftliche Selbstbiographie.* Vieweg, Braunschweig, 1962.

[2] Rife, P., *Lise Meitner and the Dawn of the Nuclear Age.* Birkhäuser, Boston, 1999.

[3] Fermi, E., Rasetti, F., and D'Agostino, O., *Ricerca Scientifica*, **5** (1934) 536. Also in [13], Vol. I, 704.

[4] Fermi, E., *Nature*, **133** (1934) 898. Also in [13], Vol. I, 748.

[5] Noddack, I., *Angewandte Chemie*, **47** (1934) 653.

[6] von Grosse, A. and Agruss, M. S., *Nature*, **134** (1934) 773.

[7] Meitner, L., Hahn, O., and Strassmann, F., *Zeitschrift für Physik*, **106** (1937) 249.

[8] Hahn, O. and Strassmann, F., *Naturwissenschaften*, **27** (1939) 11.

[9] Frisch, O. R., *What little I remember.* Cambridge University Press, Cambridge, 1979.

[10] Meitner, L. and Frisch, O. R., *Nature*, **143** (1939) 239.

[11] Frisch, O. R., *Nature*, **143** (1939) 276.

[12] Hahn, O. and Strassmann, F., *Naturwissenschaften*, **27** (1939) 89.

[13] Fermi, E., *Collected Papers, 2 vols.* University of Chicago Press, Chicago, 1962.

Two Transuranium Elements Finally Found – Neptunium and Plutonium (1940/41)

In 1789, the Berlin chemist Klaproth[1] discovered a new element in pitchblende ores from Johanngeorgenstadt in Saxony and from Sankt Joachimsthal in Bohemia. He called it *uranium* after the outermost planet then known. That planet, the seventh, had been discovered eight years earlier by Herschel[2] in England and named *uranus* after Urania, the muse of astronomy and geometry. At that time the Periodic Table was still unknown. But after it was established in 1869 and generally accepted in the years to follow, the name uranium was considered fitting for the last in the list of elements. However, the analogy was spoiled a little because since 1846 an eighth planet, *neptune*, was known.

We saw earlier (in Episodes 55 and 60) that Fermi's and Hahn's groups believed to have produced elements, situated beyond uranium in the Periodic Table, by bombarding uranium with neutrons. But, when the fission of uranium was discovered as the dominating consequence of irradiation with neutrons, the possible existence of transuranium elements, for a short time, became an open question again. Shortly after fission became known, McMillan in Berkeley found a surprisingly simple way to separate fission products from possible transuranium elements. With this method the first such element was found in 1940 and later aptly called *neptunium*.

McMillan[3] [1] grew up in Pasadena as son of a physician and took B.Sc. and M.Sc. degrees at the California Institute of Technology in his home town. Shortly after completing his Ph.D. thesis in Princeton, he won a National Research Council fellowship, which he used to work in Lawrence's new Radiation Laboratory in Berkeley. His first project there, an attempt to measure the magnetic moment of the proton with a molecular beam, was abandoned when Stern's group in Hamburg did the experiment (Episode 52). From then on McMillan became deeply involved in nuclear physics work at the Berkeley cyclotrons.

When the news on uranium fission, observed by Hahn and Strassmann, reached Berkeley early in 1939, McMillan performed a simple experiment to measure the range and thus the kinetic energy of the fission fragments. He spread a thin layer of uranium oxide on paper, placed that between two stacks of thin aluminium foils, and exposed it to neutrons produced by 8-MeV deuterons striking a beryllium target in the 37-inch cyclotron. Fission was already known to yield predominantly two nuclei of about half the mass of the uranium nucleus, flying apart back-to-back. Each of these radioactive fission products lost its energy in the aluminium foils and was stopped in one of them.

McMillan re-creating the search for neptunium at the time of the announcement of the discovery, 8 June 1940.

---

[1] Felix Klaproth (1743–1817)   [2] (Sir) Friedrich Wilhelm (William) Herschel (1738–1822)   [3] Edwin Mattison McMillan (1907–1991), Nobel Prize 1951

By measuring the radioactivity of the individual foils after irradiation, McMillan found the maximum range of the fission products in matter to be equivalent to about 2 cm in air. He repeated the experiment with cigarette paper instead of aluminium foils and filter paper as carrier of the uranium oxide instead of ordinary paper and made sure that no heavier elements (from minerals) were left in these types of paper. When studying the radioactivity in the cigarette paper, which served as fission-fragment catcher, he found, as expected, a decay curve indicating the presence of nuclei of many different half-lives. But the uranium sample (including the filter-paper carrier) clearly displayed two half-lives. One of those, about 25 minutes, had already been identified in 1937 by Meitner, Hahn, and Strassmann [2] as due to a new radioactive isotope of uranium, $^{239}_{92}$U, formed by neutron bombardment of the copious uranium isotope $^{238}_{92}$U. The other half-life was much longer, about 2 days. In a short note [3] McMillan presented his results and stressed that with his method one could easily separate the recoil activity (the fission products) from non-recoiling activity, i.e., $^{239}_{92}$U and its possible decay products. Obviously the non-recoiling activity was the place to look for transuranium elements since $^{239}_{92}$U was expected to form a nucleus with charge number 93 via β decay [2]. But McMillan left that unsaid and simply stated: 'This work is being continued.'

$_1$H																	$_2$He
$_3$Li	$_4$Be											$_5$B	$_6$C	$_7$N	$_8$O	$_9$F	$_{10}$Ne
$_{11}$Na	$_{12}$Mg											$_{13}$Al	$_{14}$Si	$_{15}$P	$_{16}$S	$_{17}$Cl	$_{18}$Ar
$_{19}$K	$_{20}$Ca	$_{21}$Sc	$_{22}$Ti	$_{23}$V	$_{24}$Cr	$_{25}$Mn	$_{26}$Fe	$_{27}$Co	$_{28}$Ni	$_{29}$Cu	$_{30}$Zn	$_{31}$Ga	$_{32}$Ge	$_{33}$As	$_{34}$Se	$_{35}$Br	$_{36}$Kr
$_{37}$Rb	$_{38}$Sr	$_{39}$Y	$_{40}$Zr	$_{41}$Nb	$_{42}$Mo	(43)	$_{44}$Ru	$_{45}$Rh	$_{46}$Pd	$_{47}$Ag	$_{48}$Cd	$_{49}$In	$_{50}$Sn	$_{51}$Sb	$_{52}$Te	$_{53}$I	$_{54}$Xe
$_{55}$Cs	$_{56}$Ba	57–71	$_{72}$Hf	$_{73}$Ta	$_{74}$W	$_{75}$Re	$_{76}$Os	$_{77}$Ir	$_{78}$Pt	$_{79}$Au	$_{80}$Hg	$_{81}$Tl	$_{82}$Pb	$_{83}$Bi	$_{84}$Po	(85)	$_{86}$Rn
(87)	$_{88}$Ra	$_{89}$Ac	$_{90}$Th	$_{91}$Pa	$_{92}$U	(93)	(94)	(95)	(96)	(97)							

Lanthanoides:	$_{57}$La	$_{58}$Ce	$_{59}$Pr	$_{60}$Nd	(61)	$_{62}$Sm	$_{63}$Eu	$_{64}$Gd	$_{65}$Tb	$_{66}$Dy	$_{67}$Ho	$_{68}$Er	$_{69}$Tm	$_{70}$Yb	$_{71}$Lu

Periodic Table as seen in the 1930s. Elements are denoted by their chemical symbol and nuclear charge number. A number in parentheses stands for an as yet unknown element.

In the 1930s, the understanding of the end of the Periodic Table was not what it is today. Although it was relatively easy to assign the correct nuclear charge number Z to each element, that was not the case for the quantum numbers of all electrons in its shell. It was assumed that the chemical similarity of the rare-earth elements or *lanthanoides* ($Z = 57, 58, \ldots, 71$) was due to the fact that the two electrons in the outer shell of all these elements had the same quantum numbers and that the elements differed only what concerned inner shells, less important for chemical properties. But it was not known that this peculiarity repeated itself for the elements with $Z = 89, 90, \ldots$, i.e., actinium, thorium, protactinium, uranium, ... Consequently, uranium (U) was placed below tungsten (W) in the Periodic Table and the hypothetical elements with $Z = 93$ and $Z = 94$ below rhenium (Re) and osmium (Os), respectively. They were referred to as eka-rhenium (Eka-Re) and eka-osmium (Eka-Os) [2];

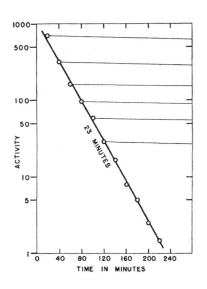

McMillan's and Abelson's proof that element 93 is a daughter product of $^{239}_{92}$U. The precipitates taken at regular intervals from a solution show an initial activity (circles) proportional to the $^{239}_{92}$U content of the solution at the time of extraction. The activity of each precipitate drops with a half-live of 2.3 days as indicated by the slightly sloped lines. Figure reprinted with permission from [5]. Copyright 1940 by the American Physical Society.

$$^{238}_{92}U + n \rightarrow \gamma + \,^{239}_{92}U$$
$$^{239}_{92}U \rightarrow \beta^- + \,^{239}_{93}Np \quad (23\ min)$$

The reactions leading to neptunium 239.

such a nomenclature had been used in nineteenth-century chemistry to refer to predicted elements in as yet empty positions of the Periodic Table.

With an improved method McMillan produced a sample of non-recoiling activity and determined the two half-lives more accurately: 23 minutes (exactly the value given by Meitner and her colleagues for) $^{239}_{92}$U and 2.3 days. It was now tempting to suspect that eka-rhenium was indeed formed in the 23-minute decay and that it decayed into eka-osmium with a half-life of 2.3 days. A homologue of rhenium, placed above it in the Periodic Table, the element technetium, had been discovered by Segrè and Perrier (see end of Episode 47) and it was natural to ask Segrè, who was in Berkeley, to undertake a search for eka-rhenium in the sample. Segrè performed the experiment by preparing a solution from the sample in which the element rhenium would dissolve, by then adding some rhenium to the solution, by precipitating the rhenium from the solution and looking for the 2.3-day activity in the precipitate, which would have been carried along with the rhenium, had it been chemically similar. The experiment was entirely negative. Nothing chemically similar to rhenium was found, but the 2.3-day activity showed properties familiar from the rare earths [4]. Was it a fission product after all?

McMillan went on to improve his method and, using very thin collodium foils, he found that the 2.3-day activity had a range of at most 0.1 mm in air. He became more convinced than ever that it could not be a fission fragment and therefore must have a nuclear charge number higher than $Z \approx 60$, typical of the rare earths. He even began to do some chemistry himself and found that the 2.3-day activity did not in every aspect show similarity to the rare earths. At that time, in the spring of 1940, Abelson[4], a former Ph.D. student of the Radiation Laboratory, returned to Berkeley for a vacation. At the Carnegie Institution in Washington he had, independently of McMillan, tried to separate the 2.3-day activity from large samples of uranium. The two decided to work together; McMillan prepared the samples, Abelson did the chemical work. It was found that element 93, suspected to be behind the 2.3-day activity, was chemically rather similar to uranium but still sufficiently different to allow a method to be found to separate it quantitatively from uranium.

With this method McMillan and Abelson demonstrated that element 93 was indeed produced in the decay of $^{239}_{92}$U. They prepared a solution containing $^{239}_{92}$U, and at intervals of 20 minutes, removed the suspected element 93 from it by precipitation. As they expected, all precipitates showed the 2.3-day activity and their initial activity, which was a measure for the amount of element 93 present immediately after precipitation, fell in unison with the concentration of $^{239}_{92}$U in the solution, i.e., by a factor of two every 23 minutes.

As already mentioned, the first transuranium element was later named *neptunium* by McMillan. The chain of reaction leading to its production was now clear: $^{239}_{92}$U was formed from $^{238}_{92}$U by neutron bombardment and that gave rise to $^{239}_{93}$Np by β decay.

The signature for neptunium was its 2.3-day β activity. It was therefore most natural to assume that, in this β decay, neptunium changed into the next transuranium element of charge number 94 and mass number 239. There were

---

[4] Philip Hauge Abelson (1913–2004)

good reasons to assume that such an element would suffer α decay.

McMillan and Abelson prepared a strong sample by extracting neptunium from 500 g of an irradiated uranium compound but were unable to detect α activity from it. They concluded that, if there was any, it had a very long half-life, a million years or more. This, as McMillan pointed out in his Nobel Lecture, was the only serious error in their note *Radioactive Element 93* [5]. It was later found that element 94 is indeed a product of neptunium β decay and that it transforms into uranium 235 by α decay with a long half-life which, however, amounts to 'only' 24 000 years.

Abelson had to return from his 'vacation' to Washington and McMillan decided to try and produce another isotope of neptunium that would give rise to another isotope of element 94, which then might decay faster and therefore could be detected more readily. He exposed uranium directly to 16-MeV deuterons from the 60-inch cyclotron. Indeed, he was able to chemically separate neptunium from the irradiated sample and found α activity in it. He could not proceed with this work, because in November 1940, he was asked to take part in radar development and had to leave Berkeley. But his research was continued by Seaborg, then an instructor in chemistry at Berkeley, who, like McMillan, lived in the Faculty Club and knew of his work on element 93.

Seaborg[5] [6,7], born in Michigan, had lived in California since he was ten, studied chemistry at the University of California in Los Angeles, and taken his Ph.D. in chemistry at Berkeley in 1937. Soon he became deeply involved in radio-chemistry through the possibilities offered by the Berkeley cyclotrons. Stimulated by McMillan's and Abelson's discovery, he asked one of his graduate students, Wahl[6], to start work on neptunium. A colleague of Seaborg as chemistry instructor, Kennedy[7], also became interested in transuranium elements. When Seaborg learned that McMillan had left, he wrote to him, asking for his consent to continue the search for element 94, which was readily granted.

On 14 December 1940, uranium oxide was irradiated with deuterons and a sample was extracted containing the neptunium produced. In this sample, a growing radioactivity, due to α particles, was detected. That was attributed to the α decay with a half-life of about 50 years (the modern value is 92 years) of an isotope of element 94 which itself was produced in the β decay of neptunium 238. In these experiments, Seaborg and Wahl did mostly chemical work. Radioactivity was measured by instrumentation constructed by Kennedy.

At that time, the Second World War was waged in Europe, the American nuclear scientists had agreed to halt the spreading of results which might be important for war; but papers were written and sent to journals for later publication. Conforming with this policy, a short note entitled *Radioactive Element 94 from Deuterons* was sent to the *Physical Review* in January 1941. The note included McMillan as one of the authors and was published in 1946 [8]. In a second note with the same title, submitted in March 1941, Seaborg, Wahl, and Kennedy reported that they had studied the chemical properties of element 94 and found them similar to those of uranium [9].

It was now clear that neither was element 93 chemically similar to rhe-

$${}^{239}_{93}\text{Np} \rightarrow \beta^- + {}^{239}_{94}\text{Pu} \quad \text{(2.3 days)}$$
$${}^{239}_{94}\text{Pu} \rightarrow \alpha + {}^{235}_{92}\text{U} \quad \text{(24 400 years)}$$

Plutonium 239 is formed in the decay of neptunium 239 and itself decays into uranium 235.

Seaborg, adjusting a Geiger–Müller counter counter during the search for plutonium at Berkeley.

$${}^{238}_{92}\text{U} + d \rightarrow 2n + {}^{238}_{93}\text{Np}$$
$${}^{238}_{93}\text{Np} \rightarrow \beta^- + {}^{238}_{94}\text{Pu} \quad \text{(2.1 days)}$$
$${}^{238}_{94}\text{Pu} \rightarrow \alpha + {}^{234}_{92}\text{U} \quad \text{(92 years)}$$

Nuclear reactions leading to the production and decay of plutonium 238.

---

[5] Glenn Theodore Seaborg (1912–1999), Nobel Prize 1951   [6] Arthur Charles Wahl (1917–2006)   [7] Joseph W. Kennedy (1916–1957)

$_1$H																	$_2$He
$_3$Li	$_4$Be											$_5$B	$_6$C	$_7$N	$_8$O	$_9$F	$_{10}$Ne
$_{11}$Na	$_{12}$Mg											$_{13}$Al	$_{14}$Si	$_{15}$P	$_{16}$S	$_{17}$Cl	$_{18}$Ar
$_{19}$K	$_{20}$Ca	$_{21}$Sc	$_{22}$Ti	$_{23}$V	$_{24}$Cr	$_{25}$Mn	$_{26}$Fe	$_{27}$Co	$_{28}$Ni	$_{29}$Cu	$_{30}$Zn	$_{31}$Ga	$_{32}$Ge	$_{33}$As	$_{34}$Se	$_{35}$Br	$_{36}$Kr
$_{37}$Rb	$_{38}$Sr	$_{39}$Y	$_{40}$Zr	$_{41}$Nb	$_{42}$Mo	$_{43}$Tc	$_{44}$Ru	$_{45}$Rh	$_{46}$Pd	$_{47}$Ag	$_{48}$Cd	$_{49}$In	$_{50}$Sn	$_{51}$Sb	$_{52}$Te	$_{53}$I	$_{54}$Xe
$_{55}$Cs	$_{56}$Ba	57–71	$_{72}$Hf	$_{73}$Ta	$_{74}$W	$_{75}$Re	$_{76}$Os	$_{77}$Ir	$_{78}$Pt	$_{79}$Au	$_{80}$Hg	$_{81}$Tl	$_{82}$Pb	$_{83}$Bi	$_{84}$Po	$_{85}$At	$_{86}$Rn
$_{87}$Fr	$_{88}$Ra	89–103	$_{104}$Rf	$_{105}$Db	$_{106}$Sg	$_{107}$Bh	$_{108}$Hs	$_{109}$Mt	$_{110}$Ds	$_{111}$Rg	112						

Lanthanoides:	$_{57}$La	$_{58}$Ce	$_{59}$Pr	$_{60}$Nd	$_{61}$Pm	$_{62}$Sm	$_{63}$Eu	$_{64}$Gd	$_{65}$Tb	$_{66}$Dy	$_{67}$Ho	$_{68}$Er	$_{69}$Tm	$_{70}$Yb	$_{71}$Lu
Actinoides:	$_{89}$Ac	$_{90}$Th	$_{91}$Pa	$_{92}$U	$_{93}$Np	$_{94}$Pu	$_{95}$Am	$_{96}$Cm	$_{97}$Bk	$_{98}$Cf	$_{99}$Es	$_{100}$Fm	$_{101}$Md	$_{102}$No	$_{103}$Lr

Modern Periodic Table

[1] Jackson, J. D. and Panofsky, W. K. H., *Biographical Memoirs (Nat. Acad. Sci.)*, **69** (1996) 214.

[2] Meitner, L., Hahn, O., and Strassmann, F., *Zeitschrift für Physik*, **106** (1937) 249.

[3] McMillan, E., *Physical Review*, **55** (1939) 510.

[4] Segrè, E., *Physical Review*, **55** (1939) 1104.

[5] McMillan, E. and Abelson, P. H., *Physical Review*, **57** (1940) 1185.

[6] Hoffman, D. C., *Biographical Memoirs (Nat. Acad. Sci.)*, **78** (2000) 235.

[7] Hoffman, D. C., Ghiorso, A., and Seaborg, G. T., *The Transuranium People*. Imperial College Press, London, 2000.

[8] Seaborg, G. T., McMillan, E. M., Kennedy, J. W., and Wahl, A. C., *Physical Review*, **69** (1946) 366.

[9] Seaborg, G. T., Wahl, A. C., and Kennedy, J. W., *Physical Review*, **69** (1946) 367.

[10] Seaborg, G. T. and Wahl, A. C., *Journal of the American Chemical Society*, **70** (1948) 1128.

nium, nor element 94 similar to osmium. Both, rather, were similar to uranium and a revision of the Periodic Table was called for. That had been hinted at already by McMillan and Abelson who wrote 'that there may be a second 'rare earth' group of similar elements starting with uranium' [5]. We know now that such a group, the *actinoides*, indeed exists. It begins not with uranium but with actinium (Ac) and comprises the elements with charge numbers $Z = 89, 90, \ldots, 103$ most of which are, of course, transuranium elements. Efforts to produce transuranium elements with ever higher charge numbers are still continuing. For decades Seaborg was the domineering figure in this field [7]. At present elements up to $Z = 118$ have been produced and analysed.

We conclude this episode by a few words on the naming of element 94. In the form of the isotope 239 it is by far the most important transuranium element, because it can be 'bred' from uranium by irradiation with neutrons in a nuclear reactor (Episode 63) and then isolated chemically. It is fissionable by neutrons and therefore used as material in both reactors and bombs. At first, because of the general secrecy which had befallen nuclear physics, there was no need for a proper name, since very few people knew about element 94. But when the properties just mentioned became apparent and a report to the Uranium Committee in Washington had to be formulated by Seaborg and Wahl (published in 1948 [10]), they decided to follow the example of McMillan and to name element 94 after the planet[8] *Pluto*, which has an orbit still outside that of neptune. Pluto had been discovered in 1930 by Tombaugh[9] at the Lowell Observatory in Flagstaff, Arizona. Having also considered *plutium* as a possible name, Seaborg and Wahl chose to call element 94 *plutonium* and to give it the chemical symbol Pu.

---

[8] In 2006 the International Astronomical Union adopted a new definition for planets; with this Pluto, because of its small mass, is no longer a 'planet' but a 'dwarf planet'.   [9] Clyde William Tombaugh (1906–1997)

Landau Explains Superfluidity (1941)

62

In Episode 58 we saw how Kapitza, under circumstances forced upon him, late in 1937 discovered superfluidity. In April 1939, he declared in a letter to Molotov that he needed help from a theoretician, namely, Landau, to go on with his work. Landau at the time was in one of Stalin's prisons and in this way Kapitza succeeded to save his life. It lies in the nature of science that results, generally, cannot be produced to order. But Landau, defying this rule, within two years created the theory of superfluidity.

Landau[1] [1,2] was born in Baku as son of the chief engineer of an oil company. He never conceded to have been a child prodigy; but he also declared that he could not remember a time when he did not know differential and integral calculus. In 1922, when he was fourteen, he enrolled at Baku University in the Department of Physics and Mathematics and also in the Department of Chemistry. Two years later he transferred to the Physics Department of Leningrad University from which he graduated in 1927, when he was not yet nineteen. He continued to work in the Physico-Technical Institute in Leningrad led by Joffe, whom we already met as fosterer of Kapitza.

Senior members of the theory group in the institute were Frenkel[2] and Fock[3]. Landau was particularly close to Gamow and Iwanenko whom we encountered already in Episodes 50 and 51, respectively. They carried the nicknames 'Dau', 'Jonny', and 'Dymus' and were called the 'three musketeers'. The name 'Dau' was given to Landau by 'Dymus' Iwanenko because his full surname, heard by a Frenchman, might be misinterpreted as *l'âne Dau* [the donkey (or ass) Dau]. Landau liked the name and continued to use it even after he severed all relations to Iwanenko. The musketeers formed the nucleus of of a group of young theoreticians who called themselves the 'jazz band' and were fond of physics and fun. Another member, the future Lady Peierls, wrote about it: 'We were putting out the *Physikalische Dummheiten* [Physical Stupidities] and read it at university seminars. In general, we sharpened our sense of humour on our teachers and at their expense.' [3]. The star of the group was Gamow who had recently returned from a highly successful foreign trip (Episode 50) and who would later settle in the United States. Soon also Landau went to Western Europe with funds provided at first by the Soviet Union and then by the Rockefeller foundation.

In his time abroad between October 1929 and March 1931 Landau visited Berlin, worked with Heisenberg in Leipzig, with Pauli in Zurich, and with Dirac in Cambridge. His longest stay was in Bohr's institute in Copenhagen. There is a charming portrait of young Landau in Copenhagen [4], which includes two newspaper interviews with rather naive statements about revolu-

Landau
as a young scientist in Copenhagen . . .

. . . and as professor in Russia.

---

[1] Lev Davidovich Landau (1908–1968), Nobel Prize 1962   [2] Yakov Il'ich Frenkel (1894–1952)   [3] Vladimir Aleksandrovich Fock (1898–1974)

Box I. Collective Excitations in a Quantum Liquid: Phonons and Rotons

A quantum liquid is described by a single quantum-mechanical wave functions comprising all atoms. The wave function can be that of the ground state (the state of the lowest possible energy) or that of an excited state. An excited state corresponds to a particular motion of part of the liquid relative to the rest and must not be confused with excitations of single atoms requiring much larger energies. Such collective excitations conveniently are considered as *quasiparticles* with particular properties travelling through the liquid. In the theory of superfluidity two types of quasiparticles are introduced.

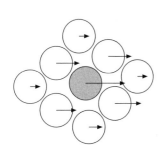

Simple visualization of a phonon ...

Phonons: A *phonon* corresponds to a tiny longitudinal sound pulse: A few atoms together carry a momentum $\mathbf{p}$, which they transmit to a neighbouring group which transmits it in turn, etc. In this way, the momentum and the corresponding energy are transported over macroscopic distances while the atoms are not. The phonon concept was introduced by Frenkel in solid-state quantum mechanics. In the classical wave theory the transported energy is proportional to the absolute value of momentum, the proportionality constant being the velocity c of sound. Similarly, the excitation spectrum of phonons, i.e., the relation between energy ϵ and momentum $\mathbf{p}$, is

$$\epsilon(\mathbf{p}) = c|\mathbf{p}| = cp \quad .$$

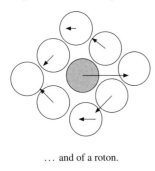

... and of a roton.

Rotons: These excitations, introduced and named by Landau, have been likened to smoke rings. Like phonons they travel in the direction given by their momentum. But in addition, there is a rotation within the ring of those atoms, which take part in the excitation at a particular time. The roton spectrum is

$$\epsilon(\mathbf{p}) = \Delta + \frac{(p - p_0)^2}{2\mu} \quad .$$

Here Δ, p_0, and μ are constants, the numerical values of which Landau extracted from measurements in 1947 [6]. A liquid with both types of excitations has a spectrum which, because of the phonons, rises linearly from the origin, drops and shows a minimum, due to the rotons, and then rises again.

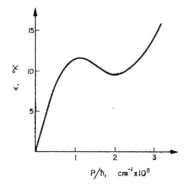

Landau's excitation spectrum of rotons. From [6].

tionary Russia. His best known work from this time is his *diamagnetism of metals* [5] in which he showed that not only the spins of free electrons contribute to the magnetic properties of a metal but also their orbits. The motion of such electrons is allowed only for discrete energy values, the *Landau levels* (see Episode 95).

At the end of the 1920s new physico-technical institutes were founded at Kharkov in the Ukraine, at Sverdlovsk (now again Jekaterinburg) in the Urals, and at Tomsk in Siberia as a result of Joffe's advice to the government to decentralize science. He argued that the greater scientific potential of Germany compared to that of France was connected to the fact that in Germany there existed research institutes in many towns whereas in France there was concentration on Paris [7]. In August 1931 Landau moved to Kharkov, where he soon established an important school of theoretical physics. With his rash and rather arrogant humour, he had made enemies in Leningrad and even alienated Joffe but the latter, knowing of Landau's talent, had proposed him for the position in Kharkov. Here Landau developed his talent as teacher. Being also a professor

---

Box II. Inducing Friction in Liquid Helium

If liquid helium flowing through a capillary is to suffer friction, it has to transmit energy to certain points, e.g., irregularities of the capillary surface. Let the liquid flow with velocity **v** and let us consider it in a coordinate frame moving along with it with the same velocity, so that in this frame the liquid is at rest. Let us now assume that an excitation is induced by collision with a surface irregularity. Its energy is $\epsilon(\mathbf{p})$ in the moving frame, **p** being the momentum of excitation in that frame. In the fixed frame the energy of the system has changed by $\epsilon(\mathbf{p}) + \mathbf{p} \cdot \mathbf{v}$. (Here $\mathbf{p} \cdot \mathbf{v}$ is a scalar product and the velocity due to **p** is taken to have an absolute value small compared to $|\mathbf{v}|$.) The excitation will readily occur if the change in energy is negative, i.e., if $\epsilon(\mathbf{p}) + \mathbf{p} \cdot \mathbf{v} < 0$. The most favourable case is that in which the direction of **p** is opposite to that of **v**, so that $\mathbf{p} \cdot \mathbf{v} = -pv$. Excitation will therefore take place provided $v > \epsilon(\mathbf{p})/p$. Now, if the function $\epsilon(\mathbf{p})/p$ has a finite minimum value, then no excitation and thus no friction can occur in liquid helium flowing with a velocity below a critical value given by that minimum,

$$v_{\text{critical}} = \min \frac{\epsilon(\mathbf{p})}{p} \neq 0 \quad .$$

This minimum is easily found to be given by the point which fulfils the condition

$$\frac{\epsilon}{p} = \frac{\mathrm{d}\epsilon}{\mathrm{d}p} \quad ,$$

i.e., the point of the excitation curve defined by $\epsilon = \epsilon(p)$ for which a straight line from the origin is tangential to the curve. It lies a little to the right and a little above the minimum of the curve. From this reasoning it becomes clear that the excitation curve of liquid helium has to possess a minimum. The existence of rotons, as postulated by Landau, explains it.

---

at the university, he lectured on practically all fields of theoretical physics and, together with his collaborator Lifshitz[4], he began to compose the monumental *Course of Theoretical Physics*. This ten-volume work, devised in detail by Landau and written down essentially by Lifshitz, was conceived as a *theoretical minimum*, the knowledge of which Landau required from those he accepted as doctoral students. Candidates, who had passed a preliminary test in mathematics, were admitted to study for the seven written tests on this 'minimum' of physics. In Landau's lifetime a total of 43 students persevered. Landau loved to rate things and people, in particular physicists. There are slightly differing reports about the scale he used, which may have changed with time. We quote here from the experience of an early student in Kharkov:

> Finally, I asked, 'What are these portraits you have here?', for in a row on the wall were small pictures of Newton, Fresnel, Maxwell, Einstein, Boltzmann, Planck, Heisenberg, Schrödinger, Dirac, Bohr, and Pauli. Landau told me that these were theoreticians of class one, Newton and Einstein even of class zero, and that all theoretical physicists can be divided into five classes. [...] I left the office, and only then I noticed on the door a plaque: 'Beware, he bites!'

---

[4] Evgeny Michailovich Lifshitz (1915–1985)

[1]   Janouch, F., *Lev D. Landau: His Life and Work.* CERN 79–03, Geneva 1979.

[2]   Khalatnikov, I. M. (ed.), *Landau – The Physicist and the Man.* Pergamon Press, Oxford, 1989.

[3]   Gorelik, G. E. and Frenkel, V. Y., *Matvei Petrovich Bronstein and Soviet Theoretical Physics in the Thirties.* Birkhäuser, Basel, 1994.

[4]   Casimir, H. B. G., *Haphazard Reality – Half a Century of Science.* Harper and Row, New York, 1983.

[5]   Landau, L., *Zeitschrift für Physik,* **64** (1930) 629. Engl. transl. in [10], p. 31.

[6]   Landau, L., *J. Phys. USSR,* **11** (1947) 91. Also in [10], p. 466, and in [11], p. 205.

[7]   Akhiezer, A. H. In [2].

[8]   Landau, L., *Zh. Eksperim. i Teor. Fiz.,* **11** (1941) 592. Engl. transl. in [10], p. 301, and in [11], p. 185.

[9]   Landau, L. and Ginzburg, V. L., *Zh. Eksperim. i Teor. Fiz.,* **20** (1950) 1064. Engl. transl. in [10], p. 546.

[10] Landau, L., *Collected Papers.* Gordon and Breach, New York, 1965.

[11] Khalatnikov, I. M., *An Introduction to the Theory of Superfluidity.* Addison-Wesley, Redwood City, 1989.

It is said that Landau placed himself in group two and a half with a later upgrade to two.

Also in Kharkov, his temper got Landau into serious trouble. In December 1936, after a row with the rector of the university, he found it wise to quit before being fired. He asked Kapitza to accept him as theoretician at his Institute of Physical Problems in Moscow. Kapitza, himself an independent character, and knowing many others from his time in England, was glad to do so. Eventually, Landau was joined by most of his former group from Kharkov. His work in Moscow on phase transitions, statistical theory of nuclei, cascade theory of electromagnetic showers, and other topics was cut short after little more than a year: In the spring of 1938 Landau was arrested as a suspected German spy. Kapitza immediately wrote a letter to Stalin, in which he states that Landau was an exceptional scientist, that he made enemies by disrespectful behaviour 'to high-ranking elders such as Academicians', but that he was not capable of 'anything dishonourable'. The letter was to no avail as was a letter written to Stalin by Bohr. A year later Kapitza finally succeeded with the letter to Molotov, which was mentioned in the introduction, and with a letter to Beria in which he formally guaranteed that 'Landau will at my Institute conduct no counter-revolutionary activities'. (Kapitza's letters are reproduced in [2].)

Now, finally, we sketch Landau's theory of superfluidity [8]. In the first section of his paper Landau formulates a theory of a *quantum liquid*, i.e., a liquid at temperatures so low that the de Broglie wavelength of its atoms becomes larger than the distance between atoms. Such a theory is needed, in particular, for helium, which stays liquid down to the lowest temperature. A quantum liquid, at least in principle, has to be described by a single wave function taking into account all constituent atoms. In the state of lowest energy, there is practically no relative motion of the atoms with respect to each other. When flowing through a capillary the liquid slips through it rather like one elastic body. At slightly higher energies, the quantum liquid can contain *collective excitations* (see Box I). These are pulses of energy and momentum travelling through the liquid. They have certain similarities to particles and are therefore referred to as *quasiparticles*. In addition to the *phonon*, already a familiar object in the quantum theory of the solid state, Landau introduced a new quasiparticle, the *roton*. With these two he obtained an energy spectrum of excitations, displaying a minimum at a finite energy. If friction occurs, excitations are formed in the liquid (see Box II). Landau was able to show that a certain flow velocity had to be exceeded for that to happen in a liquid with a spectrum of the form discussed. Below that critical velocity there is no friction and the liquid indeed is in a state of superfluidity.

In Moscow Landau re-established his unique school. Only recently two of his collaborators, Ginzburg[5] and Abrikosov[6], were awarded the Nobel Prize for work on superconductivity begun with him half a century earlier [9]. He himself accepted the prize from the hands of the Swedish ambassador when he was hospitalized in 1962. He had fallen victim to a car accident, which caused severe brain damage and the loss of short-time memory, and died six years later.

---

[5] Vitaly Lazarevich Ginzburg (born 1916), Nobel Prize 2003   [6] Alexei Alexeevich Abrikosov (born 1928), Nobel Prize 2003

Fermi Builds a Nuclear Reactor (1942)

63

The discovery of nuclear fission (Episode 60) had two important aspects: one scientific, the other technical. A nucleus could be split in two and in the process an enormous amount of energy was liberated. Lise Meitner had quite accurately estimated 200 MeV to be the energy set free in one fission process. The typical chemical binding energy of one molecule, gained, for instance, by burning one atom of carbon with two atoms of oxygen to one molecule of carbon dioxide, is on the order of 1 eV. The two numbers differ by eight orders of magnitude! A hectic activity began in many laboratories on both sides of the Atlantic. A review article [1] reports on nearly a hundred papers about fission published within less than a year after its discovery. In Europe Joliot [2], in independent work, saw fission products only days after Frisch. In the United States the news was spread by Bohr, who came on 16 January 1939 to participate in a conference and to stay on for several months at the Institute for Advanced Studies in Princeton. Only two weeks earlier, Fermi, the world expert on neutrons, had arrived in New York to work at Columbia University. Under his leadership the first nuclear reactor would be built and made to operate within less than four years. Although its construction was motivated by military reasons, we relate the story here because reactors soon became important instruments in fundamental research.

The ratio of neutrons to protons in a stable atomic nucleus depends on the position of the atom in the Periodic Table. For light nuclei the ratio is about 1; but it gradually increases, and it is nearly 1.6 for uranium. Fission products therefore have too many neutrons to be stable. Stability could be reached in two ways: by successive β decays (in each decay one neutron is changed into a proton) and by simply shedding off one or several neutrons. The former process was well known from previous neutron research but the latter process, also called *evaporation* of neutrons in the language of the liquid-drop model of the nucleus, was also expected to happen because of the high neutron excess. If these extra neutrons were set free in fission, it was clear how to produce nuclear energy on a large scale. Per fission process at least one of the extra neutrons had to provoke another fission, and so on. One would then have a *self-sustained chain reaction* in which nuclear energy was released continuously. These ideas occurred to many physicists when they first heard of fission. Experimentally they were most actively pursued by the groups of Joliot and of Fermi.

Before describing the experimental work we have to mention a remarkable reasoning made by Bohr in early February 1939 [3]. He drew attention to the fact that Meitner, Hahn, and Strassmann, in 1937, had reported on two distinctly different reactions of uranium to neutron irradiation [4]. One reaction showed a strong resonance for neutron energies around 25 eV and led, in the interpretation of 1937, to a single transuranic element. The other type, occurring for thermal as well as fast neutrons but displaying no resonance, yielded many of these new 'elements', later understood to be fission fragments. Bohr concluded that fission was not a resonance process. The resonance could be understood only as a reaction with the abundant isotope ^{238}U of uranium. (For the rare isotope ^{235}U, the cross section would exceed the upper limit which could be calculated theoretically.) In the fission process, on the other hand, the target had to be nuclei of the isotope ^{235}U of which only 0.7% is contained in uranium. The probability for fission would be particularly large for thermal neutrons. Bohr's claim that only a small fraction of natural uranium is fissionable was not generally accepted but soon proved to be true: Thermal neutrons are very effective to provoke fission of ^{235}U. The nucleus ^{238}U, through absorption of a neutron in the resonance energy region, is changed into a compound nucleus ^{239}U which, via two consecutive β decays, is transformed into ^{239}Pu, an isotope of the truly transuranic element *plutonium*, which has position 94 in the Periodic Table. During his stay in Princeton Bohr, together with Wheeler[1], published a detailed theoretical paper on *The Mechanism of Nuclear Fission* [5].

In March, in April, and in May of 1939, Joliot and two collaborators, von Halban[2] and Kowarski[3], emigrants from Austria and Russia, respectively, completed three important papers on fission [6]. In the first one they demonstrated the production of extra neutrons in uranium fission. They had placed a source of thermal neutrons in the middle of a large tank filled with a solution of uranyl nitrate and measured the density of neutrons as a function of the distance from the source. The density was also measured for a comparable solution of ammonium nitrate. It was found to decrease more slowly with distance for uranyl nitrate: Extra neutrons appeared if uranium was present. In two other papers the group showed that fission, induced by thermal neutrons, led to the production of fast neutrons and that several of such fast neutrons – they give an average number of 3.5 ± 0.7 – were produced in a single fission process.

It was now clear, at least in principle, how to achieve a self-sustained chain reaction. The fast neutrons, originating in fission processes, had to be slowed down to thermal energies by nuclei of a *moderator* material so that they could induce fission themselves. Uranium and moderator had to be grouped together in an assembly which we shall now call a *reactor* and care had to be taken to minimize the loss of neutrons. Such loss could happen in two ways: Neutrons might be captured in processes other than fission or they might simply leave the reactor. Joliot was determined to make nuclear energy available to France [7]. He procured 8 tons of uranium, mined in the Belgian Congo, through the mining company with headquarters in Brussels. It turned out that water, in spite of its many protons, is not a good moderator, since the protons also absorb neutrons to make *deuterons* ^{2}H, the nucleons of a heavy isotope of hydrogen. More suitable is heavy water, i.e., water in which the protons are replaced by deuterons. But it was available only by difficult and costly isotope separation. Larger quantities existed only in Norway, where cheap hydropower was used to produce it. Joliot convinced the French government to get 167 litres from Norway in March 1940 only days before that country was occupied by the

---

[1]   John Archibald Wheeler (1911–2008)   [2] Hans von Halban (1908–1964)   [3] Lew Kowarski (1907–1979)

German army. Work stopped when France was occupied as well. Before, the uranium was hidden in Morocco. Halban and Kowarski went to England with the heavy water. Joliot, as we saw in Episode 54, stayed in France.

In parallel to the Joliot's group in Paris, Fermi, at Columbia, studied the production and absorption of neutrons in uranium, working together with Anderson and Szilard. Anderson[4], an electrical engineer by training and then a graduate student in physics, was the first to detect fission fragments in the Unites States immediately after hearing the news brought by Bohr and soon became Fermi's close collaborator. Szilard[5] was from Hungary. In Berlin he had studied theoretical physics under Planck and von Laue and collaborated with Einstein, with whom he held a patent. In 1933 he had left Germany for England, where he had worked on nuclear physics, and had recently arrived in New York. Szilard enjoyed to have contacts to people from different spheres, including industry and politics. Through these contacts he had procured 200 kg of uranium oxide as well as a radium–beryllium neutron source containing more than 2 g of radium. The oxide was filled into longish, cylindrical cans, which were arranged in a regular pattern. The source was placed in the centre and the whole assembly put in a tank containing a 10% solution of manganese sulphate, $MnSO_4$, in water. The activity of the manganese was a measure for the neutron intensity and was found to be ten per cent higher if the uranium was in place [8]. Anderson, Fermi, and Szilard end their paper cautiously: '[More work is] required before we can conclude that a chain reaction is possible in mixtures of uranium and water.' But is was clear that the chain reaction and with it the use of nuclear energy was within reach.

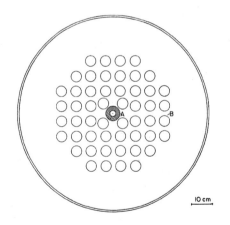

Assembly used by Anderson, Fermi, and Szilard. The neutron source A is placed in the centre of a tank filled with a solution of manganese sulphate in water; containers with uranium oxide are arranged in a regular pattern. Figure reprinted with permission from [8]. Copyright 1939 by the American Physical Society.

This work was completed in the beginning of July 1939. The political situation in Europe was sombre. Two months later, the Second World War would begin. Szilard feared that Germany might develop nuclear weapons. He persuaded Einstein to write a letter to President Roosevelt. In this letter Einstein drew attention to the work of Fermi's and Joliot's groups and he expressed the opinion that nuclear bombs could be built, of which a single one could 'destroy a whole port and some of the surrounding territory'. He therefore suggested

[4] Herbert Lawrence Anderson (1914–1988) [5] Leo Szilard (1898–1964)

that the government take a direct interest in nuclear research and increase funds for it, and finally he hinted at possible German activities in this field. (For the complete text of Einstein's letter see, for instance, [9].) The letter was signed on 2 August and was to be handed to Roosevelt by another acquaintance of Szilard's. But that happened only on 11 October. By then Europe was at war. Roosevelt appointed an Advisory Committee on Uranium, of which Szilard became a member. Research aimed at realizing the chain reaction was declared secret. Results were not published but collected in reports with restricted circulation. Some of these reports, now 'declassified', are contained in [10].

Meanwhile Fermi had identified graphite as a promising moderator because it absorbs less neutrons than water. To minimize absorption by ^{238}U he devised a *heterogeneous* assembly. Small blocks of uranium formed a lattice within a large volume of graphite. Fast neutrons from the fission of ^{235}U could then leave a block with an energy higher than that of resonance absorption by ^{238}U. In the moderator they could be slowed down below the resonance energy and enter a uranium block as thermal neutrons with a high chance to provoke fission in ^{235}U. Of course, some neutrons would still be absorbed by ^{238}U, leading to the production of plutonium. Thus, a reactor based on natural uranium is a *breeder* (the term was coined by Szilard) for plutonium. In the Bohr–Wheeler theory, plutonium was expected to have properties similar to those of the rare uranium isotope ^{235}U, namely, to be fissionable by fast as well as thermal neutrons. An explosion could be brought about by fast neutrons, i.e., bombs could be made only with essentially pure ^{235}U or plutonium. It was not clear whether enough ^{235}U could be obtained by isotope separation. But plutonium, bred in a reactor working with natural uranium, could be separated by much simpler chemical means. This, at the time, was the reason to try and build a reactor as fast as possible.

The main steps to be taken were the following: A theory of the reactor had to be developed. Properties, in particular, nuclear properties, which were unknown and could not be computed, had to be measured. Ways to purify uranium and graphite had to be found and improved since most impurities were neutron absorbers. It had to be found out how much uranium and how much graphite was needed. And, finally, these materials in the necessary quality and quantity had to be obtained and assembled. The work was done at Columbia University in New York until early 1942 and from then on in Chicago under the name Metallurgical Laboratory. More than a decade later, Fermi gave a talk on the work at Columbia [11]. The work at Chicago is described in a highly readable book by Marshall Libby[6], the only woman among the participants [12]. Group leaders under Fermi in Chicago were Anderson and Zinn[7]. Quite a number of details are contained in the official report issued by the U.S. Corps of Engineers, which became known as the *Smyth Report* and was published in 1945 [13]. Its author, Smyth[8], at the time was the chairman of the Department of Physics at Princeton University.

The way to the final reactor was marked by several intermediate experimental set-ups, called *exponential piles*. Fermi had coined the term *pile* (like 'heap') for the assembly of graphite and uranium. The goal was to realize an

---

[6] Leona Marshall Libby (born 1920)   [7] Walter Henry Zinn (1906–2000)   [8] Henry DeWolf Smyth (1898-Ŭ1986)

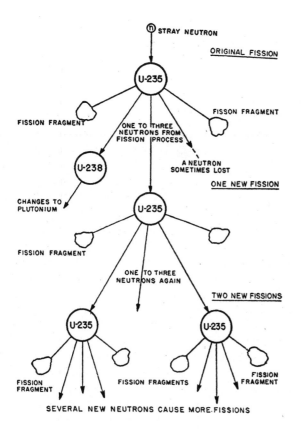

Figure in the Smyth Report, explaining the nuclear chain reaction. Figure reprinted with permission from [13]. Copyright 1945 by the American Physical Society.

assembly in which, on the average, one fission leads to k further fission processes, and so on. The chain reaction is self-sustained if the *multiplication factor k* is larger than 1. In a test assembly of graphite and uranium, containing a neutron source, the neutron intensity decreased exponentially with the distance from the source and from this decrease the factor could be computed for the ideal case of an infinitely large reactor. These experiments led to new specifications for the materials and improved reactor designs. Important for the control of a reactor was the knowledge that about 1% of the neutrons are not liberated immediately in the fission process but up to about a minute later in the transformation process of the fission fragments. Because of these *delayed neutrons* a chain reaction takes time to fully develop. The reactor can be controlled by measuring the neutron intensity and introducing or extracting neutron absorbers, made of cadmium, if required. In July 1942, for the first time an exponential pile gave a multiplication factor indicating that a self-sustained chain reaction could be achieved in a reactor of manageable size. But only several months later enough material was delivered to begin the construction.

The first reactor was assembled in provisional quarters under the west stands of the University of Chicago football field. It was made up of graphite bricks interspersed at regular intervals with uranium. About 5 tons of the uranium was in metallic form. This was placed near the centre. Uranium oxide was

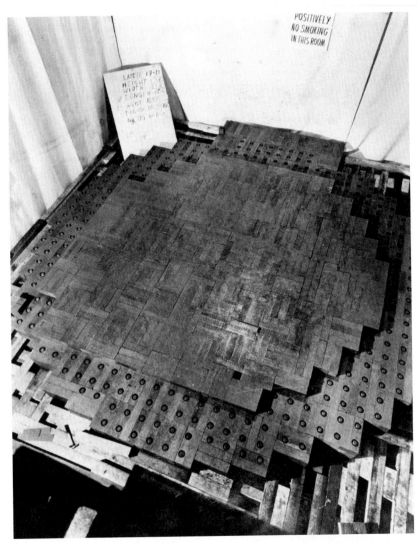

The only photograph taken during the construction of the reactor. It shows layer 19, made up of graphite bricks, already nearly covering layer 18 which contains uranium. The final assembly consisted of 57 layers. Photo: Archival Photofiles, [apf2-00502], Special Collections Research Center, University of Chicago Library.

used for the rest of the assembly. A neutron source was no longer needed. Stray neutrons, resulting, for instance, from cosmic-ray events, could start off a chain reaction. These would, however, die out, as long as the reactor was not *critical*. While layer after layer of graphite and uranium was stacked, the neutron intensity was observed to rise at first slowly with the number of layers and then very fast as the critical size was approached. Fermi has described the exciting moments in a lecture [14]:

> On the morning of December 2, 1942, the indications were that the critical dimensions had been slightly exceeded and that the system did not chain react only because of the absorption of the cadmium strips. During the morning all the cadmium strips but one were carefully removed; then this last strip was gradually extracted, close watch being kept on the intensity. From the measurements it was expected that the system would become critical by removing a length of about eight feet of this last strip.

Actually when about seven feet were removed the intensity rose to a very high value but still stabilised after a few minutes at a finite level. It was with some trepidation that the order was given to remove one more foot and a half of the strip. This operation would bring us over the top. When the foot and a half was pulled out, the intensity started rising slowly, but at an increasing rate, and kept on increasing until it was evident that it would actually diverge. Then the cadmium strips were again inserted into the structure and the intensity dropped to an insignificant level.

On 2 December the reactor was operated at a power, i.e., heat output of about 0.5 W; on 12 December the power was raised to about 200 W. The reactor was then dismantled and its components were moved to more suitable premises.

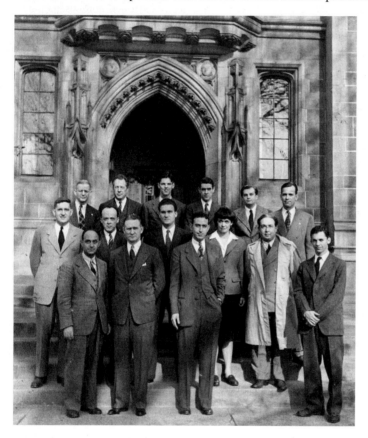

Reunion of scientists involved in the construction of the first reactor in 1946, four years after it had become critical. In front row at left: Fermi with Zinn. Anderson is at the far right. Szilard wears a raincoat. The lady is Marshall Libby.

It is typical of Fermi that after the war he did no longer concentrate on neutron research or reactor physics, both of which he had initiated. He moved on to research with mesons and observed the first excited state of the nucleon and its decay into a proton and a π meson, thus opening the field of what is now called hadron physics. In 1954 he spent part of the summer in Europe, lecturing. After his return he was diagnosed with a particularly vicious form of cancer. Fermi died in Chicago in November 1954.

[1] Turner, L. A., *Reviews of Modern Physics*, **12** (1940) 1.

[2] Joliot, F., *Comptes Rendus Acad. Sci.*, **208** (1939) 341.

[3] Bohr, N., *Physical Review*, **55** (1939) 418.

[4] Meitner, L., Hahn, O., and Strassmann, F., *Zeitschrift für Physik*, **106** (1937) 249.

[5] Bohr, N. and Wheeler, J. A., *Physical Review*, **56** (1939) 426.

[6] v. Halban, H., Joliot, F., and Kowarski, L., *Nature*, **143** (1939) 470, 680, 939.

[7] Radvanyi, P., *Les Curies*. Belin, Paris, 2005.

[8] Anderson, H. L., Fermi, E., and Szilard, L., *Physical Review*, **56** (1939) 284. Also in [10], Vol. II, 11.

[9] Segrè, E., *Enrico Fermi, Physicist*. University of Chicago Press, Chicago, 1970.

[10] Fermi, E., *Collected Papers, 2 vols.* University of Chicago Press, Chicago, 1962.

[11] Fermi, E., *Physics Today*, **November** (1955) 1. Also in [10], Vol. II, 996.

[12] Marshall Libby, L., *The Uranium People*. Crane Russak, New York, 1979.

[13] Smyth, H. D., *Atomic Energy for Military Purposes*. Princeton University Press, Princeton, 1945. Reprinted as [15].

[14] Fermi, E., *Proceedings of the American Philosophical Society*, **90** (1946) 120. Reprinted in [9]. Also in [10], Vol. II, 542.

[15] Smyth, H. D., *Reviews of Modern Physics*, **17** (1945) 351.

64 The Synchrotron: Phase Stability (1945) and Strong Focussing (1952)

Many, if not most, fundamental discoveries in nuclear physics were possible by simply using the projectiles made available by natural radioactivity. But even though, it was felt desirable to have well-controlled sources of particles of higher energies; and thus the first particle accelerators, the Cockkroft–Walton generator (Episode 50) and the cyclotron (Episode 47) were built. Part of the rapid increase in detailed knowledge on isotopes and nuclear reactions in the 1930s and 1940s was possible because of these machines. Concerning elementary particle physics, we have the reverse situation. Although some important particles were observed in a natural source, namely, the cosmic radiation, most were produced in the laboratory using artificially accelerated projectiles of ever increasing energy and intensity. All modern accelerators are based on the principle of *phase stability* invented in 1944/45. The vast majority are *synchrotrons*. Since the invention of *strong focussing* in 1952 every synchrotron design includes this feature. Let us now see how these two inventions were made, interestingly enough, each of them twice.

In Episode 47 we described how resonance acceleration was suggested by Ising, modified and demonstrated by Wideröe, and brought to fruition in the form of the *cyclotron* by Lawrence and Livingston. In the cyclotron, particles repeatedly pass through an electric field which oscillates with high frequency. Provisions are taken that the particles actually experience the field at a certain *phase*, i.e., at a particular moment of the full oscillation period. It seems obvious to choose the phase to correspond to the maximum field. However, particles arriving out of phase, i.e., a little earlier or a little later than the moment of maximum field, are accelerated less and therefore become slower than the in-phase particles and eventually drop out of phase altogether. This effect was a serious limitation of resonance acceleration. To keep it within bounds, the peak accelerating voltage had be to high, such that the number of accelerations was not too large.

McMillan lecturing on phase stability.

In September 1945, McMillan, one of the discoverers of the transuranic elements neptunium and plutonium, submitted a short paper to the *Physical Review* [1] describing the mechanism of *phase stability* with which this limitation is overcome: Particles, which are somewhat out of phase, are automatically brought back or, rather, they oscillate about the nominal phase very much like a pendulum oscillates about its equilibrium point. Phase stability (see Box I) not only ensures that groups or *bunches* of particles are kept together as if they sat in 'buckets' confining them, but also it corrects for a small energy mismatch. It is therefore possible, during an acceleration cycle, to gradually change the frequency of the accelerating electric field, the strength of the magnetic field, or even both. As McMillan pointed out in an accompanying note [2], even the

Box I. Phase Stability

A particle travelling in the machine is accelerated at particular points, for instance, the spaces between the drift tubes in Wideröe's original linear accelerator or the small region between the radio-frequency electrodes in a cyclotron. There the electric field strength E oscillates with circular frequency ω as a function of time t, $E = E_0 \sin(\omega t)$. The argument of the sine function, $\varphi = \omega t$, can be taken as an angle, growing from $0°$ to $360°$, then beginning again at $0°$, etc. Thus a particular phase φ corresponds to a particular field $E(\varphi)$ in every oscillation period. Let the accelerator be constructed to work for a particle experiencing a positive field $E_s < E_0$ in every single acceleration. This field exists for two phases, φ_s (when the field rises with time) and φ_s' (when the field falls). One of these phases is stable. Which one it is, depends on the type of accelerator.

Proton Linear Accelerator: Consider a particle of the correct energy arriving with the phase $\varphi = \varphi_s + \Delta\varphi$. If $\Delta\varphi > 0$, the particle arrives later than that of nominal phase. Compared to that it experiences a higher field, is accelerated more, travels faster and will arrive less late for the next acceleration, etc. When finally in phase, it will be somewhat too fast and, next time, will arrive before the particle of nominal phase. Now, it is accelerated less and, eventually, is brought back to the nominal phase again.

Electron Linear Accelerator: Because of their small rest mass electrons with the moderate energy of a few MeV already possess nearly the velocity of light. By further acceleration essentially their energy is increased, not their velocity. With microwave technology developed in World War II it is possible to build and energize long structures in which an electromagnetic wave travels with the speed of light. Electrons injected into the structure travel along on the crest of that wave like surfers, as envisaged by Ising in 1924. Since wave and electrons have the same speed, the particles stay *locked* to their phase.

Cyclotron and Synchrocyclotron: Particles do not travel on straight lines. In the magnetic field, which is constant in time and practically uniform in space, the path length between successive accelerations depends on the particle's energy or, more directly, its momentum. For a particle faster than nominal, the path length is increased by so much that it takes longer to reach the next acceleration point than the nominal particle. Therefore, the stable phase is φ_s' on the falling flank of the oscillating field. For energies for which the relativistic increase of the particle mass becomes appreciable the cyclotron frequency decreases. This variation is taken into account in a *synchrocyclotron* in which the radio frequency is modulated accordingly.

Synchrotron: The magnetic field is raised while the particles are accelerated to keep them on the same closed orbit. In the first generation of machines with weak focussing the spatial variation of the magnetic field was small; as in the cyclotron the stable phase was on the falling flank, changing somewhat while the particle energy increased. For modern (strongly focussing) electron synchrotrons the stable phase is on the rising flank. For proton synchrotrons during each acceleration cycle it changes its position from the falling to the rising flank. While it passes though the position of the field maximum there is no phase stability. This time, however, can be kept so short that the particles are not lost.

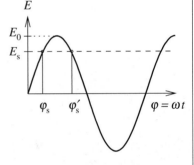

The accelerating electric field E as a function of the time t or, rather, the phase φ.

energy loss, suffered by an electron travelling on a circular orbit, is automatically corrected for. McMillan proposed the name *synchrotron* for machines employing this mechanism. In practice only ring accelerators, i.e., machines in which particles travel along the same closed orbit many times, are called synchrotrons.

In an article written four decades later [3], McMillan remembers that he was working at Los Alamos in the group that had developed the atomic bomb and was thinking about what to do on his return to Berkeley. When worrying about the limitations of the cyclotron the idea of phase stability quite suddenly occurred to him. He wrote a letter to Lawrence and discussed the topic with him when Lawrence arrived in New Mexico to attend the *Trinity Test*, the explosion of the first bomb on 16 July 1945. He also spoke to other physicists. Of these Rabi, in particular, was concerned about the importance of radiation loss if electrons were accelerated. This effect, now called *synchrotron radiation*, was known to be a limiting factor for the energy to be reached by the *betatron*, first operated successfully by Kerst[1] in 1940. It is a ring accelerator for electrons in which the accelerating electric field is not provided by a radio-frequency oscillator but by the change in time of a magnetic field; it is essentially a transformer in which the secondary winding is replaced by a ring of electrons in a vacuum pipe. Rabi asked Schwinger to do a computation on synchrotron radiation and reported the result back to McMillan who then went ahead with publication.

Early in 1946, Veksler[2] from the Lebedev Institute in Moscow, in a note to the *Physical Review* [4], stated that phase stability had been described by him already in 1944 in two papers published in Russian and early in 1945 in a publication, which had appeared in English in a Soviet journal [5]. McMillan, who had been unaware of these publications, did not hesitate to recognize Veksler's priority [6]:

> This seems to be another case of the independent occurrence of an idea in several parts of the world, when the time is ripe for the idea. The present great interest in very high energy particles furnished the need for new methods of acceleration, and the principle of phase stability applied to cyclotron-like devices is a promising solution. Since my first thoughts on the subject occurred near the beginning of July, 1945, it is clear that Veksler's discovery of the principle was earlier.

Veksler [7] was originally an electrical engineer and had worked on X rays at the All-Union Institute of Electrical Engineering. In 1936 he joined the Lebedev Institute and studied cosmic rays. There he became interested in accelerators. Already in 1946, the first 30-MeV electron synchrotron was completed at the institute. Later Veksler headed the high energy physics laboratory of the Joint Institute for Nuclear Research in Dubna. In 1957 the 10-GeV proton synchrotron in Dubna, built under his direction and called the *synchrophasotron*, was completed; for a few years it was the machine with the highest energy. In 1963 Veksler and McMillan jointly received the Atoms for Peace Award.

In the United States, work on the atomic bomb had drastically changed the style of doing nuclear physics. Physicists had learned to work in large groups

Veksler

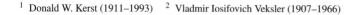

[1] Donald W. Kerst (1911–1993)   [2] Vladimir Iosifovich Veksler (1907–1966)

Box II. Magnetic Focussing – Weak and Strong

A particle of charge q and having a momentum p, oriented perpendicular to the direction of a homogeneous magnetic field B, moves on a circle of radius $R = p/qB$, see Box in Episode 47. In a ring accelerator the field is raised together with the momentum and the radius stays constant. Therefore the field has to be maintained only in a limited region, a torus around the nominal circular path. For various reasons (for instance, collisions with residual atoms in the vacuum vessel or energy loss by radiation), particles can deviate from the nominal path and have to be focussed back to it. Both vertical and radial focussing is achieved if the field in the plane changes with the distance r from the centre,

$$B = B_0(R/r)^n \quad .$$

The *field index* n is positive if the field decreases with r and negative if it increases.

Vertical Focussing: If the field changes with r, then, outside the horizontal ring plane, it necessarily has two components: B_z perpendicular to the plane and B_r in radial direction. Since the Lorentz force $\mathbf{F}_L$ acting on a charged particle in a magnetic field is perpendicular to the particle velocity $\mathbf{v}$ (which points into the plane of the figure) and to the field $\mathbf{B}$, the field component B_r gives rise to a vertical force. It focusses a particle, which is out of the plane, back to it, if the field falls off with rising r, i.e., if $n > 0$. For $n < 0$ it defocusses; the force points away from the plane.

Radial Focussing: In the ring plane a particle is influenced by two forces: the centrifugal force, with a modulus inversely proportional to the trajectory radius, $F_c \propto (1/r)$, and the Lorentz force with a modulus proportional to the field, i.e., $F_L \propto (1/r)^n$. For the nominal radius, $r = R$, both forces are equal in magnitude and balance. For $n < 1$ the centrifugal force falls more steeply with increasing r than the Lorentz force. For a particle outside the nominal radius the Lorentz force dominates, for a particle inside it is the centrifugal force. In both cases the particle is focussed back to the nominal radius.

Weak Focussing: The obvious choice for the field index is a value in the range $0 < n < 1$, which ensures vertical *and* radial focussing. Particles perform both vertical and radial oscillations about their nominal path. The amplitude of these *betatron oscillations* determines the width of the vacuum pipe in which the particles travel and over which the magnetic field has to be maintained.

Strong Focussing: The magnetic field along the circumference of the machine is composed of alternating sections with a large positive or a large negative index. In a section with $n \gg 1$ there is vertical focussing and radial defocussing, in one with $n \ll 0$ the situation is reversed. The result is an overall focussing with much reduced amplitudes in both types of betatron oscillation.

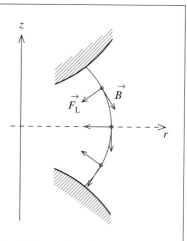

Vertical focussing: If the field drops towards the outside, $n > 0$, there is a z component of the Lorentz force bringing particles back to the nominal plane $z = 0$.

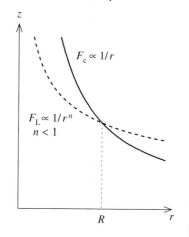

Radial focussing: For a field which decreases slowly with r or even increases, $n < 1$, there is a net force directing the particle towards the nominal radius R.

and solve many technical problems in an industrial style. Moreover, money flowed readily. McMillan was granted half a million dollars by the military to build a 300-MeV electron synchrotron in Berkeley. It reached the design energy early in 1949. Quite a while before that was achieved, proton synchrotrons were designed and proposals for funding submitted to the newly founded Atomic Energy Commission, which, in 1948, approved two large machines to be built, a 3-GeV synchrotron in the new Brookhaven National Laboratory on Long Island and a 6-GeV machine in Berkeley, later called the *Cosmotron* and the *Bevatron*, respectively. The names were chosen to allude to 'cosmic rays' and 'billion electron volts', respectively. (In the United States, at that time, it was customary to write BeV instead of the internationally used GeV.)

Lawrence standing in the iron yoke of one of the Bevatron magnets, giving an impression of its size.

These large proton synchrotrons were formidable machines, not only with respect to their scientific value, but also because of their size and mass. One reason was that the vacuum pipe had to be rather wide to accommodate the particles which perform betatron oscillations about the nominal orbit. Magnetic focussing (see Box II) is necessary because particles tend to deviate from their orbit. Oscillations arise, which were first described by Kerst and Serber[3], in studies for the betatron [8]. They found that vertical and radial focussing is achieved if the magnetic field is made to decrease rather slowly towards the outside of the machine, the slow fall-off being described by a *field index* n in the range $0 < n < 1$. The first generation of synchrotrons was built with such fields. Unfortunately, the betatron oscillations had large amplitudes calling for wide beam pipes and therefore for wide magnets. In 1952 a way out of this dilemma was found, the method of *Alternating-Gradient (AG) Focussing* or, simply, *Strong Focussing*.

The story of its discovery is told by Livingston in his small book on accelerator history [9]. Beginning shortly after the war, various efforts were made to boost again fundamental research in Europe and, in particular, cooperation among scientists. An important result was the foundation of CERN, the *Con-*

[3] Robert Serber (1909–1997)

seil Europénne pour la Recherche Nucléaire, which was to set up a large laboratory at Geneva and planned to build a 10-GeV proton synchrotron there. A visit of CERN scientists to Brookhaven and Berkeley was arranged. Livingston [10], by that time professor at the Massachusetts Institute of Technology but for the summer working in Brookhaven, prepared for the visit from Europe by attempting to extend the Cosmotron design to 10 GeV.

From the left: Courant, Livingston, Snyder, and Blewett with the model of an alternating-gradient magnet.

A cross section, perpendicular to the particle beam, through the iron core of each Cosmotron magnet had the form of the letter C, open towards the outside of the ring with the vacuum pipe placed in the gap. A magnet of such shape with flat and nearly parallel pole faces enclosing the beam pipe provides the required field with its slight fall-off towards the outside because of the asymmetry of the core. For technical reasons, given somewhat differently in [9] and in [10], Livingston, for the new design, wanted to reverse the orientation of some of the C magnets so that their gap was on the inside of the accelerator ring and that there would be an increase of the field towards the outside, corresponding to a negative field index, $n < 0$. He asked a theoretical colleague, Ernest Courant[4], one of the two physicist sons of the eminent mathematician Richard Courant, about the behaviour of a machine with alternating values of n, i.e., *alternating gradients* of the magnetic field with respect to the radial variable r. Later he recalled [9] that Courant

> ... took the problem home with him that evening. The next morning he reported, with some surprise, that preliminary calculations showed the orbits to be stable and to have even smaller transverse amplitudes than in the weak gradient of the standard synchrotron.

It was soon realized that for alternating n values of high magnitude, $|n| \gg 1$, the amplitudes became much smaller, allowing for beam pipes and magnets of much smaller cross sections and much reduced costs. When the CERN scientists arrived, the understanding of the new focussing was well advanced and

---

[4] Ernest D. Courant (born 1920)

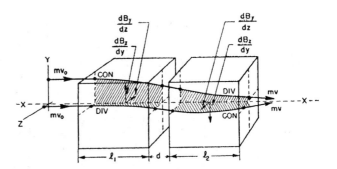

The principle of strong focussing. The magnet on the left focusses in the vertical (xy) plane and defocusses in the horizontal (xy) plane. The magnet on the right defocusses in the vertical and focusses in the horizontal plane. The overall result is focussing in both planes. Figure reprinted with permission from [11]. Copyright 1952 by the American Physical Society.

discussed with them. Indeed, CERN changed its plans to a 25-GeV machine of the new type. The CERN Proton Synchrotron began operation in 1960 and still today is used as pre-accelerator for more powerful machines. Incidentally, in the autumn of 1960, as a young Ph.D. student sent to CERN for a few months, I was allowed to help with the first exposure of a small hydrogen bubble chamber to a beam of protons from that machine.

A complete theory of the alternating-gradient principle was developed in Brookhaven together with Snyder[5]. The resulting publication [11] also includes the proposal of magnetic *quadrupole lenses* for focussing particle beams on a straight line. If the beam travels in z direction, then the field of such a lens focusses the particles in one plane, say, the xz plane, but defocusses in the plane perpendicular to it, the yz plane. With several lenses of alternating orientation again an overall focussing is reached. In an accompanying paper [12] by Blewett[6], who also pointed out that quadrupole lensing was possible with electric fields, too, this type of focussing was shown to be useful for linear accelerators.

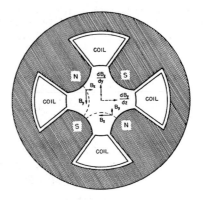

Cross section through a quadrupole magnet. Figure reprinted with permission from [11]. Copyright 1952 by the American Physical Society.

In 1953 Christofilos[7] [13], a Greek engineer on a visit to the United States, saw the paper [11] in the Brooklyn Public Library. He was born in Boston as son of Greek immigrants who had returned to Greece when he was still a child. As a schoolboy in Athens he had developed a strong interest in physics before he studied electrical engineering. He had already found the alternating-

[5] Hartland S. Snyder (1913–1962)   [6] John P. Blewett (1910–2000)   [7] Nicholas Constantine Christofilos (1916–1972)

gradient method in 1949 in his spare time but not published it. Instead, he submitted patent applications in the United States and in Greece in 1950 and sent copies to Berkeley, which, it seems, were not seen by a person able to appreciate the idea and were eventually filed. Christofilos, under the impression that his idea had been used without being cited, went to nearby Brookhaven, where he was shown around by Blewett and found that the work there was done independently. His priority was acknowledged [14], he was offered a position, and he worked on the design of the AGS, the Brookhaven 28-GeV alternating-gradient synchrotron. His US patent [15] was granted in 1956 and a settlement was reached with the Atomic Energy Commission for its usage. Later, he moved to the Livermore Laboratory to work on several original ideas in other fields of technical physics.

Section of the 25-GeV CERN Proton Synchrotron. Designed as an alternating-gradient machine, the total iron weight of its magnets is 3400 tons, only about a third of the weight of the Bevatron magnets, while it reaches four times the Bevatron energy. Photo CERN.

Under Livingston's direction the *Cambridge Electron Accelerator* (CEA) was built, a 6-GeV machine as joint research facility for Harvard and MIT; a sister machine, the *Deutsches Elektronen Synchrotron* (DESY) was built in Hamburg. Together with Blewett, Livingston wrote the standard text *Particle Accelerators* [16], which also features the famous *Livingston Plot*, from which he derived the empirical law that the maximum energy reached by the most modern accelerator shows an average 10-fold increase every 10 years. That this law held up to the present became possible by design optimization, by the ever-increasing joining of forces on the national and international scale and by new accelerator principles. We shall discuss the latest of those, colliding beams, in Episode 83.

In 1986, on the very day Livingston died, the committee in charge decided to honour him and Courant with the Fermi Award.

[1] McMillan, E. M., *Physical Review*, **68** (1945) 143.

[2] McMillan, E. M., *Physical Review*, **68** (1945) 144.

[3] McMillan, E. M., *Physics Today*, **February** (1984) 31.

[4] Veksler, V., *Physical Review*, **69** (1946) 244.

[5] Veksler, V. I., *J. Phys. USSR*, **9** (1945) 153.

[6] McMillan, E. M., *Physical Review*, **69** (1946) 534.

[7] McMillan, E. M., *Physics Today*, **November** (1966) 104.

[8] Kerst, D. W. and Serber, R., *Physical Review*, **53** (1941) 841.

[9] Livingston, M. S., *Particle Accelerators: A Brief History*. Harvard University Press, Cambridge, Mass., 1969.

[10] Courant, E. D., *Biographical Memoirs (Nat. Acad. Sci.)*, **72** (1997) 264.

[11] Courant, E. D., Livingston, M. S., and Snyder, H. S., *Physical Review*, **88** (1952) 1190.

[12] Blewett, J. P., *Physical Review*, **88** (1952) 1197.

[13] Melissinos, A. C., *Nicholas C. Christofilos: His Contributions to Physics*. In Turner, S. (ed.): *CERN Accelerator School: 5th Advanced Accelerator Physics Course, Rhodes 1993*, p. 1067. CERN, Geneva, 1995.

[14] Courant, E. D., Livingston, M. S., Snyder, H. S., and Blewett, J. P., *Physical Review*, **91** (1953) 202.

[15] Christofilos, N., U.S. Patent 2,736,799 'Focussing systems for ions and electrons', 1956.

[16] Livingston, M. S. and Blewett, J. P., *Particle Accelerators*. McGraw-Hill, New York, 1962.

65

Magnetic Resonance (1945)

Bloch

We already met Bloch as student in Zurich, witnessing Schrödinger's first public presentation of his wave equation, and in Leipzig as Heisenberg's first doctoral student, writing his thesis which laid the foundation of the quantum theory of conductors and semiconductors (Episode 44). He worked in Zurich, Utrecht, and Leiden before returning to Leipzig as Heisenberg's assistant. There he became Privatdozent with a thesis on ferromagnetism. When Hitler came to power, Bloch left Germany, spent a few months in Zurich, and then went to Rome to work in Fermi's group. Fermi, who excelled as no other physicist in both theory and experiment, advised him to do some experimental physics, because 'it was fun'. As the present episode will show, Bloch made good use of this advice.

In 1934, at the age of 28, Bloch was offered and accepted a professorship at Stanford University. There in 1936, he wrote a short note [1] in which he suggested a way to measure the magnetic moment of the free neutron. At that time this quantity was known indirectly and with poor precision from magnetic-moment measurements on the proton and on the deuteron (which is composed of a proton and a neutron), see Episode 52. Bloch proposed to study the transmission of low-energy neutrons, in fact Fermi's thermal neutrons, through an iron plate magnetized in a direction perpendicular to the line of flight of the incident neutrons. He pointed out that a neutron would experience two different forces, a strong nuclear force when encountering an iron nucleus and a much weaker electromagnetic force due to inhomogeneous magnetic fields in the iron acting on the neutron magnetic moment. The latter force is to some extent determined by the magnetization and, although it is weaker, it acts over much larger distances than the nuclear force, so that the influences of the two forces become comparable. Bloch concluded that magnetized iron can be used to produce a beam of neutrons with a preferential orientation of their magnetic moments, i.e., a beam which is partially polarized. In short, magnetized iron could serve as a *polarizer* of neutrons, very much as the Stern–Gerlach type of magnet did for atoms. Molecular beams of sufficient intensity could be well collimated along the axis of a Stern–Gerlach magnet. That was not possible with neutrons, but now magnetized iron provided a way to polarize them.

Bloch planned a resonance measurement of the neutron magnetic moment by placing two plates of magnetized iron some distance apart (the first acting as *polarizer*, the second as *analyser*) and a creating magnetic field in between. If both polarizer and analyser select the same orientation of the magnetic moment and the field between them is varied until the orientation behind the polarizer is reversed, then the intensity behind the analyser reaches a minimum. A field provoking such a reversal or *magnetic resonance* consists of a large constant field and a much smaller time-dependent field, rotating or oscillating at the

Larmor frequency (see Box I). To perform the experiment, Bloch joined forces with Alvarez, then a young member of Lawrence's group in Berkeley.

Independently, Rabi[1] devised the magnetic-resonance method with Stern–Gerlach magnets for molecular beams. He published the theory [2] in 1937 and beautiful experimental results, obtained with his collaborators [3,4], in 1938. Molecules with a particular z component of the magnetic moment, as indicated in the figure, travel from the oven O through the slit S to the detector D if no magnetic resonance occurs in region C containing the radio-frequency field. In the case of resonance, because of the magnetic moment's orientation, they are diverted in magnet B. Resonance is therefore detected as a drop in the detected beam intensity.

Rabi

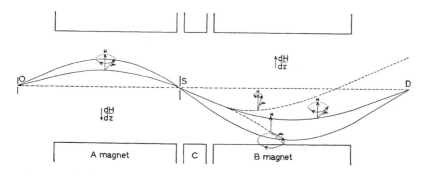

Rabi's apparatus for magnetic-resonance experiments with molecular beams. There are two Stern–Gerlach magnets A, B with the same field direction but opposite field gradients and a region C containing a homogeneous magnetic field and a radio-frequency electric field. Figure reprinted with permission from [4]. Copyright 1939 by the American Physical Society.

It took Alvarez and Bloch quite a while to make their method work. Finally, in the autumn of 1939, they measured the neutron magnetic moment with a 1% accuracy [5], using neutrons produced at a Berkeley cyclotron. The intensity of a conventional radium–beryllium neutron source had been insufficient.

At the end of the Second World War, Bloch participated in radar research at the Radio Research Laboratory at Harvard University in Boston. The Boston–Cambridge area was the centre of radar development. MIT at Cambridge ran the large Radiation Laboratory in that field. Here Purcell[2] [6], a young Assistant Professor from Harvard, worked on extending radar technology to shorter and shorter wavelengths, from 10 cm to 3 cm and later to 1.5 cm. Purcell, originally an electrical engineer who had graduated from Purdue University in 1933, had spent a year with a fellowship in Karlsruhe, where he followed courses in theoretical physics by Weizel[3] (as I did more than two decades later in Bonn). On his return to the United States he entered Harvard University as graduate student, received his Ph.D. in 1938, and stayed on as instructor in physics. It seems that Purcell and Bloch did not meet before they performed the experiments described below [7] or, if they did, only on a festive occasion, a party after the award of the Nobel Prize to Rabi became known [8].

Shortly after the end of the war, Bloch and Purcell, each with a small group of collaborators, would observe nuclear magnetic resonance in *bulk matter*, now referred to simply as *NMR*. They aimed, in particular, at detecting the resonance signals from protons. Although the magnetic moments were embedded in a rather complex environment, compared to the case of essentially

Purcell

---

[1] Isidor Isaac Rabi (1898–1988), Nobel Prize 1944   [2] Edward Mills Purcell (1912–1997), Nobel Prize 1952   [3] Walter Weizel (1901–1982)

Box I. Magnetic Moments – Units and Orders of Magnitude

From the Dirac equation it follows that the electron has a magnetic moment of one *Bohr magneton* μ_B (see Episode 43). For the discovery of a slight but important deviation from that value, see end of Episode 67. Correspondingly, the proton was expected to have a magnetic moment of one *nuclear magneton* μ_N; but a significantly larger value was observed by Stern (Episode 52). It is customary to express these facts in the form

$$\mu_e = \tfrac{1}{2} g_0 \mu_B \quad , \qquad g_0 \approx 2 \quad ,$$
$$\mu_p = \tfrac{1}{2} g_p \mu_N \quad , \qquad g_p = 5.586 \quad .$$

The numbers g_0 and g_p are called *g factors* of the electron and the proton, respectively.

The vectors of the magnetic moment μ and of the spin $\mathbf{S}$ of a particle are proportional to each other,

$$\mu = \gamma \mathbf{S} \quad .$$

The proportionality constant γ is called the *gyromagnetic ratio*. We have

$$\gamma = \gamma_e = -g_0 \mu_B / \hbar \quad \text{and} \quad \gamma = \gamma_p = g_p \mu_N / \hbar$$

for the electron and the proton, respectively.

Characteristic for the motion of a particle with charge e and mass m in a magnetic field of strength B is the *cyclotron frequency* $\omega_{cycl} = eB/m$. For a typical field of $B = 1\,\text{Tesla} = 1\,\text{T}$ one obtains

$$\omega_{cycl}^e = \frac{e}{m_e} 1\,\text{T} = 1.759 \times 10^{11}\,\text{s}^{-1} \quad \text{and} \quad \omega_{cycl}^p = \frac{e}{m_p} 1\,\text{T} = 9.579 \times 10^7\,\text{s}^{-1} \quad .$$

Electromagnetic waves in the neighbourhood of these two frequencies are produced by microwave techniques in the case of electrons and radio techniques for protons.

---

isolated atoms or neutrons in previous experiments, the apparatus needed for these measurements was much simpler. In an interesting comparison of the independent work by the two groups [8] it is pointed out that Purcell and Bloch approached the problem in complementary but equivalent ways. Purcell saw it as *resonance absorption* of quanta corresponding to the energy difference between two quantum-mechanical states. For Bloch a classical picture stood in the foreground, the change in orientation of the proton magnetic moments.

Pound[4] remembers [6] that at the end of the war most of the staff of the MIT Radiation Laboratory returned to their peace-time occupation. Some were asked to stay on. Purcell, one of them, began to contribute to a series of books, which was to preserve the achievements of the laboratory. He was particularly concerned about absorption of radar waves by water vapour in the air because this effect had rendered his 1.5-cm waves useless. That gave him the idea of another type of absorption experiment and he asked two colleagues, Torrey[5] and Pound, to help him do it in a spare-time effort. Discarded equipment from the Radiation Laboratory was moved to Harvard, where a large magnet was available, originally used for the cloud chamber in which early muon tracks

---

[4] Robert V. Pound (born 1919) [5] Henry C. Torrey (1911–1998)

Box II. Magnetic Resonance – Quantum-Mechanical and Classical Description

We consider a proton or an electron in a magnetic field $\mathbf{B} = \mathbf{B}_0 + \mathbf{B}_1(t)$, which is composed of a strong constant field $\mathbf{B}_0 = B_0\mathbf{e}_z$ in z direction and a weak field $\mathbf{B}_1(t) = B_1(\cos\omega t\,\mathbf{e}_x + \sin\omega t\,\mathbf{e}_y)$ perpendicular to the z axis and rotating around it. (For practical purposes a field $\mathbf{B}_1(t)$ oscillating in a direction perpendicular to z suffices.)

Quantum-Mechanical Description: The z component of a proton or electron spin can take the two values $S_z = \pm\frac{1}{2}\hbar$, corresponding to the z components $\mu_z = \pm\frac{1}{2}\gamma\hbar$ of the magnetic moment and to the potential energies

$$E_\pm = \mu_z B_0 = \pm\frac{1}{2}\gamma\hbar B_0 \quad .$$

Resonance transition between these two states is expected to occur if an electromagnetic field of angular frequency ω provides quanta of energy $E = \hbar\omega$ equal to the energy difference of the states, $E = \hbar\omega = \gamma\hbar B_0$, i.e., if the angular frequency of the field $\mathbf{B}(t)$ is

$$\omega = \omega_R = \gamma B_0 \quad .$$

Classical Description: Since the spin angular momentum and the magnetic moment are proportional to each other, we can consider them to be realized by a tiny magnetic needle spinning around its axis and apply the equation of motion of a spinning top under the influence of an external torque. In our case it reads $d\boldsymbol{\mu}/dt = \gamma\boldsymbol{\mu} \times \mathbf{B}$, i.e., the magnetic moment changes in a direction perpendicular to itself and to the field. If only the constant field $\mathbf{B}_0$ is present, then the moment gyrates about the field direction with the *Larmor frequency*

$$\omega_L = \gamma B_0 \quad .$$

The z component μ_z stays constant, since a field in z direction can produce changes only in the x and y components of the magnetic moment. One speaks of *Larmor precession*. The component μ_z is changed, however, by the field $\mathbf{B}_1$. The change is largest if the frequency of $\mathbf{B}_1$ is the Larmor frequency, $\omega = \omega_L$, since then the Larmor precession and the time-dependent field are in phase. In this case we have *magnetic resonance*. The vector $\boldsymbol{\mu}$ of the magnetic moment spirals around the z direction and, with a certain period, its z component oscillates between $+\mu$ and $-\mu$. For frequencies in the neighbourhood of ω_L there are still oscillations but with smaller amplitude.

Summary: In both descriptions resonance occurs for the circular frequency

$$\omega = \omega_R = \omega_L = \gamma B_0 \quad .$$

Except for a factor of $g_0/2$ or $g_P/2$, which is on the order of 1, it is equal to the *cyclotron frequency* of the electron or the proton, respectively, for which numerical values are given in Box I.

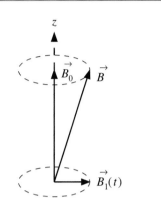

The field B is the sum of a large constant part B_0 and a small rotating part B_1.

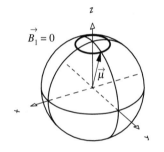

In the constant field alone the magnetic-moment vector $\boldsymbol{\mu}$, classically speaking, precesses about the field direction. Its tip describes a circle on a sphere.

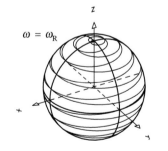

In the case of resonance the tip of $\boldsymbol{\mu}$ spirals down and up the sphere.

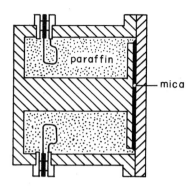

Section through Purcell's cylindrical radio-frequency cavity. Coupling with the electric circuit is achieved through the two loops. Figure reprinted with permission from [8]. Copyright 1986 by the American Physical Society.

were photographed (Episode 57). On Saturday afternoon, 15 December 1945, resonance absorption due to the proton magnetic moment was observed. The constant field B_0 was produced by the magnet, the time-dependent field B_1 by a *cavity*, filled with paraffin, and excited at its proper frequency by an external radio-wave circuit. In this circuit, the cavity can be treated as a resistance, which can be measured by standard techniques and which is determined by the energy absorption in the cavity. While the cavity frequency remained constant, the magnetic field B_0 and with it the Larmor frequency of the proton magnetic moment was varied. When the two became equal, a sharp increase in absorption was observed.

Experiments in bulk matter, like those with molecular or neutron beams, require an enrichment of one energy state. While this is achieved in beam experiments by a polarizer, interactions within the bulk matter itself lead to an enrichment. When the constant field B_0 is switched on, the two energy states are populated. The magnetic moments then interact with other moments or with the *lattice*, i.e., the ensemble of atoms, until, after some time, the *relaxation time*, a thermal equilibrium is reached. As the result of these two processes, called *spin–spin relaxation* and *spin–lattice relaxation*, respectively, the state of lower energy acquires a slightly higher population. At room temperature the difference is on the order of 1 in 10^6. But that is quite sufficient since the number of protons in the sample is on the order of 10^{23}.

The Dutch physicist Gorter[6], since the early 1930s, had studied the action of radio waves on matter. In 1936, he had (unsuccessfully) tried to measure a temperature increase caused by magnetic resonance in bulk matter [9]. In 1937, during a short visit to Columbia, in a discussion with Rabi, he contributed with his advice to the final design of the Rabi apparatus. During the war, in the Zeeman Laboratory in Amsterdam, Gorter again tried to detect NMR, this time by electronic means [10]. The result was still negative, as one now knows, because his sample had too long a relaxation time. (For Gorter's own story see [11].)

During the preparation of their experiment Purcell and his co-workers learnt of Gorter's failure. Fortunately, their sample was benevolent and on Christmas Eve 1945, a letter [12] to the *Physical Review* was completed, entitled *Resonance Absorption by Nuclear Magnetic Moments in a Solid*.

The next issue of the same journal contained a letter by Bloch, who had returned to Stanford, and two collaborators [13], completed five weeks later and describing a successful experiment on what they called *nuclear induction*. Bloch, first of all, saw magnetic resonance as a change in the direction of the proton magnetic moment. Since each moment is surrounded by its own magnetic field, there is a change of magnetic field in the sample. By virtue of Faraday's law of induction, an electric field is connected to this changing magnetic field. Together with Hansen[7] and Packard[8] he set out to measure the induced electric field. They placed a spherical container, filled with water, in a magnet providing the field B_0 parallel to the z direction. The container was surrounded by two coils: A transmitter coil, connected to a radio-frequency oscillator produced the field B_1, oscillating in the x direction. The receiver coil with its

[6] Cornelius Jacobus Gorter (1907–1980)   [7] William Webster Hansen (1909–1949)   [8] Abraham Packard (born 1921)

axis parallel to the y direction picked up the induced electric field. Half a year later, Bloch completed a comprehensive theory of nuclear induction [14], including the *Bloch equations*, which quantitatively describe the influence of the different types of relaxation on the resonance. In an adjacent paper [15] the experiment was described in detail.

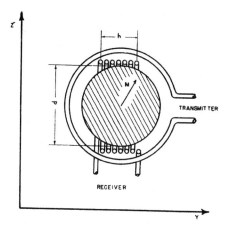

Bloch's arrangement of the sample (hatched) and two coils. Figure reprinted with permission from [15]. Copyright 1946 by the American Physical Society.

NMR was originally developed for use in nuclear physics but it has long since become indispensable in structural chemistry. Resonance frequencies for protons can be measured with relative accuracy of a few parts in 10^9. This makes them sensitive to the fields exerted on the proton by neighbouring atoms in the same molecule. Organic substances thus yield several proton resonance frequencies, each indicating a particular proton position inside a molecule. Even better known is the use of NMR imaging in medicine, which is considered the greatest contribution of physics to medical diagnostics since the days of Röntgen. Here conditions are created under which magnetic resonance occurs only within a thin straight line in the patient's body. By rotating and shifting this line and recording the resonance data, a picture providing three-dimensional information can then be reconstructed in the computer. Only recently the Nobel Prize in medicine was awarded to two pioneers of this technique, the American chemist Lauterbur[9] and the British physicist Mansfield[10].

Zavoisky

We described how NMR was discovered twice, on the east and the west coast of the Unites States, by well-equipped groups of scientists who had returned to their university laboratories after working on radar research. Magnetic resonance in bulk matter, this time the resonance of the magnetic moment of electrons, not of nuclei, was first observed a year earlier under less favourable conditions. This effect, called *electron spin resonance, ESR*, or *electron paramagnetic resonance, EPR*, was discovered during the war, in 1944, by a single scientist at the University of Kazan in the Soviet Union. Zavoisky[11] [16] was the son of a medical officer in the Russian army. From 1910 onwards, his father chose to work in a clinic in Kazan on the river Volga, because there his children would have a good education, the university being the second oldest in

---

[9] Paul C. Lauterbur (1929–2007), Nobel Prize 2003   [10] (Sir) Peter Mansfield (born 1933), Nobel Prize 2003   [11] Evgeny Konstantinovich Zavoisky (1907–1976)

[1]   Bloch, F., *Physical Review*, **50** (1936) 259.

[2]   Rabi, I. I., *Physical Review*, **51** (1937) 652.

[3]   Rabi, I. I., Millman, S., Kusch, P., and Zacharias, J. R., *Physical Review*, **53** (1938) 318.

[4]   Rabi, I. I., Millman, S., Kusch, P., and Zacharias, J. R., *Physical Review*, **55** (1939) 526.

[5]   Alvarez, L. W. and Bloch, F., *Physical Review*, **57** (1940) 111.

[6]   Pound, R. V., *Biographical Memoirs (Nat. Acad. Sci.)*, **78** (2000) 1.

[7]   Bertolotti, M., *The History of the Laser*. Institute of Physics Publishing, Bristol, 2005.

[8]   Rigden, J. S., *Reviews of Modern Physics*, **58** (1986) 433.

[9]   Gorter, C. J., *Physica*, **3** (1936) 503.

[10] Gorter, C. J. and Broer, L. J. F., *Physica*, **9** (1942) 591.

[11] Gorter, C. J., *Physics Today*, **January** (1967) 76.

[12] Purcell, E. M., Torrey, H. C., and Pound, R. V., *Physical Review*, **69** (1946) 37.

[13] Bloch, F., Hansen, W. W., and Packard, M., *Physical Review*, **69** (1946) 127.

[14] Bloch, F., *Physical Review*, **70** (1946) 460.

[15] Bloch, F., Hansen, W. W., and Packard, M., *Physical Review*, **70** (1946) 474.

[16] Kochelaev, B. I. and Yablokov, Y. V., *The Beginning of Paramagnetic Resonance*. World Scientific, Singapore, 1995.

[17] Zavoisky, E., *J. Phys. USSR*, **9** (1945) 245. Engl. transl. in [16].

Russia. Zavoisky, a radio amateur since school days, studied physics in Kazan and worked for some months at the Central Radio Laboratory in Leningrad. In 1933, he obtained his degree as *Canditate* (the Russian equivalent of a Ph.D.) with a thesis on a new way to produce 'ultrashort' electromagnetic waves, in the region of 1 m wavelength, at Kazan University and stayed there as Associate Professor. He became interested in the absorption of electromagnetic waves by matter and he learned of Gorter's attempts to detect NMR by calorimetric means. Aiming to measure that effect through observation of an energy loss in an electric circuit, he devised a set-up, similar to the one later used by Purcell, but using radio waves rather than microwaves. With two co-workers, he obtained signals in 1941 which, however, were not easily reproducible.

When the front was approaching Moscow, academy institutes were evacuated to Kazan and with them many physicists, for instance, Joffe, Kapitza, Vavilov, Frenkel, Tamm, and Landau. Zavoisky's laboratory was turned into a canteen, his apparatus was dismantled. But, eventually, he could work again and decided to search for electron magnetic resonance. Not having access to microwave components, he had to lower the required resonance frequency by drastically decreasing the strength of the constant magnetic field B_0. In 1944 Zavoisky observed absorption due to electron resonance. This discovery was the subject of a thesis which he submitted in Moscow to obtain a doctor degree (roughly equivalent to the German degree of Privatdozent). The commission in charge was not convinced that such a subtle effect could have been observed with the simple means at Zavoisky's disposal. Fortunately, Kapitza, now back in Moscow, came to Zavoisky's help and proposed to him to repeat the experiments with somewhat better equipment at his own institute in Moscow. There, early in 1945, Zavoisky was able to work at a higher magnetic field with microwaves and, within a few months, reproduced his results with better accuracy [17]. Thirty years later, he presented what remained of his apparatus to Kapitza as a birthday present, remembering, in particular, the value of two components, an American klystron and a German high-frequency cable.

ESR was, like NMR, developed into a valuable analytical tool. It is sensitive to atoms or radicals which possess an unpaired number of electrons and therefore a magnetic moment. The exact resonance frequency depends on the type of atom but also on the fields due to neighbouring atoms. The signals obtained, for instance, from defects in a semiconductor crystal, contain information on the defect type, concentration, and position.

The Pi Meson Discovered by the Photographic Method (1947)

66

More than twelve years after Yukawa had proposed the existence of a particle as carrier of the strong interaction between nucleons (Episode 57), a particle with properties expected by his theory was finally observed. The discovery was possible because important improvements had recently been made to one of the oldest techniques in the detection of ionizing radiation, the *photographic method*.

Since the last decades of the nineteenth century the light-sensitive layer of a photographic plate or film is an emulsion of gelatine with embedded fine grains of silver-halide crystals. Under action of light, some of the crystals undergo a small change: a *latent image* is formed, which is made visible by chemical development; silver halide is reduced to silver which appears black. A latent image is created not only by light but also by X rays, as found by Röntgen, and by the radiation emitted by radioactive substances. Indeed, it was in this way that Becquerel discovered radioactivity in 1896. The diffuse blackening that he observed was actually caused by β rays, because the α rays were absorbed by the black paper wrapping of the plates and because the effect of γ rays was much smaller.

About the time when Wilson developed his cloud chamber, it was recognized that α particles traversing a photographic emulsion gave rise to a chain of silver grains marking the track of the particle. The track can be studied with high precision under the microscope. Therefore, a photographic emulsion could serve as a *track chamber* similar to the cloud chamber. A charged particle, while travelling through matter, loses energy by ionization. The number of ion pairs formed per unit length depends on the charge and the velocity of the particle and on the density and type of matter. It is proportional to the square of the charge and roughly inversely proportional to the velocity squared of the particle. Thus the ionization is low for particles carrying only one elementary charge and moving with a velocity near to that of light. Such particles, e.g., electrons with an energy of a few MeV or more, are said to have *minimum ionization*. Ionization is a statistical process, i.e., ions are not distributed uniformly along the particle trajectory. To allow the formation of a droplet in a cloud chamber or a (latent) silver grain in an emulsion a certain number of ions within a small volume is required. For particles with high ionization, it is reached often along its path, they leave tracks with a high droplet density or grain density. As the ionization decreases, the density falls, until it becomes so small that tracks can no longer be recognized.

Plates available in the first third of the twentieth century yielded tracks only for α particles from radioactivity, whereas a cloud chamber was sensitive even for minimum-ionizing particles. Only when cosmic radiation was studied in

Occhialini (left) and Powell.

Lattes on Mount Chacaltaya.

detail, an advantage of emulsion was realized. It had a much larger stopping power: Because the density of emulsion is about three orders of magnitude larger than that of the cloud-chamber gas, many more interactions with nuclei are expected per unit track length. In the 1930s Blau[1] [1] in Vienna pioneered the photographic method. She was able to sensitize plates so that they would yield tracks also for low-energy protons and registered 'stars', produced by the fragments of nuclei hit by particles of the cosmic radiation [2].

Powell[2] [3], a student of C. T. R. Wilson, had built a Cockroft–Walton accelerator at the University of Bristol and had begun to study nuclear reactions in a cloud chamber, when he heard about Blau's work with emulsions. He used her method and found it advantageous, in particular, since the energy of a particle could be related to its range in emulsion. Shortly before the end of the war, Powell was joined by Occhialini[3] [4], who, together with Blackett, had built the first cloud chamber that was triggered by Geiger counters in 1932. Occhialini had left his native Italy before the war because of his opposition to fascism and had accepted a professorship São Paulo. But he lost that position when Brazil entered the war and he thereby became an 'enemy alien'. After an adventurous time, with Blackett's help, he was allowed to work in Bristol. He realized that the emulsions used as detectors might be improved by radically increasing the content of silver bromide. This was achieved shortly after the end of the war by the Ilford company. Occhialini sent photomicrographs of tracks in the new emulsion to his former student César Lattes[4] [5] in São Paulo. Lattes was excited about the possibilities of the new material. He obtained a grant from Bristol and arrived there in 1946.

Plates carrying the new Ilford emulsion were exposed to cosmic radiation. Perkins[5] of Imperial College, London, had them flown for several hours at about 10 000 m by an RAF plane. The Bristol group exposed them for much longer periods at 2800 m on the *Pic du Midi* in the French Pyrenees. Both were met with a surprise: the meson, identified by its relatively low grain density near the end of its range, finally seemed to show strong interaction and to produce disintegrations of nuclei visible as stars in the emulsion, as reported in January [6] and February [7] of 1947.

To account for the observed decay of mesons into electrons with a rather long lifetime and their lack of strong interactions, a theory had been proposed in 1940 by Tomonaga and Araki [8]: Because of their electric charge, positive mesons are repelled by nuclei and therefore are allowed to decay. Negative mesons, on the other hand, after losing their energy by ionization, are attracted by nuclei, will be captured, and then interact with them. In summary, positive mesons were expected to decay, whereas negative ones were expected to interact strongly. The stars were therefore interpreted as being caused by negative mesons, still believed to be the particles originally found by Anderson and Neddermeyer.

In May 1947, Lattes, Occhialini, and Powell, together with Powell's student Muirhead[6] reported that, while some mesons interact, others decay [9]. The charged decay products are themselves mesons in the sense that they have a

[1] Marietta Blau (1894–1970)   [2] Cecil Frank Powell (1903–1969), Nobel Prize 1950   [3] Giuseppe Occhialini (1907–1993)   [4] Cesare Mansueto Giulio Lattes (1924–2005)   [5] Donald. H. Perkins (born 1925)   [6] Hugh Muirhead (1925–2007)

mass intermediate between the electron mass and the proton mass. In their emulsion exposed on the Pic du Midi by Occhialini, they found the track of a meson, which was stopped in the emulsion, as shown by the increase of grain density towards the end point of the track. Another meson track began at the end point and also ended in the emulsion. We show a photomicrograph of this event and have to explain how such a picture is made. Tracks in a developed plate can be observed and measured with high precision under the microscope. The microscope table is movable in the horizontal x, y plane and the two coordinates x and y are recorded when a cross hair coincides with the particular point to be measured. The microscope has only a very shallow focus in z. To focus on a given point the distance between table and objective has to be adjusted; this distance is taken as the z coordinate. A series of photographs can be taken through the microscope of sections along the tracks of an event in such a way that each individual photograph is in focus. Working with scissors and glue, a mosaic is later reconstructed showing the complete event. Such mosaics or photomicrographs are used as illustrations, not for measurement.

Lattes, Occhialini, and Powell also found a second event, less complete, in which the decay meson did not stop in the emulsion. Thus, on the basis of 'one and a half' events there was evidence for the existence of two types of mesons. More events were needed to confirm the result; the low flux of cosmic radiation on the Pic du Midi was the limiting factor. In the Geographic Library in Bristol, Lattes found that near La Paz in Bolivia there existed a meteorological station on Mount Chacaltaya in the Andes at 5500 m. He proposed to go there and expose plates, 'loaded' with borax to prevent fading of the latent image before development, for one month. This was done. On his way back in La Paz Lattes developed one plate in a hurry and immediately found a complete decay event. In Bristol, the remaining plates were properly processed and scanned and many more 'double meson' events were found.

In October 1947, Lattes, Occhialini, and Powell published two papers [10] with the results obtained by the exposures on the Pic du Mici and Mt. Chacaltaya. They gave names to the two types of mesons, calling the primary the π meson and the decay product the μ meson. The tracks of all μ mesons, which ended in the emulsion and which stemmed from the decay of a π meson stopping in the emulsion, had very nearly the same length of about 600 μm. Such a μ meson, obviously, had a fixed energy and was therefore produced in the decay of a π meson into two particles. (It was found later that the other particle is a neutrino.) It was ascertained by grain-density measurements that the π meson is indeed heavier than the μ meson. Events were found in which a meson was produced in a nuclear disintegration, travelled for a short distance and produced the disintegration of another nucleus.

In the years following its discovery, it was found that the π meson, indeed, has most of the properties of the Yukawa particle. It is a boson with zero spin. It exists in three charge states, positive, negative, and neutral. (The neutral π meson, which decays rapidly into two photons, was found in 1950 [11].) In 1962, the term *hadron* was introduced by Okun to denote any strongly interacting particle. Thus the π meson is a hadron. It is now often called *pion*, for short.

The μ meson, on the other hand, does not show any strong interaction. But it

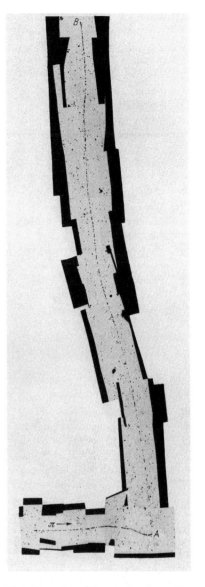

First observation of the π-meson decay. The π meson comes to rest at the point A. The muon produced there as a decay product travels through the emulsion and comes to rest at B. Reprinted by permission from MacMillan Publishers Ltd: Nature [9], copyright 1947.

[1] Rosner, R. and Strohmaier, B. (eds.), *Marietta Blau – Sterne der Zertrümmerung*. Böhlau, Vienna, 2003.

[2] Blau, M. and Wambacher, H., *Nature*, **140** (1937) 585.

[3] Powell, C. F., *Fragments of an Autobiography*. University of Bristol, 1987. http://www.phy.bris.ac.uk/history/12.%20Powell's%20Autobiography.pdf.

[4] Telegdi, V. L., *Proceedings of the American Philosophical Society*, **146** (2002) 218.

[5] Lattes, C. M. G., *My Work in Meson Physics with Nuclear Emulsions*. In Brown, L. M. and Hoddeson, L. (eds.): *The Birth of Particle Physics*, p. 307. Cambridge University Press, Cambridge, 1983.

[6] Perkins, D. H., *Nature*, **159** (1947) 126.

[7] Occhialini, G. P. S. and Powell, C. F., *Nature*, **159** (1947) 186.

[8] Tomonaga, S. and Araki, G., *Physical Review*, **58** (1940) 90.

[9] Lattes, C. M., Muirhead, H., Occhialini, G. P. S., and Powell, C. F., *Nature*, **159** (1947) 694.

[10] Lattes, C. M., Occhialini, G. P. S., and Powell, C. F., *Nature*, **160** (1947) 453, 486.

[11] Steinberger, J., Panofsky, W. K. H., and Steller, J., *Physical Review*, **78** (1950) 802.

[12] Brown, R., Camerini, U., Fowler, P. H., Muirhead, H., Powell, C. F., and Ritson, D. M., *Nature*, **163** (1949) 47, 82.

[13] Gardner, E. and Lattes, C. M. G., *Science*, **107** (1948) 270.

[14] Burfening, J., Gardner, E., and Lattes, C. M. G., *Physical Review*, **75** (1949) 382.

[15] Powell, C. F., Fowler, P. H., and Perkins, D. H., *The Study of Elementary Particles by the Photographic Method*. Pergamon, London, 1959.

interacts weakly. In particular, its decay proceeds via the weak interaction. All particles showing only weak or, at most, weak and electromagnetic interactions were called *leptons* by Møller and Pais in 1947. By this definition the μ meson is a lepton. With the years the term *meson* was reserved for hadrons. The μ meson is no longer called by its original name but is now denoted as *muon*. It decays via the weak interaction into and electron and, as was learnt later, two neutrinos (Episodes 79 and 85).

Why is it that the whole decay chain with the tracks of pion, muon, and electron was not observed in the emulsions of 1947? They were simply not sensitive enough to allow the registration of the electrons which, because of their small mass, are minimum ionizing. Emulsion with the required sensitivity became available in 1950 and then such π–μ–e events were promptly found [12].

First observation of the complete π–μ–e decay chain. Reprinted by permission from MacMillan Publishers Ltd: Nature [12], copyright 1949.

But let us return to the Bristol group. Two of its members left soon after the pion was discovered. At the end of 1947 Lattes went to Berkeley and, using the 184-inch cyclotron, was able to detect pions in photographic emulsion that were not produced by cosmic rays but by artificially accelerated particles [13,14]. In 1949 he returned to Brazil and was, successively, professor in Rio, São Paulo, and Campinas. He continued to work in the field of cosmic rays and had great influence in shaping physics research in Brazil. Occhialini, in 1948, went to Brussels. He returned to Italy in 1950 and in 1952 settled as professor in Milan. He made research with emulsion, which was relatively inexpensive, popular in post-war Italy, thus contributing to the successful concentration on elementary particle physics in this country. Powell's group continued to flourish and made further discoveries. He himself co-authored a comprehensive and beautiful book, *The Study of Elementary Particles by the Photographic Method* [15].

The Lamb Shift (1947)

In 1947, Lamb and Retherford performed an experiment on hydrogen atoms demonstrating a deviation from Dirac's relativistic theory of the electron of 1928. Although until then this theory described the observed hydrogen spectrum within the high precision achieved by experiment, it was known to be plagued by fundamental difficulties. Still more accurate data were needed to test and stimulate theoretical ideas. The measurements of Lamb and Retherford led, within a few years, to the completion of *quantum electrodynamics* (*QED*) as we now know it.

The atom of hydrogen, through the spectrum of radiation it emits or absorbs, has provided important experimental information essential for the development of quantum physics [1]. Balmer, in 1885, wrote down a simple formula, which expressed the frequencies of this spectrum as differences of two *terms*, and Bohr, in 1913, interpreted these terms (multiplied by Planck's constant h) as the different energy levels of the atom. In Bohr's theory the energy levels were $E_n = E_0/n^2$ with n, the principal quantum number, taking the values $n = 1, 2, \ldots$. Bohr was able to express the constant E_0 by fundamental constants of nature (Episode 23). The spectral lines corresponding to the transitions from states with $n > 2$ to the state $n = 2$ are in the visible part of the spectrum and called the *Balmer series*. The individual lines have names: H_α corresponding to $n = 3 \to n = 2$, H_β for $n = 4 \to n = 2$, etc.

Only two years after Balmer's work, Michelson and Morley observed that the H_α line actually consisted of two separate lines. In 1916 Sommerfeld explained this *fine structure* of the hydrogen spectrum as a relativistic effect. States with the same principal quantum number n but different angular-momentum quantum number l had slightly different energies (Episode 27). In Dirac's relativistic quantum mechanics of the electron (Episode 43), Sommerfeld's formula stayed correct (if the negative-energy states, which appear in the Dirac equation, are neglected) but the interpretation was revised. The terms are classified by four quantum numbers: In addition to n and l there is the spin quantum number $1/2$ of the electron and the quantum number j of its total angular momentum. For historical reasons, $l = 0$ is denoted by the letter S and $l = 1$ by the letter P. States with $n = 2$ can be either S or P. Since the S state has no orbital angular momentum its total angular momentum is $j = 1/2$; the state is denoted as $S_{1/2}$. There are two P states, $P_{1/2}$ and $P_{3/2}$ with angular momenta $j = 1/2$ and $j = 3/2$, respectively. In Dirac's theory, the two states $S_{1/2}$ and $P_{1/2}$ have the same energy, whereas $P_{3/2}$ lies a little higher. The experiment of Lamb and Retherford showed, however, that also the states $S_{1/2}$ and $P_{1/2}$ are different in energy. This energy difference, now called the *Lamb shift*, is minute. The frequency corresponding to a transition between the two states is only about one millionth of the frequency of the H_α

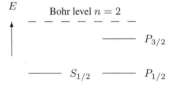

Energy levels ($n = 2$) according to the Dirac theory.

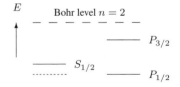

Experiment revealed that the levels $S_{1/2}$ and $P_{1/2}$ are different.

line.

As can be guessed from the smallness of the effect, the experiment was complicated. It was possible only because new techniques had recently become available. The main reason for its success is that Lamb, originally a theoretician, was aware of the importance of such a measurement and that, in the course of his career, he had become acquainted with the necessary experimental techniques, in particular, molecular beams and microwaves.

Lamb[1] [2] was born in Los Angeles as son of a telephone engineer. He obtained a B.Sc. in chemistry from the University of California in Berkeley in 1934 and decided to do his graduate studies in theoretical physics under Oppenheimer[2]. At that time Oppenheimer, who had studied at Harvard and at Cambridge, England, and taken his Ph.D. under Born at Göttingen, was already one of the leading American theoreticians and had pointed out that fundamental difficulties arose when trying to compute atomic energy levels with existing theories [3]. When Bloch arrived in nearby Stanford, Oppenheimer and Bloch set up a common seminar for their groups and often invited speakers from outside. Lamb thus met many senior physicists and learned about their subjects.

After completing his Ph.D. Lamb got a position as instructor at Columbia University in New York. He continued to work on theory but, because of Rabi's molecular beam group at Columbia, also became interested in the physics of atomic or molecular beams. One problem was the detection of the electrically neutral atoms or molecules after their passage through the apparatus. We remember that in the original Stern–Gerlach experiment the silver atoms were detected by the macroscopic silver deposit they left on a glass plate, not a very sensitive method. Some atoms could be ionized in the detector and thus yield an electric signal. It was known, at least from theoretical calculations, that some excited states of certain atoms are *metastable*. That means, they have a long lifetime, long enough to allow the atoms to travel through the apparatus in the excited state. If such an atom hits a metal surface, Lamb thought, the excitation energy it carried might serve to liberate an electron from the metal which could then be detected. Together with a visitor from Puerto Rico he published an elaboration of that idea, applied to helium atoms, in 1944 [4]. Around that time Lamb, together with other Columbia physicists, became involved in the development of microwave techniques for radar applications. Although continuing to work as a theorist he also built some apparatus himself.

Preparing lectures in 1945 Lamb, now an assistant professor, found information in the literature which became important for his experiment. He realized that the energy spacing of the hydrogen states for $n = 2$ was in the range of available microwave sources. Moreover, he found that the $S = 1/2$ state was considered to be metastable. A year later he began in earnest to plan an experiment to measure a possible substructure in the $n = 2$ state, which had been discussed in the experimental and theoretical literature for some time without conclusive results. He was joined by Retherford[3], an experienced experimentalist who had returned to Columbia for graduate studies from work on vacuum tubes at Westinghouse.

Hydrogen normally exists in the form of H_2 molecules. In a gas discharge

Lamb

---

[1]   Willis Eugene Lamb (1913–2008), Nobel Prize 1955   [2]   Julius Robert Oppenheimer (1904–1967)   [3]   Robert Curtis Retherford (1912–1981)

some of them dissociate into H atoms and of these some are excited to the $n = 2$ state. Thus it would have been relatively simple to look for the absorption of microwaves in a gas discharge of hydrogen. Such measurements, however, are influenced by the presence of free electrons and, in particular, by the violent motion of the atoms in the hot discharge vessel, which gives rise to a broadening of the resonance-absorption curve. Lamb therefore had devised a more precise, albeit much more complicated, method using a beam of atomic hydrogen.

The apparatus had four main components:

- the *dissociator* in which a beam of H atoms was produced from H_2 gas,
- the *electron bombarder* which sprayed the beam with electrons exciting some H atoms to the $n = 2$ state; of these all but those in the metastable $S_{1/2}$ state decayed to the ground state immediately,
- the *microwave region* in which the beam is exposed to microwaves,
- the *detector*, a metal plate from which metastable atoms liberate electrons; the small current of electrons is amplified by a vacuum-tube amplifier and then measured with a galvanometer.

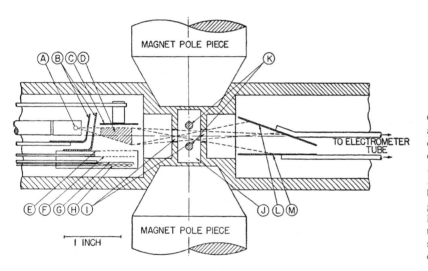

Cross section through the apparatus of Lamb and Retherford: A tungsten oven and hydrogen dissociator, B shields, C anode of electron bombarder, D bombardment region, E accelerator grid of electron bombarder, F control grid, G cathode of electron bombarder, H heater for cathode, I slits, J wave guide, K quenching wires and transmission lines, L metastable detector target, M electron collector. Figure reprinted with permission from [5]. Copyright 1950 by the American Physical Society.

If the microwave frequency ν corresponds to the energy difference ΔE between the metastable state and some nearby state, i.e., if $h\nu = \Delta E$, then the atom transits into that state and from there to the ground state; it is no longer detected. For such a frequency the detector signal decreases.

Since the microwave frequency could not be varied easily, Lamb and Retherford added a magnetic field to the microwave region. The three states $S_{1/2}$, $P_{1/2}$, and $P_{3/2}$ are subject to the Zeeman effect, which leads to level differences depending on the magnetic field strength. Resonance could therefore be achieved by gradually varying the magnetic field while the frequency remained constant and the results could be extrapolated to zero field strength.

Decades later Lamb vividly remembered the day of success [2]:

> The experiment first succeeded on Saturday, April 26, 1947, and turned
> out very much as expected, except for the location of the resonances. ... I

remember that late that night after Retherford had gone home, I went over to the laboratory to see if I could confirm the earlier results. I found that I could not manage all the knobs and data taking; so I telephoned my wife Ursula to come over to the Pupin Laboratories and help me. The two of us were able to run the apparatus and could clearly see the galvanometer spot still moved as it had done earlier that day.

Lamb and Retherford measured a pronounced absorption corresponding to a frequency of $1000\,\text{MHz} = 10^9\,\text{s}^{-1}$ at zero field, or an energy ΔE somewhat less than 10% of the level spacing between the $P_{1/2}$ and $P_{3/2}$ levels according to the Dirac theory. The result was interpreted as follows: The $S_{1/2}$ state is higher in energy by ΔE compared to the $P_{1/2}$ state. Microwaves of the right frequency make it descend to the $P_{1/2}$ state by stimulated emission of a photon with that frequency and from there to the ground state by spontaneous emission.

For three days at the beginning of June 1947, Oppenheimer organized a Conference on the Foundations of Quantum Mechanics at Shelter Island east of Long Island. It was a small exclusive circle: twenty-one physicists, one mathematician (von Neumann[4]), and one chemist (Pauling[5]). Lamb presented his results and also Rabi had news to report from his group. Kusch[6] and Foley, with the magnetic-resonance method used on molecular beams which Rabi had pioneered, had discovered another deviation from Dirac's original theory: The magnetic moment of the electron was not exactly one Bohr magneton or, in more technical terms, its g factor (see Box I in Episode 65) showed a small but significant deviation of about one-tenth of a per cent from the theoretical value $g_0 = 2$ [6–8]. These news from experiment had a very stimulating effect on the assembled theoretical elite (see Episode 71).

The first paper on the *Lamb shift* [9] was sent for publication two weeks after this conference. In the following years Lamb, together with Retherford and others, continually improved the precision of their measurement and published the results in a series of papers, all entitled *Fine Structure of the Hydrogen Atom* [5]; we only cited *Part I*.

Speaking at a conference on the history of particle physics in 1980 Lamb, obviously with some pleasure, remarked [2] that he had been in contact with the leading figures in the history of fine structure, Michelson, Sommerfeld, and Dirac. When he was still a high-school student he had met Michelson, who spent his last years at Caltech, at a chess festival in Pasadena. 'We talked a little about the game, but not about physics, and although I knew that he was famous for the measurement of the velocity of light, I did not know of any other reasons.' He never met Sommerfeld, but 'in 1950 he sent me a handwritten note that indicated that he had heard of our measurements, and he mentioned that he was the '81-year-old greatgrandfather' of the hydrogen fine-structure theory.' Lamb met Dirac repeatedly. About one occasion he remembers that Dirac asked him if he had enjoyed participating in the discovery of the fine-structure anomaly. Lamb said he had, but added that he would have had much more pleasure if instead he had discovered the Dirac equation. As Lamb recalls, after a brief pause Dirac said gently, 'Things were simpler then.'

[1] Series, G. W., *The Hydrogen Atom (An Historical Account of Studies of Its Spectrum)*. In Dassani, G. F., Inguscio, M., and Hänsch, T. W. (eds.): *The Hydrogen Atom*, p. 2. Springer, Berlin, 1989.

[2] Lamb, W. E., *The Fine Structure of Hydrogen*. In Brown, L. M. and Hoddeson, L. (eds.): *The Birth of Particle Physics*, p. 311. Cambridge University Press, Cambridge, 1983.

[3] Oppenheimer, J. R., *Physical Review*, **35** (1930) 461.

[4] Cobas, A. and Lamb, W. E., *Physical Review*, **65** (1944) 327.

[5] Lamb, W. E. and Retherford, R. C., *Physical Review*, **79** (1950) 549.

[6] Kusch, P. and Foley, H. M., *Physical Review*, **72** (1947) 1256.

[7] Foley, H. M. and Kusch, P., *Physical Review*, **73** (1948) 4126.

[8] Kusch, P. and Foley, H. M., *Physical Review*, **74** (1948) 250.

[9] Lamb, W. E. and Retherford, R. C., *Physical Review*, **72** (1947) 241.

---

[4] John von Neumann (1903–1957)   [5] Linus Carl Pauling (1901–1994), Nobel Prize 1945   [6] Polykarp Kusch (1911–1993), Nobel Prize 1955

Strange Particles (1947)

The discovery of the π meson (Episode 66) was taken as the proof of Yukawa's meson theory and therefore accepted by physicists with satisfaction. Half a year later, in December 1947, the observation of additional particles with strange, unexpected properties was announced by Rochester and Butler from the University of Manchester.

Rochester[1] was born as son of a blacksmith at Wallsend near Newcastle upon Tyne. He studied at the Armstrong College of Durham University (now the University of Newcastle), where he obtained his Ph.D. with work on spectroscopy in 1935. After two years of postdoctoral research in California, he joined the University of Manchester as assistant lecturer in 1937, the year when Blackett succeeded to the chair of William Henry Bragg at Manchester, the chair that had been Rutherford's before. Rochester soon found himself working in Blackett's particular field of interest, cosmic-ray physics. Most of basic research work was interrupted by the war but resumed in 1945. In the summer of that year Butler[2], who had obtained his Ph.D. from the University of Reading, answered an advertisement in *Nature* for a post of assistant lecturer at Manchester. Rochester was on the board of three who interviewed and accepted him.

Rochester

Early in 1946 Rochester and Butler began to reconstruct and adapt a pre-war magnetic cloud chamber of Blackett to study penetrating cosmic-ray showers. Of course, they used the method, pioneered by Blackett and Occhialini, to trigger expansion and photography of the chamber electronically by the shower itself (Episode 49). They mounted a single lead plate in the chamber to allow the cosmic-ray particles to interact. Rather soon they observed an event of a new type, which they called a *fork*: Two tracks, caused by particles of opposite electric charge, begin at a common point and continue in the downwards direction, i.e., in the direction of the cosmic rays. The event was interpreted as being caused by a neutral particle (leaving no track), which decayed in the cloud chamber into a pair of charged particles. Nine months later, another type of fork event was observed: A track, caused by a positively charged particle travelling downwards, at a certain point abruptly changed its direction by 19 degrees. It was interpreted as caused by a charged particle decaying into another charged particle and a neutral one.

Butler writes about the history of the famous paper [1] in which these events are reproduced and discussed [2]:

Butler

> When the neutral V-event was found in the autumn of 1946, its interpretation was widely discussed in the laboratory. Blackett was soon fairly convinced that it was due to a new decay process, but he would not allow immediate publication. After the discovery of the second event in

[1] George Dixon Rochester (1908–2001)   [2] (Sir) Clifford Charles Butler (1922–1999)

May 1947, urgent publication was agreed. Nevertheless, the preparation of the short paper for *Nature* took several months. Blackett took a very commanding role in the presentation of the work. Every conceivable trivial explanation of the photographs was considered and all were rejected. The precise logical order of presentation became an issue involving much discussion. Numerous drafts of the paper were prepared and rejected, usually by Blackett.

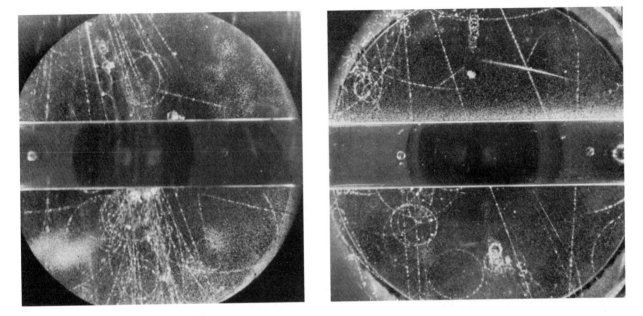

First cloud-chamber photographs of *V*-particle decays. A neutral particle (V^0) decays into two charged particles leaving tracks beginning at a common point just below the central plate in the photograph on the left. The track of a positively charged particle (V^+) changes its direction in the top-right corner of the photograph on the right, indicating that in this point the particle decayed into another positive particle and a neutral one. Reprinted by permission from MacMillan Publishers Ltd: Nature [1], copyright 1947.

The momenta of the charged particles causing the tracks could be measured from the track curvature in the magnetic field. Using formulae from the special theory of relativity (see Box), Rochester and Butler were able to estimate the mass of their new particles. Since none of the decay particles left a track with the high droplet density typical of the proton, it was rather safe to take them for π mesons. But even if their mass was taken to be that of the electron, the mass of the new particles was at least twice or thrice as high as that of the π meson. We know now that they were a neutral and a positively charged K meson, respectively.

For more than a year, no confirmation of was obtained. But in January 1949, Powell's group from Bristol reported on another exciting event they had observed in photographic emulsion exposed at an altitude of 3460 m on the *Jungfraujoch* in Switzerland [3]. A slow charged particle lost its energy in the emulsion, came to rest, and decayed into three charged particles. One of those, provoking a nuclear reaction, was identified as pion. The other two, from their ionization, had to be either pions or muons. The mass of the primary particle was estimated to be 925 electron masses under the assumption that all three decay particles were pions. That assumption was correct. The event is now taken as the first observation of the three-pion decay of the charged K meson, soon to be called τ decay. The next τ event, also in emulsion, was found at Imperial College in London [4].

Relativistic Kinematics

In the special theory of relativity (Episode 11) energy and momentum of a particle (like time and space coordinates) are combined in a *four-vector*, the *four-momentum*

$$p = (p^0, p^1, p^2, p^3) \quad ,$$

where p^1, p^2, p^3 are the components of the conventional *three-momentum* vector $\mathbf{p}$ and

$$p^0 = E/c = \sqrt{m^2 c^2 + \mathbf{p}^2}$$

is the total energy E of a particle with rest mass m, divided by the velocity c of light. The square of the four-momentum (taken according to the rules of special relativity),

$$p^2 = p \cdot p = (p^0)^2 - \mathbf{p}^2 = m^2 c^2 \quad ,$$

is equal to the square of the rest mass, multiplied by c^2.

In reactions between particles or in decays the total four-momentum is unchanged because of energy and momentum conservation. Consider the decay of particle A into particles B and C, $A \rightarrow BC$. Then

$$p_A = p_B + p_C \quad .$$

Thus, if all but one four-momenta are known, the unknown one is easily computed. In particular, if particle A is uncharged and therefore leaves no track, its rest mass, given by

$$m_A^2 c^2 = p_A^2 = (p_B + p_C)^2 \quad ,$$

can be calculated from the four-momenta, i.e., the three-momenta and the rest masses, of particles B and C.

In general, up to four unmeasured quantities can be inferred from energy–momentum conservation, not necessarily all components of the same four-vector.

Finding no new events, the Manchester group grew worried and decided to do experiments at high altitude: Rochester was to work with emulsions on the Jungfraujoch. The laboratory there was reachable by a tourist railway all year, which, however, could not accommodate Blackett's original cloud-chamber magnet. Construction of a new chamber with a take-apart magnet was begun. The laboratory on the *Pic du Midi* in the Pyrenees at 2800 m was accessible in summer by road and the Manchester cloud-chamber equipment was brought there. A new group under Butler was set up to run it. While it was installed, in November 1949, Anderson sent a telegram to Blackett, announcing the observation of 'about 30 cases of forked tracks' in his cloud chamber, exposed in Pasadena and at an altitude of 3200 m on White Mountain in California. The results of Anderson's group were published in 1950 [5]. No cloud-chamber photographs were published because of track distortions in the chamber. Anderson and Blackett coined the term *V particles* to denote all particles leaving a fork as signature in the track chamber. *V* particles could be electrically neutral, V^0, or charged, $V^\pm$.

In March 1951, the Manchester group presented results on the 36 neutral and 7 charged *V* particles observed in the first six months on the Pic du Midi [6]. The positive decay track in four of the neutral *V* particles was identified as

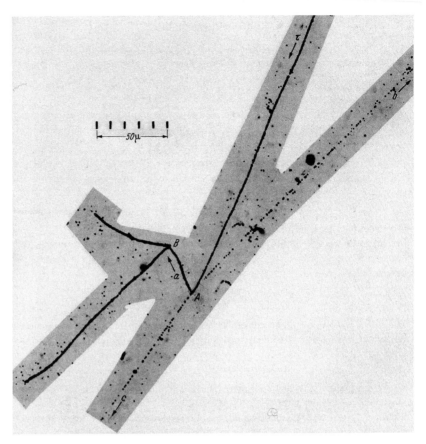

Observation of a τ event in photographic emulsion. A charged particle, labelled τ, decays at A into three pions, labelled a, b, c. The slowest one, a, disintegrates a nucleus at point B. Reprinted by permission from MacMillan Publishers Ltd: Nature [3], copyright 1949.

caused by a proton, whereas in three other cases it stemmed from a much lighter particle, presumably a pion. From this finding it had to be concluded that there were two quite different neutral V particles. In a detailed paper, *The Properties of Neutral V-Particles* [7], these two particles were called V_1^0 and V_2^0. Their decays were identified as

$$V_1^0 \to p\pi^- \quad \text{and} \quad V_2^0 \to \pi^+\pi^- \quad ,$$

respectively, and mass values for the two particles were given. Nearly at the same time, a group from Indiana, headed by Thompson[3] and using a cloud chamber especially designed to minimize distortion, also reported observing two different neutral V's [8]. Later, the group reported a very good mass separation of V_1^0 and V_2^0 [9].

In their second extensive paper, *The Properties of Charged V-Particles* [10], the Manchester group presents the results of twenty-one $V^\pm$ analysed by March 1952. Of these, most were compatible with the assumption that they were heavy mesons just as the first charged V event. In one particular event, however, a *cascade* of V decays was observed,

$$V^- \to V_{1\,\text{or}\,2}^0 + \pi^- \quad .$$

---

[3] Robert Walder Thompson (1919–1994)

V particles, eventually, were called *strange particles*. They were found to be copiously produced in interactions of cosmic rays with nuclei. It could therefore be assumed that the production process was one of *strong interaction* just as that of pions. On the other hand, their average lifetime, the time between production and decay, was rather long, around 10^{-10} s. That is a time typical of the *weak interaction*. It seems short, but that of the strong interaction is very much shorter, 10^{-21} s. It was strange, that particles like V_1^0 and V_2^0, produced by strong interaction and obviously allowed to decay into other strongly interacting particles, did not do so via the strong interaction.

Both Rochester [11] and Butler [12] describe the early years of strange-particle research as a particularly exciting time. A vivid account is also given by Peyrou [13], on whose initiative a French cloud chamber was installed on the Pic du Midi. A highlight was the international conference, devoted entirely to the new particles, held in 1953 in Bagnères de Bigorre, a resort at the foot of the French Pyrenees near the Pic du Midi. One of its aims was to order and organize the information from the various experiments. A naming scheme for particles was adopted [14]. The long-lived strange particles were divided into two groups: *Mesons*, lighter than the proton, were denoted by the letter K and *hyperons*, particles with a mass larger than the neutron's, by the letter Y.

Specific mesons and hyperons were to be named by small and capital Greek letters, respectively. Eventually four seemingly different K mesons were named in this way. We mention only the name θ for the original V_2^0 with the decay $\theta^0 \rightarrow \pi^+\pi^-$ and the name τ for the particle first seen in emulsion and observed to decay into three pions, $\tau^\pm \rightarrow \pi^\pm\pi^+\pi^-$.

As a specific hyperon the V_1^0 was named Λ^0. The choice of that capital Greek letter was obvious, since it is simply an inverted V. After the existence of the *cascade particle* was confirmed [15] it got the name Ξ^-. Yet another hyperon, the Σ^+, was observed in 1953, both in a cloud chamber [16] and in emulsion [17] and it was shown to have a negative counterpart, the Σ^- [18].

Particles were considered to belong to one family if they differed in mass by at most a few MeV like the proton and the neutron, which were taken simply as two charge states of one and the same particle, the nucleon. The θ and the τ meson had similar masses. However, there seemed to be a serious theoretical argument against taking them as one and the same particle. This *theta–tau puzzle* led to the *fall of parity* in 1957 (Episode 77).

Even before the Bagnères de Bigorre conference, two Polish physicists, Danysz[4] and Pniewski[5], working in Warsaw with nuclear emulsions that Danysz had been given during a stay in Bristol by Powell's group reported the finding of a spectacular event [19]: A fragment from a nuclear disintegration (brought about by a cosmic-ray proton) travelled a distance of about $100\,\mu$m, came to rest in the emulsion, and the 'exploded' into several particles. They correctly interpreted the fragment (now called a *hyperfragment* or *hypernucleus*) as a nucleus in which a neutron was replaced by a Λ^0 hyperon. It was thus shown that the Λ^0 has the strong interaction of a nucleon, including nuclear binding; its weak interaction, however, provokes its decay and thus the 'explosion' of the hypernucleus. For historical details of the discovery, see [20].

---

[4] Marian Danysz (1909-Ŭ1983)   [5] Jerzy Pniewski (1913–1989)

[1] Rochester, G. D. and Butler, C. C., *Nature*, **160** (1947) 855.

[2] Butler, C., *Notes Rec. Roy. Soc. London*, **53** (1999) 143.

[3] Brown, R. et al., *Nature*, **163** (1949) 82.

[4] Harding, J. B., *Philosophical Magazine*, **41** (1950) 405.

[5] Serif, A. J., Leighton, R. B., Hsiao, C., Cowan, E. W., and Anderson, C. D., *Physical Review*, **78** (1950) 290.

[6] Armenteros, R. et al., *Nature*, **167** (1951) 501.

[7] Armenteros, R. et al., *Philosophical Magazine*, **42** (1951) 1113.

[8] Thompson, R. W., Cohn, H. O., and Flum, R. S., *Physical Review*, **83** (1951) 175.

[9] Thompson, R. W., Buskirk, A. V., Etter, L. R., Karzmark, C. J., and Rediker, R. H., *Physical Review*, **90** (1953) 329.

[10] Armenteros, R. et al., *Phil. Mag.*, **43** (1952) 597.

[11] Rochester, G. In Brown, L. M., Dresden, M., and Hoddeson, L. (eds.): *Pions to Quarks*, p. 58. Cambridge University Press, Cambridge, 1989.

[12] Butler, C. C., *Journal de Physique*, **43** (1982) suppl. to No. 12, p. 177.

[13] Peyrou, C., *Journal de Physique*, **43** (1982) suppl. to No. 12, p. 7.

[14] Amaldi, E. et al., *Nature*, **173** (1954) 123.

[15] Cowan, E. W., *Physical Review*, **94** (1954) 161.

[16] York, C. M., Leighton, R. B., and Bjornerud, E. K., *Physical Review*, **90** (1953) 167.

[17] Bonetti, A., Levi Setti, R., Panetti, M. B., and Tomasini, G., *Nuovo Cimento*, **10** (1953) 345.

[18] Bonetti, A., Levi Setti, R., Panetti, M. B., and Tomasini, G., *Nuovo Cimento*, **10** (1953) 1736.

[19] Danysz, M. and Pniewski, J., *Phil. Mag.*, **44** (1953) 348.

[20] Wróblewski, A. K., *Acta Physica Polonica*, **35** (2004) 1.

69 The Transistor (1947)

Braun

Schottky

Late in 1874 Braun[1], then a young science teacher at a gymnasium in Leipzig, published a paper entitled *On the Conduction of Current through Sulphuric Metals* [1]. In this paper, now often regarded to mark the beginning of semiconductor physics, he reports that in some crystals he found a pronounced deviation from Ohm's law. In several cases the magnitude, not only the direction, of the current through a crystal changed if the voltage was reversed. He speaks of *unipolar conductivity*. In modern language, a circuit with such a behaviour is a *rectifier* since current flows more easily in one direction than in the other; in an ideal rectifier it flows in one direction only. Braun even described a case in which this effect depended on the places where the metallic leads were connected to the crystal. Three years later Braun obtained his first professorship, in Marburg; others followed. Today he is best known for *Braun's tube*, until recently the basis of oscilloscopes and TV sets, but he was awarded the Noble Prize for an essential improvement of wireless transmitters and receivers by the introduction of resonant oscillator circuits. In wireless transmission of Morse code or voice, the amplitude of a high-frequency carrier wave is modulated with the signal itself. In the receiver the signal is retrieved by rectification of the high-frequency current caused by the carrier. In the early days of wireless a *crystal detector*, designed by Braun, was usually used as rectifier in receivers. Essential for the rectifying property was the *point contact* between a suitable crystal and the sharp tip of a metal conductor.

With the development of vacuum tubes the crystal detector lost most of its importance. A *vacuum diode* is a small evacuated glass vessel containing two metallic electrodes which are electrically connected to the outside. One electrode, the cathode, electrically is brought to red heat and therefore emits electrons. If the anode, the other electrode, has a positive voltage with respect to the cathode, it collects these electrons and a current flows through the diode. If the voltage is negative, the electrons are repelled. The diode obviously is a rectifier. By mounting a third electrode, the grid, formed as a metallic mesh and placed between cathode and anode, the tube is turned into a *triode*. This can serve as *amplifier*: The voltage of the anode with respect to the cathode is kept positive and the voltage of the grid is used to determine the current through the anode.

Semiconductor rectifiers in a changed form stayed in use for low-frequency alternating currents. They became, however, better understood through work by Mott[2] [2] at the University of Bristol and, in particular, by Schottky[3] [3,4] at the Siemens Company in Berlin. A model of the electrical properties of a thin layer in the semiconductor, where it is in direct contact with the metal,

---

[1] Karl Ferdinand Braun (1850–1918), Nobel Prize 1909   [2] (Sir) Nevill Francis Mott (1905–1996), Nobel Prize 1977   [3] Arthur Walter Schottky (1886–1976)

was developed by Schottky on the basis of the band model (see Episode 44).

In a semiconductor with a surplus of donor atoms, providing electrons in the conduction band, electricity is carried by these (negatively charged) electrons; the semiconductor is said to be *n-type*. In a *p-type* semiconductor the charge carriers are (positively charged) holes in the valence band, created by electron-absorbing acceptor atoms. The relative locations (in energy) of these bands in the metal and the adjacent semiconductor, as well as the density of donor or acceptor atoms, determine the surface-layer properties. As an example, Schottky considered the interface between metal and *p*-type selenium. In the bulk of the semiconductor there is no space charge since the charges of the ionized acceptor atoms and those of the holes balance. At the interface, some electrons migrate from the metal to the semiconductor. In a thin layer near the surface they neutralize holes. There the acceptor ions form a negative space charge. The space charge in this so-called *depletion layer* leads to an upward step in potential between metal and semiconductor. By applying an external voltage in the same direction, i.e., a positive voltage of the selenium with respect to the metal, the depletion layer can be narrowed. For a negative voltage it is widened.

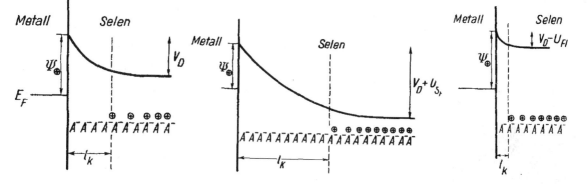

A wide depletion layer is, in essence, an insulator. Therefore, for negative external voltage the metal–selenium contact does not conduct current; for a positive voltage it does. It behaves like a vacuum diode and is now called a *Schottky diode*. In a suitably prepared *n*-type semiconductor, a surface layer, depleted of electrons and carrying a positive space charge, is obtained. A Schottky diode based on such a layer has the same properties as the one described before with the signs of all charges, currents, and voltages reversed.

Only in a few cases, such as metal and selenium, rectification was observed for a flat metal–semiconductor interface. Mostly a sharp metal tip was required. Schottky explained that by the high electric field near the tip, which produced a depletion layer of larger depth although of small lateral extension. The crystal detector, now better understood, experienced a revival in radar applications since the capacitance, inherent in vacuum tubes, inhibits their use at very high frequency. That also led to the production of semiconductors, in particular, silicon and germanium, of much improved purity.

Right after the war, a program was started at the Bell Laboratories in Murray Hill, New Jersey, aiming for improved or new electronic devices based on

Schottky's sketches of the depletion layer in selenium, a *p*-type semiconductor. Negative acceptor ions are denoted by A^-, positive holes by $\oplus$, the typical depth of the depletion layer by l_k. The electric potential, equal to the Fermi energy E_F in the metal, rises sharply by a value $\Psi_\oplus$ at the surface and then drops to a constant value. The three sketches apply to the case without external potential (left), with an additional negative potential of the semiconductor with respect to the metal (middle) and a positive potential (right). Reprinted with kind permission from Springer Science+Business Media: [4], Figures 1, 2, and 3.

semiconductors. One obvious goal was a 'semiconductor triode', which could serve as amplifier and in comparison with the vacuum triode would be smaller, consume less power, and produce much less heat because there would be no need for cathode heating. It was achieved in late 1947 when Bardeen and Brattain presented their first transistor, developed in a research group headed by Shockley [5,6].

Bardeen[4] [7] was the only person to be awarded twice the Nobel Prize in physics, the first time for the transistor, the second time for a theory of superconductivity (see Episode 78). As the two very different topics suggest, his interest and his career were divided between theory and application. The son of the dean of the medical school, he studied electrical engineering at the University of Wisconsin in Madison. He stayed on as graduate student after his B.Sc. in 1928; but two years later he followed one of his professors and worked for three years in geophysics for oil prospecting at the Gulf Oil Company in Pittsburgh. In 1933 he decided to resume graduate work, this time in theoretical physics under Wigner in Princeton. In Wigner's group the first quantitative band-model computations for real crystals were performed. Bardeen's thesis was about the work function of metals, i.e., the energy needed to liberate an electron from a metal surface. Even before that was completed in 1936, Bardeen was offered a prestigious fellowship at Harvard University, where he worked until 1938. He became assistant professor at the University of Minnesota and, in 1941, began military research for the Navy in Washington. Dissatisfied with the offer his university made for his return, he chose instead to join the Bell Laboratories and began his work at Murray Hill in October 1945.

There, Shockley[5] [8,9] had assembled an important semiconductor research group. Shockley held a B.Sc. from Caltech and a Ph.D. from MIT, obtained in 1936, and knew Bardeen from the time they overlapped in Cambridge. Already in April 1945, Shockley had drafted an idea for a semiconductor triode. He wanted to direct a current not perpendicular to the surface of a semiconductor but parallel to it, just below the surface. It would thus travel in the surface layer. If that layer could be influenced by an electric field, provided by an insulated electrode placed on the surface, the current could be influenced by that field just as the current in a triode by the grid voltage. Today *field-effect transistors*, based on this idea, are the essential components of every computer and many other devices. But at the time no effect was observed. If there was one, it was much smaller than estimated by Shockley. When Bardeen began to work at Murray Hill, he was asked by Shockley to check these theoretical estimates.

To account for the smallness of the field effect, Bardeen introduced the concept of *surface states* [10]. Localized electrons accumulate under the influence of the field on the semiconductor surface, shielding the interior from the field, and thus reducing the field effect. It was decided to put this new theory to a test. Bardeen himself sometimes helped with the measurements. At first no effect was found; but when a liquid dielectric and later an electrolyte were introduced between the electrode and the semiconductor, a field effect was observed by Brattain in the middle of November 1947.

Bardeen's and Brattain's first transistor. A gold foil is wrapped around an insulating wedge and slitted along its edge. A spring presses the wedge onto the surface of a germanium crystal thus producing two closely adjacent contacts. Photo Alcatel-Lucent/Bell Labs.

[4]   John Bardeen (1908–1991), Nobel Prize 1956 and 1972   [5]   William Bradford Shockley (1910–1989), Nobel Prize 1956

Brattain[6], too, was an old acquaintance of Bardeen; his brother had been a fellow graduate student in Princeton. He held a Ph.D. from the University of Minnesota and had been working at the Bell Laboratories since 1929. Brattain's observation started a month of experimentation with Bardeen. They varied the semiconductor material, the arrangement of electrodes and did experiments with and without electrolyte [5,6].

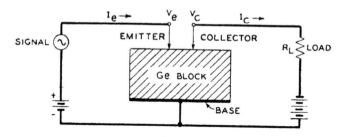

The first transistor amplifier circuit. Figure reprinted with permission from [11]. Copyright 1948 by the American Physical Society.

For the final set-up Bardeen and Brattain used a slab of n-type, 'high-back-voltage' germanium developed for use in radar detectors. The slab had a specially prepared surface near which there was an *inversion layer*, i.e., a zone where the crystal was p-type. On that layer two fine gold contacts were placed only about 50 micrometres apart. The other side of the crystal was contacted with a wide electrode and grounded. The gold contacts were realized by spreading a gold foil over a wedge of polystyrene, slitting it with a razor blade along the edge, and then pressing the wedge onto the germanium surface with a spring. Each of the two gold contacts, together with the germanium crystal and the ground electrode, formed a Schottky diode with well-known rectifying properties. One, on positive voltage, enabled current in the forward direction, i.e., into the crystal, and was called *emitter*. The crystal itself, or rather its grounded electrode, was called the *base*. The other gold contact, the *collector*, carried negative voltage and represented a rather high resistance for current in the reverse direction. Using a small voltage on the emitter and a much larger voltage on the collector the two currents were of equal magnitude.

With this arrangement it was possible to influence the current through the collector by that through the emitter. If the latter rose or fell, so did the former. What was more important, there was power amplification. For a small change of electric power in the emitter or input circuit a much larger change of power was registered in the load (for instance, a resistance) in the collector or output circuit. The device first worked on 16 December and was demonstrated to high-ranking staff of the Bell Laboratories on 23 December 1947.

According to company policy, a patent had to be filed before this work was published. Shockley tried to base it on his original idea of the field-effect transistor but it was found that a patent for such a device had already been granted in 1930 to Lilienfeld[7], a Polish-born physicist who had studied and worked in Germany and had emigrated to the United States in 1927. The patent the Bell Laboratories asked for was therefore based on the point-contact principle and filed in the names of Bardeen and Brattain. The invention was

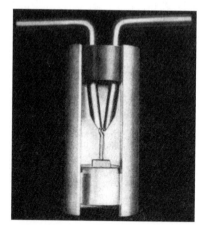

Cutaway prototype of the commercial point-contact transistor. Figure reprinted with permission from [12]. Copyright 1949 by the American Physical Society.

---

[6] Walter Houser Brattain (1902–1987), Nobel Prize 1956   [7] Julius Edgar Lilienfeld (1881–1963)

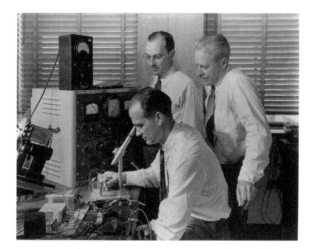

Official Bell Laboratories photograph showing Shockley in front, Bardeen behind him, and Brattain.

[1]   Braun, F., *Annalen der Physik und Chemie*, **153** (1874) 556.

[2]   Mott, N. F., *Proceedings of the Royal Society*, **A172** (1939) 27.

[3]   Schottky, W., *Zeitschrift für Physik*, **113** (1939) 367.

[4]   Schottky, W., *Zeitschrift für Physik*, **118** (1942) 539.

[5]   Riordan, M. and Hoddeson, L., *Crystal Fire – The Invention of the Transistor and the Birth of the Information Age*. Norton, New York, 1997.

[6]   Riordan, M., Hoddeson, L., and Herring, C., *Reviews of Modern Physics*, **71** (1999) S336.

[7]   Hoddeson, L. and Daitch, V., *True Genius – The Life and Science of John Bardeen*. John Henry Press, Washington, 2002.

[8]   Moll, J. L., *Biographical Memoirs (Nat. Acad. Sci.)*, **68** (1996) 303.

[9]   Shurkin, J. N., *Broken Genius – The Rise and Fall of William Shockley, Creator of the Electronic Age*. Macmillan, London, 2006.

[10] Bardeen, J., *Physical Review*, **71** (1947) 717.

[11] Bardeen, J. and Brattain, W. H., *Physical Review*, **74** (1948) 230.

[12] Bardeen, J. and Brattain, W. H., *Physical Review*, **75** (1949) 1208.

[13] Brattain, W. H. and Bardeen, J., *Physical Review*, **74** (1948) 231.

kept confidential in the laboratory until after the patent was finally filed and two short papers by Bardeen and Brattain were submitted to the *Physical Review*. The semiconductor triode was now called *transistor* because of the difference in resistance between input and output circuit. The new device was presented at a public press conference on 30 June 1948. By then the gold-foil arrangement had been replaced by two fine wires touching the germanium crystal at closely adjacent places. The transistor was enclosed in a metal casing with three leads for base, emitter, and collector.

In the first of their papers [11] Bardeen and Brattain describe the transistor and its measured characteristics, in the second [13] they try an explanation of its functioning. (A longer paper appeared in 1949 [12].) Crucial is a conductivity through holes near the two electrodes in the n-type germanium. This, they write, could be due to the narrow p-type inversion layer near the surface but also to the deformation of the valence and conductor bands of the semiconductor near the metallic emitter. The emitter current, to some extent, is carried by holes and these holes, in turn, are essential for the flow of a collector current.

Bardeen's and Brattain's *point-contact transistor* opened the semiconductor age. But it was soon replaced by a more practical device, the *junction transistor*. Immediately after the internal presentation of the point-contact transistor in December 1947, Shockley had begun to design it but did not share his thought with others for quite a while. His device was a small block of n-type germanium with a narrow p-type zone across its middle and metallic contacts on its two n-type ends (the emitter and collector) and the p-type zone (the base). Again were two diodes, but now inside the semiconductor, at the junction between n-type and p-type germanium. Through a current in the emitter diode, electrons were introduced in the p-type base and enabled a current in the collector diode (see also Episode 86).

Shockley's, or rather Lilienfeld's, original idea of a *field-effect transistor* took longer to materialize. It has two essential advantages over the junction transistor. Extremely little power is needed to influence the current through the transistor and it allows miniaturization in the extreme: millions of transistors and their connections can be placed on one surface of a small silicon crystal.

The Shell Model – A Periodic Table for Nuclei (1949)

The Periodic Table of the elements was conceived on the basis of the observed chemical similarities between elements long before the reality of atoms was generally accepted. In 1914 Moseley demonstrated that the position number of an element in the Periodic Table is equal to the number Z of elementary charges in the nuclei of its atoms (Episode 24) and in 1925 Pauli explained the periodicity by enumerating the electrons fitting into successive shells surrounding an atomic nucleus (Episode 35). When, in particular, through the work of Aston (Episode 22), it became apparent that there are, in general, several different isotopes per element, a search for systematics in the wealth of isotopes began and with it the history [1,2] of the *nuclear shell model*.

This was a problem of nuclear physics since isotopes of the same element differ only by their nuclei. A first attempt was made late in 1927 by Beck[1] in Vienna, who compiled a comprehensive table of known isotopes, ordered according to their nuclear charge number Z and their atomic mass number A, and pointed out a number of regularities [3]. The paper, written more than four years before the discovery of the neutron, is remarkable not so much for its results but for two nearly prophetic statements:

Beck in 1938

- The simplest assumption one can make in this respect is to imagine nuclei, in analogy to the electron hull of atoms, to be built up of shells.
- The regularities [observed in the table of isotopes] allow to hope that the Pauli principle and the spin will prove of value as guideposts also in the field of nuclei.

Beck became Heisenberg's first assistant in Leipzig. When the Nazis came to power, after a veritable odyssey with the most different stations, he arrived in South America and had a great influence on the development of theoretical physics in Argentina and Brazil. I vividly remember meeting him in Rio in 1980 and enjoying his Austrian charm when he took a colleague and me out for dinner and later for a glass of wine in his apartment in Copa Cabana.

With the discovery of the neutron it became clear that a nucleus of charge number Z and mass number A was composed of Z protons and $N = A - Z$ neutrons and it became customary to arrange different nuclei or *nuclides*, as they are often called in this context, in a chart with N as abscissa and Z as ordinate. But how to recognize closed shells in such a chart if there were any? In an atom with a closed shell the electrons are tightly bound; the ionization energy, i.e., the energy needed to remove one electron, is particularly large. In analogy, the binding energy of a nucleus could be taken as a measure for its stability. It can be expressed as its *mass defect*, i.e., the mass of Z protons plus

---

[1] Guido Beck (1903–1988)

the mass of N neutrons minus the mass of the nucleus itself. There are several other measures of stability of which we mention but one. It is the observation of many stable (non-radioactive) nuclides for particular values of Z and N. Since protons and neutrons differ in their isospin quantum number I_3 (Episode 51), one could expect separate shells for protons and neutrons. Stable nuclei, in which neutrons formed a closed shell, might then exist with a particularly large variety of proton numbers. Conversely, a closed proton shell might allow for the attachment of neutrons in different numbers.

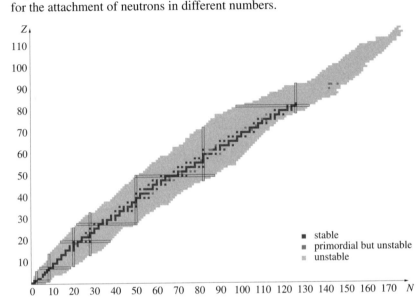

Modern chart of nuclides. The number Z of protons is plotted against the number N of neutrons. Nuclides are classified in three groups. *Stable* nuclides do not decay. *Primordial but unstable* nuclides exist since the formation of the earth but decay. *Unstable* nuclides can only be produced artificially or through the decay of other nuclides. Magic proton numbers are indicated by horizontal, magic neutron numbers by vertical bars.

The extraordinary stability of the α particle, which is the nucleus of the helium isotope $_2^4$He, was soon explained. It contains two protons with opposite spin orientation and, likewise, two neutrons with antiparallel spins. Obviously, the two protons are in two different quantum states as are the two neutrons. One might speak of a closed proton shell in the sense of the Pauli principle, as well as of a closed neutron shell.

The shell structure of atoms had already been apparent to Bohr in 1913. Electrons are far away from each other and from the nucleus. Each can be treated as a single particle in the electric field of the nucleus, which is only little influenced by the presence of the other electrons. A nucleon, i.e., a proton or a neutron, within a nucleus, on the other hand, is very near to the other nucleons and experiences forces from all. Moreover, the form of these forces was unknown. Nevertheless a *single-particle model* was constructed, in which each nucleon experiences a global potential, due to all others, which pulls it towards the centre of the nucleus. The model was worked out in some detail by Elsasser[2]; a detailed study of the empirical material was done by Guggenheimer. Both were refugees from Germany who in 1933 in Paris recognized their common interest in this field.

Guggenheimer [4] found evidence for closed shells with 50 and 82 neutrons. Depending on details of the model, Elsasser could compute two different series

---

[2] Walter Maurice Elsasser (1904–1991)

of numbers corresponding to closed shells [5,6]. Here the matter rested until 1948. The single-particle model went out of fashion with the formulation of Bohr's model of the *compound nucleus* (Episode 55). No progress was made except for an important compilation of the terrestrial and cosmic abundance distribution of nuclides by the geochemist Goldschmidt[3] in Norway which appeared in 1938. The search for closed shells in nuclei was resumed only after the Second World War independently by Goeppert-Mayer in the United States and by Haxel, Jensen, and Suess in Germany.

Maria Goeppert-Mayer[4] grew up in Göttingen as daughter of the paediatrician Friedrich Göppert who represented the sixth generation of a line of university professors in the family. In 1924 she began to study mathematics in Göttingen; but when quantum mechanics was developed at her own university, her interest shifted to physics. In 1930 she obtained her doctor degree for work in theoretical physics done under Born. Göttingen then attracted researchers from all over the world. Among them was the American chemical physicist Edward Mayer[5] her future husband, who worked with Franck. They married in 1930 and went to Baltimore when Edward Mayer became professor at Johns Hopkins University. Maria Goeppert-Mayer also did research there without having a salaried position. That repeated itself when the couple moved to Columbia University in New York. Only after the war they both became professors in Chicago.

Goeppert-Mayer with Fermi

In April 1948, Goeppert-Mayer completed a careful study of the empirical material on nuclides. The abstract of her paper [7] simply reads:

> Experimental facts are summarized to show that nuclei with 20, 50, 82, or 126 neutrons or protons are particularly stable.

Already a year earlier, Suess in Hamburg had published his findings, entitled *On Cosmic Nuclear Frequencies* [8] which he summarized by the identification of distinguished numbers for neutrons (N) and protons (Z) at

$$N:\ 20 \quad - \quad 28\ 50\ (58) - 82$$
$$Z:\ (20)\ 26\ \text{or}\ 28\ 50 \quad - \quad 74\ 82$$

Together with the numbers 2 and 8, known from the α particle and the particularly stable nucleus of the most common oxygen isotope $^{16}_{8}$O, respectively, it thus seemed that the neutron and proton numbers

$$2, 8, 20, 28, 50, 82, 126$$

were very special. They became called *magic numbers* in America (the name seems to be due to Wigner) and *distinguished numbers* (ausgezeichnete Zahlen) in Germany.

Jensen

Suess[6] [9], a Viennese, had studied physical chemistry at the university of his home town, where his father and before him his grandfather were professors of geology. Even before he received his Ph.D. in 1936 he did experiments with heavy water which had recently been discovered. He remained interested in the subject and during the war served as scientific advisor to Norsk Hydro,

---

[3] Victor Moritz Goldschmidt (1888–1947)   [4] Maria Goeppert-Mayer (1906–1972), Nobel Prize 1963   [5] Joseph Edward Mayer (1904–1983)
[6] Hans Eduard Suess (1909–1993)

n	j	$M =$ $2j + 1$	$N_n =$ $\sum M$
1	1/2	2	**2**
2	3/2	4	*6*
	1/2	2	**8**
3	5/2	6	*14*
	3/2	4	
	1/2	2	**20**
4	7/2	8	**28**
	5/2	6	
	3/2	4	
	1/2	2	*40*
5	9/2	10	**50**
	7/2	8	
	5/2	6	
	3/2	4	
	1/2	2	*70*
6	11/2	12	**82**
	9/2	10	
	7/2	8	
	5/2	6	
	3/2	4	
	1/2	2	*112*
7	13/2	14	**126**
	11/2	12	
	⋮	⋮	

Constructing the Magic Numbers

An electron in the atomic hull is characterized by the principal quantum number n, its orbital angular-momentum number ℓ, the magnetic quantum number (or z component of angular momentum) m, and the spin quantum number $s = 1/2$ with the possible z components $m_s = \pm 1/2$. For a given value of n, ℓ can take n values, $\ell = 0, 1, \ldots, n - 1$. For a given value of ℓ, m can take $2\ell + 1$ values, $m = -\ell, -\ell + 1, \ldots, \ell$. Therefore, up to and including the principal quantum number n, there are $N_n = 2(1 + 3 + \cdots + 2n - 1) = 2n^2$ states differing in at least one quantum number. Although he used a different nomenclature and spin was not yet known, this, essentially, was Pauli's explanation of closed atomic shells for $N_n = 2, 8, 18, \ldots$ electrons.

This reasoning remains valid also for protons or neutrons bound by a central force within the nucleus, but only as long as one accepts ℓ and s as separate quantum numbers. For strong spin–orbit coupling they are replaced by a single quantum number $j = \ell + s$ of total angular momentum with $M = 2j + 1$ different z components. One speaks of a multiplicity of M states. From the table one easily sees that there are $N_n = n(n + 1)(n + 2)/3 = 2, 8, 20, 40, 70, \ldots$ states up to and including n.

The energy of each state can be computed for an assumed potential by solving a Schrödinger equation. For large values of j the energy is significantly lowered. A distinct gap in energy appears between states with the highest and those with the second highest value of j. Those with the highest j have energies like states with the next lower value of n and are considered to belong to the next lower shell. With this effect the number N_n is increased by $2n + 2$ and one obtains the sequence $N_n = 6, 14, 28, 50, 82, \ldots$.

The empirical *magic numbers* **2, 8, 20, 28, 50, 82, 126** are three numbers of the first type, followed by four of the second.

which produced heavy water in Norway, then under German occupation. In 1942, he and Jensen visited Goldschmidt in Oslo to discuss his abundance distributions of nuclides. After the war, when experimental work was difficult in Germany, Suess based most of his analysis on them. At that time both Suess and Jensen worked at the University of Hamburg.

Jensen[7] [10] was born in Hamburg, where his father was gardener in the Botanic Garden. He studied in Freiburg and Hamburg. For his doctoral dissertation under Lenz in Hamburg, he used quantum-statistical methods to describe the electron cloud of heavy atoms and their binding in a crystal. It has been pointed out [10] that this approach, which he later extended, was complementary to the single-particle concept used in the shell model of the nucleus. During the war Jensen worked and published on the theory of isotope separation by diffusion. Repeatedly he discussed the distinguished numbers with Suess and also with Haxel.

Haxel[8] had studied in Munich and Tübingen, where he took his Ph.D. under Geiger. In 1936 Geiger accepted a professorship in Berlin and Haxel became Privatdozent in his institute. During the war, he also participated in the German nuclear-energy effort. In 1947 he moved from Berlin to the Max-Planck Institute for Physics in Göttingen. Haxel, at first quite independently,

Goeppert-Mayer and Jensen

---

[7] Johannes Hans Daniel Jensen (1907–1973), Nobel Prize 1963   [8] Otto Haxel (1909–1998)

had worked on distinguished numbers using different nuclear data than Suess. Haxel, the experimental nuclear physicist, and Suess, the physical chemist, discussed their findings and sought for an explanation. As Suess put it much later [11], they agreed 'to publish whatever came into our minds together'.

One day, after consulting textbooks on quantum mechanics, Suess constructed a table similar to the one in our Box. But, of course, the values of the quantum number j of total angular momentum were in ascending order $1/2, 3/2, \ldots$. Then he realized that he would get the higher magic numbers $28, 50, 82, 126$ if he included in a seemingly closed shell of principal quantum number n also the states of highest j belonging to $n + 1$. He showed this play of numbers to Jensen, who went through the data by himself and the next day, according to Suess [11] declared: 'Well, if there is something to it – if the scheme you drew up means something – it would mean that there is a strong spin–orbit coupling.' (See Box.) Soon thereafter, Jensen visited Bohr's institute in Copenhagen for the first time after the war and was surprised that Bohr, the originator of an apparently competing model, was very interested and encouraging. A series of three short papers with rotating author names [12–14] was completed by Haxel, Jensen, and Suess in February and April 1949, showing the table reproduced in our Box, pointing out strong spin–orbit coupling, and drawing various consequences for the properties of nuclei.

Also in February 1949, Maria Goeppert-Mayer submitted a letter [15] to the *Physical Review*. It, too, is based on strong spin–orbit coupling and contains, essentially, the same table. The last sentence reads: "Thanks are due to Enrico Fermi for his remark, 'Is there any indication of spin–orbit coupling?', which was the origin of this paper." After this remark, as she later explained [16], she knew how to proceed and construct the table. With some pride, she remembered that Fermi, who had been sceptical at first, soon taught her theory in his class in nuclear physics.

Although the nuclear shell model in its new form not only explained the magic numbers but also many properties of nuclides, a reason for the strong spin–orbit coupling was not apparent. Therefore Oppenheimer told Jensen: 'Maria and you are trying to explain magic by miracles.' Jensen was fond of this bon mot; he quoted it in his Nobel Lecture [17] and, occasionally, in conversations. Goeppert-Mayer and Jensen first met in 1950. Together they wrote a book on the shell model which appeared in 1955.

After the war, Bothe held the only physics chair at the University of Heidelberg. He succeeded in having a chair in theoretical physics established for Jensen, and when Jensen arrived for the winter semester 1948/49 he welcomed him as the 'first theoretician in Heidelberg since Kirchhoff'. Jensen not only founded an important school of theoretical physics but had great influence on the flourishing of all physics in Heidelberg. A second institute of experimental physics was established for Haxel. Other fields, such as applied and high-energy physics followed. From the beginning Jensen strove and soon succeeded to re-establish scientific contacts with colleagues abroad. On Jensen's suggestion, the university acquired a spacious villa overlooking Heidelberg Castle and situated in a park-like garden (in which he often worked himself) to house Theoretical Physics and the Physics Library. It is there that I met him frequently in his later years.

[1] Johnson, K. E., *American Journal of Physics*, **60** (1992) 1644.

[2] Johnson, K. E., *Physik Journal*, **2 (12)** (2003) 53.

[3] Beck, G., *Zeitschrift für Physik*, **47** (1928) 407.

[4] Guggenheimer, K. F., *Journal de Physique*, **5** (1934) 253.

[5] Elsasser, W. M., *Journal de Physique*, **4** (1933) 549.

[6] Elsasser, W. M., *Journal de Physique*, **5** (1934) 389.

[7] Goeppert Mayer, M., *Physical Review*, **74** (1948) 235.

[8] Suess, H. E., *Zeitschrift für Naturforschung*, **2A** (1947) 311, 604.

[9] Waenke, H. and Arnold, J. R., *Biographical Memoirs (Nat. Acad. Sci.)*, **87** (2005) 1.

[10] Dosch, H. G. and Stech, B., *J. H. D. Jensen – Leben und Werk.* http://www.thphys.uni-heidelberg.de/home/info/historie_dir/jensen_dosch_stech.html.

[11] Suess, H. Discussion remark in [18].

[12] Haxel, O., Jensen, J. H. D., and Suess, H. E., *Naturwissenschaften*, **35** (1949) 376.

[13] Suess, H. E., Haxel, O., and Jensen, J. H. D., *Naturwissenschaften*, **36** (1949) 153.

[14] Jensen, J. H. D., Suess, H. E., and Haxel, O., *Naturwissenschaften*, **36** (1949) 155.

[15] Mayer, M. G., *Physical Review*, **75** (1948) 1969.

[16] Goeppert Mayer, M., *Nobel Lectures, Physics 1963–1970*, p. 20. Elsevier, Amsterdam, 1972.

[17] Jensen, J. H. D., *Nobel Lectures, Physics 1963–1970*, p. 40. Elsevier, Amsterdam, 1972.

[18] Bethe, H. A., *The Happy Thirties*. In Stuewer, R. H. (ed.): *Nuclear Physics in Retrospect*, p. 1067. University of Minnesota Press, Minneapolis, 1977.

71 Quantum Electrodynamics and Feynman Diagrams (1949)

Quantum electrodynamics or, as it is often simply called, QED, is the theory in which the electron (and other charged leptons) as well as the electromagnetic field are consistently described in the framework of relativistic quantum mechanics or, as to use a more appropriate name, of a relativistic quantum field theory. It took nearly a quarter of a century to complete it [1], about the same time it took from the first appearance of the quantum idea in 1900 to the creation of quantum mechanics by Heisenberg, Dirac, and Schrödinger in 1925/26. Once completed, QED became the model *quantum field theory*. Such theories now successfully describe not only the electromagnetic but also the weak and the strong interaction between particles. In all of them *Feynman diagrams*, also called *Feynman graphs*, play a central role in visualization and computation. Feynman and his diagrams will also be in the centre of this episode.

Feynman[1] [2] was one of the most colourful scientists of the twentieth century. His delightfully direct way of speaking is still evident in the famous *Feynman Lectures* and in his autobiographical stories [3], both edited by friends on the basis of tape recordings. Feynman's father had come from Russia to the United States as a boy, attended high school, and later a homoeopathic medical institute but had chosen to work as salesman. Feynman grew up in Far Rockaway just outside New York City on Long Island. Already as a schoolboy he was interested in doing and understanding technical things, such as 'fixing radios by thinking' [3]. In the fall of 1935, he entered the Massachusetts Institute of Technology, initially intending to become a mathematician, then an electrical engineer; but soon he settled for physics. He was a brilliant student and was granted his B.Sc. after three years instead of the regular four. Thereafter he moved to Princeton for graduate studies under Wheeler, who had just become assistant professor there at the age of twenty-six. Feynman's occupation with QED dates back to his undergraduate days.

The existing theory was plagued with infinities. In fact, they appeared even in classical electrodynamics. The electric field of a point charge contains an infinite amount of energy. Lorentz therefore attributed a finite extension, the *classical electron radius*, to the electron. Also, an accelerated charge seems to possess an abnormally high mass since energy was not only needed to overcome inertia but also to power the radiation it emits. This effect could be accounted for, if one assumed that the field of the charge acts back on the charge itself. Planck in 1900 and, in particular, Einstein in 1905 had described the electromagnetic field in a box as a set of quanta; but in non-relativistic quantum mechanics only massive particles, usually the electrons, were considered

Feynman

---
[1] Richard Phillips Feynman (1918–1988), Nobel Prize 1965

quanta, whereas the field was treated as classical. In 1927, Dirac introduced a method he called *second quantization*, which allowed to describe the creation and annihilation, i.e., the emission and absorption of light quanta. With the advent of the positron also the creation and annihilation of electron–positron pairs had to be explained. Now new difficulties were encountered and eventually resolved by *renormalization*, the distinction between the 'bare' charge of the electron and that observed in experiment and, likewise, the distinction between its 'bare' and its observed mass. In a few words this highly technical mathematical procedure is usually explained as follows. In the strong field near a solitary electron 'virtual' electron–positron pairs exist. There is not enough energy for them to be 'real', i.e., to be detected as free particles, but, because of the uncertainty relation, they can exist for a short time before they annihilate again. They change the charge distribution near the electron (the effect is called *vacuum polarization*) and hence the observed charge. Similarly, the problem of the particle mass was altered in second quantization.

It was Feynman's dream to formulate classical electrodynamics in an alternative way in which these difficulties would simply not show up and then turn it into a quantum theory. He resorted to drastic measures. A charge was forbidden to act on itself and the field was abolished and replaced by action at a distance as in Newton's original description of gravitation. The action had to be retarded because of relativity. Although his original theory was still flawed, together with Wheeler he indeed worked out an alternative electrodynamics. When trying to construct his quantum theory, Feynman found that he could not base it on the Hamiltonian formalism of classical mechanics (as Heisenberg and Schrödinger had done) but that he had to start out from the Lagrangean formalism. Fortunately there was a paper by Dirac on the subject. Feynman now developed his *path-integral* formulation of quantum mechanics in which he followed all paths that a particle, described quantum mechanically, could take and superimposed the contributions of all paths. He also used diagrams in a coordinate system, spanned by space and time axes, to symbolize such paths. Such a system, now named *spacetime* had been introduced by Minkowski (Episode 11), who had called it *world*. A trajectory in spacetime is still named *world line*.

As Feynman tells in his Nobel Lecture [4], one day (in the fall of 1940) Wheeler called him by phone and asked, 'Feynman, I know why all electrons have the same charge and the same mass.' 'Why?' 'Because, they are all the same electron!' Wheeler had realized that an arbitrary world line for an electron could be decomposed into sections going forward in time and others going backwards. The latter could be attributed to a positron going forward in time (see Box II). A similar, more detailed, use of world-line diagrams was first published in 1941 by Stückelberg[2], professor at the University of Geneva.

Feynman obtained a Ph.D. with a thesis on his path-integral method and joined the theoretical group working on the atomic bomb at Los Alamos. Bethe who led that group made sure that Cornell, his own university, offered a professorship to Feynman before another one did. Only after the war, at Cornell, did Feynman return to his theory. In 1948 he published it as a complete, alternative

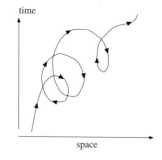

Sketch of Wheeler's world line for a single electron.

---

[2] Baron Ernst Carl Gerlach Stückelberg von Breidenstein zu Breidenbach und Melsbach (1905–1984)

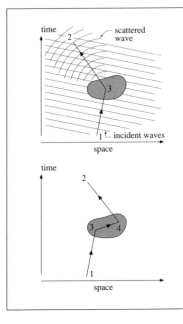

Box I. Feynman Diagrams for the Schrödinger Equation

Feynman introduces diagrams to symbolize formulae. His simplest case is the scattering of an electron by an electromagnetic field existing in the shaded scattering region. The incident electron starts at point 1 in spacetime (space coordinate $\mathbf{x}_1$, time coordinate t_1) very far from the scattering region and can therefore be described by a plane wave. Scattering is thought to take place at point 3, giving rise to a scattered spherical wave. Of interest is the probability (in technical terms: the scattering cross section) to observe the particle at point 2. This is computed by adding up the amplitudes for scattering at all possible points 3. The result obtained is said to be of *first order*.

A computation done in *second order* also includes the sum over all amplitudes for which scattering occurs successively at two points 3 and 4, provided that $t_4 > t_3$.

We have sketched here the contributions in first and second order to a *perturbation series* which, in principle, can be extended to any order. The contribution of every order contains a power of Sommerfeld's fine-structure constant $\alpha \approx 1/137$; the power is rising with the order. Therefore results in low order usually are quite accurate.

way to formulate non-relativistic quantum mechanics [5].

Already in the previous year, at the beginning of June 1947, an exclusive physics conference with only 23 participants had been organized on Shelter Island close to the East of Long Island by the National Academy of Sciences. The conference was dominated by theoreticians. We mention only Bethe from Cornell, Oppenheimer from the Institute for Advanced Studies, and Weisskopf[3] from MIT and, of the young generation, Schwinger from Harvard and Feynman. There were a few experimentalists, among them Lamb, who reported on his and Retherford's still unpublished observation of the *Lamb shift* (Episode 67). Lamb's talk was the highlight of the meeting because he presented unambiguous and quantitative deviations between experiment and Dirac's treatment of the hydrogen atom (Episode 43), which took into account relativity but not second quantization.

Theoretical physicists now had a checkpoint for testing improved theories. The first to take advantage of it was Bethe, who performed a rough but ingenious calculation of the Lamb shift right after the meeting on a train ride from New York City to Schenectady in upstate New York. On returning to Cornell, Bethe gave a lecture on this work, in which he made it clear that a more elegant way of dealing with infinities than his was needed. Feynman told him, after the lecture [4]: 'I can do that for you, I'll bring it in for you tomorrow.' Although he did not have the correct answer the next day, Feynman used the insight and experience he had gained from his alternative ways of doing quantum mechanics and electrodynamics and in using diagrams to develop a set of rules for calculations in QED. He did this, as he explains, mainly by guessing the answers (a way open probably to Feynman only) and checking them against

---

[3] Victor Frederick Weisskopf (1908–2002)

**Box II. Feynman Diagrams for the Dirac Equation.
Positrons Depicted as Electrons Moving Backwards in Time**

In Dirac's formalism an electron is described by a *spinor* with four components which takes into account the two different orientations of the electron spin as well as the possibility of positive- and negative-energy states. Dirac looked at states of negative energy as unoccupied states (positively charged 'holes') in a 'sea' of infinitely many states of negative energy. In the formulae these states are described as if they were ordinary electrons of positive energy travelling backwards in time. That is seen best by looking at the current density of a charged particle, a *four-vector* in the sense of special relativity, which is proportional to the electric charge of the particle times its four-momentum with the components $E/c, p_x, p_y, p_z$, i.e., the energy (divided by the speed of light) and the components of momentum. The current density remains unchanged if both the particle charge and the four-momentum change their sign. A negative energy can be turned positive by inverting the signs of charge and momentum, i.e., of charge and direction of motion.

The Dirac equation for the first-order scattering of an electron by a field has solutions of positive energy, symbolized by the trajectory $1 \to 3 \to 2$. But it has also solutions of negative energy which are turned by the argument above into ones of positive energy and symbolized by $1 \to 3 \to 2'$. Of course, there is no such thing as a particle moving backwards in time. Since $t_{2'} < t_3$ a particle has to travel from $2'$ to 3. If the particle described by $1 \to 3$ is an electron, that travelling the path $2' \to 3$ is a positron. While the graph $1 \to 3 \to 2$ symbolizes the scattering of an electron at 3, the graph $1 \to 3 \to 2'$ corresponds to the annihilation of an electron and a positron into energy of the electromagnetic field at 3.

In second order there are contributions from scattering at points 3 and 4 but we can now have both $t_4 > t_3$ and $t_4 < t_3$. In the first case the three pieces of the graph all correspond to the electron. In the second case the line $4 \to 3$ is a positron line and the graph should be interpreted as follows: Still while the original electron travels on its path from 1 to 3 an electron–positron pair is created at 4; the positron annihilates with the original electron at 3 and the electron of the pair continues to 2. For the computation both graphs can be taken as one single electron line. Feynman, in his original paper [6], writes:

> Following the charge rather than the particles corresponds to considering this world line as a whole rather than breaking it up into pieces. It is as though a bombardier flying low over a road suddenly sees three roads and it is only when two of them come together and disappear again that he realizes that he has simply passed over a long switchback in a single road.

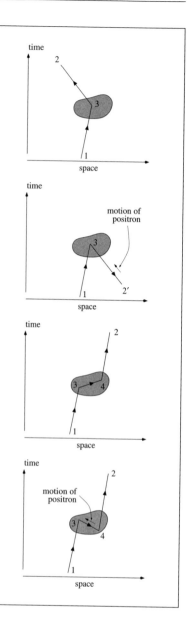

what was already known. Results in second quantization had been obtained in the 1930s for some processes such as the elastic scattering of two electrons by Møller[4] and that of a positron by an electron by Bhabha[5].

Feynman devised diagrams which, at first sight, seem to be trajectories of particles depicting a process in spacetime but which stand for formulae. In fact, each graphical component of a diagram stands for a factor. Their product is a complex quantity called the matrix element or amplitude of the diagram. Its absolute square, apart from simple factors, is the cross section describing

---

[4] Christian Møller (1904–1980) [5] Homi Jehangir Bhabha (1909–1966)

Box III. Feynman Diagrams for QED Processes

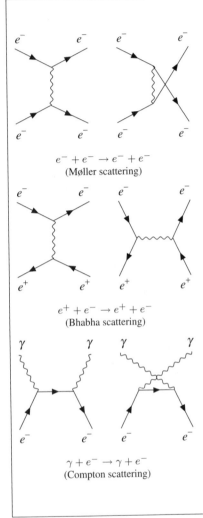

$e^- + e^- \rightarrow e^- + e^-$
(Møller scattering)

$e^+ + e^- \rightarrow e^+ + e^-$
(Bhabha scattering)

$\gamma + e^- \rightarrow \gamma + e^-$
(Compton scattering)

Because in QED Feynman diagrams represent formulae rather than world lines in spacetime, their orientation is no longer important and they are usually presented to be read like ordinary text: particles come in from the left, react, and exit to the right. Each component of a diagram stands for a factor in the amplitude or matrix element of a process. There are three types of components:

- An *outer line* depends on the type of incoming or outgoing particle and on its four-momentum.

- A *vertex factor* depends on the type of particles whose lines meet in one point and on the nature of the coupling between them. In the diagrams for QED, it is proportional to the elementary charge e.

- A *propagator* describes the exchange of a particle and is symbolized by an inner line. It depends on the nature of this particle and its four-momentum.

Conservation of energy and momentum, i.e., conservation of four-momentum, holds for the reaction as a whole and also at each vertex. The mass squared of an exchanged particle computed from the energy E and the momentum $\mathbf{p}$ of a propagator, $M^2 = E^2/c^4 - \mathbf{p}^2/c^2$, in general, is not equal to the square m^2 of its rest mass (c, as usual, is the speed of light). The exchange particle is said to be *off mass shell*. That is allowed by the uncertainty relation for the short time the particle exists in that state. The propagator is the smaller, the farther off shell the particle is. A photon propagator ($m^2 = 0$) is said to be *time-like* for $M^2 > 0$ and *space-like* for $M^2 < 0$.

All diagrams, i.e., amplitudes have to be added before the absolute square is taken, which, up to simple factors, is the cross section for the process. Diagrams are shown here describing in lowest order three different scattering processes.

Møller scattering: If the initial electrons are A and B and there are two detectors D and F, one cannot distinguish the case $A \rightarrow D, B \rightarrow F$ from $A \rightarrow F, B \rightarrow D$. Therefore two diagrams contribute, both with a space-like photon propagator.

Bhabha scattering: Experimentally indistinguishable is the case of 'simple' scattering with the exchange of a space-like photon (left) from the annihilation of the initial electron–positron pair into a time-like photon from which a new pair is created (right).

Compton scattering: There are two diagrams with an electron propagator corresponding to the absorption of the initial-state photon before or after the final-state one is emitted.

the probability for the process to occur. Before taking the absolute square, the amplitudes of all diagrams, having the same initial and the same final states, have to be added. The number of vertices in a diagram defines its *order*. In general an infinity of diagrams of different order can contribute to a process. But, since the product of two vertex factors is proportional to Sommerfeld's fine-structure constant $\alpha \approx 1/137$, higher orders are quickly suppressed. Results take the form of a sum with contributions of different orders, called a *perturbation series*, which can be cut off at some order depending on the accuracy required. In many cases results in lowest order are satisfactory. We do not present the *Feynman rules* explicitly but show (see Box III) the Feynman diagrams of lowest order for three simple reactions in QED.

Schwinger who had also attended the Shelter Island Conference, in contrast to Feynman, set himself the goal to formulate QED by starting from first principles. Schwinger[6] [7] was born and grew up in New York City. His parents had come to the United States as children from different parts of Eastern Europe and his father had become a well-to-do clothes designer. Schwinger's unusual gifts soon became apparent and Rabi saw to it that he could study at Columbia University very early. He published his first research paper when he was only sixteen and obtained his B.Sc. in 1936. Schwinger worked with Teller on neutron scattering in 1937 and soon completed a doctoral thesis on that subject but worked on various topics before taking his Ph.D. in 1939. After a two years' stay at Berkeley in Oppenheimer's group, he became instructor in 1941 and assistant professor in 1942 at Purdue University. In 1941 he was invited to lecture at the prestigious Summer School of Theoretical Physics at the University of Michigan. It is there that he first met Weisskopf who had done pioneering work on renormalization. As part of the war effort he did theoretical work on radar at MIT but also on nuclear physics. In these years Schwinger was a confirmed night worker, arriving at his office after dinner (which was his breakfast) and staying until dawn. Although this habit led to raised eyebrows, he was appointed associate professor at Harvard in 1945 and full professor two years later.

Schwinger

At Harvard Schwinger assembled a group of extraordinary graduate students and held close contact with Weisskopf's group at MIT. Together with Weisskopf he travelled by train to New York to proceed from there to Shelter Island. Both knew already about the Lamb shift and agreed that a calculation similar to that actually done a few days later by Bethe in his train ride was possible. Schwinger began his intensive work on QED in the fall of 1947. We cannot possibly attempt to describe his work in detail and only point out that he made it his principle to exploit symmetry properties as much as possible: He insisted that *Lorentz covariance* and *gauge invariance* should be apparent (manifest, as he called it) throughout his calculations. (Lorentz covariance is the way physical quantities transform under Lorentz transformation in special relativity. The word *gauge* denotes the freedom one has in modifying the mathematical expression for the electromagnetic field without changing its physical effects. For instance, a constant value can be added at every point in space to a static potential. The importance of the gauge concept in quantum mechanics was realized in the late 1920s.)

Tomonaga

Both Schwinger and Feynman presented their work at a conference at Pocono Manor in Pennsylvania which, after ten months, followed the one at Shelter Island. Whereas Schwinger argued in a mathematical language familiar to the theorists present, among whom were Bohr, Dirac, and Oppenheimer, Feynman with his 'intuitive' approach and his diagrams had severe difficulties to make himself understood. In personal talks, however, the two compared notes, exchanged techniques, and became convinced that they both were right.

On his return to Princeton from the Pocono Conference, Oppenheimer found a letter from Tomonaga[7], drawing his attention to research on QED Tomonaga had done with his group in isolation from international science during and immediately after the war in Japan. Tomonaga's work was not unlike Schwinger's but preceded it. Oppenheimer sent a copy of that letter to the par-

Weisskopf (left) and Dyson in 1952 on a boat to Copenhagen.

---

[6] Julian Schwinger (1918–1994), Nobel Prize 1965   [7] Sin-Itiro Tomonaga (1906–1979), Nobel Prize 1965

[1] Schweber, S. S., *QED and the Men Who Made It: Dyson, Feynman, Schwinger, and Tomonaga*. Princeton University Press, Princeton, 1994.

[2] Mehra, J., *The Beat of a Different Drum – The Life and Science of Richard Feynman*. Clarendon Press, Oxford, 1994.

[3] Feynman, R. P., *"Surely You're Joking, Mr. Feynman!" – Adventures of a Curious Character*. Norton, New York, 1985.

[4] Feynman, R. P., *Nobel Lectures, Physics 1963–1970*. Elsevier, Amsterdam, 1972.

[5] Feynman, R. P., *Reviews of Modern Physics*, **20** (1948) 367.

[6] Feynman, R. P., *Physical Review*, **76** (1949) 749.

[7] Mehra, J. and Milton, K. A., *Climbing the Mountain – The Scientific Biography of Julian Schwinger*. University Press, Oxford, 2003.

[8] Tomonaga, S., *Physical Review*, **74** (1948) 224.

[9] Schwinger, J., *Physical Review*, **74** (1948) 1439.

[10] Schwinger, J., *Physical Review*, **75** (1949) 651.

[11] Schwinger, J., *Physical Review*, **76** (1949) 790.

[12] Dyson, F., *Disturbing the Universe*. Harper and Row, New York, 1979.

[13] Dyson, F., *Physical Review*, **75** (1949) 486.

[14] Dyson, F., *Physical Review*, **75** (1949) 1736.

[15] Feynman, R. P., *Physical Review*, **76** (1949) 769.

[16] Schwinger, J. In Brown, L. M. and Hoddeson, L. (eds.): *The Birth of Particle Physics*, p. 329. University Press, Cambridge, 1983.

[17] 't Hooft, G. and Veltmann, M. J. G. CERN 79-9, Geneva, 1973.

ticipants of the Pocono meeting and urged Tomonaga by telegram to write 'a summary of present state and views for prompt publication [in the] *Physical Review*'. That summary [8] soon appeared. It contains references to earlier papers in the English-language journal *Progress in Theoretical Physics* published in Japan and is accompanied by a note by Oppenheimer.

After the Pocono Conference Schwinger began to write three extensive papers, all entitled *Quantum Electrodynamics* [9–11]. In July/August of 1948 he lectured again at the Michigan Summer School in Ann Arbor. The lectures were based in part on his first two as yet unpublished papers. In the audience was Dyson[8] [12], who had gone to Winchester, one of England's renowned Public Schools, graduated from Cambridge and was at the time working towards a Ph.D. at Cornell, a degree he would later pride himself to never have taken. Dyson, a brilliant mathematician already at Winchester, was deeply interested in QED and had regularly served as discussion partner to Feynman in Cornell. Feynman, who was to spend the summer at Los Alamos, had taken Dyson along in his car on the latter's way to Michigan and they had talked physics most of the time [12]. At Ann Arbor, Dyson had the chance to learn all about Schwinger's approach 'straight from the horse's mouth'.

Dyson, who at the time was probably the only person really familiar with both Schwinger's and Feynman's theories, successfully constructed the missing bridge between the two and showed that the latter can be derived from the former. He wrote a paper, entitled *The Radiation Theories of Tomonaga, Schwinger, and Feynman* [13]. That was a somewhat delicate undertaking since Dyson's paper appeared when Feynman still had not published his method and only the first of Schwinger's papers was printed. His second paper on the subject [14] is considered to contain the definite treatment of renormalization in QED.

Feynman's two papers on QED were completed in April and May 1949. In the first one, *The Theory of Positrons* [6], he carefully explains the meaning of his diagrams beginning with their application to the Schrödinger equation (Box I). Application to the Dirac equation (Box II) yields an interpretation of the positron in which Dirac's original hole theory is no longer needed. The second paper, *Space-Time Approach to Quantum Electrodynamics* [15], contains the Feynman rules and explains their usage. By these rules, computations for specific problems are simplified so much that Schwinger, much later, said [16]:

> Like the silicon chip of more recent years, the Feynman diagram was bringing computation to the masses.

Originally, Schwinger's presentation of QED was considered as more fundamental. But one can also take the Feynman rules to define the theory. Indeed, a quarter of a century after its creation we even find the following statement by leading theorists [17]:

> Few physicists object nowadays to the idea that diagrams contain more truth than the underlying formalism …

---

[8] Freeman John Dyson (born 1923)

Glaser's Bubble Chamber (1953)

For at least a decade, the late 1950s and all through the 1960s, most discoveries in particle physics were made with a new kind of detector, the *bubble chamber*, exposed to beams of particles produced by accelerators. The bubble chamber was invented by Donald Glaser who was searching for a track detector adapted to the needs of high-energy particle physics and, only a little later, forged by Luis Alvarez and his group into an indispensable tool.

Glaser[1] was born in Cleveland, Ohio and obtained a B.Sc. degree in physics and mathematics from the Case Institute of Technology there in 1946. For graduate studies he moved to the California Institute of Technology. For his thesis, supervised by Anderson, he used a system of two cloud chambers and a magnet and measured the momentum distribution of charged cosmic-ray particles. In 1949 he accepted a position in teaching and research at the University of Michigan in Ann Arbor, becoming a professor in 1957.

Glaser valued cloud chambers in which a charged particle leaves a clear track that can be measured with high precision. The discovery of *V particles* in 1947 (Episode 68) had underlined its importance for detecting events with a complicated structure. But Glaser also realized that the small density of the cloud-chamber filling, a gas–vapour mixture, was a drawback. Interactions of high-energy particles are rarely observed in the gas. To compensate this, material of high density, e.g., copper or lead plates, was placed outside or within a cloud chamber. Reactions could then take place in this material, but particle tracks were only visible outside it. Photographic emulsions offered a way out but were by no means ideal. It was difficult to fill a large volume, say several litres, with emulsion. Moreover, in contrast to the cloud chamber, emulsion is continuously sensitive. That is a disadvantage in the intense particle beams of accelerators.

Because of all this, Glaser undertook a systematic search for possible high-density materials that could exist in a metastable state such that the small amount of energy, deposited by a charged particle ionizing the material, would suffice to make it locally return to the stable state. The supersaturated vapour in the cloud chamber is such a material. A few ions initiate the formation of a droplet, the energy needed for its formation is already present in the vapour. After considering several other possibilities, Glaser opted for the complement of the supersaturated vapour, namely the superheated liquid. Suppose a liquid is kept in a closed vessel at a temperature above its boiling point at atmospheric pressure. Of course, the pressure is then higher than atmospheric. If the pressure is suddenly reduced, the liquid is in the superheated state and, sooner or later, will start boiling.

What Glaser hoped for, was that bubbles would first form along the path

Glaser

---

[1] Donald Arthur Glaser (born 1926), Nobel Prize 1960

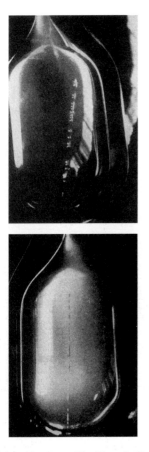

Particle tracks observed by Glaser in his first small, ether-filled chamber. Figure reprinted with permission from [3]. Copyright 1953 by the American Physical Society.

of an ionizing particle if one happened to traverse the liquid. Bubble formation, in which part of the liquid turns into vapour, is a case of *phase transition*. So Glaser studied the theory of that process. He decided that diethyl ether, the substance used for narcosis in former times, might have the right properties. At atmospheric pressure it boils at $34.6°C$. Glaser estimated that at about $140°C$ it could be sensitive to minimum-ionizing particles. Before himself experimenting with ether he looked up the literature and found a paper by physical chemists [1] who, already in 1924, had studied several superheated liquids in closed small glass vessels. The authors stressed the 'capricious' behaviour of ether at $130.5°C$: The time between the pressure drop, i.e., the setting-in of superheating, and the moment of boiling varied erratically between 7 seconds and more than two minutes; a list of about 25 such time intervals was presented. Glaser realized that these time intervals followed what is known as the *Poisson distribution* in mathematical statistics. That showed that boiling began after random and independent time intervals. Only the average time was well defined. (Rutherford and Geiger had used the same method in 1910 to conclude that individual α decays occur independently of each other.) Moreover, Glaser was able to estimate that the average time interval, 60 seconds, was about equal to the time in which the first cosmic-ray particle would have passed through the superheated ether. He repeated the experiments with superheated ether in small glass vessels and demonstrated that boiling indeed was started by cosmic rays and also by radiation from a radioactive source. These findings were published in a short note in 1952 [2].

What remained to be done, was to produce and photograph bubble tracks along a particle's path. To this end Glaser used a small glass chamber containing $3\,cm^3$ of ether, fitted it with an expansion mechanism, and mounted Geiger counters above and below it. If, after expansion, both counters in coincidence signalled the passage of a particle, a few microseconds later a flashlight was triggered. Pictures of the chamber taken in the light of that flash indeed showed strings of bubbles forming particle tracks. In April 1953, Glaser presented a summary of his results illustrated by photographs at a meeting of the American Physical Society in Washington and, a few weeks later, sent it for publication [3].

The importance of Glaser's invention was realized immediately. Other groups began developing their own chambers. Glaser himself, of course, continued his efforts. A comprehensive report of 1955 [4] contains details of bubble growth and presents pictures of a 6-inch chamber, filled with isopentane and exposed to a proton beam from the Brookhaven Cosmotron. Together with two colleagues he developed a chamber filled with the liquefied noble gas xenon [5], whose nuclei have the high charge number $Z = 54$. Therefore, there is a high probability that a photon is converted into an electron–positron pair in the chamber and can thus be registered. Glaser also participated in a number of early bubble-chamber experiments performed at Brookhaven. In 1960, only seven years after seeing the first bubble tracks, Glaser received the Nobel Prize. As is the custom, after the ceremonies in Stockholm, he was invited for a number of talks in festive surroundings in Europe. One such occasion was in Düsseldorf. Wolfgang Paul from Bonn, who was to give the opening address, took me along because for my diploma thesis under him I had built a 1-litre isopentane bubble chamber and the electronics to run it at the Bonn 500-MeV electron synchrotron. In this way I came to shake Glaser's hand and that of the young lady he had just married.

Glaser and his young wife in Düsseldorf with Leo Brandt, my father.

Glaser accepted a professorship at the University of California in Berkeley in 1959. In 1962, he changed his field of research from physics to molecular biology in which he was deeply interested since his graduate studies.

Let us now turn to the work of Alvarez' group. We already met Alvarez[2] [6] in Episode 65, measuring the neutron magnetic moment. He had earned his Ph.D. under Compton and had become a member of Lawrence's cyclotron group in the 1930s. During the war he worked on both radar and the atomic bomb and after the war he invented and built the first *Alvarez-type* linear accelerator for protons. He headed a large research group at Berkeley and was not only an excellent physicist but also a gifted and experienced organizer. Alvarez tells in his Nobel Lecture [7] how he learned about the first bubble chamber. During the 1953 Physical Society Meeting in Washington, which was already mentioned, he found himself together with other physicists at a table in a hotel garden.

> A young chap ... was seated at my left and we were soon talking of our interests in physics. He expressed concern that no one would hear his 10-min contributed paper, because it was scheduled as the final paper of the Saturday afternoon session, and therefore the last talk to be presented at the meeting. ... I admitted that I wouldn't be there, and asked him to tell me what he would be reporting. And that is how I heard first hand from Donald Glaser how he had invented the bubble chamber, and to what state he had brought its development. ... He showed me photographs of bubble tracks in a small glass bulb, about 1 centimeter in diameter and 2 centimeters long, filled with diethyl ether. ... I was greatly impressed by his work, and it immediately occurred to me that this could be the "big idea" I felt was needed in particle physics.

The same night, in his Washington hotel room, Alvarez discussed the strategy he had for using the bubble chamber with a colleague from Berkeley and told him that he hoped they 'would get started on the development of a liquid-hydrogen chamber, much larger than anything Don Glaser was thinking about'.

Alvarez

---

[2] Luis Walter Alvarez (1911–1988), Nobel Prize 1968

Liquid hydrogen is an ideal chamber filling, since the partners for strong inter-actions of incoming particles are then known to be protons, the nuclei of hydro-gen atoms. The chamber should be designed for use at the Bevatron, then being completed in Berkeley, which would be the most powerful accelerator in the world. The chamber had to be large enough to allow a good fraction of incom-ing particles to react in the first third of the chamber and the reaction products to be detected. In particular, the production and decay of *strange particles* was to be studied. Therefore the chamber had to be long enough to allow these par-ticles to decay and their decay products to leave well-measurable tracks. The chamber had to be embedded in a strong magnetic field to allow momentum measurement from track curvature. Moreover, methods had to be developed to perform measurements on a very large number of bubble-chamber pictures.

Alvarez, with his well-equipped and superbly staffed group at UCRL, the *University of California Radiation Laboratory* (now named the *Lawrence Berke-ley National Laboratory*, LBNL) in Berkeley, achieved all this in just six years. The first two were used for preliminary studies. It should be noted that liquid hydrogen is by no means easy to handle. Hydrogen, if mixed with oxygen or air, is highly explosive. It becomes liquid only at very low temperatures. In a bubble chamber it is normally under overpressure at a temperature of about 27 K, i.e., $-246°C$. Amounts of liquid hydrogen such as those Alvarez was thinking about had not been handled before. Within less than a year the Berke-ley group photographed the first tracks in a 1.5-inch hydrogen bubble cham-ber [8]. (That overheated hydrogen was radiation sensitive had been demon-strated earlier by a group in Chicago [9].)

Glaser's early chambers were smooth all-glass vessels. That assured that bubbles, likely to form at edges or other irregularities, would occur only along particle tracks. It was now recognized that for use with accelerators chambers of a more robust construction, made of metal with ports closed by glass win-dows could be used. An accelerator yields particles at a well-determined time. If the chamber is expanded just before and a flash is triggered just after that time, a few extra bubbles, formed at gasket edges or the like, do not matter. With this in mind, in rapid succession, bubble chambers with ever-increasing radius, ranging from 2.5 to 15 inches, were built, including the first chamber placed in a magnetic field, and several of these chambers were used for long physics runs.

In April 1955, two years after meeting Glaser, Alvarez was able to specify in a detailed document, *The Bubble Chamber Program at UCRL*, what had been in his mind all along. His chamber was to have a sensitive volume of oblong shape, 72 inches long, 20 inches wide, and 15 inches deep. It was to be illuminated and photographed through a single glass window able to bear a force of 100 tons. It was to be placed in a magnetic field of 1.5 Tesla. The magnet would weigh about 100 tons and would consume 2 to 3 megawatts of electric power. Semiautomatic machines would have to be developed for the measurement of the pictures taken and computers would be required and would have to be programmed for the analysis of the data so obtained. The bottom line: 2.5 million dollars were needed.

Alvarez writes how Lawrence, the inventor of the cyclotron and founder of UCRL, helped to obtain that money from the Atomic Energy Commission [7]:

Alvarez with various bubble chambers built in his group. In front the glass window of the 72-inch chamber.

... we talked in one day to three of the five Commissioners: Lewis Strauss, Willard Libby [later a Nobel Laureate] and the late John Von Neumann, the greatest mathematical physicist then living. That evening, at a cocktail party at Johnny Von Neumann's home, I was told that the Commission had voted that afternoon to give the laboratory the 2.5 million dollars we had requested. All we had to do now was build the thing and make it work!

The Berkeley 72-inch hydrogen bubble chamber. The actual chamber is situated in its magnet between the two platforms. Photographs are taken from above. Photo: Lawrence Berkeley National Laboratory.

That was achieved within less than four years. The 72-inch hydrogen bubble chamber began to operate in March 1959. Using the chamber at the Bevatron, the Alvarez group reaped a rich harvest in particle physics. Here we only name two examples, which are discussed later together with theoretical ideas: the confirmation of the *strangeness* concept (Episode 74) and the discovery of new short-lived particles, which was important for the proposal of the *quark hypothesis* (Episode 87).

Particles from the Bevatron interacting in the 72-inch chamber. Photo: Lawrence Berkeley National Laboratory.

[1] Kenrick, F. B., Gilbert, C. S., and Wismer, K. L., *Journal of Physical Chemistry*, **28** (1924) 1297.

[2] Glaser, D. A., *Physical Review*, **87** (1952) 665.

[3] Glaser, D. A., *Physical Review*, **91** (1953) 762.

[4] Glaser, D. A. and Rahm, D. C., *Physical Review*, **97** (1955) 474.

[5] Brown, J. L., Glaser, D. A., and Perl, M. L., *Physical Review*, **102** (1956) 586.

[6] Alvarez, L. W., *Alvarez: Adventures of a Physicist*. Basic Books, New York, 1987.

[7] Alvarez, L. W., *Nobel Lectures, Physics 1963–1970*, p. 241. Elsevier, Amsterdam, 1972.

[8] Wood, J. G., *Physical Review*, **94** (1954) 731.

[9] Hildebrand, R. H. and Nagle, D. E., *Physical Review*, **92** (1953) 517.

It remains to be told why the bubble-chamber era came to an end in the 1970s. We give only two of the reasons: The bubble chamber had one important drawback and electronic detectors were improved considerably. Concerning the drawback, the lifetime of the *latent image* of a bubble chamber is much shorter than that of a cloud chamber. Therefore, if the presence of an interesting event is signalled by counters and the chamber is then expanded, the image is gone. A bubble chamber cannot be triggered by counters. That is unimportant for frequently occurring events but makes the chamber of little use for very rare events.

Concerning electronic detectors I want to relate what I heard Kowarski say at CERN in the early 1960s. At that time Kowarski, who had worked with Joliot in early fission experiments (Episode 60), headed the Data Division and supervised the development of a machine designed to automatically measure bubble-chamber pictures and transfer the data to an electronic computer. He said that the initial process in the bubble chamber was electric, the formation of ions along the path of a particle. That was followed by the thermodynamic process of bubble formation, the optics of photography, the chemistry of development, the electromechanics and optics of measurement, the encoding of data on punched cards and, finally, the entry of these data into a computer, where they were dealt with in electric processes. He expected that eventually, also for complicated events with many tracks that at the time could be resolved only in the bubble chamber, detector systems would be available that used electric signals throughout, from the initial ionization to the final computing. Because of the advances in detector and computer technology that has indeed happened.

The Maser (1954)

In one of his very influential papers [1] Einstein identified three fundamental types of interaction between matter and light, *absorption*, *spontaneous emission*, and *stimulated emission*: By absorbing a light quantum a molecule (or atom) performs a transition to a state of higher energy; it can spontaneously, i.e., without being prompted to do so, return to a state of lower energy by emitting a quantum; finally, it can also emit a quantum and transit to a lower energy, if stimulated by another light quantum. In the latter case the stimulating and the emitted quantum have the same wavelength and also the same phase. It is this quality which made the *maser* and, later, the *laser* (Episode 82) instruments of outstanding qualities.

Stimulated emission is not observed in ordinary matter in thermal equilibrium. It is hidden under the absorption because states of lower energy are more frequently populated than those of higher one. The more frequent states absorb quanta, the less frequent ones may undergo stimulated emission, overall an absorption results (see Box). The special case of *population inversion* was demonstrated for the first time by Purcell and Pound [2] in 1950 in the context of nuclear magnetic resonance. We remember from Episode 65 that in a constant magnetic field, aligned parallel to the z direction, two states develop, for which the z component of the magnetic moment is negative or positive, the energy being smaller in the first case. The state with the smaller energy is the more frequent one. If such a system is prepared and the direction of the field suddenly reversed, the state with the higher energy becomes the one which is more frequent. Purcell and Pound performed a magnetic-resonance experiment with a sample of lithium fluoride, having a particularly long relaxation time, which showed the typical resonance absorption. After the field reversal, emission was observed instead. While relaxation within the sample took place, emission decreased and absorption gradually reappeared. To describe the state of population inversion, Purcell and Pound coined the term *negative temperature*, because, only if the temperature is negative, such a state can be described by a formula of thermodynamics. The term is suggestive but also misleading, because temperature is, first of all, a measure of the average energy of the constituents of a sample, which, also for population inversion, is by no means negative.

The idea to use population inversion as operating principle of an amplifier (with the stimulating radiation as input signal and the sum of stimulating and stimulated radiation as output) appeared in print in three independent publications. In 1952 Weber[1] presented it at an engineering conference and published it in the conference proceedings [3]. The paper did not attract the attention it deserved, probably because it was not published in a widely read physics jour-

---

[1] Joseph Weber (1919–2000)

Absorption vs. Stimulated Emission

Let us consider one type of molecules and, in particular, two states with energies E_1 and E_2 (E_1 being smaller than E_2). These states are occupied by n_1 and n_2 molecules, respectively. Light quanta of the right energy $E = E_2 - E_1$ can provoke absorption ($E_1 \rightarrow E_2$) or stimulate emission ($E_2 \rightarrow E_1$). The probabilities for the two processes are proportional to the *populations* n_1 and n_2, respectively. If the molecules are in thermal equilibrium at (absolute) temperature T, then the ratio of the two populations is

$$\frac{n_2}{n_1} = \exp\left(-\frac{E}{kT}\right) = \exp\left(-\frac{E_2 - E_1}{kT}\right) \quad , \tag{1}$$

where k is Boltzmann's constant. Since n_2 is smaller than n_1, absorption exceeds stimulated emission: Absorption is the only net effect observed.

Stimulated emission can be made to dominate if, by particular experimental techniques, a *population inversion*, i.e., n_2 larger than n_1, can be achieved. Such a situation can formally still be described by (1), however, only with a negative value of the temperature T.

Townes

nal. Weber is now best remembered for his pioneering work in constructing detectors for the still unobserved gravitational waves, predicted by the general theory of relativity. In a letter to *Physical Review* of May 1954 [4], a small group from Columbia University in New York, led by Townes, not only published the idea but also described an apparatus based on that idea and presented first results obtained with it.

Townes[2] [5] was born in Greenville, South Carolina. At the tender age of 19 he became both a Bachelor of Science in Physics and a Bachelor of Arts in Modern Languages at Furman University in Greenville. After that he continued to pursue his interest in both languages and science and obtained a Master of Arts degree from Purdue University in 1936 and a Ph.D. in Physics from the California Institute of Technology in 1939. He joined the Bell Telephone Laboratories, where he worked on radar during the war. Like Purcell he was intrigued by the absorption of microwaves by water vapour. In 1947, following a suggestion by Rabi, he was appointed to the faculty of Columbia University, concentrating his work on the microwave spectroscopy of molecules. During a visit to Washington in 1951 to attend a meeting on millimetre waves, it occurred to Townes that population inversion could be reached with molecular-beam techniques. He did not mention this idea at the meeting but, at Columbia, set out to realize it together with Gordon[3], a new graduate student, and Zeiger[4], a postdoctoral scientist on a scholarship from industry.

After unsuccessful trials with millimetre waves, they decided to use 1.5-cm waves because the technology was available from radar research. Ammonia (NH_3) was chosen as a suitable molecule with a transition between two energy states in the region of that wavelength. It had been Townes' idea to reach population inversion by producing the two states in a molecular beam, separating molecules in different states as in the Stern–Gerlach experiment and, finally,

---

[2] Charles Hard Townes (born 1915), Nobel Prize 1964   [3] James P. Gordon (born 1928)   [4] Herbert J. Zeiger (born 1925)

removing most molecules in the lower state. Rather than by an inhomogeneous magnetic field, as in the classic experiments by Stern or Rabi, this was achieved by an inhomogeneous electric field, because molecules in the two states, in an electric field, develop electric dipole moments of opposite sign. Following ideas developed by Paul in Göttingen together with Friedburg and later with Bennewitz in Bonn [6,7], an electric quadrupole lens was used. (Paul was a master of particle lensing which, eventually, earned him a Nobel Prize, Episode 80.) The quadrupole lens consists of four metallic rods, surrounding the beam axis and connected alternately to two different voltages. With this lens or 'focusser', as it was called by Townes' group, molecules in the upper energy state are focussed, i.e., concentrated on the beam axis. Those in the lower state are defocussed, i.e., removed from the beam.

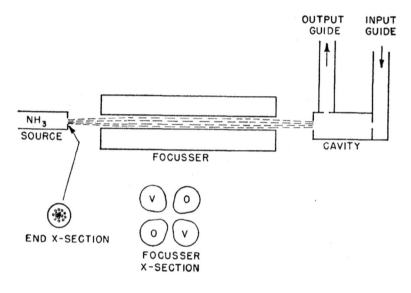

Principle of operation of the first maser. Ammonia molecules in two slightly different energy states pass a focusser which enriches those in the higher state. They enter a cavity with an input of stimulating radiation. The output is a sum of stimulating and stimulated radiation. Figure reprinted with permission from [4]. Copyright 1954 by the American Physical Society.

The apparatus had three essential parts, all enclosed in a vacuum vessel: a container (the *source*) filled with ammonia and releasing molecules through many fine parallel channels, the *focusser* creating the two energy states and allowing only one to pass, and the *cavity* into which the selected molecules enter. The cavity is a conducting vessel in which standing electromagnetic waves in a certain frequency range can reside. Input and output of waves is achieved by conducting tubes, called wave guides. In the presence of the selected ammonia molecules and for an input with a frequency corresponding to the transition between the two states, the apparatus worked as amplifier: The output exceeded the input. In the detailed paper, which followed their letter [4], Gordon, Zeiger, and Townes write [8]:

> The device utilizes a molecular beam in which molecules in the excited state of a microwave transition are selected. Interaction between these excited molecules and a microwave field produces additional radiation and hence amplification by stimulated emission. We call an apparatus using this technique a "maser," which is an acronym for "microwave amplification by stimulated emission of radiation."

The *maser* proved to be an instrument of extraordinary qualities. It can not

Basov (left) and Prokhorov.

Bloembergen

only be used as amplifier but also as oscillator, i.e., transmitter of electromagnetic waves. If the molecular beam is sufficiently intense, then one spontaneous emission will set off continuous, self-sustained, stimulated emission. From the outside the maser can be seen as a piece of electronic equipment with conventional input and output lines. Inside, however, the essential constituents are not macroscopic, man-made components but molecules. These are all identical with their well-defined energy states and transition frequencies, determined by quantum mechanics. A new field, *quantum electronics*, was opened up by the advent of the maser. Radiation from a maser oscillator is *coherent*, i.e., it possesses a well-defined phase like that of a conventional electronic oscillator. That is not the case for the radiation emitted by atoms and molecules under normal circumstances. This radiation, for instance, the one observed in emission spectroscopy, is the *incoherent* superposition of photons from spontaneous emission occurring independently in many molecules. Because of its unique properties the maser became an instrument of unprecedented precision, limited only by the uncertainty principle of quantum mechanics.

The third independent publication [9] of the maser idea, full of theoretical details (see also [10]), appeared in a Soviet journal. It was written by Basov[5] and Prokhorov[6] from the Lebedev Institute in Moscow. Unlike Townes, the two Russian scientists had no radar experience but had become interested in microwaves as a means for spectroscopy. Prokhorov's thesis work, suggested by Veksler, one of the inventors of the synchrotron, had been the study of *synchrotron radiation*, i.e., the electromagnetic radiation emitted by electrons on a curved trajectory, for instance, the orbit in an accelerator. For Basov the work on the maser eventually became the subject of his Ph.D. thesis. Basov and Prokhorov also came to the conclusion to produce population inversion in a molecular-beam set-up and opted for a molecule with a large electric dipole moment: caesium fluoride, CsF. The plans were discussed at a Soviet-internal conference in 1952 and the first paper was submitted to the journal in December, 1953, but appeared in print only ten months later, after the letter by Townes' group. A few months after reading that letter, Basov and Prokhorov had their first maser in operation.

As we have seen, the first masers were of the *two-level* type. Molecules in two energy levels were prepared in a beam and those with the lower energy were physically removed from it. A great step forward was done with the proposal and the realization of the *three-level* solid-state maser, a step comparable to that from magnetic resonance in molecular beams to the one in bulk matter. The proposal [11] was made in 1956 by Bloembergen[7], then Professor of Applied Physics at Harvard University. Bloembergen had studied physics under the difficult conditions of the war in his native Netherlands. Right after the war he was accepted as graduate student at Harvard, where he joined Purcell's group which, only a few weeks before his arrival, had produced nuclear magnetic resonance in condensed matter. He went back to Holland for some time but returned to Harvard in 1949 working, in particular, in solid-state spectroscopy. This is what Bloembergen proposed: With respect to many of their energy levels, impurities in a crystal behave like isolated atoms or atoms in a gas of very low pressure. Impurity atoms with three suitable energy levels E_1, E_2, E_3 are selected, where $E_1 < E_2 < E_3$. Under normal circumstances the populations of these states decrease with increasing energy, $n_1 > n_2 > n_3$. By continuously submitting the sample to radiation of the frequency corresponding to the energy $E_{31} = E_3 - E_1$ the population n_3 can be increased at the expense of n_1. This procedure later was called *pumping*.

Three-level system: Population inversion of the levels E_3 and E_2 can be obtained by irradiating the sample with the pumping frequency $\nu_{31} = E_{31}/h$. That enables the observation of stimulated radiation of the frequency $\nu_{32} = E_{32}/h$.

Under circumstances discussed by Bloembergen, one can achieve $n_3 > n_2$, i.e., population inversion of the two levels E_2 and E_3. Stimulated emission, required by a maser, is then possible. Maser activity is at the energy $E_{32} = E_3 - E_2$. Bloembergen noted that the energies E_2, E_3, via *Zeeman effect*, can depend on an external magnetic field and that, by varying that field, the maser energy can by *tuned*.

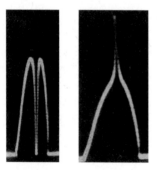

Early maser signal obtained by Bloembergen's group. Both oscilloscope curves show a frequency spectrum. The broad line is that of the cavity. Without pumping there is a sharp absorption line due to the material in the cavity (left). Pumping results in a population inversion; the absorption turns into emission (right). Figure reprinted with permission from [12]. Copyright 1958 by the American Physical Society.

Bloembergen and his group reported the successful experimental realization of such a maser in December 1957 [12]. The first maser operating on Bloembergen's principle was built a year earlier by a group at the Bell Telephone Laboratories [13].

[1] Einstein, A., *Mitteilungen der Physikalischen Gesellschaft Zürich*, **16** (1916) 47. Also published as [14].

[2] Purcell, E. M. and Pound, R. V., *Physical Review*, **81** (1951) 279.

[3] Weber, J., *Transactions of the Institute of Radio Engineers*, **3** (1953) 1.

[4] Gordon, J. P., Zeiger, H. J., and Townes, C. H., *Physical Review*, **95** (1954) 282.

[5] Townes, C. H., *How the Laser Happened – Adventures of a Scientist*. Oxford University Press, Oxford, 1999.

[6] Friedburg, H. and Paul, W., *Naturwissenschaften*, **38** (1951) 159.

[7] Bennewitz, H. G. and Paul, W., *Zeitschrift für Physik*, **139** (1954) 489.

[8] Gordon, J. P., Zeiger, H. J., and Townes, C. H., *Physical Review*, **99** (1955) 1264.

[9] Basov, N. G. and Prokhorov, A. M., *Zh. Eksperim. i Teor. Fiz.*, **27** (1954) 431.

[10] Basov, N. G. and Prokhorov, A. M., *Zh. Eksperim. i Teor. Fiz.*, **30** (1956) 560. Engl. transl. [15].

[11] Bloembergen, N., *Physical Review*, **104** (1956) 324.

[12] Artman, J. O., Bloembergen, N., and Shapiro, S., *Physical Review*, **109** (1958) 1392.

[13] Scovil, H. E. D., Feher, G., and Seidel, H., *Physical Review*, **105** (1957) 762.

[14] Einstein, A., *Physikalische Zeitschrift*, **18** (1917) 121.

[15] Basov, N. G. and Prokhorov, A. M., *Soviet Physics JETP*, **3** (1956) 426.

74 Strangeness – A New Quantum Number (1955)

In Episode 68 we saw how, beginning in 1947, new particles were observed which were considered 'strange', because they were obviously produced by strong interactions but decayed only via the weak interaction. The first important step towards an understanding of these *strange particles* was done by Pais in 1952 with his idea of an *associated production*.

Pais[1] [1] was born in Amsterdam and obtained bachelor degrees in both physics and mathematics from the university of his home town. He then became a graduate student of Uhlenbeck in Utrecht who had greatly impressed him with guest lectures given in Amsterdam. When Uhlenbeck left for the United States the month before the Second World War began, Pais continued under his successor Rosenfeld[2]. He managed to complete his Ph.D. in 1941 just before, in occupied Holland, the award of degrees to Jews was prohibited. Pais, and also his parents, survived the final war years in hiding. His sister and her husband were killed in a camp in Poland. Pais writes how easily, in his solitude, he could concentrate on theoretical physics. Occasionally Kramers, who was professor in Leiden but whose university had been closed, came for a discussion. When the war had ended, Pais published the results of his lonely research and, on Rosenfeld's suggestion, wrote letters of application to Bohr in Copenhagen and to Pauli in Princeton. He was accepted by both, decided to go to Copenhagen first, where he worked closely with Bohr for the first half of 1946, then moved on to Princeton. He was to stay there for a long time at the Institute for Advanced Studies. Of course, he met Einstein, who was a member of the institute since 1932. Pais now is known in particular for his excellent biographies of Einstein and Bohr. He himself considered his contribution to the understanding of strange particles to be his best work [1].

In a paper on which he worked late in 1951 [2], Pais introduced a new quantum number to explain the queer behaviour of strange particles. He later summarized its essentials as follows [3]:

> The selection rule I proposed, the 'even–odd rule', can in the simplest form be stated as follows. Assign a number 0 to all 'old' particles (π, N, γ, leptons) and a number 1 to the new particles Λ, K (the others were not there yet). In any process, first add these numbers for the initial-state particles, then for the final-state particles, the numbers being n_i and n_f. Then in all strong and electromagnetic processes n_i and n_f must be both even or both odd; in weak decays of the new particles one shall be odd, the other even. Thus $\pi^- + p \rightarrow \Lambda + \pi^0$ is strongly forbidden, $\pi^- + p \rightarrow \Lambda + K^0$ is strongly allowed. In general new particles come in pairs, a mechanism later named 'associated production' (I do not know

Pais

---

[1] Abraham Pais (1918–2000)   [2] Léon Rosenfeld (1904–1974)

who coined this term). Electromagnetic decays like $K^0 \rightarrow 2\gamma$ are forbidden. $\Lambda \rightarrow p + \pi^-$, $K^0 \rightarrow 2\pi$, etc., proceed by weak interaction. Later I learned that several Japanese colleagues had been considering a series of options which included the idea of a strong interaction selection rule [4–7].

Although the introduction of a new quantum number was premature, Pais' idea that strange particles can be produced strongly only in pairs was confirmed experimentally two years later. A beam of negative π mesons, produced by the Cosmotron accelerator at Brookhaven, was allowed to enter a cloud chamber. Two events of the type predicted by Pais were observed, the associated production of a Λ^0 hyperon and a K^0 meson in collisions of a π^- meson with a proton [8].

When Pais was writing his paper on associated production, the office next to his was occupied by a twenty-two year physicist who had recently obtained a Ph.D. from the Massachusetts Institute of Technology. Pais writes [1] that it did not take him long to recognize that Gell-Mann was unusually bright. The two discussed Pais' paper at length. Soon after, Gell-Mann left to assume a position as instructor at the University of Chicago.

In the course of his distinguished career, Gell-Mann was to achieve the deepest insight into the structure of matter since the development of the periodic system of elements and its explanation by Bohr and Pauli. As a first step, in Chicago in 1953, he applied Heisenberg's idea of *isospin* to the new particles [9]. We remember (Episode 51) that the nucleon was assigned an isospin quantum number of $I = \frac{1}{2}$ in order to account for its two different charge states, the proton (p) with the third component of isospin $I_3 = +\frac{1}{2}$ and the neutron (n) with $I_3 = -\frac{1}{2}$. Since the π meson occurs in three charge states, π^+, π^0, π^-, it was natural to assign it $I = 1$ and $I_3 = +1, 0, -1$, respectively. Thus, what concerned 'normal' particles, nucleons (which are fermions, i.e., have half-integer spin) also have half-integer isospin, and mesons (which have integer spin and therefore are bosons) also have integer isospin. What Gell-Mann postulated for strange particles was the opposite: Strange mesons (bosons) have half-integer isospin, hyperons (fermions) integer isospin. A particular feature of Gell-Mann's postulate was that K mesons came in doublets: a doublet of *particles*, K^+, K^0, with $I_3 = +\frac{1}{2}, -\frac{1}{2}$, and a doublet of *antiparticles*, $K^-, \bar{K}^0$, with $I_3 = -\frac{1}{2}, +\frac{1}{2}$. Thus, the K^- meson was taken to be the antiparticle of the K^+ meson; the two did not belong to the same isospin multiplet, contrary to the case of π mesons. That could also explain the fact that experimentalists had observed the K^- much less frequently than the K^+.

Gell-Mann postulated a difference in behaviour for the three types of interaction between particles:

- Strong interactions conserve isospin, $\Delta I = 0$.
- Electromagnetic interactions can change isospin by an integer value.
- Weak interactions do not respect isospin at all. (It was found that very frequently $|\Delta I| = \frac{1}{2}$.)

With these postulates Gell-Mann was able to explain the observations then available. As an example we consider the reaction

$$\pi^- + p \rightarrow \Lambda^0 + K^0 \quad . \tag{1}$$

Early observation of associated production in a cloud chamber. The photograph (one of a stereoscopic pair) shows the stopping of the track of an incoming π^- meson. Only neutral particles are produced in the interaction of the pion with a proton. Two neutral particles are seen to decay into pairs of charged ones, a Λ^0 into a proton (track 1a) and a π^- (2a), and a K^0 into a π^+ (1b) and a π^- (2b). Figure reprinted with permission from [8]. Copyright 1954 by the American Physical Society.

For the left-hand side we find $I_3 = -1 + \frac{1}{2} = \frac{1}{2}$ and for the right-hand side $I_3 = 0 + \frac{1}{2} = \frac{1}{2}$. Thus, in the strong process of associated production, isospin is conserved. For the weak decay of the Λ^0 hyperon,

$$\Lambda^0 \to p + \pi^- \quad , \tag{2}$$

however, it is not ($I_3 = 0$ for the initial, $I_3 = \frac{1}{2} - 1 = -\frac{1}{2}$ for the final state).

At this point we would like to introduce (or to recall) generic names for particle families indicating the type of interactions they can participate in:

- *Hadrons* are all particles with strong interaction. The proton and all particles with a proton as heaviest final decay product are called *baryons*; the term is due to Pais. Baryons comprise nucleons and hyperons. All other hadrons are *mesons*.
- *Leptons* are all particles with weak and possibly electromagnetic interactions, like the electron, the muon, and the neutrino.
- The *photon* interacts electromagnetically, but neither strongly nor weakly.

Pais remembers [1] that as a manuscript Gell-Mann's paper [9] still had the title *Isotopic Spin and Curious Particles* and that only the 'stodgy editors' of the *Physical Review* replaced 'Curious' by 'New Unstable'. He also mentions that Gell-Mann sometimes used the term 'queerness' in describing the new particles.

When Pais was invited to present a talk on these particles at a conference, to be held in June 1954 in Glasgow, he asked Gell-Mann to co-author a joint report. This report [10], in its final form completed just after the conference, contains the following information on hyperons, based on recent experiments and mostly (except for the particle names) already presented in an earlier unpublished report by Gell-Mann.

Gell-Mann

- The Λ^0 has no charged partners. Its isospin assignments are $I = I_3 = 0$.
- Two somewhat heavier charged hyperons are given the names Σ^+ and Σ^- and form an isospin triplet with their neutral partner Σ^0 which is predicted to exist.
- The cascade particle, by now confirmed, is given the name Ξ^-. In order to explain its weak decay, $\Xi^- \to \Lambda^0 + \pi^-$, it is assigned a half-integer isospin, contrary to the original assumption that all strange fermions have integer isospin. A neutral partner, the Ξ^0 with $I_3 = \frac{1}{2}$, is predicted to exist and to form an isospin doublet with the Ξ^-, having $I_3 = -\frac{1}{2}$.

Both predicted particles were found in due course, the Σ^0 in 1956 and the Ξ^0 in 1959.

If one plots the various long-lived mesons and baryons in diagrams with the electric charge Q on one axis and the particle mass m on the other, one notes a distinctive shift of the average charge with mass. With increasing mass, the average charge tends towards positive values for mesons and towards negative values for baryons. As Pais remembers [1], he remarked to Gell-Mann, when they wrote their report, that this fact should be expressed by a new quantum number. Gell-Mann, then, thought that isospin was a sufficient characterization.

In 1955, Gell-Mann did introduce a new quantum number [11]. He notes that, for 'ordinary' mesons, i.e., pions, the charge number Q is identical to the third component I_3 of isospin, $Q = I_3$, whereas for ordinary baryons it is not. Now, if baryons are given the *baryon number* $B = 1$ (and their antiparticles $B = -1$) whereas mesons get $B = 0$, then a common relation for the charge number can be written down for all ordinary particles, mesons and nucleons, namely, $Q = I_3 + B/2$. The term $B/2$ is equal to the 'centre of charge' of a charge multiplet. For pions the centre of charge is zero but for nucleons (or antinucleons) it is 'displaced' from zero by $B/2 = 1/2$ (or $B/2 = -1/2$). Gell-Mann found that the relation held also for strange particles if he wrote the charge displacement in the form $B/2 + S/2$. This defines the number S for all charge multiplets. The resulting simple formula reads

$$Q = I_3 + \frac{B}{2} + \frac{S}{2} \quad . \tag{3}$$

Concerning the symbol S, Gell-Mann writes [11]:

> We propose to identify all known hyperons and K-particles as members of displaced multiplets and to account for some of their properties in that way. Since we have $S = 0$ for ordinary particles and $S \neq 0$ for "strange" ones we refer to S as "strangeness".

From this scheme it follows that the K mesons K^+ and K^0 have $S = 1$. The Λ and Σ hyperons have $S = -1$, the cascade particle Ξ has $S = -2$. The strangeness of an antiparticle is the strangeness of the corresponding particle multiplied by -1.

Strangeness, like electric charge, is conserved in strong and electromagnetic interactions. But it is not conserved in weak interactions. In the *strong* production process (1), the total strangeness is the same in the initial and in the final state, $S_i = S_f = 0$. In the *weak* decay (2), however, it is not, $S_i = -1, S_f = 0$. From this scheme a large number of predictions followed, which are discussed in Gell-Mann's paper.

The relation (3) is now called the Gell-Mann–Nishijima formula. Nishijima[3] from the Osaka City University had, independently, put forward much the same scheme as Gell-Mann [12]. Instead of *strangeness* he spoke of η charge. Relation (3) is often written in the form

$$Q = I_3 + \frac{Y}{2} \quad , \qquad Y = B + S \quad . \tag{4}$$

The quantity Y, the sum of the baryon number B and the strangeness S, is called *hypercharge*.

In his Nobel Lecture [13] Alvarez proudly writes: 'My research group eventually confirmed all of Gell-Mann's and Nishijima's early predictions, many of them for the first time, and we continue to be impressed by their simple elegance.' As a particularly beautiful example, we mention the production and decay of the Ξ^0 hyperon [14]. The 15-inch Berkeley hydrogen bubble chamber was exposed to a beam of negative K mesons produced by the Bevatron accelerator and the following chain of events was observed in it. In a collision of a K meson with a proton of the bubble-chamber filling two particles were

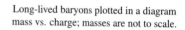

Long-lived baryons plotted in a diagram mass vs. charge; masses are not to scale.

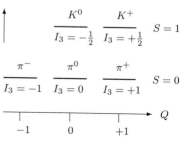

Long-lived mesons plotted in a diagram mass vs. charge; masses are not to scale.

---

[3] Kazuhiko Nishijima (born 1926)

[1]   Pais, A., *A Tale of Two Continents – A Physicist's Life in a Turbulent World*. Oxford University Press, Oxford, 1997.

[2]   Pais, A., *Physical Review*, **86** (1952) 663.

[3]   Pais, A., *Inward Bound – Of Matter and Forces in the Physical World*. Oxford University Press, Oxford, 1986.

[4]   Nambu, Y., Nishijima, K., and Yamaguchi, Y., *Progress in Theoretical Physics (Japan)*, **6** (1951) 615, 619.

[5]   Aizu, K. and Kinishita, T., *Progress in Theoretical Physics (Japan)*, **6** (1951) 630.

[6]   Miazawa, H., *Progress in Theoretical Physics (Japan)*, **6** (1951) 631.

[7]   Oneda, S., *Progress in Theoretical Physics (Japan)*, **6** (1951) 633.

[8]   Fowler, W. B., Shutt, R. P., Thorndike, A. M., and Whittemore, W. L., *Physical Review*, **93** (1954) 861.

[9]   Gell-Mann, M., *Physical Review*, **92** (1953) 833.

[10] Gell-Mann, M. and Pais, A., *Theoretical Views on the New Particles*. In Bellamy, E. H. and Moorhouse, R. G. (eds.): *Proc. Glasgow Conf. on Nuclear and Meson Physics*, p. 342. Pergamon, Oxford, 1955.

[11] Gell-Mann, M., *Nuovo Cimento Suppl.*, **4** (1956) 848.

[12] Nishijima, K., *Progress in Theoretical Physics (Japan)*, **13** (1955) 285.

[13] Alvarez, L. W., *Nobel Lectures, Physics 1963–1970*, p. 241. Elsevier, Amsterdam, 1972.

[14] Alvarez, L. W. et al., *Physical Review Letters*, **2** (1959) 215.

produced, a Ξ^0 and a K^0. Both, being uncharged, did not leave tracks and decayed, the Ξ^0 into two other neutral particles, $\Xi^0 \rightarrow \Lambda^0 + \pi^0$, the K^0 into two oppositely charged pions, $K^0 \rightarrow \pi^+ + \pi^-$. One of the neutral decay particles of the Ξ^0, the Λ^0, also decayed in the chamber into two charged particles, $\Lambda^0 \rightarrow p + \pi^-$. The momenta of all charged particles in this chain,

$$
\begin{array}{ccc}
K^- + p \rightarrow \Xi^0 & & + K^0 \\
\quad \llcorner \Lambda^0 + \pi^0 & & \quad \llcorner \pi^+ + \pi^- \\
\qquad \llcorner p + \pi^- & &
\end{array} \quad ,
$$

were measured. (For the original proton, at rest in the chamber, the momentum was known to be zero.) From the laws of special relativity those of all neutral particles could then be computed. The analysis of this single photograph was a triumph both for the bubble-chamber method and for the strangeness scheme. It is easily seen that the total strangeness is conserved in the strong production process (it is $S = -1$). In each of the three weak decays, on the other hand, it changes by one unit. The strangeness $S = -2$ of the Ξ^0 in the first stage of its decay cascade changes to $S = -1$ and in the second stage to $S = 0$.

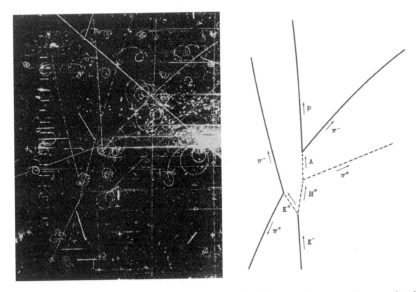

The production and decay of the Ξ^0 hyperon observed by Alvarez and his group. Figure reprinted with permission from [14]. Copyright 1959 by the American Physical Society.

Antimatter (1955)

In Episode 64 we described the invention of a new type of particle accelerator, the synchrotron, and mentioned that the US Atomic Energy Commission, in 1948, approved funds to build two large synchrotrons, one accelerating protons to an energy of 3 GeV (the Cosmotron) at Brookhaven and one of 6 GeV (the Bevatron) at Berkeley. The latter energy was chosen because that would suffice for the production of *antiprotons* if such particles existed. It was generally expected there would be an antiproton, i.e., a particle with the proton mass but of opposite (negative) charge as partner of the proton in the same way as there was the positron as antiparticle of the electron (Episode 49). (That expectation was, however, by no means a certainty. Because it had an anomalous magnetic moment the proton was assumed to have a substructure; it might not be an elementary particle in the same sense as the electron was.)

Just like electrons and positrons, protons (p) and antiprotons ($\bar{p}$) were expected to be created in pairs, for instance, in a collision of two protons,

$$p + p \rightarrow p + p + p + \bar{p} \quad .$$

If the two protons of the initial state have equal and opposite momenta, then their joint centre of mass is at rest and so is that of the four particles in the final state. The energy to create the $p\bar{p}$ pair which, according to Einstein's famous formula, is equal to twice the proton mass m_p multiplied by the square of the velocity c of light, $2m_pc^2$, and has to be taken out of the kinetic energy of the initial protons. If the kinetic energy is larger, the particles in the final state can still be given momentum. Thus the minimum energy in the *centre-of-mass system* is $E_{\min}^{\mathrm{cm}} = 2m_pc^2$. In practice, the experiment had to be done in the *laboratory system*, in which one of the initial protons is at rest as part of a piece of matter, the *target*. Only the other proton carries kinetic energy gained in the accelerator. In the laboratory system the centre of mass is not at rest. More energy is needed to produce a $p\bar{p}$ pair because these particles now have to have momentum, so that the momentum of the centre of mass is the same in the initial and in the final state. A simple computation shows that the minimal kinetic energy is $E_{\min}^{\mathrm{lab}} = 6m_pc^2$. Therefore, since $m_pc^2 = 0.938\,\mathrm{GeV}$, protons hitting the target with an energy of 6 GeV were expected to produce proton–antiproton pairs.

Several groups conceived and set up experiments to detect antiprotons while the Bevatron was being built. The one first meeting with success was Segrè's. We already met Segrè[1] [1] as Fermi's first student in Rome (Episodes 53, 55) and as professor in Palermo discovering the element technetium. In 1938, he went to Berkeley as research associate in the Radiation Laboratory and became professor of physics at the University of California in Berkely in 1946. His

Segrè

Chamberlain

---

[1] Emilio Gino Segrè (1905–1989), Nobel Prize 1959

three collaborators, Chamberlain[2], Wiegand[3], and Ypsilantis[4], had all been his students. Chamberlain and Wiegand began their graduate studies in Berkeley and, during the war, worked under Segrè on the atomic-bomb project both at Berkeley and at Los Alamos. In 1946 Chamberlain went to Chicago, where he earned his Ph.D. under Fermi, and in 1948 returned to Berkeley. Wiegand obtained his doctor's degree for work under Segrè at Berkeley. Ypsilantis, originally a chemist and the youngest in the group, got his degree at Berkeley in 1955. Together with Wiegand he had realized a scheme to produce a polarized beam of protons, i.e., of protons with a definite spin orientation.

Surrounding Edward Lofgren (centre), who was in charge of the Bevatron, are (from the left) Segrè, Wiegand, Chamberlain, and Ypsilantis.

The experiment was designed to use electronic detectors only. Its aim was, of course, to unambiguously identify particles with the mass m_p of the proton but with the electric charge $-e$, opposite to the charge of the proton. The mass of a particle cannot be measured directly but can be computed from the measurements of two other quantities. Momentum and velocity of the particle were chosen.

The momentum of a particle of known charge is measured quite precisely by its deflection in a magnetic field. Negative particles produced by protons falling on a copper target in the Bevatron entered a system of two bending magnets M1 and M2, deflecting the particles, and two magnetic lenses Q1 and Q2, made of three quadrupole magnets each and focussing them. The momentum of the particles reaching the counter S2 was then well defined to be $p = 1.19 \, \text{GeV}/c$ within an error of 2%.

The velocity v of a fast particle is conveniently expressed as the fraction $\beta = v/c$ it is of the velocity c of light. For antiprotons of the selected momentum the numerical value is $\beta_{\bar{p}} = 0.78$ (slightly reduced to $\beta = 0.765$ by passage through counters); for the lighter pions it is $\beta_{\pi} = 0.99$. Velocity measurements had to reliably distinguish between these two values. One way to measure the velocity was the so-called *time-of-flight method*. The particles

[2] Owen Chamberlain (1920–2006), Nobel Prize 1959   [3] Clyde Wiegand (1915–1996)   [4] Thomas John Ypsilantis (1928–2000)

traversed the two scintillation counters S1 and S2 which were a distance of 40 feet (12.2 m) apart. The time of flight between the two counters was thus 40 nanoseconds (ns) $= 40 \times 10^{-9}$ s for a pion and 51 ns for an antiproton. The two counters were connected with cables of identical lengths to an electronic decision-taking circuit and to an oscilloscope on which the time difference between the signals from S2 and S1 could be displayed. Both with the circuit and on the oscilloscope, the time of flight could be measured with a precision of 1 ns.

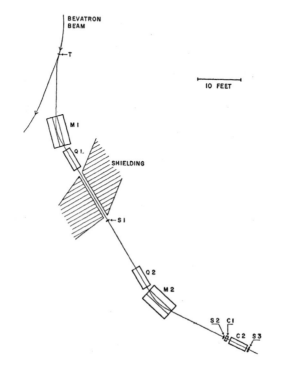

The system of magnets and counters used to identify the antiproton. Figure reprinted with permission from [2]. Copyright 1955 by the American Physical Society.

A second, independent velocity measurement was provided by a *Cherenkov counter*. As told in Episode 56, a charged particle, traversing a medium faster than the speed of light in that medium, emits light under an angle with respect to its path, which is characteristic of the velocity. The counter C2 was specially designed so that it detected light only within a narrow angular range, corresponding to a range in β between 0.75 and 0.78, comprising the antiproton velocity but not that of pions. Another Cherenkov counter, the *guard counter* C1, was of the threshold type. A material (quartz) was employed that yielded signals only for particles with β larger than 0.79, i.e., for pions but not for antiprotons. A final scintillation counter, S3, was placed behind the two Cherenkov counters. The decision-taking circuit selected events with signals in S1 and S2 (having a time difference of about 51 ns between them, indicating an antiproton), a signal in C2 (another indication for an antiproton), and a signal in S3 (assuring that the particle traversed the Cherenkov counter C2 along its axis, essential for the proper functioning of that counter). For these events the oscilloscope was triggered and automatically photographed; it displayed the signals of S1 and S2 and (somewhat later in time and with reversed

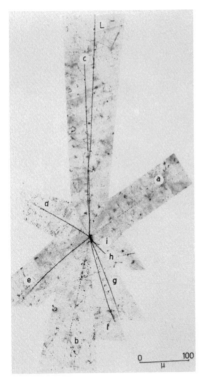

Oscilloscope traces showing (from left to right) pulses of the counters S1, S2, and C1. The three different cases correspond to the passage of (a) a pion, (b) an antiproton, and (c) an accidental coincidence. The time difference between the first two pulses is 40 ns in (a) and 50 ns in (b). Figure reprinted with permission from [2]. Copyright 1955 by the American Physical Society.

Antiproton annihilation event observed in emulsion. The track marked L is that of the incoming antiproton. Figure reprinted with permission from [3]. Copyright 1956 by the American Physical Society.

sign) the signal from the guard counter C1, if there was one. Photographs with S1 and S2 signals, showing the proper time difference, and no guard-counter signal were accepted as due to antiprotons. If there was a signal from the guard counter, the seemingly correct time difference could have been caused not by a single antiproton but by two independent pions. The ratio of antiprotons to pions traversing the apparatus was about 1 in 30 000. Even less had been expected, therefore the precaution against faked times of flight.

The experiment began early in August 1955. On 21 September first evidence for antiprotons was obtained. On the basis of 60 unambiguous events, the group submitted a letter entitled *Observation of Antiprotons* to the *Physical Review* near the end of October 1955 [2].

An antiproton can be created in a $\bar{p}p$ pair together with a proton and such an antiparticle–particle pair can also annihilate, preferentially into pions. It is characteristic of an annihilation that the particles in the final state carry more kinetic energy than those in the initial state, because the rest mass of the initial particles is, at least in part, transformed into kinetic energy. This feature can be used to identify annihilation events. In particular, if an antiproton comes to rest, there is no kinetic energy in the initial state but still annihilation with a release of energy is possible. Segrè's group, in conjunction with a group in Rome led by Amaldi, searched for annihilation events in photographic emulsion. Slow antiprotons, produced by the Bevatron, were stopped in the emulsion, which was developed and scanned for annihilation events. One such event was found in Rome still in 1955 [3]. An antiproton entered a nucleus and annihilated with one of its nucleons. With the energy released, pions were created and the other nucleons were dispersed, the protons from the nucleus and the charged pions leaving tracks in the emulsion.

A year after the existence of the antiproton was ascertained, *antineutrons* were detected by another group in Berkely [4]. That was more difficult, because they carry no electric charge and therefore leave no signals in scintillation or Cherenkov counters. An antineutron can, however, be detected through the particles appearing in its annihilation. The way used to create antineutrons in a controlled way was the *charge-exchange reaction*

$$\bar{p} + p \rightarrow \bar{n} + n \quad .$$

If the incoming antiproton has sufficient energy, annihilation need not hap-

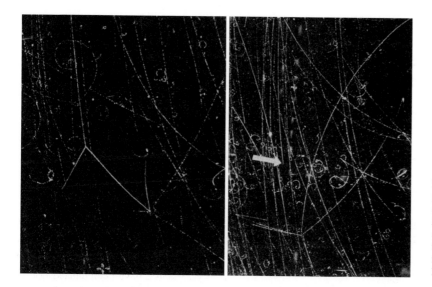

Interactions of antiprotons observed in a propane bubble chamber. *Left:* Scattering and subsequent annihilation of an antiproton. *Right:* Charge exchange of an antiproton and subsequent annihilation of the resulting antineutron. Figure reprinted with permission from [5]. Copyright 1958 by the American Physical Society.

pen; but the negative charge of the antiproton can be transferred to the proton. Both particle and antiparticle become neutral and fly apart again. In the experiment, antiprotons (identified through magnetic deflection and an elaborate time-of-flight system) were allowed to fall on a converter target, where charge exchange could take place. Particles from the target could enter a Cherenkov counter with lead glass as active medium by traversing two scintillation counters. Annihilation of an antineutron in the lead glass would lead to a large signal in the Cherenkov counter. If, simultaneously, the two scintillation counters gave no signal (showing that no charged particle had entered the Cherenkov counter) and an antiproton was known to have fallen on the converter target, it was assured that a large Cherenkov signal, indeed, was caused by an annihilating antineutron.

With the advent of bubble chambers, registration of complicated events became possible. To conclude this episode we show two photographs taken in a bubble chamber filled with propane (a compound composed of carbon and hydrogen atoms) by an extension of the original antiproton crew [5]. The beam of negative particles entering the chamber contained only few antiprotons, which nevertheless were often identified by characteristic interactions. In both photographs, particles enter from the top. In the first one, an antiproton is seen to scatter elastically off a proton from the chamber filling. It is deflected to the right side of the picture while the hit proton flies to the left. Further along on its path the antiproton hits a carbon nucleus and produces an *annihilation star* consisting of five tracks left by protons from the carbon and pions produced in the annihilation.

In the second photograph, near the arrow, a track left by an antiproton suddenly ends. At this point charge exchange happened, the neutron and the antineutron leaving no tracks. Further down in the photograph, the annihilation of the antineutron is recorded by the tracks of five charged pions produced in it.

[1] Segrè, E., *A Mind Always in Motion – The Autobiography.* University of California Press, Berkely, 1993.

[2] Chamberlain, O., Segrè, E., Wiegand, C., and Ypsilantis, T., *Physical Review,* **100** (1955) 947.

[3] Chamberlain, O., Chupp, W. W., Goldhaber, G., Segrè, E., Wiegand, C., Amaldi, E., Baroni, G., Castagnoli, C., Franzinetti, C., and Manfredini, A., *Physical Review,* **101** (1956) 909.

[4] Cork, B., Lambertson, G. R., Piccioni, O., and Wenzel, W. A., *Physical Review,* **104** (1956) 1193.

[5] Agnew, L., Elioff, T., Fowler, W. B., Gilly, L., Lander, R., Oswald, L., Powell, W., Segrè, E., Steiner, H., White, H., Wiegand, C., and Ypsilantis, T., *Physical Review,* **110** (1958) 994.

76   The Neutrino Finally Observed (1956)

In Episode 46 we told how, in 1930, Pauli postulated the existence of the *neutrino* in order to save the conservation laws of energy, momentum, and angular momentum in β decay. Fermi, in the successful theory of weak interactions he developed in 1933 (Episode 53), made use of the creation as a pair of an electron and a neutrino. With the advent of the positron as antiparticle of the electron it became clear that the neutrino which was pair-created with a positron, for instance, in the β decay of an artificially produced isotope which emits positrons, is the antiparticle of the original Pauli neutrino. By convention the original neutrino was called an *antineutrino*, $\bar{\nu}$, in order to describe the pair of particles created in β decay as one particle, the electron, and one antiparticle, the antineutrino. Using this notation the decay of a neutron (n) into a proton (p) and such a pair is written as

$$n \rightarrow p + e^- + \bar{\nu} \quad . \tag{1}$$

It was of great interest to obtain direct experimental evidence for the existence of the (anti)neutrino, for instance, through the observation of the process

$$\bar{\nu} + p \rightarrow n + e^+ \quad . \tag{2}$$

This reaction is called the *inverse β decay* since it can be seen as the original decay reaction (1), read from the right to the left side with the electron brought (with its charge reversed) to the other side of the arrow. In 1934 Bethe and Peierls [1] estimated the cross section for reaction (2) for antineutinos from the β decay of natural radioactive sources to be ridiculously small, around 10^{-44} cm^2; for the concept of interaction cross section (see Box I in Episode 16). They concluded that a neutrino would, on the average, travel a distance about 10^{19} metres or a hundred light-years in ordinary matter and concluded 'that there is no practically possible way of observing the neutrino.' Although the cross-section estimate was essentially correct, the neutrino was observed through reaction (2) in 1956 by Reines, Cowan, and co-workers. That was possible because the fission products in a nuclear reactor emit so many neutrinos that some do interact in the experimental set-up. (If a single neutrino yields an interaction in 10^{19} metres of detector, one out of 10^{19} neutrinos interacts within a single metre.)

Reines[1] was born in Paterson, New Jersey, as the youngest of four children. His parents were immigrants from a small Russian town. When he was a child his father ran a general store in a rural region of the State of New York. The family later returned to New Jersey, and Reines studied at the Stevens Institute of Technology in Hoboken. There he graduated in engineering in 1939 and

---

[1] Frederick Reines (1918–1998), Nobel Prize 1995

took a M.Sc. in mathematical physics in 1941. For graduate studies he moved
to New York University and wrote a theoretical Ph.D. thesis on nuclear fission.
Before he completed that in 1944, Reines moved to Los Alamos in New Mexico to join the Manhattan Project, as the effort to build the atomic bomb was
called. Another young theorist there was Feynman, who later remembered that
Bethe, the leader of theoretical physics at Los Alamos, got to like him because
of his unabashed way in discussions [2]: 'I got a notch up on account of that,
and I ended up as a group leader under Bethe with four guys under me.' One
of these 'guys' was Reines who would soon become a group leader himself.

When the war ended, many scientists left Los Alamos to return to their
universities. Reines belonged to those who stayed on. He became director of Operation Greenhouse which performed and analysed experiments on
atomic-bomb explosions on the Eniwetok and Bikini atolls in the Pacific and
in Nevada. In 1951 he asked for a 'leave in residence', so that he could think
about some fundamental physics to do [3]:

> I moved to a stark empty office, staring at a blank pad for several months
> searching for a meaningful question worthy of a life's work. It was a
> very difficult time. The months passed and all I could dredge up out of
> the subconscious was the possible utility of a bomb for direct detection
> of neutrinos. After all such a devise produced an extraordinarily intense
> pulse of neutrinos ...

He consulted Fermi, who agreed that 'the bomb' would be a good source of
neutrinos; but neither Reines nor Fermi knew how to build a detector of the
required size. Only a little later Reines convinced Cowan to join forces with
him and track down the neutrino.

Reines

Cowan[2] had a B.Sc. in chemical engineering. He had been an officer in the
United States Army during the war, stationed in England since 1942. Leaving
active service in 1946, he took advantage of the G.I. Bill which provided financial support for the university studies of soldiers. He obtained a Ph.D. in 1949
from Washington University in St. Louis and then joined the scientific staff at
Los Alamos.

In February 1953, Reines and Cowan submitted a letter to the *Physical Review*, outlining the experiment they proposed to do [4]. The envisaged source

---

[2] Clyde Lorrain Cowan (1919–1974)

of neutrinos was now a reactor and a way to detect reaction (2) was described. To distinguish the rare process (2) from possible background, both particles in the final state, the positron and the neutron, had to be clearly identified. The positron, after losing its kinetic energy by ionization, annihilates with an electron; as a result two gamma rays appear, each one carrying the energy of 0.511 MeV corresponding to the mass of one electron. The annihilation can also take place while the positron is still 'in flight', in which case the photon energy is higher. The neutron was to be slowed down in a substance rich in hydrogen (Episode 55); a slow neutron is captured with high efficiency by a cadmium nucleus, which then emits several gamma rays. The 'signature' for a neutrino event was the simultaneous appearance of two annihilation photons followed a short time (several microseconds) later (needed by the neutron to slow down in repeated collisions with protons) by photons emitted by an excited cadmium nucleus.

Gamma rays, i.e., high-energy photons, can be detected in several ways. A convenient method, new at that time, was to use a *liquid scintillator*. Solid scintillators were long before known. Already Röntgen and Rutherford had used light emitted by zinc sulphide to detect X rays and α particles. A high-energy photon in matter produces a cascade of electrons and photons of lower energy. These particles can excite certain *scintillating* molecules to emit visible light. If such molecules are present in a transparent liquid, the light can be collected in photomultipliers and transformed into an electronic signal. The height of that signal is a measure for the original photon energy.

The first detector consisted of a single large volume of liquid serving several purposes at a time. It contained protons as targets for the neutrino reaction (2), it had scintillating properties, and it contained cadmium to absorb the neutrons. Light was collected by 90 photomultipliers in optical contact with the liquid. This detector was installed near a reactor of the plutonium factory at Hanford, Washington. The results reported in 1953 [5] indicated that reaction (2) was observed. They were, however, not statistically significant in a strict sense; other explanations were still possible though rather unlikely.

Reines and Cowan redesigned their apparatus. A large flat rectangular tank containing 200 litres of water and some cadmium salt served as interaction volume for the incident neutrinos, for the moderation and capture of the neutrons, and for the annihilation of the positrons. It was placed between two wider tanks filled with liquid scintillator and fitted with photomultipliers for the detection of photons. Coincident signals from both tanks were required to identify a positron annihilation. If another coincidence, typical for a neutron capture, appeared within the appropriate delay, then the event was accepted as caused by a neutrino reaction.

The new neutrino detector was installed near a reactor at the Savannah River Site, a recently built plutonium factory. The whole apparatus was a 'club-sandwich' [7] made of three 'bread' layers (the scintillator tanks) and two 'meat' layers (the interaction volumes). It was installed in an underground room of the reactor building and completely enclosed by paraffin and lead shielding to protect it from particles other than neutrinos. Events with the required signature triggered two oscilloscopes on which the signals from the three scintillator tanks were displayed and automatically photographed for

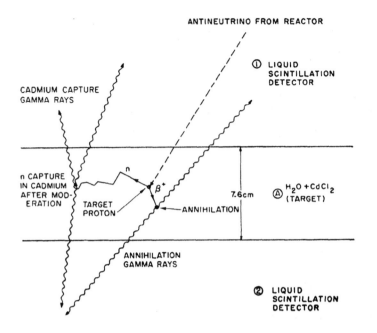

ANTINEUTRINO FROM REACTOR

① LIQUID
SCINTILLATION
DETECTOR

CADMIUM CAPTURE
GAMMA RAYS

n CAPTURE
IN CADMIUM
AFTER MOD-
ERATION

TARGET
PROTON

n

β⁺

ANNIHILATION

7.6cm

Ⓐ $H_2O + CdCl_2$
(TARGET)

ANNIHILATION
GAMMA RAYS

② LIQUID
SCINTILLATION
DETECTOR

Principle of the Savannah River experiment. Figure reprinted with permission from [6]. Copyright 1960 by the American Physical Society.

later inspection and analysis. When the reactor was in operation, the estimated flux of neutrinos through the detector was 1.2×10^{17} per square metre and second.

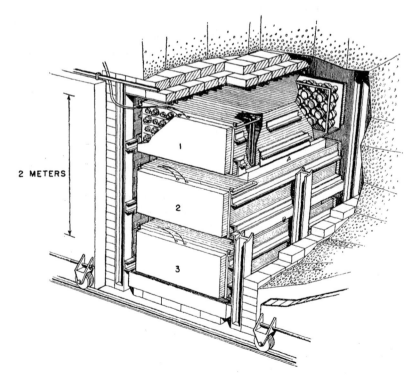

2 METERS

1

2

3

The neutrino detector at Savannah River. The tanks 1, 2, and 3 contain liquid scintillator and are viewed by 110 photomultipliers each. In the two gaps between them are the target tanks *A* and *B* containing water and cadmium chloride. Figure reprinted with permission from [6]. Copyright 1960 by the American Physical Society.

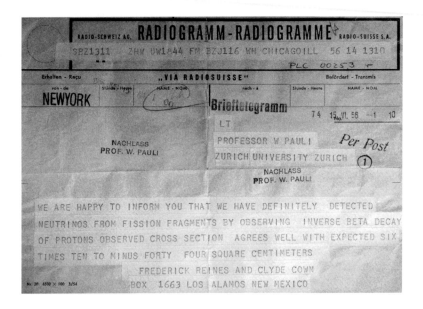

The telegram received by Pauli. From CERN Pauli Collection.

The existence of the (anti)neutrino was unambiguously established in 1956. Short reports were published in July [7] and September [8]. A detailed paper appeared later [6]. Here are some of the results quoted in [8]:

The experiment was run for a total time of nearly 1400 hours, including periods in which the reactor was on and others in which it was off. During reactor-on time, an average of 2.88 ± 0.22 events per hour were registered. The background of accidental coincidences, which might have faked true neutrino interactions, was on the order of 5% of the registered events. A number of very detailed checks was performed: The frequency of neutrino events was found to be proportional to the density of protons in the target tank. The first signal showed all the characteristics of a positron–electron annihilation. The delayed signal (as shown by its size and its dependence on the density of cadmium atoms) was due to a neutron capture. Other particles coming from the reactor did not contribute significantly, since additional shielding had no influence on the event rate.

Also, the absolute event rate coincided with the theoretical expectations. With the detailed knowledge of nuclear properties and reactor physics it had been possible to compute the energy spectrum of the neutrinos. Weak interaction theory gave an average cross section of 6×10^{-44} cm^2 for the inverse-β-decay reaction (2) for these neutrinos, somewhat more favourable than what Bethe and Peierls had estimated two decades earlier.

In his Nobel Lecture Reines [3] recalls how the news were received by the great masters. A telegram was sent to Pauli, who answered by a night letter:

Thanks for message. Everything comes to him who knows how to wait.
                                                                    Pauli

When Bethe was confronted with his and Peierls's statement that there was no way to observe the neutrino, he simply said:

Well, you shouldn't believe everything you read in the papers.

[1] Bethe, H. A. and Peierls, R. E., *Nature*, **133** (1934) 532.

[2] Feynman, R. P., *"Surely You're Joking, Mr. Feynman!"* – *Adventures of a Curious Character*. Norton, New York, 1985.

[3] Reines, F., *Nobel Lectures, Physics 1991–1995*, p. 202. World Scientific, Singapore, 1997.

[4] Reines, F. and Cowan, C. L., *Physical Review*, **90** (1953) 492.

[5] Reines, F. and Cowan, C. L., *Physical Review*, **92** (1953) 830.

[6] Reines, F., Cowan, C. L., Harrison, F. B., McGuire, A. D., and Kruse, H. W., *Physical Review*, **117** (1960) 159.

[7] Cowan, C. L., Reines, F., Harrison, F. B., Kruse, H. W., and McGuire, A. D., *Science*, **124** (1956) 103.

[8] Reines, F. and Cowan, C. L., *Nature*, **178** (1956) 446.

Parity – A Symmetry Broken (1957)

In October 1957, I attended my first conference, the annual meeting of the German Physical Society which was held in Heidelberg. I was in my sixth semester and had just begun to work on my diploma thesis under Paul in Bonn. The highlight was a talk by Weisskopf who reported on the 'fall of parity'. I had never heard of parity, let alone that so far it had been tacitly taken to be conserved. Mirror-image symmetry or, as it was technically called, parity invariance had been taken for granted so much that it was not even mentioned in most text books. When it was found not to hold, Dirac was asked for a comment. His answer was [1]: 'I never said anything about it in my book.'

In physics *symmetries* under transformations, *unobservable quantities*, and *conservation laws* are closely linked to one another. We give a simple example. For an isolated system in classical mechanics it can be easily shown, that the total momentum of the system does not change with time, if the Hamiltonian function, i.e., the total energy of the system, is unaltered although another origin of the coordinate system is chosen. A change in origin corresponds to the transformation of every position vector $\mathbf{r}$ to $\mathbf{r}' = \mathbf{r} + \Delta\mathbf{r}$. Invariance under such a translation or, as it is also called, symmetry of the physical laws describing a system is therefore equivalent to a conservation law, that of momentum. But it is also equivalent to the fact that, for that physical system, an absolute spatial position is unobservable. Similarly, there are other transformations in space and time pointing out other unobservables and conservation laws. The experimental facts that absolute time, absolute orientation, and absolute rest are unobservable are equivalent to the conservation of energy, of angular momentum, and of four-momentum squared, respectively.

The transformations in space and time, mentioned so far, are continuous; but they can also be discrete (see Box I). One is space reflection, the transformation in which the space coordinates are multiplied by a minus sign, $x, y, z \rightarrow -x, -y, -z$. It corresponds to a reflection of the original system at the origin. While in classical physics reflection symmetry does not lead to a conservation law, in quantum mechanics it can. Let us discuss the symmetry properties under reflection a wave function $\Psi(\mathbf{r})$ can possess. We define a *parity operator* P_{op} which turns $\Psi(\mathbf{r})$ into $\Psi(-\mathbf{r})$, i.e., $P_{\mathrm{op}}\Psi(\mathbf{r}) = \Psi(-\mathbf{r})$. Now, if the wave function is *symmetric* ($\Psi(\mathbf{r}) = \Psi(-\mathbf{r})$) or *antisymmetric* ($\Psi(\mathbf{r}) = -\Psi(-\mathbf{r})$) under reflection, then it obeys the simple equation $P_{\mathrm{op}}\Psi(\mathbf{r}) = P\Psi(\mathbf{r})$ with $P = +1$ or $P = -1$, respectively. It is an *eigenvalue equation* like the Schrödinger equation (Episode 39). The eigenvalue P is called the eigenparity or simply *parity* of the state. A state need not have a definite parity.

The full wave function of the electron in the hydrogen atom (in Episode 39 we only showed the radial, not the angular part) has a parity depending on the angular-momentum quantum number ℓ; it is $(-1)^{\ell}$, i.e., $P = +1, -1, +1, \ldots$

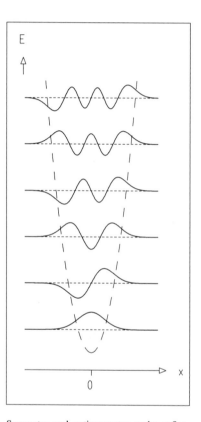

Symmetry and antisymmetry under reflection. The curves represent the wave functions of the harmonic oscillator in one dimension, i.e., a particle in a potential with parabolic form, shown by the long-dashed curve. The energies of the different states are indicated by the short-dashed horizontal lines which, in the plot, serve as zero lines for the wave functions. Beginning with the lowest-energy state, under reflection ($x \rightarrow -x$) the wave functions are symmetric, antisymmetric, symmetric, etc. The states have parity $+1, -1, +1, \ldots$. (The energies are $E_n = \hbar\omega(n + 1/2)$; $n = 0, 1, 2, \ldots$, where ω is the circular frequency of the oscillator.)

Box I. Spacetime Symmetries and Conservation Laws

In Box II at the end of Episode 11 we stated that under a *Lorentz transformation* from a spacetime coordinate system x, y, z, t to another system x', y', z', t' the square of a four-vector is *invariant*,

$$s^2 = (ct)^2 - x^2 - y^2 - z^2 = (ct)^2 - (\mathbf{r})^2$$
$$= (ct')^2 - (x')^2 - (y')^2 - (z')^2 = (ct')^2 - (\mathbf{r}')^2 \quad .$$

The transformations which leave s^2 invariant can have rather different forms. In particular, we distinguish two types:

1. **Continuous transformations**

   - *Translation in space*, $\mathbf{r}' = \mathbf{r} + \Delta\mathbf{r}$;
     symmetry implies conservation of momentum.
   - *Translation in time*, $t' = t + \Delta t$;
     symmetry implies conservation of energy.
   - *Rotation in space*, e.g., x', y', z' rotated with respect to x, y, z by an angle φ around the common axis $z = z'$;
     symmetry implies conservation of angular momentum.
   - *Lorentz transformation* (Box I in Episode 11);
     symmetry implies conservation of four-momentum squared.

2. **Inversions**

   - *Space reflection*, $\mathbf{r}' = -\mathbf{r}$;
     symmetry implies conservation of parity.
   - *Time inversion*, $t' = -t$.

---

for $\ell = 0, 1, 2, \ldots$. It was known from the study of optical spectra that transitions between two states with absorption or emission of radiation was accompanied by a change in angular momentum, $\Delta\ell = \pm 1$. From the symmetry properties of the electromagnetic field in such transitions it is known that the photon has parity -1. Parity is a *multiplicative quantum number*, i.e., the parity of a composite system is the product of the parities of its constituents. Thus in a radiative transition the total parity is conserved; e.g., in a transition $\ell \rightarrow \ell + 1$ the initial state has $P = (-1)^\ell$ and the final state $P = (-1)^{\ell+1}(-1) = (-1)^{\ell+2} = (-1)^\ell$.

Parity conservation was first found in electromagnetic interactions but it was quite natural to take it to be as generally valid as the conservation laws resulting from symmetries under continuous transformations. Indeed, it was found to hold generally if an *intrinsic parity* was assigned to each type of particle. Only the parity of one particle relative to another one can be measured; therefore the intrinsic parities of some particles are defined by general agreement, in particular, those of the proton and the neutron are set to be $P_p = P_n = +1$. Reactions between nucleons and pions were found to conserve parity if the pion was given the intrinsic parity $P_\pi = -1$.

The assumption of parity conservation proved useful in the derivation of particle quantum numbers from the experimental study of reactions between

---

Box II. Vector, Axial Vector, Scalar, and Pseudo Scalar

The following quantities are distinguished by their behaviour under space reflection:

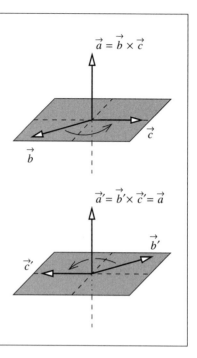

$$\vec{a} = \vec{b} \times \vec{c}$$

- An ordinary (or polar) *vector* **v**, just like the position vector **r**, simply inverts its direction, $\mathbf{v} = -\mathbf{v}'$. Examples are **b** and **c** in the figure.

- An *axial vector* defined as the vector product of two polar vectors, $\mathbf{a} = \mathbf{b} \times \mathbf{c}$, remains unchanged, i.e., $\mathbf{a}' = \mathbf{b}' \times \mathbf{c}' = \mathbf{a}$. An axial vector defines a *handedness*, indicated by the direction of **a** and the arrow in the form of an arc. It can be taken to symbolize a right-handed screw: If turned as shown by the arrow it moves forward along **a**. Because it describes a rotation also the spin is an axial vector. Conservation of parity implies that there is no preferred handedness.

$$\vec{a}' = \vec{b}' \times \vec{c}' = \vec{a}$$

- A *scalar s* is a number which stays unchanged. An example is the scalar product $s = \mathbf{a} \cdot \mathbf{b}$ of two vectors.

- A *pseudo scalar* does change sign. An example is the scalar product of a vector **d** with an axial vector **a**, which is the mixed product of three vectors,

$$p = \mathbf{d} \cdot \mathbf{a} = \mathbf{d} \cdot (\mathbf{b} \times \mathbf{c}) \quad , \qquad p' = \mathbf{d}' \cdot \mathbf{a}' = -\mathbf{d} \cdot \mathbf{a} = -p \quad .$$

p is positive if **d** and **a** include an angle smaller than $90°$ and negative otherwise.

---

particles until a severe problem turned up which was soon called the θ–τ *puzzle*. All strange mesons (Episode 68) were denoted by the letter K and different K mesons were distinguished by small Greek letters. Two such mesons, the θ^+ and the τ^+ with the decays

$$\theta^+ \to \pi^+ + \pi^0 \quad , \qquad \tau^+ \to \pi^+ + \pi^+ + \pi^- \quad ,$$

were apparently different, because they seemed to have different parities. Here is the reason, cut a little short: In neither decay is there an angular momentum; the parity of the τ^+ was computed to be the product of three pion parities, $P(\tau^+) = (-1)^3 = -1$, and, by the same token, $P(\theta^+) = (-1)^2 = +1$. When more accurate experiments were done it turned out that both the masses and the lifetimes of θ and τ were identical within the small measurement errors. Why should there be two particles with identical properties apart from their parities? The puzzle was intensively discussed at the Sixth Rochester Conference on High Energy Nuclear Physics at Rochester, New York, in April 1956. In an introductory talk Yang reported on a rather desperate solution he and Lee had constructed, postulating the existence of particle pairs with opposite parities but otherwise identical properties for odd strangeness numbers, not only for mesons but also for hyperons [2].

Yang[1] was already professor at the Institute for Advanced Studies in Princeton. In autobiographic notes woven into the commentaries of his *Selected Papers* [3] Yang, son of a professor of mathematics, tells of the good education in physics he received at the university in Kumming in spite of miserable ex-

---

[1] Chen Ning Yang (born 1922), Nobel Prize 1957

ternal conditions during the Sino-Japanese War. A fellow student, four years younger, was Lee[2]. The two young men came to know each other only when in 1946 both, on Chinese fellowships, became graduate students of Fermi in Chicago. Yang received his Ph.D. in 1948 with a thesis on nuclear physics. Lee got his in 1950 with work on the nuclear processes in white dwarf stars. Yang went to Princeton in 1949 and Lee to Berkeley in 1950. From 1950 to 1953 also Lee was on the staff of the Institute for Advanced Studies and worked closely with Yang. When Lee joined the faculty at Columbia University in 1953, mutual visits, twice a week, were arranged so that their intense collaboration could proceed.

Lee (left) and Yang.

Among the participants of the 1956 Rochester Conference was Feynman who shared a room with Martin Block, an experimentalist he had not met before. Block asked him why the rule of parity conservation was taken so seriously; what if it did not hold? After a discussion which lasted for the better part of the night, Block convinced Feynman to raise the question at the conference, claiming that nobody would listen if he himself would do it. In the conference proceedings we read [4]:

> *Feynman* brought up a question of Block's: Could it be that the θ and τ are different parity states of the same particle which has no definite parity, i.e., that parity is not conserved. That is, does nature have a way of defining right or left-handedness uniquely? Yang stated that he and Lee looked into that matter without arriving at any definite conclusions ... Perhaps one could say that parity conservation, or else time inversion invariance, could be violated. Perhaps the weak interactions could all come from this same source, a violation of spacetime symmetries ...

In the middle of April 1956, Yang moved to Brookhaven for some months and continued his twice-a-week meetings with Lee from there. One day in late April or early May, they had lunch together in a Chinese restaurant in Manhattan when Yang declared that one should dissociate the discussion of the symmetry for the production process of strange particles from that of the decay processes. Computations and a study of the literature showed that there was excellent evidence for parity conservation in strong interactions but that existing experimental data could not at all exclude parity violation in weak interactions. Specific experiments were called for to settle the question.

In a paper, entitled *Question of Parity Conservation in Weak Interactions* [5], completed in June 1956, Lee and Yang describe their findings and propose such experiments both in elementary particle physics and in nuclear β decay. They point out that certain quantities would, on the average, not be zero, if parity conservation was violated. These are *pseudo scalars*, which can be formed by three vectors or a vector and an axial vector (see Box II). The momentum of a particle is a vector, its spin an axial vector. The scalar product of both, essentially, is the projection of the spin on the particle's direction of motion, now called *helicity*. Observation of a preferred helicity, for instance, of the muons from the decay of pions ($\pi^+ \rightarrow \mu^+ + \nu$) would signal parity violation.

Another experiment Lee and Yang proposed became particularly famous, the study of the β decay of a cobalt nucleus into an excited nickel nucleus, an electron and an antineutrino, $^{60}\text{Co} \rightarrow \ ^{60}\text{Ni}^* + e^- + \bar{\nu}_e$. In this case the

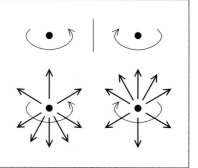

Box III. Mirror-Inverted Beta Decay Does Not Happen

Space reflection in which all three coordinates are inverted can also be thought of as rotation by $180°$ about some axis, followed by a simple mirror reflection on a plane perpendicular to that axis. For symmetry under rotation parity invariance is, therefore, equivalent to mirror invariance: If a certain process is observed, it must be possible to make the mirror image of that process happen.

The mirror image of a spinning ^{60}Co nucleus is a nucleus spinning the other way (upper figure). If we add to these two sketches the directions of some decay electrons, the mirror symmetry is broken, since electrons are emitted preferentially against the ^{60}Co spin direction (lower figure).

pseudo scalar of interest is the scalar product of the ^{60}Co spin with the average direction of motion of the decay electron. The idea of this experiment was born in a discussion between Lee and Wu, an experimental colleague at Columbia.

Wu[3], often respectfully called Madame Wu, had studied at Nanking and had come to Berkeley in 1936, where she received her Ph.D. in 1940. After teaching at Smith College and at Princeton University, she had joined Columbia University in 1944. During the war, she had worked on the atomic-bomb project and later she had done extensive experiments on β decay. Wu tells [6] how she pointed out to Lee that it had recently been possible to orient the spins of ^{60}Co nuclei in the direction of a magnetic field and that this technique should be used in an experimental test of parity conservation.

Madame Wu was eager to do the experiment herself. Polarization of ^{60}Co required temperatures as low as a hundredth of a Kelvin, which could be reached with paramagnetic demagnetization of certain crystals such as cerium–magnesium nitrate. If a thin layer containing radioactive ^{60}Co was grown on such a crystal, after cooling as much as 65% of these nuclei could be aligned along an external magnetic field. Wu knew that Ambler[4] had pioneered this technique in Oxford and was now working in a well-equipped low-temperature laboratory at the National Bureau of Standards in Washington. She readily got his agreement to cooperate with his group. For the experiment (see Box IV) the crystal with the ^{60}Co source was protected from heat by a housing of several larger crystals leaving an opening only in the direction of the detector for decay electrons. Source, housing, and detector were mounted in an evacuated glass tube which was immersed in a liquid-helium bath. After demagnetization cooling, a coil producing the aligning field was brought into position and the electron rate was registered while the sample slowly (in a time period of about 8 minutes) warmed up and lost its polarization. For polarized nuclei a strong asymmetry was detected: electrons are emitted preferentially in the direction opposite to the ^{60}Co spin. Were parity conserved there would be no preferred direction.

^{60}Co has a spin of $5\hbar$, ^{60}Ni one of $4\hbar$; the difference is carried away by the electron and the antineutrino, each of which has a spin of $\hbar/2$. The spin of the decay electron has to be parallel to the ^{60}Co spin, its preferential direction of

Wu

---

[3] Chien Shiung Wu (1912–1997)   [4] Ernest Ambler (born 1923)

Box IV. The Wu Experiment

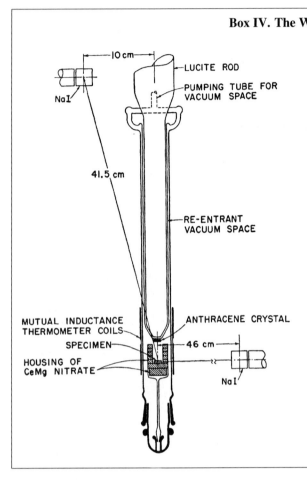

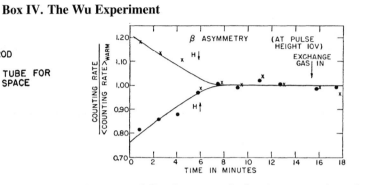

Apparatus (left): On a crystal of cerium magnesium nitrate (specimen), which can be cooled to extremely low temperature (0.01 K), there is a thin layer of radioactive cobalt. After cooling a magnetic field orients the spin of the ^{60}Co nuclei parallel or antiparallel to the vertical axis. Electrons from the β decay are detected if they hit the anthracene crystal in which they produce scintillation light. The Lucite rod serves as light guide allowing amplification and detection of the light signals outside the low-temperature region and the magnetic field. Two sodium-iodide crystals (NaI) detect γ rays emitted by the daughter nucleus ^{60}Ni* and allow an independent measurement of the ^{60}Co polarization. *Results* (above): The electron counting rate was observed for two orientations of the magnetic field, i.e., two orientations of the ^{60}Co spin, while the specimen was allowed to warm up and gradually lose it polarization. Emission is suppressed in the field direction and enhanced opposite to it. Both figures reprinted with permission from [7]. Copyright 1957 by the American Physical Society.

flight was found to be antiparallel to it. Therefore the electron preferentially has negative helicity; it is preferentially *left-handed*. The antineutrino, on the other hand, is *right-handed*. We shall see in Episode 79 that Lee and Yang, on the basis of Madame Wu's experiment, successfully argued that the antineutrino can only be a right-handed particle and its antiparticle, the neutrino, only be left-handed.

Pauli was eagerly awaiting the results of Wu's experiments. Always sceptical until thoroughly convinced, he opposed the idea of symmetry breaking in one type of interaction only. Those were not yet the times of instant global communication and so it happened that still on 17 January 1957 Pauli wrote from Zurich to Weisskopf [8]:

> I don't believe that the Lord is a weak left-hander and I am ready to bet a very high sum that the experiment will give a symmetric angular distribution of electrons. I do not see any logical connection between the strength of an interaction and its mirror invariance.

Two days earlier, on 15 January, a press conference had been held at Columbia University, chaired by Rabi, in which not only Wu's results [7] were presented but also those of a second experiment [9], which is less known but not less remarkable. The next morning the front-page headline of the *New York Times* read 'Basic Concept in Physics Upset in Tests'.

The story of the second experiment is told by Garwin[5], one of its authors [10]. He held a Ph.D. from the University of Chicago and at the time was directing a project for IBM at Poughkeepsie. But he was keeping close contacts with his colleagues in physics at Columbia when he was in New York City. On the evening of 4 January 1957, he got a telephone call from Lederman who worked at the NEVIS Laboratory in Irvington-on-the-Hudson, where the Columbia Physics Department had constructed and was now running a 385-MeV synchrocyclotron. Lederman knew already of Wu's results and wanted to measure the helicity of muons from pion decay predicted by Lee and Yang.

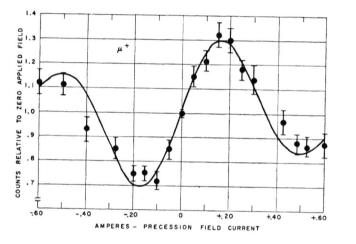

Positron counting rate as a function of coil current, observed in the experiment of Garwin, Lederman, and Weinrich. Figure reprinted with permission from [9]. Copyright 1957 by the American Physical Society.

Garwin and Lederman designed an ingenious experiment and began measurements that very evening. Positive pions from the synchrocyclotron were stopped in carbon and then allowed to decay. The decay muons, polarized along the line of flight of their parent, were stopped in a carbon target. A coil was wound around the target producing a magnetic field perpendicular to that line of flight. The muon magnetic moment and the muon spin could precess in the field and with it the asymmetry of the positron direction from the muon decay. Positrons were registered by counters under a fixed angle after allowing a time delay of about one microsecond for precession from the moment the muons came to rest. The angle of precession was determined by the magnetic field, i.e., by the electric current in the coil. Varying that current Garwin, Lederman, and Weinrich, a graduate student, measured the asymmetry of the muon β decay and demonstrated parity non-conservation. But they also measured for the first time the spin and the magnetic moment of the muon. The experiment was completed after only four days. The paper [10] was printed directly after that by Wu and her collaborators whose priority is graciously acknowledged.

---

[5] Richard L. Garwin (born 1928)

[1] Polkinghorne, J. C., *At The Feet of Dirac*. In Kursunoglu, B. N. and Wigner, E. P. (eds.): *Paul Adrien Maurice Dirac*. Cambridge University Press, Cambridge, 1987.

[2] Lee, T. D. and Yang, C. N., *Physical Review*, **102** (1956) 290.

[3] Yang, C. N., *Selected Papers*. Freeman, San Francisco, 1983.

[4] Ballam, J. (ed.), *High Energy Nuclear Physics: Proceedings of the 6. Annual Rochester Conference*, p. VIII 27. Interscience, New York, 1957.

[5] Lee, T. D. and Yang, C. N., *Physical Review*, **104** (1956) 254.

[6] Wu, C. S., *The Discovery of Nonconservation of Parity in Beta Decay*. In Novick, R. (ed.): *Thirty Years Since Parity Nonconservation*. Birkhäuser, Boston, 1988.

[7] Wu, C. S., Ambler, E., Hayward, R. W., Hoppes, D. D., and Hudson, R. P., *Physical Review*, **105** (1957) 1413.

[8] Pauli, W., *Wissenschaftlicher Briefwechsel*, vol. IV, part IV, A:1957, p. 82. Springer, Berlin, 2005.

[9] Garwin, R. I., Lederman, L. M., and Weinrich, M., *Physical Review*, **105** (1957) 1415.

[10] Garwin, R. L., *Demonstration of Parity Nonconservation in the π–μ–e Chain*. In Novick, R. (ed.): *Thirty Years Since Parity Nonconservation*. Birkhäuser, Boston, 1988.

Superconductivity Explained by Bardeen, Cooper, and Schrieffer (1957)

In May 1950, Bardeen received a phone call from Bernard Serin from Rutgers University in New Brunswick which alerted him to a new experimental development in superconductivity. Bardeen was still working at the Bell Laboratories, where, a little more than two years ago, he and Brattain had invented the first transistor (Episode 69). Since Shockley, the leader of their research group, was pursuing the development of the next type of transistor on his own, Bardeen was looking for a new research topic. The phone call made him return to superconductivity, a field on which he had tried his hand nearly ten years earlier at Harvard. Before we can discuss the work of Bardeen and his group [1,2] we have to mention at least a few of the advances in superconductivity since its discovery in 1911 (Episode 18).

In 1933, Meissner[1] and his assistant Ochsenfeld[2] at the Physikalisch-Technische Reichsanstalt in Berlin found an unexpected effect [3]: Superconductors expel magnetic fields. Its discovery had important experimental and theoretical consequences. Since the superconducting material decreases the external field, it is diamagnetic and since it, in fact, reduces it to zero, it is said to show *perfect diamagnetism*. Diamagnetism is much easier to measure than conductivity and therefore the Meissner–Ochsenfeld effect is often used as experimental indication for superconductivity.

A theoretical description of superconductivity, stimulated by the Meissner–Ochsenfeld effect, was formulated in 1934 by the brothers Fritz London[3] and Heinz London[4]. They had fled from Nazi Germany and worked in Oxford, helped by the initiative of Lindemann to support refugee scientists. The London brothers were able to formulate an equation, now called the *London equation* [4], which supplements Maxwell's equations of electrodynamics in the case of superconductors in much the same way as they are supplemented by Ohm's law for ordinary conductors (see Box I). In this way superconductivity could be well described although not explained. The London equation could be derived from a model, called the *London theory*, in which it was assumed that electrons in a superconductor can be of two kinds: One permits superconductivity, the other does not. In this theory, the energies of the two kinds are separated by a small but significant gap. Superconducting electrons become normally conducting only if they gain at least the gap energy.

Since electrons are particles of spin $\frac{1}{2}\hbar$, i.e., fermions, they have to obey the Pauli principle and Fermi–Dirac statistics (Episode 41). For a free Fermi gas at zero absolute temperature all energy states up to the *Fermi energy* E_{F}

Meissner   Ochsenfeld

Heinz London (left) and Fritz London.

[1] Fritz Walther Meissner (1882–1974)   [2] Robert Ochsenfeld (1901–1993)   [3] Fritz London (1900–1954)   [4] Heinz London (1907–1970)

Box I. Meissner–Ochsenfeld Effect and London Equation

Consider the two experiments sketched on the margin. In the column on the left, a sample is made superconducting by cooling it below the critical temperature. Then a magnetic field is turned on which, however, cannot enter the interior of the sample. That is not surprising since no electric field can exist in the superconductor. By Faraday's law of induction for zero electric field there cannot be a change in magnetic field: the latter has to remain zero in the superconductor. This is brought about by currents, induced on the surface of the superconductor, which generate a magnetic field cancelling the external field inside the sample and deforming it outside it.

The experimental result, sketched in the right-hand column, the *Meissner–Ochsenfeld effect*, by which a magnetic field is expelled from a sample when this becomes superconducting, is not so easily explained. To describe it, an extra equation is needed in addition to those of classical electrodynamics. That is the *London equation*

$$\mathbf{j} = -\frac{1}{\mu_0 \lambda^2} \mathbf{A} \quad ,$$

which relates the electric current density $\mathbf{j}$ in the superconductor to a vector potential $\mathbf{A}$, from which the magnetic field is computed by differentiation. (For a superconductor the London equation replaces Ohm's law $\mathbf{j} = \sigma \mathbf{E}$, which relates $\mathbf{j}$ to the electric field $\mathbf{E}$ by the conductivity σ; in a superconductor σ becomes infinite.) A significant current flows only within a thin layer near the surface. The layer thickness is characterized by the *penetration depth* λ; μ_0 is the magnetic field constant of electrodynamics.

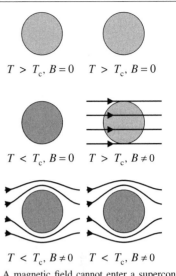

$T > T_c, B = 0 \qquad T > T_c, B = 0$

$T < T_c, B = 0 \qquad T > T_c, B \neq 0$

$T < T_c, B \neq 0 \qquad T < T_c, B \neq 0$

A magnetic field cannot enter a superconducting sample (left). If there is a field while the sample is still normally conducting, it is expelled as soon as the temperature drops below the value T_c at which superconductivity sets in (right).

---

are occupied; all states of higher energy are empty. An electron possessing the Fermi energy has a momentum of modulus p_F, the *Fermi momentum*. In momentum space, spanned by the three components p_x, p_y, p_z of momentum, all electrons with the Fermi energy lie on the *Fermi surface*, which is simply the sphere of radius p_F. All others lie inside it. At finite temperature some states outside the Fermi surface are occupied, leaving empty states inside it; the rigid surface of the sphere becomes somewhat diffuse. For electrons in a conductor or superconductor the same concept is used. An electric current is carried by states near the Fermi surface; it leads to an asymmetry favouring momenta pointing in one direction. The gap in the energy of states, needed for the description of superconductivity, therefore had to lie at the Fermi energy.

The new experimental fact, of which Bardeen was told on the phone by one of its discoverers in May 1950, was the *isotope effect*. Through the war activities in nuclear physics, samples of several elements had become available in the form of pure or greatly enriched isotopes. It had now been found independently at the National Bureau of Standards [5] and at Rutgers University [6] that the critical temperature, below which the sample is superconducting, falls as the atomic mass of the mercury isotope increases. Now it was clear that the atoms of the superconductor play and important role and Bardeen outlined a theory [7,8] in which an electron exchanges energy and momentum with a group of atoms, extended over a macroscopic region of about a millionth of a centimetre. Such an exchange of energy and momentum with the lattice of atoms can be described quantum mechanically as the exchange of a quasiparticle, first studied and named *phonon* by Frenkel (see Box I in Episode 62).

Bardeen

The energy gap, Bardeen suggested, appears because electrons near the Fermi surface either lose or gain energy by phonon exchange with the lattice and thus are suppressed below or lifted above the Fermi energy, respectively. A similar proposal had been made little earlier by Fröhlich[5] even before the isotope effect was found. Fröhlich, a student of Sommerfeld and another refugee from Nazi Germany, was professor at Liverpool, then on leave at Purdue University in Indiana.

Bardeen preferred to work on basic research at a university and was soon offered a professorship in electrical engineering and physics at the University of Illinois in Urbana where he moved with his family in 1951. He undertook a concentrated study of the literature and wrote a comprehensive review article on the theory of superconductivity for the *Handbuch der Physik* which appeared in 1956. Feeling that additional skill in modern theoretical methods was needed, he invited Cooper, who had been recommended by C. N. Yang, to join his group. Cooper[6], a New Yorker, had studied physics at Columbia University, obtained his Ph.D. there in 1954, and worked at the Institute for Advanced Studies in Princeton. He arrived at Urbana in September 1955 and began his work by delivering a series of seminar talks on quantum field theory; quantum electrodynamics had only recently been completed. Cooper shared the office with Bardeen; closely working with them was Schrieffer, a graduate student who had helped Bardeen to proofread his *Handbuch* article and, inspired by it, had chosen superconductivity as subject for his Ph.D. thesis. Schrieffer[7] was born in Illinois. He had studied at MIT and had gone to Urbana for graduate studies. The group was striving, as many had before them, for a microscopic theory of superconductivity, i.e., a full quantum-mechanical description of electrons in the superconductor.

An important step towards this goal was done in 1955 when Cooper studied the interaction of just two electrons near the Fermi surface [9]. He found that for opposite momenta of the two electrons their total energy is lowered by their interaction through phonons in spite of the electrostatic repulsion (see Box II). Such a *Cooper pair* is described by one wave function extending over a rather large region in space. A minimum amount of energy is needed to break it up.

What was needed in the end was the construction of a single wave function for all electrons contributing to superconductivity taking into account this pairing. Bardeen suggested this task as subject for his thesis to Schrieffer, who finally succeeded while Bardeen was in Sweden to receive his first Nobel Prize. The full wave function of the superconducting state is *rigid*, i.e., it is extended in space and can move through the superconductor unaffectedly, unless a minimal amount of energy is absorbed by the state. At sufficiently low temperature, however, that amount is not available.

Bardeen, Cooper, and Schrieffer published their theory, the *BCS theory* as it came to be called, in 1957, first as a short letter [10] and half a year later as an extensive paper [11], which also includes comparisons with experimental data. The theory explained the known phenomena of superconductivity. Needless to say that soon simpler phenomenological theories like the London theory could be derived from it.

Cooper

Schrieffer

---

[5] Herbert Fröhlich (1905–1991)   [6] Leon Neil Cooper (born 1930), Nobel Prize 1972   [7] John Robert Schrieffer (born 1931), Nobel Prize 1972

The existence of Cooper pairs, the most important feature of the theory, was established in 1961 by Doll[8] and Näbauer[9] [12] in an institute in Herrsching, Bavaria, headed by Meissner, and simultaneously with a complementary method by Deaver[10] and Fairbank[11] [13] at Stanford University. They measured magnetic *flux quantization*, predicted by Fritz London in 1948 [14].

Box II. Cooper Pairs

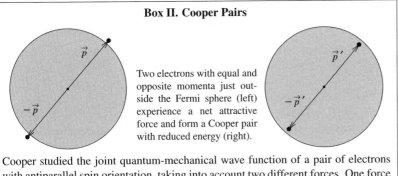

Two electrons with equal and opposite momenta just outside the Fermi sphere (left) experience a net attractive force and form a Cooper pair with reduced energy (right).

Cooper studied the joint quantum-mechanical wave function of a pair of electrons with antiparallel spin orientation, taking into account two different forces. One force is the electrostatic repulsion between the electrons, modified ('screened') inside the lattice structure of the superconductor. The second is due to the exchange of a phonon between the two electrons, i.e., their indirect interaction through exchange of energy and momentum with the atomic lattice. Let us assume that both electrons have the same energy, slightly higher than the Fermi energy E_F. Their momenta, therefore, have the same modulus slightly above the Fermi momentum p_F and can differ only in direction. Cooper found that for opposite or very nearly opposite directions the two electrons form a bound state with a total energy less than $2E_F$. It is separated by a tiny energy gap from other allowed states of the pair. The joint wave function of a Cooper pair has a macroscopic spatial extension on the order of a thousandth of a millimetre exceeding by far atomic dimensions. No net electric current is carried by a Cooper pair with exactly equal and opposite momenta $\mathbf{p}'$ and $-\mathbf{p}'$, respectively; but that is easily changed by increasing slightly the magnitude of one momentum and decreasing that of the other.

As we saw, a magnetic field cannot enter a superconductor; but that statement holds true only below a certain critical field (which depends on temperature). Once that is surpassed, the field enters either abruptly (in superconductors of type I) or gradually (in those of type II). In the latter case small ring-like persistent currents appear, not unlike the electric current on a Bohr orbit in an atom. They give rise of a flux of magnetic field, which is quantized in much the same way as the magnetic moment due to an orbiting electron. The magnitude of the *flux quantum*, the smallest value of the magnetic flux of such a persistent current, is h/q; q is the electric charge carrying the current and h is Planck's constant. In both experiments the quantization was clearly observed with $q = 2e$, the charge of two electrons, that is, the charge of a Cooper pair. (The smallest possible, elementary flux quantum is $\Phi_0 = h/e$.)

[1] Hoddeson, L., *Journal of Statistical Physics*, **103** (2001) 625.

[2] Hoddeson, L., Braun, E., Teichmann, J., and Weart, S., *Out of the Crystal Maze – Chapters from the History of Solid-State Physics*. Oxford University Press, Oxford, 1992.

[3] Meissner, W. and Ochsenfeld, R., *Naturwissenschaften*, **21** (1933) 787.

[4] London, F. and London, H., *Proceedings of the Royal Society*, **A149** (1935) 71.

[5] Maxwell, E., *Physical Review*, **78** (1948) 477.

[6] Reynolds, C. A., Serin, B., Wright, W. H., and Nesbit, L. B., *Physical Review*, **78** (1948) 487.

[7] Bardeen, J., *Physical Review*, **79** (1948) 167.

[8] Bardeen, J., *Physical Review*, **80** (1948) 567.

[9] Cooper, L. N., *Physical Review*, **104** (1956) 1189.

[10] Bardeen, J., Cooper, L. N., and Schrieffer, J. R., *Physical Review*, **106** (1957) 162.

[11] Bardeen, J., Cooper, L. N., and Schrieffer, J. R., *Physical Review*, **108** (1957) 1175.

[12] Doll, R. and Näbauer, M., *Physical Review Letters*, **7** (1961) 51.

[13] Deaver, D. S. and Fairbank, W. M., *Physical Review Letters*, **7** (1961) 43.

[14] London, F., *Physical Review*, **74** (1948) 652.

---

[8] Robert Doll (born 1923)   [9] Martin Näbauer (1910–1962)   [10] Bascom Sine Deaver (born 1930)   [11] William Martin Fairbank (1917–1989)

Weak Interaction Better Understood – The V − A Theory (1957)

Fermi's 'theory of β rays' (Episode 53) or, as it came to be called, the theory of weak interaction, was decisively advanced by two experimental discoveries. The first was the realization that the muon observed in cosmic rays (Episode 57) was not the meson proposed by Yukawa. Studies of its decay led to the notion of *universal Fermi interaction*. The second discovery was that of parity non-conservation (Episode 77). Still in 1957 it led to a formulation of the theory of weak interaction much more precise than the one originally proposed by Fermi in 1933.

Pontecorvo[1] [1,2], whom we briefly met in Episode 55 as a student of Fermi in Rome, had joined Joliot's group in Paris in 1936. After the occupation of France, he fled first by bicycle to Toulouse and then by train to Lisbon, from where he reached the United States. For some years he worked in oil prospecting, developing a neutron activation method, which gave information about the composition of the bedrock surrounding a borehole. In 1943 he joined the staff of the newly founded Anglo-Canadian nuclear research laboratory in Montreal and contributed to the design af a reactor at Chalk River, Ontario. When the reactor went critical, on 22 July 1947, Pontecorvo was present in the control room. In 1949 Pontecorvo moved to Bristol and in 1950 he suddenly showed up with his family in the Soviet Union. From then on he worked at the Joint Institute of Nuclear Research at Dubna near Moscow. Under the impression of fascism in Italy he had joined the clandestine Italian communist party in 1936 and was a fervent believer in communism. Of course, the media and some historians suspected him to be a spy. But no serious accusations have ever been voiced against him.

In June 1947, Pontecorvo had sent an important note, less than a page long, to the *Physical Review* [3]. (For his reminiscences about it see [4].) The paper begins from the by then well-established fact that the muon is not the Yukawa meson because it does not show the required strong interaction with matter. Pontecorvo points out that, on the other hand, the probability for the capture of negative muons by nuclei is similar to that of *K capture* of electrons (see end of Episode 53) if allowance is made for the mass difference between muon and electron. If one assumed the muon to possess a spin of $\hbar/2$ (rather than zero as required for the Yukawa meson), one could describe muon capture, for instance, that by a proton yielding a neutron and a neutrino,

$$\mu^- + p \to n + \nu \quad ,$$

by using Fermi's theory of β decay, including the Fermi coupling constant. Since in the Fermi theory all interactions involved four fermions, Pontecorvo

Пшую Понтекорви

Pontecorvo

---

[1] Bruno (Maximovich) Pontecorvo (1913–1993)

proposed as decay products of the muon an electron and two neutrinos (which we denote, more specifically, as a neutrino and an antineutrino),

$$\mu^- \to e^- + \nu + \bar\nu \quad , \qquad \mu^+ \to e^+ + \nu + \bar\nu \quad .$$

This decay, if it existed, was particularly interesting because it did not involve nucleons, only leptons, particles which show weak and, possibly, electromagnetic but no strong interaction.

Before Pontecorvo's note, still in the spirit of the meson theory, the muon had been assumed to decay into two particles only, an electron and a neutrino. Experimentally, a two-particle decay has a simple signature. If the decaying particle is at rest, the two decay particles have equal and opposite momenta and each has a fixed, discrete energy. Thus for the decay $\mu^+ \to e^+ + \nu$ at rest, the energy spectrum of the positrons would be discrete and contain only one energy. For a three-body decay, on the other hand, the spectrum is continuous.

In 1949, Steinberger, as his thesis work under Fermi in Chicago, measured the energy spectrum of decay electrons from cosmic-ray muons [5]. His results excluded a simple two-body decay of the muon and were identified by theoreticians as evidence for the decay into an electron and two neutrinos [6].

Fermi had devised his theory of β rays in analogy to quantum electrodynamics. It was therefore natural to now use Feynman diagrams to describe it, to compare it to the more precise experimental data then available and to generalize it, if necessary. The central idea in Fermi's theory was the pair creation of an electron and an antineutrino. It is convenient to consider one member of the pair as incoming, the other one as outgoing. That is always possible, since, in a Feynman diagram, an outgoing particle is equivalent to an incoming antiparticle. Diagrams can therefore be drawn, composed of two continuous lines, each with a section for an incoming and an outgoing fermion with a difference in electric charge between incoming and outgoing particle. If we denote by a, c and b, d the particles corresponding to these lines, the diagram represents the process $a + b \to c + d$. The most general form, describing this process, contains five terms, each with its own coupling constant (see Box). (Fermi had been aware of this fact but chosen to consider only one term, the vector term V, most similar to the case of quantum electrodynamics.) For several years, available data on β decays were analysed in order to obtain numerical values of the five coupling constants but no clear picture was arrived at.

Usually, a symmetry or conservation law simplifies a theory. For the weak interaction the contrary was true; it was simplified by parity violation. When, in January 1957, they knew the result of the Wu experiment, Lee and Yang postulated that there exist only left-handed neutrinos and right-handed antineutrinos [7]. (Such an assumption was possible only if the neutrino mass, like the photon mass, was assumed to be exactly zero. In this case neutrinos always have the speed of light and it is impossible for an observer to 'overtake' it. Were that possible, then, for that observer, the neutrino would change from left-handed to right-handed.) This theory has interesting properties:

- Of the four components of the Dirac spinor, representing the neutrino wave function, only two are different from zero. One speaks of the *two-component theory* of the neutrino.

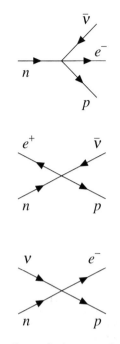

Feynman diagram for the neutron decay $n \to p + e^- + \bar\nu$ (top) and diagrams derived from it by replacing an outgoing particle by an incoming antiparticle, $e^+ + n \to \bar\nu + p$ (middle) and $\nu + n \to e^- + p$ (bottom).

The General Form of the Weak Interaction if Parity were Conserved

Fermi's bilinear form for the interaction Hamiltonian reflects itself in the form of the matrix element $\mathcal{M}$,

$$\mathcal{M} = \sum_{\ell=1}^{5} g_\ell j_\ell^{(ac)} j_\ell^{(bd)} + \text{Hermitian conjugate} \quad,$$

where the expressions

$$j_\ell^{(ac)} = (\bar{\Psi}_c O_\ell \Psi_a) \quad, \qquad j_\ell^{(bd)} = (\bar{\Psi}_d O_\ell \Psi_b) \quad,$$

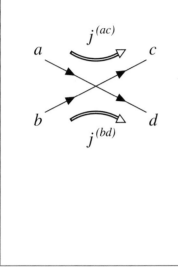

which describe the transitions $a \to c$ and $b \to d$, respectively, are called charge-changing currents or, slightly misleading, simply *charged currents*. Each is bilinear in the Dirac spinors (for instance, Ψ_a, Ψ_c) of an incoming and an outgoing particle ($\bar{\Psi} = \Psi^{*+}\gamma_0$, where the * stands for complex conjugation of all four components of the spinor, where the $^+$ transforms a column spinor into a row spinor, and where γ_0 is one of the Dirac gamma matrices, Episode 43). There are five different ways to combine two spinors so that $|\mathcal{M}|^2$, as required, becomes invariant under Lorentz transformations. They differ in the choice of the operator O_ℓ, a four-by-four matrix built from the unit matrix and Dirac matrices. Depending on the choice of O_ℓ the charged currents have different transformation properties in spacetime, denoted as scalar (S), pseudo scalar (P), (four-)vector (V), axial vector (A), and tensor (T), see also Box II in Episode 77. In this general form of the weak interaction there are five coupling constants g_ℓ.

- The mirror image of the (left-handed) neutrino, constructed by application of the parity operator P on the wave function, does not exist. It would be a right-handed neutrino, since P reverses the direction of motion (a vector) but not that of spin (an axial vector).
- Also the state obtained by applying the so-called *charge-conjugation* operator C on the neutrino wave function does not exist. The operator C reverses the sign of the electric charge and of other charge-like quantum numbers and turns a particle into the corresponding antiparticle. Space-oriented quantities like momentum and spin are unchanged.
- However, the combined application of C and P changes the wave function of a neutrino into that of an antineutrino and vice versa.

In the two-component theory the neutrino ν is always left-handed, its spin (wide arrow) is antiparallel to its momentum (slim arrow). The operations of parity P and charge conjugation C lead to unphysical states shown in square brackets. Only the combined operation CP yields the right-handed antineutrino.

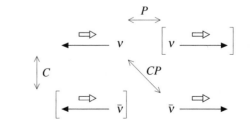

Lee and Yang had introduced projection operators P_L and P_R which changed a general four-component spinor into one which was only left-handed or only right-handed, respectively. They look a little frightening if written in terms of the unit matrix and the matrix γ_5, which is the product of all four Dirac matrices; but they are quite simple four-by-four matrices. In April 1957, Lee presented a talk on the new neutrino theory at the Seventh Rochester Conference on High Energy Nuclear Physics. This, by the way, was the last of a series of annual meetings held at Rochester, New York. It then became a large international conference organized every second year in a different country. Like the year before, when he had raised the possibility of parity violation for his room mate Block, Feynman participated. This time he stayed with his sister in nearby Syracuse. Lee had given him a copy of the paper beforehand. In his inimitable way Feynman tells [8] (the narration is repeated with more details in [9]) how he told his sister in the evening that he could not understand the paper by Lee and Yang and how she replied, 'what you mean is *not* that you can't understand it, but that you didn't *invent* it. You didn't figure it out your *own* way from hearing the clue. What you should do is imagine you're a student again, and take this paper upstairs, read every line of it, and check the equations. Then you'll understand it very easily.'

Feynman took the advice and now found the paper 'very obvious and simple'. Moreover, he extended the idea by Lee and Yang as follows. Leptons with mass, i.e., the electron and the muon, are represented by spinors, which can be seen as superpositions of left-handed and right-handed states. Instead of the full spinor Ψ he used only the left-handed part $P_{\mathrm{L}}\Psi$ in the expression for the weak interaction. Feynman was not scheduled to speak at the conference but a kind colleague let him have five minutes of his own time to sketch the idea and predictions obtained from it. His work aroused little interest, and he himself went to Brazil for a few months.

On his return to Caltech, colleagues told him about problems with the interpretation of recent experiments. Feynman quotes them like this [8]: 'The situation is so mixed up that even some of the things they've established for *years* are being questioned—such as the beta decay of the neutron is S and T. It's messed up. Murray [Gell-Mann] says it might even be V and A.' At that point Feynman jumped up, exclaiming: 'Then I understand EVVVVVERYTHING!' He had realized that, if the operator P_{L} was applied not only to lepton spinors but to all spinors (also to those of the neutron and the proton in β decay), all terms corresponding to scalar (S), pseudo scalar (P), and tensor (T) coupling would vanish automatically and only vector (V) and axial vector (A) coupling would contribute to weak interactions. Two coupling constants remained, but these were related to one another. A relative minus sign appears between the contributions of V and A. Therefore the theory of left-handed coupling is called $V − A$ (V minus A) theory. In this way there was only a single independent constant which could be taken to be the original Fermi coupling constant G_{F}. That was a constant of nature given by experiment. Still that night, Feynman used his new theory to compute the half-life of the muon and found it to be 9% off the experimental value. When he reported the success to his colleagues the next morning, he learned that he had used an obsolete numerical value of G_{F}. Taking the correct value, he found that his result was good to 2%, which was also the estimated experimental error.

Feynman was very fond of this achievement. He considered the decipherment of the exact form of the weak interaction as the discovery of a law of

[1] Bilenky, S. M. et al. (eds.), *B. Pontecorvo, Selected Scientific Works.* Editrice Compositori, Bologna, 1997. Also contains biographical essays on Pontecorvo.

[2] Bonolis, L., *American Journal of Physics*, **73** (2005) 487.

[3] Pontecorvo, B., *Physical Review*, **72** (1947) 246.

[4] Pontecorvo, B. M., *Recollections on the establishment of the weak-interaction notion.* In Brown, L. M., Dresden, M., and Hoddeson, L. (eds.): *Pions to Quarks*, p. 367. Cambridge University Press, Cambridge, 1989.

[5] Steinberger, J., *Physical Review*, **75** (1949) 1136.

[6] Tiomno, J. and Wheeler, J. A., *Reviews of Modern Physics*, **21** (1949) 144.

[7] Lee, T. D. and Yang, C. N., *Physical Review*, **105** (1957) 1671.

[8] Feynman, R. P., *"Surely You're Joking, Mr. Feynman!" – Adventures of a Curious Character.* Norton, New York, 1985.

[9] Mehra, J., *The Beat of a Different Drum – The Life and Science of Richard Feynman.* Clarendon Press, Oxford, 1994.

[10] Feynman, R. P. and Gell-Mann, M., *Physical Review*, **109** (1958) 193.

[11] Henley, M. and Lustig, H., *Biographical Memoirs (Nat. Acad. Sci.)*, **76** (1998) 2.

[12] Sudarshan, E. C. G., *Midcentury Adventures in Particle Physics.* In Brown, L. M., Dresden, M., and Hoddeson, L. (eds.): *Pions to Quarks*, p. 485. Cambridge University Press, Cambridge, 1989.

[13] Sudarshan, E. C. G. and Marshak, R. E., *Physical Review*, **109** (1958) 1860.

nature comparable to (although not quite as important as) the Dirac equation or Maxwell's equations and he valued it higher than his contributions to quantum electrodynamics. He soon found that he was not the sole discoverer but that did not diminish his pleasure. Gell-Mann, who had been on vacation when Feynman returned to Caltech, had done similar work. The two decided to write a common paper on the theory [10] which, therefore, is also known as *Feynman–Gell-Mann theory.*

Gell-Mann and Feynman

The $V − A$ theory had already been worked out in Rochester, even before Feynman had his inspiration there, by Marshak[2] [11], the initiator of the Rochester Conferences, and Sudarshan[3], his doctoral student. Marshak was born in the Bronx, New York City, as son of poor Russian immigrants from Minsk. Highly gifted and widely interested, he studied at the College of the City of New York and at Columbia University and, at the age of twenty-two, completed his Ph.D. under Bethe at Cornell. He joined the staff of the University of Rochester in 1937 and, after war work (mainly at Los Alamos), transformed the physics department there into a leading centre of research. In doing so, he attracted promising graduate students from abroad, such as Sudarshan who came from the Tata Institute in Bombay. Much later Sudarshan gave a detailed account of the events in 1957/58 [12]. Together with Marshak he had worked out the theory and it was agreed that he would present it at the conference when Marshak, the conference chairman, decided that as a graduate student Sudarshan could not be an official delegate and thus not give a talk. Marshak himself gave a principal lecture on another topic. Their common work was first presented by Marshak at a conference in Padua and Venice. A publication in a regular journal [13] appeared only several months after that by Feynman and Gell-Mann.

---

[2] Robert Eugene Marshak (1916–1992)   [3] Ennackal Chandy George Sudarshan (born 1931)

Keeping Ions in a Trap (1958)

An important element of the first maser (Episode 73) was an electrostatic 'focusser' which acted on the electric dipole moment of neutral molecules. It was based on the use of multipole fields pioneered by Paul in Göttingen and brought to perfection in his first years in Bonn.

Paul[1] grew up in Munich where his father was professor of pharmaceutical chemistry at the university. In his later years, Paul told me about the visit he paid to Sommerfeld before he began with his studies of physics. 'Since you don't intend to become a theoretician', Sommerfeld said, 'I advise you to attend the *Technische Hochschule* [now Technical University]. There you will learn how to calculate. (Da lernen Sie wenigstens rechnen.)' Sommerfeld, himself a former mathematician, knew of the different ways in which mathematics was (and still is) taught at classical and technical universities in Germany. In 1934 Paul moved from Munich to the Technical University of Berlin. He began to do research under Kopfermann[2], a former student of Franck and a specialist in hyperfine spectroscopy. Paul obtained his diploma in 1937 in Berlin and then followed Kopfermann to Kiel where the latter had become professor at the university. For his doctoral thesis, the measurement of nuclear moments of beryllium, which was completed in 1939, he developed an atomic beam as the best light source for optical spectroscopy. After a short spell of military service Paul rejoined Kopfermann in 1940 who, by then, had moved to Göttingen. He became Privatdozent there in 1944.

In 1951, together with Friedburg, Paul published his first paper on multipole focussing [1]. A magnetic moment experiences a force in an inhomogeneous magnetic field. (This fact was the basis of the Stern–Gerlach experiment, Episode 30.) If the direction of an atomic beam is along the z axis and if the force acts towards the axis and rises linearly with the distance from it, then diverging beam particles are brought back to the axis; they are focussed. Since the force is proportional to the gradient of the field, the absolute value of the field has to rise quadratically, and that of its potential with the third power. Friedburg and Paul pointed out that a sextupole field has these properties. It can be realized by six pole shoes, oriented parallel to the z axis and alternating in polarity: north, south, north, etc. The direction of the force depends on the component of the magnetic moment with respect to the magnetic field. Depending on its sign, particles are focussed or defocussed. For the maser the method was applied to electric dipole moments. Since these are proportional to the (electric) field, only a quadrupole field was required.

In 1952 Paul became professor of physics at the University of Bonn and was followed there by several members of his Göttingen group. In 1953, together

Paul

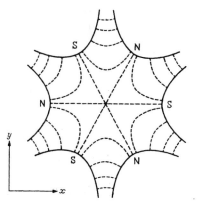

Form of the pole shoes (continuous lines) and field lines (broken lines) of a magnetic sextupole. Reprinted with kind permission from Springer Science+Business Media: [1], Fig. 1.

---

[1] Wolfgang Paul (1913–1993), Nobel Prize 1989   [2] Hans Kopfermann (1895–1963)

with Steinwedel he wrote a paper, less than two pages long, on the possible application of time-dependent electric quadrupole fields [2]. The patent application of the same year [3] contains more details. This time the focussing of ions, i.e., charged particles, in an electric quadrupole field was discussed. The field could be realized by four metallic rods parallel to the beam. If the field is constant in time, focussing occurs only in the plane containing the two rods with a voltage repelling the beam particles. The rods in the perpendicular plane attract, i.e, defocus, the beam. Therefore Paul and Steinwedel give the voltage between pairs of rods a periodic time dependence. In this case, the field pattern rotates with high frequency about the beam axis. Particles successively experience focussing and defocussing. The result can be an overall focussing effect.

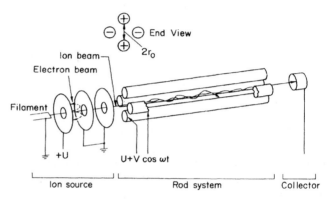

Paul's quadrupole mass spectrometer or mass filter. From [4]. © Nobel Foundation. Reprinted with permission.

Paul and Steinwedel note that the situation is analogous to the principle of *strong focussing* in alternating-gradient synchrotrons (Episode 64). There the alternation is brought about by the geometry, here by the high-frequency time structure of the field. In Paul's institute in Bonn research concentrated on two rather different areas, which were connected at this very point, low-energy atomic and molecular physics with ions and high-energy particle physics with accelerators. With his group he designed and set up in Bonn the first alternating-gradient machine in Europe, a 500-MeV electron synchrotron. (As Paul's student, for my diploma in 1959, I built a bubble chamber and the electronics to operate it at that machine and registered the first pions produced by it and decaying in my chamber.)

In his Nobel Lecture [4] Paul demonstrated the focussing of a quadrupole field with a mechanical analogue model: A Perspex plate, machined to a saddle form, is rotated about the vertical symmetry axis; a small steel ball stays permanently near the centre.

Ions are transmitted through the system of rods if certain conditions are met. These are determined by the specific charge, i.e., the quotient q/m of charge q and mass m of the ion and the form of the field. That is given by the distance r_0 of the rods from the axis and the voltage between pairs of rods, $U + V\cos(\omega t)$, which is the sum of a constant term and a term oscillating with the circular frequency ω. The four parameters U, V, r_0, ω can be chosen in such a way that only a narrow band of specific charges q/m is focussed and passes the apparatus. Ions outside the band end their paths on the rods. The

Snapshot of Paul in 1960.

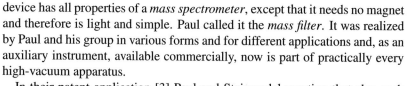

FORSCHUNGSBERICHTE

DES WIRTSCHAFTS- UND VERKEHRSMINISTERIUMS

NORDRHEIN-WESTFALEN

Herausgegeben von Staatssekretär Prof. Dr. h. c. Dr. E. h. Leo Brandt

Nr. 415

Prof. Dr.-Ing. Wolfgang Paul
Dr. rer. nat. Otto Osberghaus
Dipl.-Phys. Erhardt Fischer

Physikalisches Institut der Universität Bonn

Ein Ionenkäfig

Title page of the 1958 research report by Paul, Osberghaus, and Fischer entitled *An Ion Cage* and edited by my father, Leo Brandt.

device has all properties of a *mass spectrometer*, except that it needs no magnet and therefore is light and simple. Paul called it the *mass filter*. It was realized by Paul and his group in various forms and for different applications and, as an auxiliary instrument, available commercially, now is part of practically every high-vacuum apparatus.

In their patent application [3] Paul and Steinwedel mention that also such quadrupole fields can be produced that confine ions not near a line but near a point in space. An instrument achieving this is the now famous *Paul trap*. Paul himself had named it an *ion cage*. Its principles are given in some detail in the patent. The first ion cage was built in Bonn by Paul, Osberghaus, and Fischer. It is described in a now very rare research report of the Ministry of Economy and Transport of North Rhine-Westphalia [5]. This series of reports was edited by my father, who, in the difficult post-war years, had liberated ministry funds to support basic research. More details about the first cage and its properties are given in Fischer's doctoral thesis [6].

The quadrupole field for a Paul trap is produced by three electrodes with hyperbolic surfaces. Two, the mushroom-shaped top and bottom electrodes, form a two-sheet hyperboloid, the ring-shaped middle electrode a single-sheet hyperboloid. A radio-frequency voltage is applied between the outer electrodes and the ring. Focussing alternates between the axial coordinate (z) and the radial coordinate r, the distance from the vertical axis (see Box). The imprisonment of an ion in the cage is easily understood by looking at the computer simulation of a trajectory, projected on the horizontal symmetry plane. It consists of pieces curved towards the centre (in times in which the field is focussing in r and defocussing in z) and pieces curved outwards (defocussing in r, focussing in z). The frequency of this inward–outward oscillation is, of course, that of the external field. But there are other characteristic frequencies, in particular, that for completing a full turn around the origin. They are typical of the ion's specific charge and were detected as follows. A second radio-frequency circuit was connected to the cage; its frequency could be tuned over a wide range. Absorption of energy from this circuit at particular frequencies

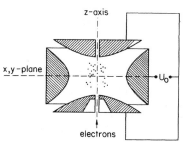

Cross section through a Paul trap. The trap has rotational symmetry about the z axis. The time-dependent voltage $U_0 = U + V\cos(\omega t)$ is applied between the outer electrodes and the ring electrode. From [4]. © Nobel Foundation. Reprinted with permission.

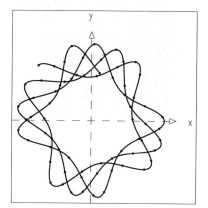

Computer simulation of an ion trajectory in a Paul trap, projected onto the xy plane.

Quadrupole Fields Focussing in Two and in Three Dimensions

Consider an electric potential Φ that is quadratic in the Cartesian coordinates x, y, z,

$$\Phi = \Phi_0(t) \frac{1}{2r_0^2}(\alpha x^2 + \beta y^2 + \gamma z^2) \quad,$$

with the time-dependent factor

$$\Phi_0(t) = U + V\cos(\omega t) \quad.$$

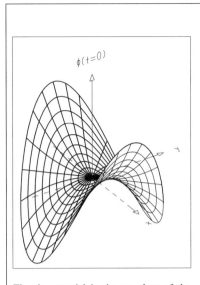

$\phi(t=0)$

The laws of electrodynamics (here Poisson's law) require $\alpha + \beta + \gamma = 0$.
For the *quadrupole mass spectrometer* one chooses $\alpha = 1, \beta = -1, \gamma = 0$ and obtains the potential

$$\Phi = \Phi_0(t) \frac{1}{2r_0^2}(x^2 - y^2) \quad,$$

which has the form of a saddle rotating about the z axis. Ions of the 'right' specific charge are confined to a region near the axis.
For the *Paul trap* one takes $\alpha = 1, \beta = 1, \gamma = -2$ and, with $r = \sqrt{x^2 + y^2}$ denoting the distance from the z axis, the potential is

$$\Phi = \Phi_0(t) \frac{1}{2r_0^2}(r^2 - 2z^2) \quad.$$

Electric potential in the xy plane of the quadrupole mass spectrometer. As a function of time, the saddle form rotates about the z axis. In the Paul trap the potential is similar if plotted over the rz plane.

It depends only on r and z and, in a plane spanned by these quantities, also has saddle form. Therefore it confines ions of the 'right' specific charge in a region near the coordinate origin.

---

signalled resonance with ions of such proper frequencies. Thus, the ion cage served as a very sensitive mass spectrometer, able to detect tiny traces of atoms or molecules in its residual gas.

More than two decades after its conception and first realization, the Paul trap became one of the principal instruments in the new field of *quantum optics*. That was made possible by the advent of the laser, which offered the possibility to 'cool' ions and to visually observe single ions. As the name suggests, to cool ions means to remove most of their energy. In the field of the trap an ion is similar to an electron in the field of an atomic nucleus. It can be in one of many energy states. By stimulating it with radiation in the right frequency range it can be made to transit into a state of lower energy, emitting the surplus energy in the form of radiation. A single ion, laser-cooled in this way, stays practically at rest in the centre of the trap.

An ion, except for one or possibly more missing or additional electrons, is just an atom with an optical spectrum. One of its optical frequencies can be excited by laser light and the emitted light can be observed. A sophisticated lens system backed by a sensitive electronic camera looking through the upper electrode of a Paul trap can produce the image of a single ion. Such a system was employed by Walther[3] and his group in Munich to produce and photograph 'crystals', formed by a few positive magnesium ions ($^{24}Mg^+$) [7]. Walther's group also constructed a storage ring for cold ions which is, essen-

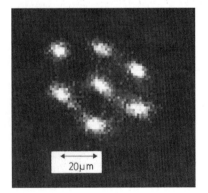

Crystal formed of seven $^{24}Mg^+$ ions in a Paul trap. Figure reprinted with permission from [7]. Copyright 1987 by the American Physical Society.

20µm

---

[3] Herbert Walther (1935–2006)

tially, a quadrupole mass spectrometer bent to form a closed ring. Ions are confined within a narrow ring and through their mutual electrostatic repulsion form regular spiral patterns [8]. Also a *linear Paul trap* has been realized, consisting of a quadrupole mass spectrometer with an additional repulsive field at either end. Rows of a few ions can be stored like pearls strung along its axis. It is considered possible that these may serve as registers in possible future *quantum computers* [9].

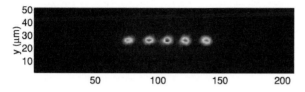

String of five Ca^+ ions in a linear Paul trap. Figure reprinted with permission from [9]. Copyright 2000 by the American Physical Society.

Paul devoted much of his energy to provide research opportunities to young scientists from Germany and abroad. We only mention his contributions to the founding and leading of DESY, the large accelerator centre in Hamburg, and his presidency, from 1979 to 1989, of the Alexander von Humboldt Foundation for the promotion of international research cooperation. In his later years Paul, with a few collaborators, returned to magnetic sextupoles. A storage ring for 'ultracold' neutrons with a typical velocity of $10\,m/s$ was constructed and tested in the late 1970s. A precision measurement of the lifetime of free neutrons [10] became possible in 1989 when neutron beams of higher intensity became available to fill the ring. I remember well that Paul was particularly pleased with this experiment, not only because his two sons were his collaborators but also because it allowed the measurement of the neutron's gravitational mass. Magnetic forces attract the neutrons to the symmetry plane of the storage ring but gravitation pulls them downwards. The equilibrium orbit is therefore some distance (on the order of a centimetre) below the symmetry plane. In his Noble Lecture Paul proudly points out that the 'neutron storage ring represents a balance with a sensitivity of $10^{-25}\,g$'.

We cannot conclude this episode without at least sketching an elegant and most accurate measurement made by Dehmelt and co-workers who employed a different type of trap. Dehmelt[4] grew up in Berlin, where he attended the *Gymnasium zum Grauen Kloster* with a tradition going back to the sixteenth century. In the war years he served at both the eastern and western front with some leave to study physics in Berlin. Only in 1946 he could resume his studies, this time in Göttingen. He attended lectures by Heisenberg and Kopfermann and the advanced laboratory courses offered by Paul and obtained both his diploma and his Ph.D. with experimental work on nuclear quadrupole moments done in Kopfermann's institute.

Dehmelt was invited in 1949 as postdoctoral associate to Duke University and two years later went to the University of Washington in Seattle, rising through the ranks from visiting assistant professor in 1952 to full professor in 1961. From Osberghaus, who had been his fellow graduate student in Göttingen, Dehmelt received a copy of the research report [5] on the Paul trap and employed the novel apparatus for precision spectroscopy in his group. But he

Dehmelt

---

[4] Hans Georg Dehmelt (born 1922), Nobel Prize 1989

[1]   Friedburg, H. and Paul, W.,
      Naturwissenschaften, **38** (1951)
      159.

[2]   Paul, W. and Steinwedel, H.,
      Zeitschrift für Naturforschung,
      8 A (1953) 448.

[3]   Paul, W. and Steinwedel, H.,
      German Patent 944 900; filed
      1953.

[4]   Paul, W., *Nobel Lectures, Physics
      1981–1990*, p. 601. World
      Scientific, Singapore, 1993. Also
      in [14].

[5]   Paul, W., Osberghaus, O.,
      and Fischer, E., *Forschungs-
      berichte des Wirtschafts- und
      Verkehrsministeriums Nordrhein-
      Westfalen*, **Nr. 415** (1958).

[6]   Fischer, E., *Zeitschrift für Physik*,
      156 (1959) 1.

[7]   Diedrich, F., Peik, E., Chen,
      J. M., Quindt, W., and Walther,
      H., *Physical Review Letters*, **59**
      (1987) 2931.

[8]   Waki, I., Sassner, S., Birkl, G.,
      and Walther, H., *Physical Review
      Letters*, **68** (1992) 2007.

[9]   Nägerl, H. C. et al., *Physical
      Review*, **A 61** (2000) 023405.

[10] Paul, W., Anton, F., Paul, L., Paul,
      S., and Mampe, W., *Zeitschrift für
      Physik*, **C 4562** (1989) 25.

[11] Van Dyck, R. S., Schwinberg,
      P. B., and Dehmelt, H. G.,
      Physical Review, **D34** (1986) 722.

[12] Van Dyck, R. S., Schwinberg,
      P. B., and Dehmelt, H. G.,
      Physical Review Letters, **38**
      (1977) 310.

[13] Van Dyck, R. S., Schwinberg,
      P. B., and Dehmelt, H. G.,
      Physical Review Letters, **59**
      (1987) 26.

[14] Paul, W., *Reviews of Modern
      Physics*, **62** (1990) 513.

also used it as a starting point for a new instrument, which he named the *Penning trap*. In 1936 Penning[5], in the Netherlands, had used a magnetic field to curl up the trajectories of electrons in a glow discharge. By thus lengthening their paths he could maintain such discharges at much lower pressures than previously possible. In the Penning trap, designed for the confinement and observation of a single electron, a static voltage is applied to the electrodes of a Paul trap which provides a harmonic force in axial (z) direction pulling the electron towards the centre. This field defocusses in the radial (r) direction; but the defocussing is more than compensated by a strong axial magnetic field forcing the electron to circle the z axis. In practice the magnetic field is made so high that the electron is essentially confined to the z axis and oscillates about the origin in the electrostatic field.

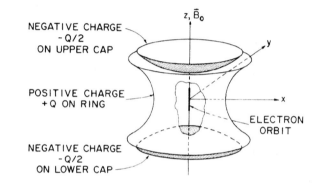

The Penning trap used by Dehmelt's group. Figure reprinted with permission from [11]. Copyright 1986 by the American Physical Society.

A single electron in a well-known field is reminiscent of the hydrogen atom. Dehmelt called his system *geonium* 'because ultimately the electron is bound to the earth via the trap structure and the magnet' [12]. The energy levels of this atom can be computed in quantum electrodynamics. They depend on the magnetic moment of the electron, expressed by the *gyromagnetic ratio* or *g factor*. The Dirac equation (Episode 43) yields exactly $g = 2$. But at the same time as the Lamb shift a slight deviation from that value was observed, which could be explained in the framework of quantum electrodynamics. Using high-frequency techniques that we shall not discuss, Dehmelt's group was able to measure the g factor of the electron with ever increasing precision. It has become customary to express the deviation from 2 by the quantity $a = (g - 2)/2$. In 1987, a was measured [13] as

$$a = 0.001\,159\,652\,188\,4 \pm 0.\,000\,000\,000\,004\,3 \quad,$$

that is, with a relative accuracy of 4 parts in 10^9. A cumbersome theoretical calculation in quantum electrodynamics agrees within this tiny error with the experiment. This makes quantum electrodynamics the best-tested fundamental theory.

---

[5] Frans Michel Penning (1894–1953)

The Mössbauer Effect (1958)

A simple and most instructive experiment, shown in many lectures for beginners in physics, is the demonstration of *resonant absorption* and *resonant fluorescence* (also called *resonant scattering*) with yellow sodium light. A beam of light produced by a lamp, in which a glow discharge is maintained in sodium vapour, is directed through an evacuated glass globe containing some metallic sodium. The beam falls on a screen behind the globe, producing a bright yellow patch. As the globe is heated and thus fills with sodium vapour, the yellow patch loses intensity and at the same time the incident beam becomes visible as a yellow glow within the glass globe. Excited sodium atoms in the lamp perform a transition to the ground state by emitting light quanta of energy E_0. These, in turn, can excite atoms in the globe and thus be absorbed from the beam. The excited atoms return to their ground state, again by emitting quanta of the energy E_0 and causing a visible glow where the incident light beam passes.

Like the atomic hull also the atomic nucleus possesses a ground state and excited states; transitions between them are accompanied by the absorption and emission of photons, albeit in a very different energy range. In nuclear transitions energies are typically 10^5 eV compared to a few eV for atomic ones. While being emitted from the atom or nucleus, the quantum imparts on it a recoil and thus loses part of its own energy. The effect is negligible for visible light but becomes important for γ rays because in the latter case the energy E_0 is so much higher (see Box). By the loss of the recoil energy E_R to the nucleus, the γ energy is shifted from E_0 to $E_\gamma = E_0 - E_R$. When a photon of that energy hits a nucleus of the kind which emitted it, then again it transfers a mechanical energy E_R. All in all only the energy $E_0 - 2E_R$ remains to excite the nucleus and that is usually insufficient.

In spite of these difficulties, in 1950 Moon[1] in Birmingham performed the first resonant scattering experiment with γ rays from the mercury isotope ^{198}Hg [1]. The excited state was created by irradiating gold (^{197}Au) with neutrons in a reactor, creating ^{198}Au which subsequently decayed to excited ^{198}Hg. In Moon's experiment the source was placed on a steel rotor which was spun so fast that it approached the target with a speed up to 700 metres per second. In this way the energy lost by recoil in the source and the target was compensated by the kinetic energy given to the fast-moving source.

Another way to achieve resonance was to broaden the width of the emitted and the absorbed line by thermal motion. Because of the uncertainty relation the γ energy is not exactly E_0 but extends over a small region around that value characterized by the natural line width Γ (see Box). Due to the thermal motion of the atoms the emitting nuclei have a kinetic energy and thus a velocity by

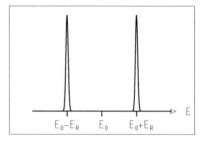

The emission line is shifted to $E_0 - E_R$, the absorption line to $E_0 + E_R$. Since both lines are sharp, the resonance is lost.

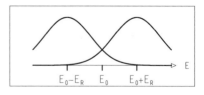

Resonance is partly re-established by thermal Doppler broadening, which leads to some overlap of the broadened lines.

---

[1] Philip Burton Moon (1907–1994)

Facts Relevant to the Mössbauer Effect

Recoil Energy In the radiative transition of an atom or a nucleus (at rest) from an excited state to the ground state the energy E_0 is set free. This energy is shared by the radiated photon (energy E_γ) and the recoiling system of mass M (energy E_M) which can be the atom, the nucleus, or, in the case of the Mössbauer effect, a whole crystal. From energy and momentum conservation one easily obtains the *recoil energy*

$$E_{\mathrm{R}} = E_M = E_0 - E_\gamma = \frac{E_\gamma^2}{2Mc^2} \quad.$$

Here c is the speed of light and therefore Mc^2 the relativistic rest energy of the recoiling system.

Natural Line Width If $\Delta t = \tau$ is the mean lifetime of the excited state, then, according to Heisenberg's uncertainty relation $\Delta E \, \Delta t = \hbar/2$, the energy of that state is not exactly given but has an uncertainty ΔE, called the *width* Γ,

$$\Gamma = \hbar/\tau \quad.$$

This is the *natural width* of the spectral line centred around the energy E_0 (or rather E_γ, if there is a recoil).

Doppler Effect If a source sending radiation moves with the velocity v relative to the detector receiving it, then the frequency ν_{D} observed in the detector is changed with respect to the frequency ν_{S} in the system in which the source is at rest. The relative shift in frequency, which for quanta is also a shift in energy, is

$$\frac{|\nu_{\mathrm{D}} - \nu_{\mathrm{S}}|}{\nu_{\mathrm{S}}} = \frac{|E_{\mathrm{D}} - E_{\mathrm{S}}|}{E_{\mathrm{S}}} = \frac{\Delta E_{\mathrm{Doppler}}}{E_{\mathrm{S}}} = \frac{v}{c} \quad.$$

Numerical Values for Two γ-Ray Lines (assuming recoil on a single nucleus)

Isotope	E_0	Γ	Γ/E_0	E_{R}
^{191}Ir	129×10^3 eV	3.5×10^{-6} eV	2.71×10^{-11}	5×10^{-2} eV
^{57}Fe	14.4×10^3 eV	4.6×10^{-9} eV	3.20×10^{-13}	3×10^{-3} eV

---

which the line is shifted through the Doppler effect. Since there are velocities in all directions, an overall *Doppler broadening* results, which increases as the temperature rises. Thus the broadened emission line (centred at $E_0 - E_{\mathrm{R}}$) and the absorption line (centred at $E_0 + E_{\mathrm{R}}$) can have some overlap and in that region the resonance condition is met. Obviously, it was expected that there was more resonant absorption at high than at low temperature. This was the situation in 1953 when Mössbauer entered the field.

Mössbauer[2] was born in Munich, where he also went to school and began his physics studies at the Technical University in 1949. He chose to do his theses for the diploma and the doctorate under Maier-Leibnitz[3], who had been a student of Franck in Göttingen and a collaborator of Bothe in Heidelberg. Maier-Leibnitz suggested that Mössbauer do resonant spectroscopy using thermal Doppler broadening. In a talk given decades later, Mössbauer reports interesting details of his early work [2]. For his diploma he built a system of

---

[2] Rudolf Ludwig Mössbauer (born 1929), Nobel Prize 1961   [3] Heinz Maier-Leibnitz (1911–2000)

12 proportional counters giving a combined efficiency of 5% for γ rays and chose to study the 129-keV radiation of the iridium isotope ^{191}Ir because it was suitable for the thermal method and was available from the reactor centre at Harwell in England. (At that time, Allied Regulations did not allow reactors in Germany.) Moreover, he read publications which Maier-Leibnitz had given him, in particular, the work of Moon and a theoretical paper by Lamb entitled *Capture of Neutrons by Atoms in a Crystal* [3]. Lamb had studied the excitation of nuclei in crystals by neutrons, taking into account the quantum-mechanical properties of the crystal.

Mössbauer

On completion of the diploma, Maier-Leibnitz arranged for Mössbauer to do the actual measurements at the Department of Nuclear Physics in the Max-Planck Institute of Medical Research in Heidelberg. The department had been created for Bothe, see end of Episode 34. Mössbauer began his work there in 1955, the year of Bothe's death. At the Max-Planck Institute the equipment, the workshop, and the technical know-how were superior to that in his Munich university institute. Mössbauer was soon convinced that he had to abandon his original detector system in favour of a modern sodium-iodide (NaI) crystal, backed by a photomultiplier that provided nearly 100% efficiency. He decided to effectuate the temperature change by cooling rather than heating since it was 'much simpler to build a cryostat rather than a furnace' [2]. The cryostat was a simple Dewar vessel, cooled with liquid nitrogen, which could accommodate either an iridium absorber or a (non-resonant) platinum absorber used for comparison. The source could also be cooled in this way but, alternatively, left at room temperature. Measurements were performed with the source at 88 K and at 303 K.

What Mössbauer found, to his utmost surprise, was not a smaller but a higher absorption for the lower source temperature. At first he looked for some 'dirt effect'. When he found none he went for help to Jensen, the theoretician in Heidelberg whom we met in Episode 70. Jensen told him that, if the effect was real, an explanation must be hidden in Lamb's paper. Mössbauer frankly admits that he did not know much quantum mechanics at the time. Nevertheless, he succeeded in adapting Lamb's theory to the excitation of nuclei by high-energy photons. He showed that the photon spectra for emission and for absorption do not only consist of recoil-shifted and Doppler-broadened humps but also contain a very sharp line of the natural width Γ at the energy E_0. Some of the photons are affected neither by recoil nor by thermal Doppler effect and their proportion rises as the temperature decreases. The atoms in the crystal oscillate about their average positions. In quantum mechanics the oscillation energy is quantized, i.e., it can be increased or decreased only in finite steps or quanta. The probability that an oscillator absorbs an additional quantum increases with the number it already possesses. At low temperature that number is small and often a single oscillator cannot absorb the recoil energy. Instead, the γ ray recoils against the crystal as a whole and, since that possesses a mass so much larger than that a single atom has, the recoil energy E_R becomes entirely negligible.

Mössbauer completed this work in 1957. He moved back to Munich, described measurements and theory in his thesis, and sent a paper to the *Zeitschrift für Physik* [4]. He obtained his Ph.D. and prepared to do neutron work at the

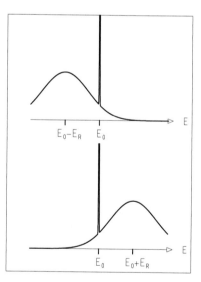

Mössbauer's explanation for the unexpected outcome of his experiment: If the emitting and absorbing nuclei are embedded in crystals, then some of the emission (top) and absorption (bottom) is not affected by recoil or Doppler effects. Emission and absorption lines of natural width appear at E_0. Their intensity increases as the temperature falls. (The broad 'background' under the sharp lines is now better understood and shows more structure than originally assumed by Mössbauer and shown here.)

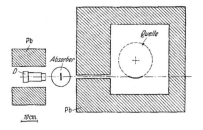

Mössbauer's arrangement with a rotating source (Quelle) to measure the natural line width. Reprinted with kind permission from Springer Science+Business Media: [5], Fig. 1.

first German research reactor which was being constructed near Munich under the direction of Maier-Leibnitz. (The Allied Restrictions had finally been lifted.) Later he remembered what happened when he received the printed version of his paper [2].

> One still reads the first publication of oneself, while later publications are of no interest. During this reading it occurred to me, that I had not performed the main experiment: It should be possible to measure the sharp resonance lines by using the linear Doppler effect.

Immediately, he travelled back to Heidelberg, where, fortunately, his apparatus still existed.

With cogwheels from a toy shop Mössbauer hastily constructed a turntable on which he placed his source to give it a velocity relative to the absorber. The arrangement was similar to that of Moon but the velocity was very much smaller, only about a centimetre per second. The Doppler shift due to this small velocity was sufficient to detune the resonance between source and absorber and thus to significantly reduce the observed absorption. Varying the velocity in small steps Mössbauer swept the source line over the absorber line and thereby measured an absorption dip corresponding to twice the natural line width. He published his results in a short note [5], dated 13 August 1958, and in a somewhat longer paper [6] in which he also summarized his theory.

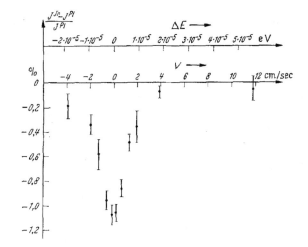

Intensity dip behind the absorber as a function of the velocity v of the source (in cm/s) as measured by Mössbauer. The upper scale indicates the Doppler shift given to the γ quanta emitted by the source (1 unit corresponds to 10^{-5} eV). Reprinted with kind permission from Springer Science+Business Media: [5], Fig. 2.

While Mössbauer's thesis, which he had sent to experts including Moon, and his first paper went largely unnoticed, his direct line-width measurement did not. Within a week after publication of the note [5], he received 260 requests for reprints. It was with this measurement, that *Mössbauer spectroscopy* was established. Suitable γ lines, now called *Mössbauer lines*, have very small natural widths (see Box). They present the best-defined radiation available, allowing measurements of very high precision. Not unlike nuclear magnetic resonance (NMR), Mössbauer spectroscopy has become a widely used tool in nuclear and solid state physics, in chemistry, and archaeology. Mössbauer himself called the phenomenon that enabled such precision measurements the *recoilless resonant absorption*; the scientific public soon replaced this cumbersome expression with *Mössbauer effect*.

The west end of the Jefferson Physical Laboratory of Harvard University. The building was erected in 1884, including an inner tower, extending from the basement to the penthouse, which is isolated from the rest of the building. This tower housed the Pound–Rebka experiment.

In 1960 Pound and Rebka at Harvard University used the extraordinary precision, offered by that effect, to perform the first terrestrial experiment verifying the basis of the general theory of relativity. That is Einstein's postulate of the equivalence of gravitation and acceleration. On the basis of this *equivalence principle*, in 1911, Einstein had predicted an effect, now called *gravitational redshift*, which we discussed in Episode 26. If a photon of energy E travels a distance h against a gravitational field which would exert an acceleration g on a massive body, it loses the energy $\Delta E = Egh/c^2$ (c is the speed of light). At the surface of the earth one has $g = 9.81\,\mathrm{m/s^2}$ and therefore, for a height h of 1 m, a relative energy shift of $\Delta E/E = 1.09 \times 10^{-16}$.

The height available to Pound and Rebka in their laboratory was 22.6 m, resulting in a shift of $\Delta E/E = 2.56 \times 10^{-15}$. It was effectively doubled by doing experiments with upward- and with downward-directed radiation. This corresponds to only about 2% of the width of the Mössbauer line of the iron isotope $^{57}\mathrm{Fe}$ which they decided to use. Rebka and Pound modulated the source velocity and measured absorption at positions right and left of the maximum, where the absorption curve is steepest. From the difference between the two, the shift could in principle be obtained. Instead, the difference was exactly compensated by a superimposed small constant velocity v of the source provided by a hydraulic system. Einstein's prediction ($v = gh/c$ or $7.4 \times 10^{-5}\,\mathrm{cm/s}$ for the height of the tower) was verified with a relative error of 10% [7]. In a second experiment by Pound and Snider the accuracy was raised to 1% in 1964 [8]. For a historical narrative by Pound on these experiments see [9,10].

[1] Moon, P. B., *Proceedings of the Physical Society*, **64** (1951) 76.

[2] Mössbauer, R. L., *Hyperfine Interactions*, **126** (2000) 1.

[3] Lamb, W. E., *Physical Review*, **55** (1939) 190.

[4] Mössbauer, R. L., *Zeitschrift für Physik*, **151** (1958) 124.

[5] Mössbauer, R. L., *Naturwissenschaften*, **45** (1959) 538.

[6] Mössbauer, R. L., *Zeitschrift für Naturforschung*, **14 A** (1959) 211.

[7] Pound, R. V. and Rebka, G. R., *Physical Review Letters*, **4** (1960) 337.

[8] Pound, R. V. and Snider, J. L., *Physical Review Letters*, **13** (1964) 539.

[9] Pound, R. V., *Physics in Perspective*, **2** (2000) 224.

[10] Pound, R. V., *Physics in Perspective*, **3** (2001) 4.

The Laser (1960)

Schawlow

In Episode 73 we described the development of an instrument with unusual properties and of outstanding precision, the *maser*, which was first realized by Townes and co-workers. It generates microwaves (electromagnetic waves with a wavelength around 1 cm) which are coherent and of extremely well-defined and stable frequency. In 1958, in a very influential paper, entitled *Infrared and Optical Masers* [1], Schawlow and Townes discussed possibilities to extend the maser principle to much shorter wavelengths.

Schawlow[1] [2] was born in Mount Vernon in the state of New York. His family moved to Toronto when he was only three. After high school he wanted to become a radio engineer. There were no scholarships in engineering, but Schawlow obtained one in physics, a subject sufficiently close to the one he thought he liked best. After obtaining a master's degree from the University of Toronto, he worked for several years in industry on radar equipment. After the war he returned to the university. For his Ph.D. he did research on high-precision spectroscopy. In 1949, with a fellowship from industry, he joined Townes' group at Columbia University in New York. Schawlow stayed only for two years before he accepted a position at the Bell Telephone Laboratories. But in these two years he not only did important work but also established a lasting friendship with Townes and got married to one of Townes' sisters. The two scientists continued to work together and to jointly write a book on microwave spectroscopy.

It was originally thought that population inversion, needed for the functioning of a maser, could not be achieved for energy-level differences corresponding to infrared or visible light. But when Townes began to critically review the situation in the second half of 1957, he came to the conclusion that such an inversion was not excluded, if a suitable environment could be provided to contain a high enough density of both the pumping and the stimulating radiation. In a typical three-level maser a sample is used with three carefully chosen energy levels with energies $E_1 < E_2 < E_3$. The sample is placed in a microwave cavity resonating at two frequencies $\nu_{13} = (E_3 - E_1)/h$ and $\nu_{23} = (E_3 - E_2)/h$. The cavity is excited at the *pumping frequency* $\nu_{\text{pump}} = \nu_{13}$, transitions $E_1 \rightarrow E_3$ are provoked until the population n_3 of the level E_3 is larger than the population n_2 of the level E_2. Now a spontaneous transition $E_3 \rightarrow E_2$, giving rise to radiation of frequency ν_{32}, can trigger off stimulated emission at that *emission frequency* $\nu_{\text{em}} = \nu_{23}$. Depending on the particular system, there may be more energy levels involved. But there is always an emission frequency ν_{em} and usually a pumping frequency ν_{pump} with $\nu_{\text{pump}} > \nu_{\text{em}}$; in some cases pumping is achieved by other means than irradiation with electromagnetic waves.

---

[1] Arthur Leonard Schawlow (1921–1999), Nobel Prize 1981

A cavity for electromagnetic waves serves the same purpose as an organ pipe for sound waves. It houses standing waves of one or several wavelengths, directly connected to the dimensions of the cavity. Microwaves with typical wavelengths of around 1 cm require cavities which are quite easily built. But what could be an efficient 'cavity' for light with wavelengths more than ten thousand times shorter? Even if there were ways to make such a tiny box, it would not offer the place to contain the sample. Townes at first thought of a rectangular box with mirrored walls to contain light waves of various frequencies. He discussed this idea with Schawlow whom he saw regularly because, at the time, he also did consulting for the Bell Laboratories.

Schawlow proposed to use only two parallel mirrors. Such an arrangement was well known in optics under the name *Fabry–Pérot* interferometer. It singles out highly directional standing waves perpendicular to the surfaces of the mirrors. All other radiation leaves through the open sides. Also waves with particular wavelengths are selected, namely, those for which half a wavelength fits an integer number of times between the two mirrors. Stimulated emission of waves of such a direction and wavelength was then favoured, because stimulating radiation with just these properties was kept enclosed between the two mirrors. If one of the mirrors was made a little transparent or if a small hole was made in its reflective coating, then part of the stimulated radiation could leave and be used. It would be monochromatic, coherent, and essentially parallel.

As a specific example, Schawlow and Townes discussed a column of potassium vapour enclosed between two mirrors and illuminated from the side by a lamp providing intense light in the region of the pumping frequency. At Columbia university Townes and his group tried to realize an optical maser based on potassium vapour. Townes decided that the work with Schawlow fell into his consulting activities with Bell Labs and suggested that they take out a patent for the optical maser, which they did.

The paper [1] by Schawlow and Townes inspired work by groups and individuals in many places. A particular case is that of Gould[2], then a graduate student at Columbia, working only a few doors away from Townes but not a member of Townes' group. He invented the term *laser* (light amplification by stimulated emission of radiation) for the optical maser; at any rate he was the first to use it in print. Gould was more interested in inventing and patenting than in experimenting and publishing. He applied for patents on the laser. In a 'patents war', raging over thirty years, Gould and the companies representing his interests finally won a number of important points (and important money) [3,4].

The first laser was built by Maiman[3] [5] at the Hughes Research Laboratories in Malibu, California. Maiman had studied at the University of Colorado and earned a Ph.D. at Stanford University with a thesis on microwave spectroscopy. His work in industrial laboratories began at Lockheed Aircraft. At Hughes, another aircraft company, he achieved important advances in solid-state maser technology. For instance, he reduced the weight of a maser by a factor of thousand. At that time, ruby was a popular maser material. Ruby

Maiman

[2] Gordon Gould (1920–2005)   [3] Theodore Harold Maiman (1927–2007)

Maiman's laser opened up.

is a crystal of aluminium oxide Al_2O_3 with a small admixture of chromium Cr. Levels of the isolated chromium atoms dispersed in the crystal provided maser activity. While a pure Al_2O_3 crystal appears transparent like glass, ruby is pink or bright red, depending on the content of chromium: Luminescence in the frequency region of red light is provoked by shining light of higher frequency on it. Maiman decided he should try to reach population inversion in energy levels of chromium corresponding to the emission of this red light.

In September 1959, Maiman attended the first international conference on quantum electronics held at a resort hotel in the Catskills in the state of New York. Here, Schawlow gave good reasons why ruby would probably not be suitable as a laser material. But Maiman was not discouraged. Back in his laboratory he first measured important optical properties of ruby and constructed his simple but efficient laser which first worked on 16 May 1960. Maiman sent a short paper, sketching his success, to *Physical Review Letters*; but the editors, swamped by papers on the maser, refused it without even asking referees to report on its value. Thus Maiman published in *Nature*, the venerable London journal [6]. At first, Maiman had only made sure that stimulated emission indeed occurred. He had not looked out for a bright spot of red light on the wall, now considered so typical for the ruby laser. The bright spot was seen and the coherence of the light was ascertained by Schawlow and his group soon thereafter [7]. About half a year after his letter to *Nature*, Maiman, together with four collaborators, submitted a comprehensive paper on his laser, its properties and the properties of its light to the *Physical Review* [8]. We quote the short paragraph describing the apparatus:

> The material samples were ruby cylinders about $\frac{3}{8}$ in. in diameter and $\frac{3}{4}$ in. long with flat ends parallel to within $\lambda/3$ at 6943 Å [the wavelength of the emitted light]. The rubies were supported inside the helix of a flash tube, which in turn was enclosed in a polished aluminum cylinder (see Fig.); provision was made for forced air cooling. The ruby cylinders were coated with evaporated silver at each end; one end was opaque and the other was either semitransparent or opaque with a small hole in the center.

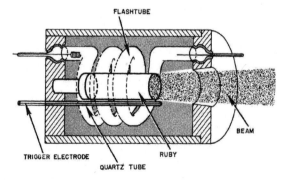

Astonishingly enough, the first laser was a very simple device: A slab of ruby with two parallel end faces, silvered to serve as mirrors, and placed inside a conventional flash tube as used for photography which provides the pumping radiation. Indeed it was much simpler than any maser had been.

An early spectacular application of the ruby laser was the precision measurement of the distance between earth and moon. On 21 July 1969, Neil Armstrong and Edwin Aldrin, the first men to reach the moon, placed a ten-by-ten-inch reflector made of one hundred so-called corner-cube prisms on the moons surface. Corner cubes have the property to reflect light exactly back to where it came from. Within about two weeks, two teams at the Lick Observatory of the University of California and at the McDonald Observatory of the University of Texas, respectively, succeeded in measuring the time it took a light pulse to travel from a place on earth to the reflector on the moon and back. That time could be converted into distance with the accuracy of about one inch. Three more reflectors were deployed on the moon, including a French one brought there by an unmanned Soviet craft. Experiments of *lunar laser ranging* with these reflectors still continue and yield valuable information on the moon's motion.

An ever-increasing activity set in to develop and manufacture lasers of different properties. The first laser used a solid-state *medium*, ruby. It was operated in a *pulsed* mode and was *radiation pumped* by the flash tube. It yielded light of *low intensity* at *fixed frequency* determined by the properties of the chromium impurities within the crystal. All these properties were varied and extended. Laser media can be gaseous, liquid, and solid. Often they are specially tooled like the semiconductor laser diodes (Episode 86) used in compact-disc players. Lasers can operate in continuous mode or in pulsed mode. Laser light pulses can be made as short as 10^{-15} s. Pumping can be done by methods other than radiation. In the laser diode it is achieved by simply passing an electric current through it. There are lasers with a tunable frequency and, last but not least, lasers of high intensity. Since laser light is created parallel, it can be easily focussed onto a tiny spot, which is then irradiated by a very high power per area. This is one of the reasons why lasers are successfully used as tools in widely differing fields such as micro-machining and medicine.

For some applications, for instance, eye surgery, it is important that laser light traverses transparent media without being absorbed. Light needed to do work at the retina is simply entered through the iris. Schawlow enjoyed to demonstrate this particular property in his own way: He had a blue balloon, shaped like Mickey Mouse, inside a larger transparent balloon. When 'shooting' with a laser, mounted in a mock-up ray gun, on that set of balloons, Mickey collapsed while the outer balloon remained unharmed.

The laser has become a multi-billion dollar business and, therefore, seems to be a good example for the 'useful' industrial application of basic research, something required more often than not by governments and other research-funding institutions. However, it took nearly half a century from Einstein's postulation of stimulated emission in 1916 to the realization of the first laser in 1960. Steps on the way were the discovery of quantum mechanics, needed for quantitative calculations, and the development of vital experimental methods: molecular beams, magnetic resonance, and the maser. Although neither molecular beams nor magnetic resonance were used to build the laser, they were needed for the maser, its forerunner.

[1] Schawlow, A. L. and Townes, C. H., *Physical Review*, **112** (1958) 1940.
[2] Chu, S. and Townes, C. H., *Biographical Memoirs (Nat. Acad. Sci.)*, **83** (2003) 1.
[3] Bertolotti, M., *The History of the Laser*. Institute of Physics Publishing, Bristol, 2005.
[4] Townes, C. H., *How the Laser Happened – Adventures of a Scientist*. Oxford University Press, Oxford, 1999.
[5] Maiman, T., *The Laser Odyssey*. Laser Press, Blaine, WA, 2000.
[6] Maiman, T. H., *Nature*, **187** (1960) 493.
[7] Collins, R. J., Nelson, D. F., Schawlow, A. L., Bond, W., Garret, C. G. B., and Kaiser, W., *Physical Review Letters*, **5** (1960) 303.
[8] Maiman, T. H., Hoskins, R. H., D'Haenens, I. J., Asawa, C. K., and Evtuhov, V., *Physical Review*, **123** (1961) 1151.

83   Particle–Antiparticle Colliders (1961)

Panofsky in the tunnel of the two-mile linear accelerator at SLAC.

One morning in December 1959, I was taking the final oral exam for my diploma in the office of my advisor Paul in Bonn. One of the topics was particle accelerators. After all, Paul had built a 500-MeV strong-focussing electron synchrotron in Bonn which I had used in my diploma thesis work. We even touched on the new technique of *storage rings*, of which I had only learnt the day before in a seminar given by Panofsky. While we talked, Panofsky came in to say good-bye to his host. Kindly, he also shook my hand. At that time Panofsky[1], professor at Stanford University, was director of the Stanford High Energy Physics Laboratory (HEPL) and preparing to found the Stanford Linear Accelerator Center (SLAC). Largely due to him, Stanford developed into a leading centre of particle physics.

I remember well that Panofsky began to give his seminar in German but after a while switched to English, because he 'hadn't learnt physics in German'. Later I was told that he was born in Berlin and grew up in Hamburg, where his father was a renowned historian of art. When he was fifteen, the family emigrated to the United States. He studied in Princeton and took his Ph.D. at Caltech in 1942. Having worked on the atomic bomb at Los Alamos, he moved to Berkeley in 1945 and to Stanford in 1951.

The Princeton–Stanford electron storage rings. There are two synchrotrons which are filled (via magnets and inflectors) with electrons from a linear accelerator. Energy loss due to synchrotron radiation is compensated by radio-frequency cavities. Since the electrons circulate in the same sense (clockwise) in both rings, they can collide in the shared section of the beam pipe in the centre. Reactions are registered by counters surrounding that region. Photo: Stanford University.

In his seminar in Bonn, Panofsky reported on an experiment under construction, in which a beam of 500-MeV electrons was to collide head-on with another electron beam of the same energy. On his slides the arrangement looked somewhat like two copies of our own synchrotron placed side by side. Where they touched, they had a common beam pipe and there the electrons from the

---

[1] Wolfgang K. H. Panofsky (1919–2007)

Colliding Beams vs. Fixed Target

We consider collisions between two particles of equal rest mass m. In a conventional 'fixed-target' experiment one, the target particle, is at rest in the laboratory; the other has energy E_{lab}. Part of that energy is used simply to make the centre of mass of the two particles move in the laboratory and is lost for the study of the interaction proper. In a 'colliding-beam' experiment both have energy E and equal but opposite momenta, and the centre of mass is at rest. A simple relativistic calculation, assuming that the rest energy mc^2 is small compared to E_{lab} and to E, shows that the useful interaction energy is

$$E^{(\text{int})}_{\text{fixed target}} = \sqrt{2mc^2 E_{\text{lab}}} \quad , \qquad E^{(\text{int})}_{\text{coll}} = 2E$$

in the two cases. The small table below lists the beam energy E of three colliders and the energy E_{lab}, needed to reach the same interaction energy in a fixed-target experiment.

Collider	Particles	E	Equivalent E_{lab}
Stanford–Princeton (this episode)	$e^- e^-$	0.5 GeV	978 GeV
S$\bar{p}p$S (Episode 96)	$\bar{p}p$	300 GeV	192×10^3 GeV
LEP (Episode 98)	$e^+ e^-$	100 GeV	39×10^6 GeV

Besides the energy, the key figure of a collider is its *luminosity* L. It is the rate (in events per second) for a reaction of unit cross section ($1\,\text{cm}^2$). Luminosities vary greatly between different machines but a good value is $L = 10^{32}\,\text{cm}^{-2}\,\text{s}^{-1}$. That seems to be a large number. However, in high-energy e^+e^- physics, a typical cross section is only $\sigma = 10^{-34}\,\text{cm}^2$, corresponding to an average of 1 event in 100 seconds. Prerequisite for high luminosity is the precise collision of very narrow beams of high intensity.

Particle 2 is at rest in a *fixed-target* experiment.

In a *colliding-beam* experiment both particles have equal and opposite momenta.

two machines could collide. Panofsky explained that such an experiment, if it worked, had one clear advantage over conventional set-ups, in which a beam of accelerated particles falls on a stationary target: The energy useful for the study of interactions is much larger (see Box).

The first to write about colliders was Wideröe, who had already stimulated Lawrence (Episode 47). He patented the idea in 1953 but did not follow it up technically. Independently this idea, quite naturally, was discussed among particle physicists. In 1956 Kerst, the inventor of the betatron, and collaborators from the Midwestern Universities Research Association in Champaign, Illinois, published a note [1], discussing the possibility of colliding beams with a new type of proton accelerator the association was trying to construct. That same year, at an accelerator conference in Geneva, O'Neill[2], a young professor at Princeton, also propagated colliding-beam facilities for protons, proposing that they could be added to 25-GeV synchrotrons, then under construction at CERN and Brookhaven. (That, indeed, was done at CERN in the late 1960s.) Moreover, he proposed colliding beams of electrons as a means to experimentally test quantum electrodynamics at very high energy corresponding to very small distance. In 1957, he convinced Panofsky to attempt the latter experi-

---

[2] Gerald K. O'Neill (born 1927)

Richter

Touschek (on the left).

ment as a joint Princeton–Stanford project. Three young physicists were recruited to work with O'Neill, among them Richter, who would reap a Nobel Prize with a collider experiment at Stanford (Episode 93).

Richter[3], a New Yorker, had studied at MIT and taken his Ph.D. there in 1956 with an experiment on the creation of pions in the collision of high-energy photons with protons. The photons were produced with the help of an electron synchrotron. Through this work with electrons and photons Richter got interested in testing quantum electrodynamics. He obtained a position at Stanford and successfully tested QED in an experiment of electron–positron pair production at the 700-MeV electron linear accelerator there. It was natural that he was keen to work on O'Neill's project [2].

The principal problem of a colliding-beam experiment is to obtain a reasonable reaction rate; the figure of merit is called *luminosity* (see Box). Instead of being a material object the target is itself a particle beam with a density lower by many orders of magnitude. In practically all colliders the particle beams are therefore made to circulate in rings, each ring being a synchrotron (Episode 64). Beam particles, not suffering a collision with a particle in the other beam, simply continue to circulate and get more chances to collide. Such colliders are *storage rings*. In a conventional synchrotron electrons stay for a few milliseconds, while in a storage ring they have to remain for hours. That posed quite a few severe problems, many of which are touched upon by Richter in his overview on the history of colliding beams [3]. New vacuum technologies had to be developed to reduce the loss of beam particles by collision with the rest gas in the beam pipes. The long-time behaviour of the beams had to be understood and the guiding and focussing fields had to be adapted to cope with various types of beam instabilities. This pioneering work, full of surprises and compared by Richter to the voyages of Odysseus [3], took many years. The final publication on the test of QED was published in 1971 [4].

The first electron–positron collider was built in Italy by Touschek and his group. Touschek[4] [5] was born in Vienna. Because his mother was Jewish he was expelled from the University of his home town in 1940. Helped by senior physicists, including Sommerfeld, he moved to Hamburg where he attended lectures at the university without being inscribed and for a long time lived in the flat of Lenz, one of the physics professors. To sustain himself he took on various jobs and, for some time, worked with Wideröe on the construction of a betatron. At the beginning of 1945 Touschek was arrested on racial grounds. Fortunately, in spite of dramatic happenings, he survived the time as prisoner. Only in 1946, he became officially a student of physics in Göttingen and obtained his diploma still the same year. In 1947 Touschek went with a British stipend to Glasgow, where he worked with an electron synchrotron, that was being built, and used it for studying meson production. Thus his thesis work was quite similar to that Richter would do a few years later. Together with his Ph.D. Touschek got an appointment as lecturer in natural philosophy at Glasgow. In 1952, he moved to Rome, where he had obtained a research position at the local section of the Istituto Nazionale di Fisica Nucleare (INFN), led by Amaldi, who had been one of Fermi's first students.

---

[3] Burton Richter (born 1931), Nobel Prize 1976   [4] Bruno Touschek (1921–1978)

The INFN operates the Laboratori Nazionali di Frascati in the small town of Frascati, beautifully located among vineyards not far from Rome. In March 1960, Touschek gave a seminar there in which he proposed a collider for electrons and positrons. Because of the symmetry between these particles they could circulate in the same ring but in opposite directions. Touschek stressed, in particular, the physics interest of such collisions. An electron and a positron can annihilate into a quantum-mechanical state corresponding to a meson and yield, in an unprecedented way, information about the strong interaction. At that time the Frascati laboratory had a 1.1-GeV electron synchrotron serving experimental groups from all over Italy. In his enthusiasm Touschek asked for the immediate conversion of that machine into a collider. That was, of course, impractical but Touschek was provided with the means and a group of three physicists to perform a proof-of-principle experiment. The result was a small machine named AdA for Anello di Accumulazione (accumulation ring). Its history is related in detail by a member of the original team [6].

Electrons and positrons of 200 MeV could be stored in the AdA ring [7,8] which had a diameter of 1.6 metres. The injection system was simple. There were two metal targets on opposite positions within the ring, a little outside of the nominal orbit. One was bombarded by photons from the Frascati synchrotron to produce electron–positron pairs. Some electrons were captured in the ring. They had to have just the right energy and lose enough of it by synchrotron radiation not to hit the target again in one of the first turns. Positrons were removed by the magnetic field. Once filled with electrons, AdA, which was mounted on rails, was moved perpendicular to the photon beam. The second target was irradiated and positrons were stored. AdA began to operate in February 1961, less than a year after Touschek's seminar. Stored electrons of either sign were recorded, but they were too few to yield a measurable collision rate. Therefore, in 1962, AdA was transferred to the Laboratoire de l'Accélérateur Linéaire (LAL) in Orsay near Paris, which had a high-intensity linear accelerator. There, in 1963, precise measurements of the beam lifetimes were made [9]. Also a particular beam instability was observed and soon analysed by Touschek, which allowed countermeasures to be taken. It is now known as *Touschek effect*. AdA had a very low luminosity because of the rather primitive injection. Touschek and his group could only detect the most frequent electron–positron collision, that giving rise to the emission of a single *bremsstrahlung* photon, $e^+ + e^- \rightarrow e^+ + e^- + \gamma$.

AdA raised the interest in electron–positron colliders and such machines were designed and realized in the following years in several laboratories: ACO in Orsay, ADONE (for 'big AdA') in Frascati, DORIS at the Deutsches Elektronen-Synchrotron (DESY) in Hamburg and VEPP-2 at the Institute of Nuclear Physics in Novosibirsk. The most successful of all, SPEAR at Stanford, was built under the direction of Richter and began operation only in 1972, three years after ADONE. The US Atomic Energy Commission had refused to provide construction funds although proposals had been submitted every year from 1964 to 1970 [3]. Finally a way was found to build SPEAR out of the ongoing budget of SLAC. The complex machine became the model of all later electron–positron colliders and was completed in only a year and a half.

AdA mounted on its movable support. Photo: INFN-LNF/SIS Photovideo.

[1] Kerst, D. W. et al., *Physical Review*, **102** (1956) 590.

[2] O'Neill, G. K., *Storage Rings for Electrons and Protons*. In Kowarski, L. (ed.): *International Conference on High Energy Accelerators and Instrumentation*, p. 125. CERN, Geneva, 1959.

[3] Richter, B., *The Rise of Colliding Beams*. In Hoddeson, L., Brown, L., Riordan, M., and Dresden, M. (eds.): *The Rise of the Standard Model*, p. 261. Cambridge University Press, Cambridge, 1997.

[4] Barber, W. C., O'Neill, G. K., Gittelman, B., and Richter, B., *Physical Review D*, **3** (1971) 2796.

[5] Amaldi, E., *The Bruno Touschek Legacy*. Yellow Report CERN 81-19, Geneva, 1981.

[6] Bernardini, C., *Physics in Perspective*, **6** (2004) 156.

[7] Bernardini, C. et al., *Nuovo Cimento*, **18** (1960) 1293.

[8] Bernardini, C. et al., *Nuovo Cimento*, **23** (1962) 202.

[9] Bernardini, C. et al., *Physical Review Letters*, **10** (1963) 407.

<table>
<tr><td>

84

</td><td>

Nonlinear Optics (1961)

</td></tr>
</table>

The laws of *reflection* and *refraction* of light were, at first, found empirically and later derived from various theories of light and matter. In his doctoral thesis of 1875 and, in more detail, in his book of 1895 [1] Lorentz gave a derivation from Maxwell's equations and a model of matter which agreed with experiments until 1961. His model was based on the assumption (by no means generally accepted at the time) that matter consists of molecules and that these may exist in charged form as ions. He pointed out that the existence of ions had proved necessary to explain electric conduction in gases and continued [1]:

> As it seems to me, nothing stands in the way of the assumption, that also the molecules of ponderable dielectric bodies contain such particles, which are bound to certain equilibrium positions and can only displaced therefrom by external electric forces; just herein would then rest the "dielectric polarisation" of such bodies.

Lorentz also assumed that the displacement was strictly proportional to the electric force. As we shall see, this proportionality is the foundation of *linear optics*. When, for very high electric fields, proportionality gave way to more complicated relations, the field of *nonlinear optics* emerged.

With our knowledge of the atomic structure of matter Lorentz' model can easily be rephrased (see Box). The essential fact remains: An electric field induces electric dipole moments in a dielectric medium. The macroscopic dipole moment per unit volume is called the dielectric polarization. If the field oscillates in time, as does that of a light wave, the dipoles oscillate with the same frequency and act as tiny Hertzian antennas emitting radiation (light) of the same frequency. The frequency of light remains unchanged in the Lorentz theory.

In the year following the invention of the laser (Episode 82) Franken and his group demonstrated that laser light could double its frequency while passing through quartz. They focussed the beam of red light from a ruby laser onto a plate of crystalline quartz, placed a spectrograph behind the plate, and, for a single laser pulse, registered the optical spectrum of the light, emerging from the quartz, on a photographic plate. The plate showed an intensive spot at the wavelength of red ruby laser light (694.6 nm), but also a weak one at exactly half that wavelength, i.e., twice the frequency of the incident light: Besides

Figure from the paper *Generation of Optical Harmonics* by Franken and his group. The original caption reads: 'A direct reproduction of the first plate in which there was an indication of second harmonic. The wavelength scale is in units of 100 Å [10 nm]. The arrow at 3472 Å indicates the small but dense image produced by the second harmonic. The image of the primary beam at 6946 Å is very large due to halation.' The small dot, to which the arrow was intended to point, somehow got lost in the editing process and never appeared in *Physical Review Letters*. Figure reprinted with permission from [2]. Copyright 1961 by the American Physical Society.

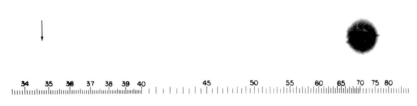

red light there also was blue light. The experiment proved that, under the conditions of laser light, the dipoles of the medium are made to oscillate not only with the frequency of the incident light but also with twice that frequency [2].

Franken[1] [3] was born in New York City and studied at Columbia University, earning his Ph.D. in 1952 with magnetic-resonance research under Kusch. He continued to work successfully in this field, first at Stanford University and, beginning in 1956, at the University of Michigan in Ann Arbor. Franken recognized that a laser could produce an extraordinarily high field in a dielectric. The laser light can easily be focussed on a small spot with a diameter of, say, ten wavelengths. Moreover, it is polarized (the electric field vector has a well-defined direction) and coherent (all light waves have the same phase). Within one laser pulse the spot experiences a very high flux of energy per unit area and unit time, which is proportional to the square E_0^2 of the field-strength amplitude. Franken's estimate was $E_0 \approx 10^7 \, \text{V/m}$ for his laser. This field is still some orders of magnitude lower than a typical electric field inside an atom. But Franken hoped that it was strong enough for the atomic dipoles to become anharmonic rather than harmonic oscillators, i.e., to radiate at the fundamental frequency ω of the incoming light and at the frequencies of *overtones* or *harmonics*. What his group observed was the second harmonic.

Already in the original paper [2] the small intensity of blue with respect to red light was explained by a *phase mismatch*. The velocity of light $v = c/n$ in a medium is equal to the speed c of light in vacuum divided by the *index of refraction* n. Now every medium shows *dispersion*, i.e., n depends on the frequency ω and so does v. The dipoles, excited by the incident red light, form themselves a wave pattern, propagating together with the red light with velocity $v_r = c/n_r$. The blue light, they emit, travels at a different velocity $v_b = c/n_b$. Thus a phase shift between the emitters of blue light and the wave of blue light develops; the blue-light intensity rises and falls periodically. The length of this period is on the order of $\ell = \lambda_b n_b / (n_b - n_r)$, where λ_b is the wavelength of the blue light. This effect severely limits the blue-light intensity. Moreover, the intensity depends critically on the optical thickness of the crystal: Blue light will appear outside only if there is an intensity maximum very near the surface through which the radiation exits.

Franken

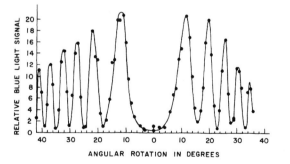

The effect of phase mismatch as demonstrated by Terhune's group. Shown is the intensity of blue light leaving a thin quartz platelet, which has an inclination with respect to the incident laser beam of red light. The intensity goes through maxima and minima as the optical thickness of the platelet is varied by changing the angle of inclination. Figure reprinted with permission from [4]. Copyright 1962 by the American Physical Society.

---

[1] Peter Alden Franken (1928–1999)

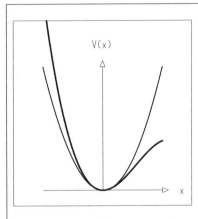

Potential energy $V(x)$ for a harmonic oscillator (*thin line*) and an anharmonic oscillator. Only if the anharmonic potential is asymmetric (as in the figure) there will be a nonvanishing susceptibility $\chi^{(2)}$. That requires a crystal without inversion symmetry, i.e., a crystal not symmetric under the inversion $x \to -x, y \to -y, z \to -z$.

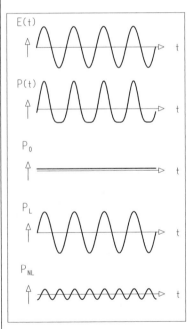

Time dependence of the electric field strength E and the polarization P. The latter is the sum of a constant term P_0, a linear term P_L, oscillating with the frequency ω of the electric field, and a nonlinear term P_{NL} oscillating with 2ω.

Linear and Nonlinear Dielectric Polarization

Harmonic Oscillator. Consider a particle of mass m bound by a linear force $F = -ax$ to its equilibrium position $x = 0$. Its potential energy, $V = (a/2)x^2$, is quadratic in x. Suppose there is some small energy loss (damping). If there is an external harmonically oscillating force $F = F_0 \sin \omega t$, the particle oscillates with the (circular) frequency ω of that force (driven oscillator).

Anharmonic Oscillator. If the force is not linear, i.e., the potential is not quadratic in x, the driven system can no longer be described by an oscillation of a single frequency. However, for sufficiently small x every binding force is well approximated by a quadratic potential. Therefore the anharmonic nature becomes apparent only for large amplitudes in x, i.e., for a large external force.

Linear Polarization. In a dielectric medium the role of the particle is played by an electron (or several electrons), carrying charge Q, in an atom. The binding force is the electrostatic force from the rump (nucleus and other electrons) of that atom plus the force from neighbouring atoms. The external force is provided by an electric field, $E = E_0 \sin \omega t$, acting on Q, i.e., $F = QE$. It leads to a time-dependent dipole moment $d = xQ$ of the atom. The dipole moment per unit volume is called the *dielectric polarization P* of the medium. For moderate fields it is proportional to the field strength,

$$P = \varepsilon_0 \chi E \quad .$$

The proportionality constant χ is called *dielectric susceptibility*; ε_0 is the electric field constant.

Nonlinear Polarization. For very high fields the simple proportionality is lost. The susceptibility becomes a function of the field strength, $\chi = \chi(E)$, which may be developed into a power series,

$$P = \varepsilon_0 \chi(E) E = \varepsilon_0 (\chi^{(1)} E + \chi^{(2)} E^2 + \chi^{(3)} E^3 + \cdots) \quad .$$

Typical numerical values,

$$\chi^{(1)} \approx 1 \quad , \qquad \chi^{(2)} \approx 10^{-12} \, \mathrm{m \, V^{-1}} \quad , \qquad \chi^{(3)} \approx 10^{-21} \, \mathrm{m^2 \, V^{-2}} \quad ,$$

show that very high fields are needed to obtain observable nonlinear effects. We now consider the time dependence of P and find (up to the second order in E) the relation

$$P(t) = \varepsilon_0 (\chi^{(1)} E_0 \sin \omega t + \chi^{(2)} E_0^2 \sin^2 \omega t) \quad ,$$

which can be rewritten as

$$P(t) = \frac{\varepsilon_0}{2} \chi^{(2)} E_0^2 + \varepsilon_0 \chi^{(1)} E_0 \sin \omega t - \frac{\varepsilon_0}{2} \chi^{(2)} E_0^2 \cos 2\omega t = P_0 + P_L(\omega) + P_{NL}(2\omega) \, .$$

The polarization is seen to be a sum of a constant term P_0, a *linear* term $P_L(\omega)$, oscillating with ω, and a *nonlinear* term $P_{NL}(2\omega)$, oscillating with 2ω.

Tensor Structure of χ. Although the electric field and the polarization are vectors, no vector symbols were used in the equations above, which actually stand for more complicated relations. In a crystal the two vectors, in general, are not parallel. $\chi^{(1)}$ is a tensor with 9 elements, because each of the three components of the vector $\mathbf{P}$ depends on all three components of the vector $\mathbf{E}$. The tensor $\chi^{(2)}$ has 27 elements because every component of $\mathbf{P}$ depends on all 9 products $E_x E_x, E_x E_y, \ldots, E_z E_z$ of field-strength components.

Within months after Franken's pioneering work, phase mismatch was demonstrated experimentally by a group under Robert Terhune at the Scientific Laboratory of the Ford Motor Company in Dearborn, Michigan [4]. They changed the optical thickness of a thin quartz plate by varying its inclination with respect to the incident red light and found that the intensity of the outgoing blue light displayed pronounced minima and maxima.

A method, avoiding phase mismatch and now widely used for frequency doubling, was presented by Terhune's group in the same paper [4] and, independently, by Joseph Giordmaine at Bell Telephone Laboratories in Murray Hill, New Jersey [5]. It uses a crystal with pronounced *birefringence*. If a beam of monochromatic light enters such a crystal, it is split in two. These beams are called *ordinary* ray (o) and *extraordinary* ray (e), respectively. The polarizations, i.e., the planes in which the electric field strengths oscillate, are different for the two rays and perpendicular to each other. For the ordinary ray, the refractive index n_o is independent of the direction of the ray within the crystal. The refractive index n_e for the extraordinary ray depends on the angle between the ray direction and a particular direction within the crystal, called its *optic axis*. Graphically, the angular dependence of n_e can be shown as follows. Consider an xyz coordinate system with z the direction of the optic axis, and an ellipsoid, rotationally symmetric about this axis with its centre in the origin. A straight line, drawn from the origin, after a length n_e, intersects the ellipsoid. Its direction, given by the polar angle θ and the azimuthal angle ϕ in the coordinate system, is that of the extraordinary ray; its length $n_e(\theta, \phi)$ is the refractive index for that ray. For the refractive index of the ordinary ray, obviously, the ellipsoid becomes a simple sphere of radius n_o. In a crystal of potassium dihydrogen phosphate (KDP) there is a polar angle θ_0, for which the refractive index n_e for blue light equals the index n_o for red light.

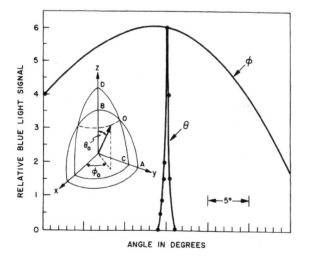

Phase matching in a KDP crystal as demonstrated in [4]. The inset shows an octant of the ellipsoid describing the refractive index $n_{e,b}$ of blue light and an octant of the sphere for the refractive index $n_{o,r}$ of red light. Phase matching occurs for a polar angle θ_0, at which ellipsoid and sphere intersect, i.e., for $n_{e,b} = n_{o,r}$. This is verified by the intensity measurement of blue light as a function of theta, which sharply peaks at $\theta = \theta_0$. With θ kept fixed at that angle, there is also a weaker dependence on the azimuthal angle ϕ with a maximum at ϕ_0. Figure reprinted with permission from [4]. Copyright 1962 by the American Physical Society.

Terhune and his group measured a very pronounced intensity peak for blue light at the angle θ_0. Keeping that angle fixed, they also found a dependence on the azimuthal angle ϕ. The maximum at $\phi = \phi_0$ is explained by considering the tensor structure of the susceptibility $\chi^{(2)}$ (see Box). All three axes of the coordinate system are defined by the KDP crystal. Under the direction, defined by (θ_0, ϕ_0), in addition to phase matching, the ordinary ray is polarized exactly in the xy plane and the extraordinary ray perpendicular to it. Making use of that special direction, frequency doubling is obtained in KDP with high efficiency.

The theory of nonlinear optics became the domain of Bloembergen and his group at Harvard University. We met Bloembergen in Episode 73 as the inventor of the three-level principle for the maser, which later became the operating principle of many lasers. While waiting for one of the first commercial lasers to be delivered, he decided to extend Lorentz' theory to include nonlinear or anharmonic response to the external electromagnetic field. In his Nobel Lecture [6], delivered in 1981, he says admiringly about Lorentz:

> If he had admitted some anharmonicity, he could have developed the field of nonlinear optics seventy years ago. It was, however, not experimentally accessible at that time, and Lorentz lacked the stimulation from stimulated emission of radiation.

A light wave travels in the direction of its *wave vector* $\mathbf{k}$. Its modulus is $k = \omega/c$ in vacuum and $k = n\omega/c$ in a medium with refractive index n. Maxwell's equations demand certain continuity conditions for the electric field strength and the dielectric polarization to be fulfilled at the surface between different media. Let the surface be the xy plane of a coordinate system and let the vector $\mathbf{k}$ be parallel to the xz plane. It has only two components: k_z, perpendicular to the surface, and k_x parallel to it. Because of the continuity conditions for the fields, the vector k_x does not change at the surface; k_z can then be computed from k and k_x. In this way the wave vectors of the reflected and refracted waves are easily found, given that of the incident wave. The intensities of the reflected and refracted waves are also obtained from the continuity conditions for the fields. All this was done by Lorentz for linear optics. His results were the same as obtained in 1823 by Fresnel, who treated light as an elastic wave in the ether or other media.

Example of field continuity in linear optics. *Left:* The xy plane is the surface of a dielectric, which occupies the region $z < 0$; there is vacuum at $z > 0$. The wave vector $\mathbf{k}$ of the incident light lies in the xz plane, as do the wave vectors $\mathbf{k}^{\mathrm{R}}$ of the reflected wave and $\mathbf{k}^{\mathrm{T}}$ of the refracted wave. The surface represents the component E_y of the electric field strength, $E_y = E_y(x, y)$, at a fixed moment in time. It is continuous at $z = 0$. The wiggly pattern at $z > 0$ is due to the interference of incident and reflected wave. The simple plane wave at $z < 0$ is the refracted wave. *Right:* The wave vectors $\mathbf{k}, \mathbf{k}^{\mathrm{R}}, \mathbf{k}^{\mathrm{T}}$ in the xz plane. All three have the same x component.

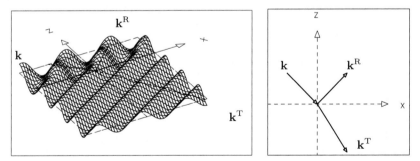

The essential point in the extensive work of Bloembergen and his collaborators [7,8] is the incorporation of the dielectric polarization as a source of radiation with a frequency different from that of the incident radiation. It causes refracted and reflected waves of the new frequency. We present here only an illustration of a situation which, as far as possible, is analogous to the linear case discussed above. It is a special case of an example in [8]. The coordinate system is as before and also the incident (red) light wave of frequency ω, as before, is polarized in the y direction. But now the medium at $z < 0$ has also nonlinear properties. The incident wave suffers refraction and reflection at this surface according to the laws of conventional, linear optics. The refracted light causes a dielectric polarization in the medium, which radiates linearly with ω and nonlinearly with 2ω. The nonlinear polarization is a vector field oriented parallel to the y direction. It can be described by a wave propagating with the wave vector $\mathbf{k}^S(2\omega)$. Since it is caused directly by the refracted red light, this wave vector is simply twice the wave vector of the latter, $\mathbf{k}^S(2\omega) = 2\mathbf{k}^T(\omega)$.

To understand the properties of the (blue) light of frequency 2ω, we consider two more wave vectors. One of them, $\mathbf{k}^R(2\omega) = 2\mathbf{k}^R(\omega)$, is parallel the wave vector $\mathbf{k}^R(\omega)$ of the reflected red light. The other one, $\mathbf{k}^T(2\omega)$, is the wave vector which would describe refracted blue light in the medium if the incident light were blue instead of red. It is not parallel to that of the refracted red light because of the dispersion in the medium (the change of refractive index with frequency). The continuity conditions at the surface now enforce the following for the electric field strength of the blue light, which is also parallel to the y direction: In the medium it is a superposition of waves with the wave vectors $\mathbf{k}^S(2\omega)$ and $\mathbf{k}^T(2\omega)$. It displays a beat structure, because there is a phase mismatch. The structure is complicated further since the wave vectors $\mathbf{k}^S(2\omega)$ and $\mathbf{k}^T(2\omega)$ are not parallel. There is also a blue light wave in the vacuum with the wave vector $\mathbf{k}^R(2\omega)$. This frequency-doubled 'reflected' light was observed by Bloembergen a year after he predicted it [9].

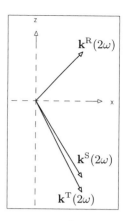

The wave vectors $\mathbf{k}^S(2\omega)$, $\mathbf{k}^T(2\omega)$, and $\mathbf{k}^R(2\omega)$ in the xz plane. Similar to the linear case, all three vectors have the same x component.

[1] Lorentz, H. A., *Versuch einer Theorie der electrischen und optischen Erscheinungen in bewegten Körpern*. Brill, Leiden, 1895.

[2] Franken, P. A., Hill, A. E., Peters, C. W., and Weinreich, G., *Physical Review Letters*, **7** (1961) 118.

[3] Gibbs, H. M., Meystre, P., and Wright, E. M., *Physics Today*, **October** (1999) 105.

[4] Maker, P. D., Terhune, R. W., Nisenoff, M., and Savage, C. M., *Physical Review Letters*, **8** (1962) 21.

[5] Giordmaine, J. A., *Physical Review Letters*, **8** (1962) 19.

[6] Bloembergen, N., *Nobel Lectures, Physics 1981–1990*, p. 12. World Scientific, Singapore, 1993.

[7] Armstrong, J. A., Bloembergen, N., Ducuing, J., and Pershan, P. S., *Physical Review*, **127** (1962) 1918.

[8] Bloembergen, N. and Pershan, P. S., *Physical Review*, **128** (1962) 606.

[9] Ducuing, J. and Bloembergen, N., *Physical Review Letters*, **10** (1963) 474.

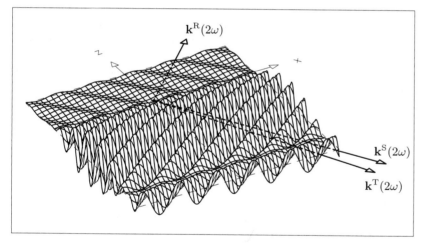

The surface represents the component E_y of the frequency-doubled (blue) light wave for the special case discussed in the text. In the nonlinear material (the region $z < 0$) the field is a superposition of waves with the wave vectors $\mathbf{k}^S(2\omega)$ and $\mathbf{k}^T(2\omega)$. The beat structure of the field is well visible, in particular, at the left margin. There is also a 'reflected' wave with wave vector $\mathbf{k}^R(2\omega)$ in the vacuum (the region $z > 0$).

Nonlinear optics is by no means limited to frequency doubling, though that is a useful practical application. Sums and differences of various frequencies can be formed. Moreover, the new field of *nonlinear spectroscopy*, exploiting nonlinear absorption phenomena, has developed.

There is More than One Kind of Neutrino (1962)

In Episode 79 we mentioned that the muon decays into an electron and two neutrinos and that this fact was ascertained, in particular, on the basis of measurements which Steinberger performed for his thesis and published in 1949. A dozen years later he was the senior scientist in a group demonstrating that these two neutrinos are of different nature.

Steinberger[1] [1] was born as son of the cantor and Hebrew teacher of the Jewish community in Bad Kissingen, Franconia. Already in 1934, when he was thirteen, his parents felt that he and his elder brother should be spared the rising Nazi oppression and accepted an offer to place them in American foster families. The brothers moved to Chicago where the family was reunited in 1938 after parents and younger brother also had been able to leave Germany. That same year Steinberger began to study chemical engineering on a scholarship he had won. When that ran out, for a year, he secured a job as laboratory aid and attended night classes. Supported by another scholarship he obtained a bachelor's degree in chemistry at the University of Chicago in 1942. America having entered the war, Steinberger joined the U. S. Signal Corps and soon found himself working on radar at MIT, gradually turning into a physicist. In 1945 he did some months of active duty, some of them under Lieutenant Leon Lederman. In 1946 Steinberger became a member of Fermi's group in Chicago; among his fellow graduate students were Yang and Lee.

After obtaining his Ph.D. in 1948, Steinberger worked for a year in theoretical physics under Oppenheimer at the Institute for Advanced Studies in Princeton and computed the decay $\pi^0 \rightarrow \gamma + \gamma$ of the still unobserved neutral pion into photons. In 1950, working at Lawrence's Radiation Laboratory in Berkeley and returning to experiment, he was able to detect that decay and thus reveal the existence of the π^0 meson. Still in 1950, Steinberger accepted a position as assistant professor at Columbia University in New York. Columbia, at its NEVIS Laboratory, possessed a cyclotron which was then the best source of pions. These were made available as an external beam, designed and built by Lederman, Steinberger's former army lieutenant.

Lederman[2] had spent several years in active army service and was still a graduate student. A native of New York City, he had received his education in his home town and got a degree in chemistry from the City College of New York in 1943. He obtained his Ph.D. in 1951 and stayed on at Columbia. At the end of Episode 77 we briefly described the beautiful experiment by which, in four days, he demonstrated parity violation and determined the muon spin.

In 1956, Steinberger performed the first bubble-chamber experiment at

---

[1] Jack Steinberger (born 1921), Nobel Prize 1988   [2] Leon M. Lederman (born 1922), Nobel Prize 1988

Brookhaven which yielded new information on strange particles. The chamber had been built together with three graduate students including Schwartz.

Schwartz[3] was born in New York and attended the Bronx High School of Science before studying physics at Columbia. He obtained his Ph.D. in 1958 and continued to experiment at Brookhaven as a member of the Columbia group. In his Nobel Lecture [2] Schwartz recalls a physics discussion over coffee at Columbia in 1959 at which Lee deplored the fact that there seemed to be no means to test weak-interaction theory at high energies. Experiments were restricted to β decay, in which only little energy, a few MeV, is released, and to neutrino interaction at reactors (Episode 76). The neutrinos in such experiments are of low energy since they also are produced in β decay. In the evening it occurred to Schwartz that it might be possible to use neutrinos produced with the help of the new particle accelerators, which were then being constructed or planned. Their flux would be much smaller than that of reactor neutrinos but that could be made up by the increased energy. Schwartz phoned Lee at home who responded with enthusiasm.

In the $V-A$ theory (Episode 79) the matrix element for neutrino–nucleon interaction is independent of energy; but since the number of available quantum-mechanical states rises with energy, so does the interaction cross section: it increases as the square of the available energy in the centre-of-mass system. The increase could not, however, continue indefinitely. Simple quantum-mechanical considerations limit the cross section. (Its maximum value is $\sigma_{\max} = \pi \lambda^2 / 2$, where $\lambda = \hbar / p^*$ is, essentially, the de Broglie wavelength of the neutrino in the centre-of-mass system, p^* is the neutrino momentum in that system.) That limit is reached only for laboratory energies of the neutrino of 300 GeV, by far out of reach at that time, but it pointed at a defect of the theory. In fact, ever since Yukawa's work (Episode 57) the possibility had been considered to transmit weak interaction by exchange of a boson, then referred to as *intermediate boson* and now called weak boson or *W boson*. In the language of Feynman diagrams that meant that the four fermion lines no longer met in a single point in spacetime but that there appeared an internal line, representing an electrically charged boson and connecting two pairs of fermion lines. This inner line or propagator contributes a factor to the matrix element of the reaction which, essentially, is proportional to $1/(q^2 - M_W^2)$, where M_W is the mass of the intermediate boson and q is the four-momentum transfer, i.e., the difference in four-momentum between an incoming and an outgoing fermion. Such a theory was consistent with experiment if M_W was so large that $M_W^2 \gg q^2$ and the factor became a constant, $-1/M_W^2$. With increasing values of q^2 the factor would rise; the cross section would reach a maximum for q^2 around M_W^2 and then drop off again.

Low-energy experiments thus could not decide about the existence of an intermediate boson; but there was a subtle point to which attention had been drawn by Feinberg[4] from Brookhaven. He considered two cases:

1. There is only one kind of neutrino ν; the intermediate boson can couple to pairs of leptons such as (e^-, ν) and (μ^-, ν).
2. There are two neutrinos, ν_e and ν_μ, of different kind which are related

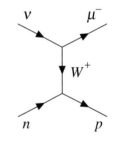

Diagram for the reaction

$$\nu + n \rightarrow \mu^- + p$$

including a weak boson W.

[3] Melvin Schwartz (1932–2006), Nobel Prize 1988   [4] Gerald Feinberg (1933–1992)

to the electron and the muon, respectively. The W boson couples only to lepton pairs of one kind, i.e., (e^-, ν_e) and (μ^-, ν_μ).

In the first case the muon can decay into an electron and a photon, $\mu^- \to e^- + \gamma$. The muon transforms (for the very short time allowed by the uncertainty relation) into a boson W^- and a neutrino. In the process the charged boson suffers acceleration and emits a photon. It can then recombine with the neutrino to form an electron. In the second case that process is forbidden. The decay $\mu^- \to e^- + \gamma$ had been searched for but was not observed although the experiment [3] had been sensitive to detect it if case 1 applied. Feinberg concluded [4] that this hinted at the existence of two different neutrinos.

Early in 1960, a note by Schwartz appeared, describing the feasibility of a high-energy neutrino experiment [5]. It was directly followed by a paper by Lee and Yang on possible physics questions to be answered with the new method [6]. Schwartz envisaged an experiment of the following kind.

An intense proton beam from an accelerator hits a target, producing pions predominantly moving in the forward (the original proton) direction. A certain fraction of the charged pions decays, $\pi^\pm \to \mu^\pm + \nu$, yielding high-energy neutrinos also moving predominantly forward. At a distance of 10 metres from the target a shielding wall was placed, another 10 metres thick. It was to absorb all strongly interacting particles, such as pions, protons, and neutrons and also all charged leptons, i.e., electrons and muons. But neutrinos could penetrate it and reach a detector. Schwartz estimated that for an existing accelerator, yielding 3-GeV protons, about one neutrino interaction per hour would result, provided the detector contained 10 tons of material. He concluded that the experiment was impractical with existing machines but could be done with accelerators planned or under construction. In a note, added in proof, Schwartz states that his attention had been drawn to a similar proposal arrived at, independently and somewhat earlier, by Pontecorvo in Russia [7].

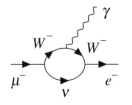

Diagram allowing the decay $\mu^- \to e^- + \gamma$.

Lederman (left), Steinberger (centre), and Schwartz.

At Brookhaven a new proton accelerator, the *Alternating Gradient Synchrotron* (AGS) was under construction. Lederman, Schwartz, and Steinberger, now all faculty members at Columbia, assembled a group preparing for a neutrino experiment there. After first considering a large bubble chamber, they settled for the recently invented *spark chamber* because of its higher mass. A spark chamber is an assembly of parallel metal plates interspersed by gas-filled gaps. If a charged particle traverses it and high voltage is applied to the plates (with a sign alternating from plate to plate), the ions produced by the particle in the gas initiate sparks where the particle passed. Photographs reveal particle tracks as chains of sparks.

Ten spark-chamber modules, each consisting of 9 aluminium plates and weighing a ton, were constructed. The whole detector consisted of two rows of five modules each, one on top of the other. It was surrounded by anti-coincidence counters. Trigger counters were placed between the individual modules. The shielding, 13.5 m thick, was of iron, partly armour plates from a scrapped battle ship.

Schwartz with the spark chamber.

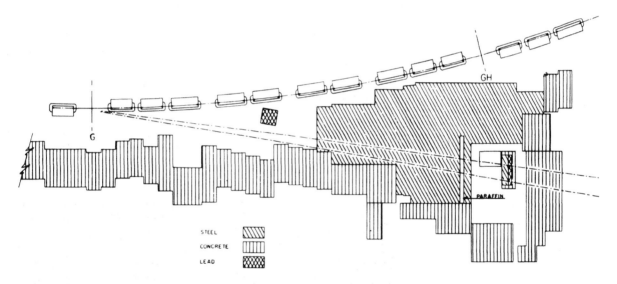

STEEL

CONCRETE

LEAD

During the data taking high voltage was applied to the spark chambers a few microseconds after three conditions were met simultaneously: (a) Trigger counters indicated one or more charged particles within the detector, (b) the absence of signals from the anti-coincidence counters made sure that no charged particles entered from the outside, (c) the timing was right for the synchrotron to produce particles. The ionization in the gas gaps persisted over a few microseconds; sparks were generated and a photograph taken.

Neutrinos created at accelerators in the decay of π mesons and, to a lesser extent, of K mesons were created together with muons. If there were two kinds of neutrino, they would be muon neutrinos. A muon neutrino, hitting a nucleon, could produce a muon, e.g., $\nu_\mu + n \rightarrow p + \mu^-$, but not an electron. If there was only one neutrino, electrons and muons would be produced in equal proportion. The signatures of a muon and an electron in the chamber are

Plan of the Brookhaven neutrino experiment. The target (within the synchrotron) is near point G, the detector (enclosed by shielding) at the lower right. Figure reprinted with permission from [8]. Copyright 1962 by the American Physical Society.

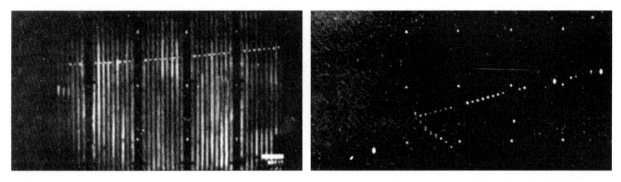

Single-muon event (left) and vertex event (right). Both figures reprinted with permission from [8]. Copyright 1962 by the American Physical Society.

Particle	L_e	L_μ
e^-, ν_e	+1	0
e^+, $\bar\nu_e$	−1	0
μ^-, ν_μ	0	+1
μ^+, $\bar\nu_\mu$	0	−1

Assignment of lepton numbers.

[1] Steinberger, J., *Learning About Particles – 50 Privileged Years.* Springer, Berlin, 2005.

[2] Schwartz, M., *Nobel Lectures, Physics 1981–1990*, p. 467. World Scientific, Singapore, 1993.

[3] Lokanathan, S. and Steinberger, J., *Physical Review*, **98** (1955) 240.

[4] Feinberg, G., *Physical Review*, **110** (1958) 1482.

[5] Schwartz, M., *Physical Review Letters*, **4** (1960) 306.

[6] Lee, T. D. and Yang, C. N., *Physical Review Letters*, **4** (1960) 307.

[7] Pontecorvo, B. M., *Zh. Eksperim. i Teor. Fiz.*, **37** (1959) 1751. Engl. transl. in [9].

[8] Danby, G., Gaillard, J. M., Goulianos, K., Lederman, L. M., Mistry, M., Schwartz, M., and Steinberger, J., *Physical Review Letters*, **9** (1962) 36.

[9] Pontecorvo, B. M., *Soviet Physics JETP*, **10** (1959) 1236.

quite different. A muon, which loses energy only very gradually by ionization, leaves one long track. An electron through successive *bremsstrahlung* and pair-creation events causes an electromagnetic shower resulting in a short and wide group of sparks.

The large majority of events with the characteristics of having been induced by high-energy neutrinos and which showed tracks with a minimum total energy clearly contained muons. They were either *single-muon events*, in which only the muon left a visible track, or *vertex events* with an additional track from the recoil proton or even several hadron tracks. Against a total of 56 such events there stood 8 'shower' events which, however, could be explained as caused by background reactions of various kinds. In their paper [8] the group concluded, that

> the most plausible explanation for the absence of electron showers, and the only one which preserves universality [i.e., equal coupling of electrons and muons in weak interactions], is then that $\nu_\mu \neq \nu_e$; i.e., that there are at least two types of neutrinos.

This result was verified, a little later, by experiments at the newly built proton synchrotron at CERN in Geneva and could be best expressed by assigning two different *lepton numbers*, L_e and L_μ, to every lepton, see Table. These are additive quantum numbers, like electric charge or baryon number. In all processes the sum over L_e over all particles and, likewise, the sum over L_μ is conserved. (Or so it seemed until very recently, see Episode 100.) For instance, in the decay of the negative muon, $\mu^- \to e^- + \bar\nu_e + \nu_\mu$, these sums are 1 for L_μ and 0 for L_e both before and after the reaction.

The paper [6], which Lee and Yang wrote before any high-energy neutrino experiment was done, touched on many more topics which, eventually, would become accessible. These are, in particular, the possible existence of so-called weak neutral currents in addition to the charged ones discussed so far and a search for the weak bosons. Both were eventually found, the neutral currents in 1973 (Episode 91) and the bosons in 1983 (Episode 96).

Semiconductor Heterostructures – Efficient Laser Diode Proposed (1963) and Built (1970)

86

The essential elements of the first transistor, the *point-contact* transistor discussed in Episode 69, are the two transition regions (also called contacts, interfaces, or junctions) between a metal and a semiconductor, both having the property of a rectifier. Already in 1948 Shockley applied for the patent [1] on a quite different and later widely used device, the *junction transistor*. Here the interfaces are between regions of the same semiconductor material (originally germanium, now usually silicon), doped by acceptor atoms on one side of the junction and by donor atoms on the other side. Such interfaces are now called *homojunctions*. An interface between two different crystals is a *semiconductor heterojunction*. The new possibilities, inherent in heterojunctions, were first understood theoretically. Intense experimental efforts then led to the development of new devices, particularly in the field of optoelectronics. Furthermore, special devices containing heterojunctions allow fundamental research, such as studies of an electron gas in two, one, or even zero space dimensions.

On page 193 in Episode 44, we showed the energy distributions of electrons and holes in an intrinsic semiconductor, in which the *Fermi energy* E_F is halfway between the upper edge E_v of the valence band and the lower edge E_c of the conduction band. In (acceptor-doped) *p*-type material E_F lies between E_v and the energy level of the acceptors, i.e., only a little above E_v. In (donor-doped) *n*-type material it is a little below E_c. In the transition region of a *pn* junction some holes drift to the *n* side and some electrons to the *p* side of the interface, forming positive space charge on the former and negative space charge on the latter. The width of the space-charge region is on the order of 1 μm. Because of the space charge a step in energy appears in both the valence and the conduction band over this width. The band gap, i.e., the separation of the two bands, stays unchanged over the step. This step makes the *pn* junction a rectifier or *diode*.

Shockley's junction transistor consists of an *emitter* (*p*-type), a thin *base* (*n*-type), and a *collector* (*p*-type). It can be seen as a *pn* rectifier followed by an *np* rectifier and therefore allows no current, whatever sign the voltage between emitter and collector has. The situation changes if also the base is connected to an external circuit. The emitter–base diode can be made conducting and holes enter the base. Here they are *minority carriers* (electrons are the *majority carriers*) and are not hindered at all by the base–collector diode. The principle of the junction transistor is the control of the collector current I_C by the base current I_B. There is a snag here: The base current is not only carried by holes coming from the emitter but, to some extent, also by electrons leaving toward the emitter. The amplifier properties of the transistor are drastically improved

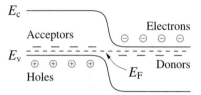

Band diagram of a *pn* homojunction. The horizontal coordinate is the position perpendicular to the junction. Shown are the band edges E_v and E_c. (E_F is the Fermi energy.) The step in both bands is caused by a characteristic feature of the *pn* junction, the space-charge layer. Since the vertical coordinate is the potential energy of an electron, electrons are inclined to move 'downhill' along the edge E_c of the conduction band, while holes seek to move 'uphill' along the edge E_v of the valence band. Obviously, the motion of both electrons and holes is hindered by the step. By applying an external voltage between the ends of the *pn* junction the step can be lowered or increased, thus enabling or disabling a current: The *pn* junction acts as a rectifier or *diode*.

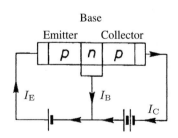

A *pnp*-junction transistor with its circuit.

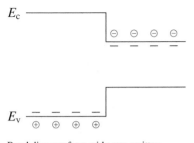

Band diagram for a wide-gap emitter.

Kroemer

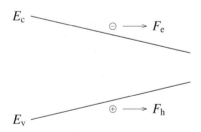

Quasi-electric forces, sketched by Kroemer for converging bands. In this special case the forces on electrons and holes act in the same direction.

if this electron current is reduced. That can be achieved if the step size in the emitter–base junction is larger for electrons than for holes, and this requires different band gaps in the two regions. An emitter with lower valence-band edge E_v but a higher conduction-band edge E_c, compared to the values in the base, is ideal. In this case holes from the emitter are accelerated at the upward edge of E_v towards the base and electrons from the base experience a repulsive force at the edge of E_c. Claim number 2 in Shockley's patent application [1] reads:

> A device as set forth in claim 1 [the junction transistor] in which one of the separated zones is of a semiconductive material having a wider energy gap than that of the material in the other zones.

Shockley did not elaborate on possible realizations of such a device. Ways to improved transistors, including one with a wide-gap emitter, were pointed out by Kroemer, who was unaware of Shockley's claim.

Kroemer[1] was born as son of a municipal civil servant in Weimar, the town of Goethe and other heroes of German classic. In 1948 he began his studies of physics at the university of nearby Jena, where Schiller had been professor of history. It was the time of the Berlin blockade. Kroemer felt that there was severe political suppression in East Germany. In the short autobiography for the Nobel Foundation he writes: 'Every week, some of my fellow students had suddenly disappeared, and you never knew whether they had fled to the West, or had ended up in the German branch of Stalin's Gulag, like the uranium mines near the Czech border. During the Berlin airlift, I was in Berlin as a summer student at the Siemens company, and I decided to go West via one of the empty airlift return flights.' Kroemer continued his studies in Göttingen. For his diploma and his doctorate under Sauter he did theoretical work; his doctoral dissertation dealt with aspects of the recently invented transistor.

Having received his Ph.D. before reaching the age of 24, Kroemer joined the Central Telecommunications Laboratory, the Fernmeldetechnisches Zentralamt FTZ, of the German postal service as the only theoretician in its small semiconductor research group. There he wrote a series of articles on transistor theory. One includes the proposal of a wide-gap emitter [2]. In 1957, now working in Princeton at the laboratories of the Radio Corporation of America (RCA), Kroemer repeated this proposal in more detail [3,4]. He stressed that it should be possible to produce semiconductor crystals with a bandgap varying continuously with position, for instance, an alloy of several semiconductors such as germanium and silicon with varying composition. Such a crystal is now said to possess a *graded gap*. The forces on electrons and holes are determined by the slopes of the conduction-band and valence-band edges, respectively. Kroemer called them 'quasi-electric' and pointed out [3]:

> They present a new degree of freedom for the device designer to enable him to obtain effects with the quasi-electric fields that are basically impossible to obtain with ordinary circuit means involving only "real" electric fields.

At this stage we have to introduce more semiconductors besides crystals of germanium and silicon. The latter elements owe their semiconductor proper-

---

[1] Herbert Kroemer (born 1928), Nobel Prize 2000

ties to the fact that they occupy column IV of the Periodic Table. Their outermost shell contains four electrons. (The Pauli principle would allow eight, at most.) Regular crystals with similar structure can also be built of equal amounts of atoms from column III (with three valance electrons) and from column V (with five). They are called III–V semiconductors. Similarly there also are II–VI semiconductors. Fortunately, the *lattice constant* is very nearly equal for gallium-arsenide (GaAs) and for aluminium-arsenide (AlAs) crystals (about 0.567 nm). But the band gaps of the two crystals are quite different. Because of the nearly identical lattice constants, one of these crystals can be grown on the other without appreciable *lattice mismatch*. At the interface an abrupt change of bandwidth appears. Moreover, crystals (also called solutions) of the type $Al_xGa_{1-x}As$, aluminium–gallium arsenide, can be grown in which the ratio of aluminium to gallium atoms is x to $1 - x$ and which have a band gap, determined by x, between that of GaAs and that of AlAs.

III	IV	V
B	C	N
Al	Si	P
Ga	Ge	As
In		Sb

Elements in columns III, IV, and V of the Periodic Table, which are important in semiconductor physics.

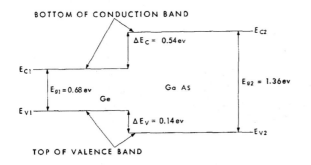

Band diagram of Anderson's Ge–GaAs heterojunction. From [5]. Reprinted with permission.

At first AlAs was not considered because it is unstable in moist air, nor was $Al_xGa_{1-x}As$. But GaAs was studied. It so happens that its lattice constant is also nearly identical to that of germanium. In 1960, Richard Anderson at the IBM Research Center in Poughkeepsie, New York, succeeded in growing a thin film of germanium on a crystal of gallium arsenide and thus created a structure with an abrupt change of band gap (not Kroemer's graded gap) [5]. That was done by exposing a GaAs crystal to a hot gas, containing germanium-iodide molecules. On the cooler crystal surface these molecules dissociate; germanium atoms are deposited on the GaAs lattice, forming a germanium crystal matching that of GaAs. In Anderson's paper the term *heterojunction* is coined, which was later generalized to *heterostructure*. In the following years and decades refined methods for the production of heterostructures were developed. Procedures similar to Anderson's now come under the heading *chemical vapour deposition* (CVD). Deposition from a suitable liquid, *liquid phase epitaxy* (LPE), is also possible. The most versatile process is *molecular beam epitaxy* (MBE), pioneered by Cho[2] at the Bell Telephone Laboratories in Murray Hill, New Jersey [6,7]. Wide molecular beams from different effusion cells (ovens) can be directed onto a substrate. The exposure of the substrate to the beams, carrying different atoms or molecules, is regulated by shutters. Multiple layers of many different thin crystals can be deposited, some of which may only be a few atomic layers thick.

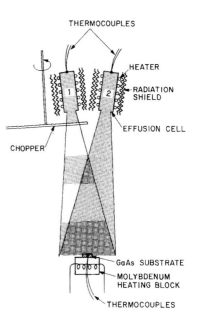

Early apparatus for molecular beam epitaxy by Cho. Reprinted with permission from [7]. Copyright 1971 American Institute of Physics.

---

[2] Alfred Y. Cho (born 1937)

Kroemer had moved to Varian Associates in Palo Alto, California, when in March 1963 he attended a review talk, presented by a colleague, on the semiconductor-junction laser or *laser diode*, which had been invented a few months earlier at the General Electric Research Laboratory in Schenectady, New York [8]. The first laser diode operated with a *pn* homojunction of GaAs, highly doped on either side. If a very high current is drawn through the junction, then there can be so many electrons and holes in the junction region that the effective Fermi level for electrons is inside the conduction band and that for holes inside the valence band. In other words, there are more electrons in the lowest states of the conduction band than in the highest states of the valence band. This is *population inversion*, the requirement for laser operation (Episode 82). In the transition of an electron from the conduction band to the valence band, an electron–hole recombination, the energy difference is radiated off as a photon. Light emission (but not laser operation) also occurs without population inversion, in this case one speaks of a *light-emitting diode* (LED). The required high current made it necessary that laser diodes were cooled to very low (liquid nitrogen) temperatures. Moreover, continuous laser action was impossible; cooling-off periods were needed between laser bursts.

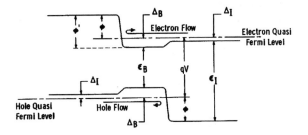

Kroemer's band diagram of a double-heterostructure laser. Reprinted with permission from [9], © 1963 IEEE.

Alferov

When Kroemer asked the speaker about the chances to achieve continuous operation at room temperature, he was told that experts thought it to be fundamentally impossible. Pointing at the properties of his wide-gap emitter, Kroemer protested. He wrote a paper [9] and a patent application [10], describing a *double-heterostructure laser*. A thin low-gap zone, which he called *base*, for recombination and stimulated emission is sandwiched between two wide-gap regions, which emit holes from one side and electrons from the other side into the base. Both types of charge carriers are, to a large extent, trapped in the base and accumulate, thus leading to population inversion and laser action, even at moderate currents through the device which, electrically, is still a diode.

Kroemer wanted to develop his laser diode at Varian but was refused the necessary resources. Seven years later, reports of double-heterostructure lasers with continuous operation at room temperature were published by Alferov's group in Leningrad [11] and, one month later, by a group at the Bell Labs in Murray Hill [12].

Alferov[3] was born in the Vitebsk. His father, originally a docker in St. Petersburg, served as non-commissioned officer with the Czar's Life-Guard Hussars in the First World War and as officer of the Red Army in the ensuing civil war; later he graduated from an industrial academy and managed Soviet

[3] Zhores Ivanovich Alferov (born 1930), Nobel Prize 2000

technical enterprises in various regions of the country. Alferov's beloved elder brother Marx fell in the Second World War. He himself studied at the Ulyanov Electrotechnical Institute in Leningrad, from which he graduated in December 1952. The following month he joined the Physico-Technical Institute, founded by Joffe (Episode 58), in the same town. The institute, of which Alferov later became director, now bears Joffe's name; but Joffe had been dismissed in 1950. Alferov's first work was to help with the development of germanium diodes and transistors based on pn homojunctions. The group succeeded in March and displayed the first transistor radios to the higher Soviet authorities in May 1953.

Alferov and a colleague, independently, obtained a patent on a double-heterostructure laser [13]. The group at the Joffe Institute first tried heterojunctions of the type GaP_xAs_{1-x}–$GaAs$. These showed laser action but, because of lattice mismatch, also only at the temperature of liquid nitrogen. One day, at the end of 1966, Alferov was told that a colleague had prepared samples of $Al_xGa_{1-x}As$ which had been sitting in a desk drawer for two years without decomposing and thus, in contrast to pure AlAs, were stable.

Alferov's group achieved liquid phase epitaxy of $Al_xGa_{1-x}As$ on a substrate of GaAs and constructed light-emitting diodes, based on a single, ideally lattice-matched, $Al_xGa_{1-x}As$ pn heterojunction [14]. The same achievement was reported by a group from the IBM Watson Research Center in Yorktown Heights, New York, a few weeks later [15].

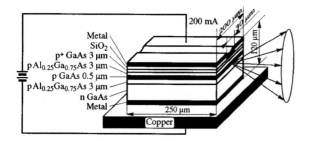

Alferov's sketch of the first semiconductor laser with continuous operation at room temperature. The double heterostructure is formed by the three middle layers, the laser-active zone is the 0.5 μm GaAs layer in the centre. The outside layers of p-doped GaAs (near the top) and n-doped GaAs (near the bottom) provide holes and electrons, respectively. From [16]. Reprinted with permission.

The race for the first double-heterostructure laser was on. They were built, not only in Leningrad, but at first required too much current. As mentioned, Alferov and his group were first to have a semiconductor laser, operating continuously at room temperature. In all, the structure comprised five different semiconductor layers plus external metal contacts. Laser action took place in the central GaAs layer. It also served as the optical cavity of the laser. Its interfaces with neighbouring layers and with the surrounding air reflect most of the light, but some was allowed to pass to the outside in a direction perpendicular to that of the electric current [11].

Modern life is all but unthinkable without double-heterostructure lasers. They are, for example, essential for the optical writing and reading of digital mass-storage media such as the compact disc (CD) and digital video disk (DVD). Even more important, they provide the radiation for telephone, television, and Internet communication via glass fibre.

In 1978 an elegant way to prepare a two-dimensional electron gas in a heterostructure was found at the Bell Labs in Murray Hill [17]. In the conduction

[1] Shockley, W., US Patent 2,569,347; filed June 26, 1948.

[2] Kroemer, H., *Archiv der elektrischen Übertragung*, **8** (1954) 499.

[3] Kroemer, H., *RCA Review*, **18** (1957) 332.

[4] Kroemer, H., *Proceedings of the IRE*, **45** (1957) 1535.

[5] Anderson, R. L., *IBM Journal of Research and Development*, **4** (1960) 283.

[6] Cho, A. Y., *Journal of Vacuum Science and Technology*, **8** (1971) S 31.

[7] Cho, A. Y., *Applied Physics Letters*, **19** (1971) 467.

[8] Hall, R. N. et al., *Physical Review Letters*, **9** (1962) 366.

[9] Kroemer, H., *Proceedings of the IEEE*, **51** (1963) 1782.

[10] Kroemer, H., US Patent 3,309,553; filed August 16, 1963.

[11] Alferov, Z. I. et al., *Fizika i Tekhnika Poluprovodnikov*, **4** (1970) 1826. Engl. transl. in *Soviet Physics – Semiconductors*, **4** (1970) 1573.

[12] Hayashi, I. et al., *Applied Physics Letters*, **17** (1970) 109.

[13] Alferov, Z. I. and Kazarinov, R. F., Soviet Patent 181 737; priority as of March 30, 1963.

[14] Alferov, Z. I. et al., *Fizika i Tekhnika Poluprovodnikov*, **1** (1967) 1579. Engl. transl. in *Soviet Physics – Semiconductors*, **1** (1968) 1313.

[15] Rupprecht, H., Woodall, J. M., and Petit, G. D., *Applied Physics Letters*, **11** (1967) 81.

[16] Alferov, Z. I., *Physica Scripta*, **T 68** (1996) 32.

[17] Dingle, R., Störmer, H. L., Gossard, A. C., and Wiegmann, W., *Applied Physics Letters*, **33** (1978) 665.

[18] Störmer, H. L., Dingle, R., Gossard, A. C., Wiegmann, W., and Sturge, M. D., *Solid State Communications*, **29** (1979) 705.

band of a double heterostructure there is a potential well in one direction (x). If the well is narrow enough, the possible eigenstates of an electron are well separated in energy. At very low temperature only the ground state can be assumed. Electrons are deprived of moving in x; motion is allowed only in the yz plane. If the crystal in the well is perfect, i.e., devoid of impurities and dislocations, the ideal case of a free, two-dimensional electron gas is approached. One needs donors to have electrons in the well, and these are themselves impurities. By precise molecular beam epitaxy the layers neighbouring the well layer can be doped with donor atoms. The electrons, they provide, move to the well which itself stays free of impurities, leading to the formation of a near-perfect two-dimensional electron gas. The separated space charge of donors and electrons leads to a distortion of the band edges. The well bottom becomes convex with triangle-shaped zones at its boundaries. Such a zone appears also in a single heterojunction, doped on one side. It, too, serves as a quantum well, useful for the preparation of a two-dimensional electron gas [18].

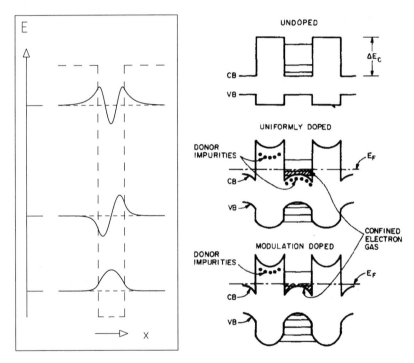

Left: The three lowest eigenstates of an electron in a potential well. The long-dashed broken line indicates the potential as function of the position x. The horizontal bars at the energy scale are the energy eigenvalues. The wave functions for the different eigenvalues are shown as continuous lines. *Right:* This figure sketches the band diagram of a double heterostructure, realizing a potential well in the x direction. Shown are three situations: an undoped structure (top), a structure with uniform donor doping and electrons as well as donor impurities present in the well region (middle), and one in which only a neighbouring layer of the well is doped (bottom). In the last case there are electrons in the well but no donor impurities. The distortion of the edges of the conduction band *CB* and of the valence band *VB* is due to the excess of negative charge in the well and that of positive charge outside. Reprinted with permission from [17]. Copyright 1978 American Institute of Physics.

Three Quarks – Order in the Wealth of New Particles (1964)

In the early 1960s an unexpected inflation was observed in the number of strongly interacting elementary particles or *hadrons*, as they came to be called. Most of these were found with the help of bubble chambers, many by the group of Alvarez in Berkeley, see Episode 72. As an example, we briefly discuss the first observation of a hyperon, which is now called the Σ^* or $\Sigma(1385)$ [1]. A hydrogen bubble chamber was exposed to a beam of negative K mesons and events were selected in which a Λ^0 hyperon and two pions of opposite charge were produced. Accumulations were observed at a fixed pion energy (in the centre of mass) for both pions. The fixed energy of a pion (say, the π^+) corresponds to a fixed energy of the system of the other two particles (Λ^0, π^-) and to a fixed rest mass or *invariant mass* of that system. This fact had to be interpreted as the existence of a new hyperon and the observation of its decay,

$$\Sigma^{*+} \rightarrow \Lambda^0 + \pi^+ \quad , \qquad \Sigma^{*-} \rightarrow \Lambda^0 + \pi^- \quad .$$

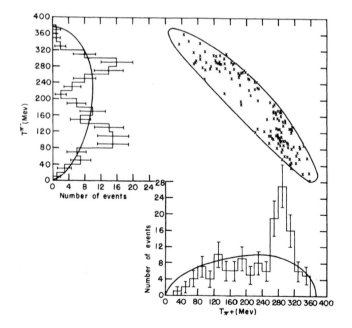

Discovery of the Σ^* hyperon in the analysis of 141 events of the type

$$K^- + p \rightarrow \Lambda^0 + \pi^+ + \pi^-$$

observed by Alvarez' group for a fixed K-meson energy. A *Dalitz plot* is constructed by entering every event as a point in a plane, spanned by two of the energies (in the centre-of-mass system of the interaction) of the particles in the final state. (The energy of the third particle is then defined as well.) Bands in the energy of the positive and the negative pion, centred around 280 MeV, are clearly visible. Projected onto the two energy axes the bands show up as well-defined maxima. Figure reprinted with permission from [1]. Copyright 1960 by the American Physical Society.

In contrast to the strange particles known until then, the Σ^* did not travel any measurable distance in the chamber. But its mean lifetime $\Delta t = \tau$ could be determined from the uncertainty relation $\Delta E \, \Delta t = \hbar$, where $\Delta E = \Gamma$ is the energy uncertainty or *width* of the relevant energy or mass distribution.

Gell-Mann

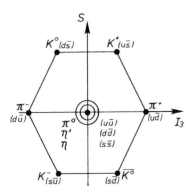

The nine pseudoscalar mesons, i.e., mesons with spin $J = 0$ and parity $P = -1$, appear as an octet and a singlet in $SU(3)$. Two elements of the octet as well as the singlet appear in the centre of the diagram. The η and η' mesons were predicted to exist by Gell-Mann and Ne'eman in 1961 and experimentally observed still in 1961 and in 1964, respectively. (Note that the three meson states π^0, η, η' are superpositions of the three quark–antiquark states $(u\bar{u}), (d\bar{d}), (s\bar{s})$.)

With $\Gamma \approx 60\,\text{MeV}$ one obtains $\tau \approx 10^{-23}$ s. Such short-lived states were at first called *resonances*; the first resonance, now called the $\Delta(1236)$, was already observed in 1952 by Fermi and his group [2] in pion–proton scattering. It became clear that resonances are particles which can decay via the strong interaction, whereas particles with a longer lifetime like the Λ^0 with $\tau \approx 10^{-10}$ s cannot. (Strangeness is conserved in the decay $\Sigma^{*\pm} \to \Lambda^0 + \pi^{\pm}$ but not in the decay $\Lambda^0 \to p + \pi^-$.) Particles decaying via the electromagnetic interaction have intermediate lifetimes. (The neutral pion, which decays into two photons, $\pi^0 \to \gamma + \gamma$, has the lifetime $\tau \approx 10^{-16}$ s.)

In quick succession further strange and non-strange short-lived particles were found. We only mention explicitly the mesons $K^*(892)$ [3], ρ [4,5], and ω [6]. Symbols of more particles will appear in the diagrams to be discussed below. Their masses and the quantum numbers (spin J, parity P, strangeness S, isospin I, and its third component I_3) were determined by methods we shall not go into. Of course, it became increasingly difficult to consider all these hadrons as *elementary* particles.

In 1964 the wealth of hadrons was ordered by considering them to be composed of only three different types of particles, called *quarks*, and their antiparticles. The name was coined by Gell-Mann who was not alone in this search for order but only he contributed at all three stages of this search.

Gell-Mann[1] [7] was born in New York. His father, originating from Galicia in the East of the Austro-Hungarian Empire, had studied in Vienna and Heidelberg and come to the United States before the First World War. Gifted in languages, for some time before the great depression, he had run his own language school. Gell-Mann himself was a child prodigy; at the age of 14 he completed high school and immediately began to study physics at Yale University on a scholarship, graduating four years later in 1948. He became a graduate student under Weisskopf at MIT, obtaining his doctorate early in 1951, and moved on to the Institute of Advanced Studies in Princeton. It was there that we met him in Episode 74, working with Pais and later, when he had moved to Chicago, introducing the *strangeness* quantum number. That was the important first stage on the way that led to quarks. From then on hadrons with the same baryon number, spin, and parity were placed in a set and the regularities of these sets were studied. In 1954 Gell-Mann went to Columbia University as Visiting Associate Professor. In 1955 he became Associate Professor and in 1956 Full Professor at the California Institute of Technology in Pasadena where, in 1957 together with Feynman, he co-authored the V − A theory of weak interactions (Episode 79).

The second stage on the way to quarks was what Gell-Mann called the *eightfold way*. Symmetries are an important feature of physics. They are properties observed when transformations are applied to physical systems. The mathematical theory of transformations is called *group theory* (see Box). In earlier episodes we already discussed features of angular momentum, spin, and isospin, which are described by a *special unitary group* called $SU(2)$, without mentioning that term. In 1961, Gell-Mann combined isospin and strangeness to form a quantity which he called *unitary spin* and studied its properties under

---

[1] Murray Gell-Mann (born 1929), Nobel Prize 1969

A Little on Group Theory, in Particular, on the Group $SU(3)$

A *group* is a set G of elements with the following properties. (a) For each pair a, b of elements a *product* $d = ab$ is defined which is element of G. (b) For this multiplication the *associative law* holds, $(ab)c = a(bc)$. (c) There is a *unit element* e such that $ea = ae = a$ for all elements a. (d) For each element a there is an *inverse element* a^{-1} with the property $aa^{-1} = a^{-1}a = e$.

A *representation* of a group is the allocation of a matrix $D(a)$ to each element a in such a way that the property of the product is preserved by the matrices, $D(ab) = D(a)D(b)$. Many groups used in physics, in particular, those mentioned below, are *Lie groups*, called after the Norwegian mathematician Lie[2], who taught in Christiania (now Oslo) and Leipzig. Their matrix representations describe continuous transformations; in the neighbourhood of the identity transformation they are given by the unit matrix plus the linear combination of m linear independent matrices called *generators*. Such matrices may operate in a vector space spanned by the *state vectors* of a quantum-mechanical system. Here we are interested in the *special unitary groups* $SU(2)$ and $SU(3)$ for which $m = 3$ and $m = 8$, respectively.

The generators of the simplest (or fundamental) representation of $SU(2)$ are the *Pauli matrices* operating in a vector space of 2 dimensions (see Box II in Episode 43). One of these is diagonal; used in an eigenvalue equation it yields the eigenvalue s_z. Representations in higher dimensions are general operators of angular momentum. The group $SU(2)$ also describes isospin; here the isospin quantum number I_3 plays the role of s_z. Eigenvalues characterize possible quantum-mechanical states. In the case of $SU(2)$ they show up graphically as points on an axis, labelled I_3.

The 8 *Gell-Mann matrices* are generators of the fundamental representation of $SU(3)$; 2 of them are diagonal, i.e., there are 2 eigenvalues. They can be taken to be I_3 and the strangeness quantum number S. Physical states are depicted as points in an I_3, S plane. The eigenstates of the fundamental representation form a *triplet*, referred to as **3** of *quarks*, named *up* (u), *down* (d), and *strange* (s). In an equivalent way, the corresponding *antiquarks* $\bar{u}, \bar{d}, \bar{s}$ form an *antitriplet* $\bar{\mathbf{3}}$. Representations in spaces of higher dimension can be constructed in a systematic way from the fundamental one. By tensor multiplication and subsequent 'reduction' to a sum, *multiplets* are formed, representing sets of hadrons. (Multiplication and summation in this context are symbolized by $\otimes$ and $\oplus$, respectively.) The elements of a multiplet are transformed only among themselves by the matrices forming the elements of a representation. Following these rules, mesons, made up of a quark and an antiquark, are grouped in singlets and octets ($\mathbf{3} \otimes \bar{\mathbf{3}} = \mathbf{1} \oplus \mathbf{8}$). Baryons, consisting of three quarks, form singlets, octets and decuplets ($\mathbf{3} \otimes \mathbf{3} \otimes \mathbf{3} = \mathbf{1} \oplus \mathbf{8} \oplus \mathbf{8} \oplus \mathbf{10}$).

If the symmetry were *perfect*, then particles within a multiplet would differ only in the quantum numbers I_3 and S; since it is somewhat *broken*, they also differ in mass.

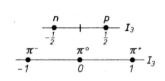

$SU(2)$ multiplets.
Top: The fundamental doublet for $I = 1/2$, i.e., $I_3 = \pm 1/2$, here identified with neutron n and proton p.
Bottom: The triplet for $I = 1$ with $I_3 = -1, 0, +1$, here identified with the three π mesons.

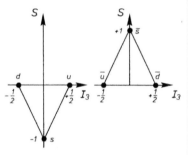

Fundamental triplets of $SU(3)$.
The quarks u, d, s (left) and the antiquarks $\bar{u}, \bar{d}, \bar{s}$ (right).

the group $SU(3)$ of transformations [8].

Gell-Mann found that he could construct the group $SU(3)$ from a set of 8 matrices, now called the *Gell-Mann matrices*. He studied *multiplets* of this group and found that he could assign the existing hadrons to elements of such multiplets. Within a multiplet all elements have identical properties, except for the quantum numbers I_3 and S, in particular, they have the same spin J and the same parity P. Since the $SU(3)$ symmetry is not perfect, they also have

---

[2] Marius Sophus Lie (1842–1899)

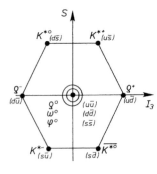

The nine vector mesons (spin $J = 1$ and parity $P = -1$) form an octet and a singlet. (While the neutral vector-meson states ρ^0 and ω^0 are superpositions of the states $(u\bar{u})$ and $(d\bar{d})$, the state ϕ^0 is essentially pure $(s\bar{s})$.)

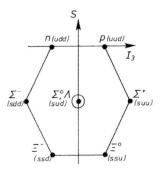

The octet of baryons with spin $J = 1/2$ and parity $P = +1$.

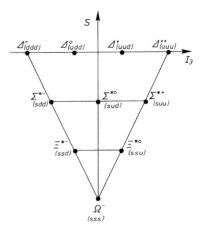

The decuplet of baryons with spin $J = 3/2$ and parity $P = +1$.

different masses.

The *pseudoscalar mesons* which have $J = 0$ and $P = -1$ or, for short, $J^P = 0^-$, were grouped in an octet comprising the three pions π^-, π^0, π^+, the kaons K^0, K^+, and their antiparticles $\bar{K}^0, K^-$. Gell-Mann postulated the existence of an additional meson with $I_3 = S = 0$ as the eighth member of the octet. Such a meson, now called η, was found still in 1961 [9]. In addition to the octet, $SU(3)$ predicted a singlet with the same quantum numbers. A particle, fitting that description and called η', was first observed in 1964 [10]. We have to add here that the physical particles π^0, η, η' which occupy the same position in an I_3, S diagram, need not be quantum-mechanical eigenstates of $SU(3)$ but can be superpositions of such states. That applies also to multiply occupied places in diagrams of other particle families.

Likewise, the *vector mesons* with $J^P = 1^-$ were grouped in an octet and a singlet. The equivalent of the π meson, which had recently been discovered, was called ρ by Gell-Mann, 'being the succeeding letter of the Greek alphabet'. The octet member with $I_3 = S = 0$ he called ω. Both names stuck. The singlet, now called φ, was found in 1962 [11].

The meson diagrams contain both particles and antiparticles because a meson is turned into its antiparticle by inverting the signs of charge and strangeness. For baryons also the baryon number has to be inverted; neutron n and antineutron $\bar{n}$ are different. Gell-Mann constructed an octet of the best-known baryons, those with $J^P = 1/2^+$, namely, $p, n, \Sigma^-, \Sigma^0, \Lambda, \Sigma^+, \Xi^-, \Xi^0$. The baryon octet was found to display some regularity concerning the particle masses. While the masses are nearly equal for a fixed value of S, they increase in two sizable steps as S changes from 0 to -1 and from -1 to -2. Other possible baryon multiplets were hinted at.

Completely independently and unbeknownst to Gell-Mann, essentially the same work, using the group $SU(3)$ and culminating in the proposal of particle multiplets, was done by a hitherto unknown Israeli physicist in London. Ne'eman[3] was born in Tel Aviv and had studied mechanical engineering at the Technion in Haifa. He pursued a military, a scientific, and a political career, usually two of them in parallel. When he was military attaché in London, in addition to his official obligations he studied physics at Imperial College in South Kensington close to the Embassy. The work on unitary symmetry was done as part of his Ph.D. thesis under Salam. Ne'eman's paper [12] was submitted for publication several weeks before Gell-Mann's. A few years later the two scientists edited *The Eightfold Way*, a collection of early papers in the field by themselves and others [13]. In 1961 Ne'eman left the Army with the rank of a colonel. He founded the physics department of Tel Aviv University in 1965 and was president of the university for the first half of the 1970s. In the 1980s and early 1990s he was member of the Knesset, the Israeli parliament, and twice served as minister of science.

In July, 1962, both Gell-Mann and Ne'eman attended the International Conference on High-Energy Physics at CERN in Geneva. (This, by the way, was the first international conference I witnessed. As a junior CERN scientist, I did not participate officially but was allowed to attend two of the less-frequented

---

[3] Yuval Ne'eman (1925–2006)

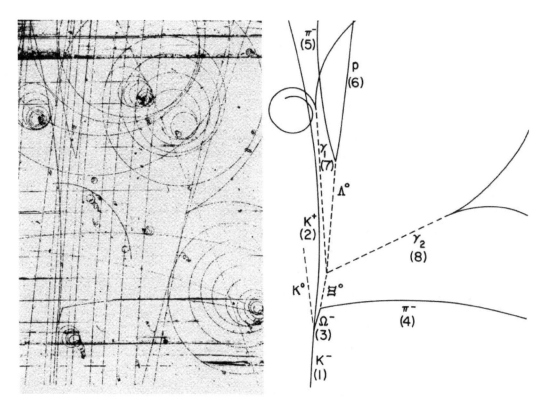

sessions, not the one now to be mentioned.) Speaking in the discussion following a report on the strong interaction of strange particles, Gell-Mann proposed a new kind of multiplet, a decuplet, for baryons with $J^P = 3/2^+$ [14]. Only for four of its members, $\Delta^-, \Delta^0, \Delta^+, \Delta^{++}$ with $S = 0$, the values of spin and parity were well established. But Gell-Mann assumed that three resonances with $S = -1$ ($\Sigma^{*-}, \Sigma^{*0}, \Sigma^{*+}$) and two with $S = -2$ (Ξ^{*-}, Ξ^{*0}) would fit in; they certainly did what concerned their regular increase in mass with falling strangeness. Having said this, Gell-Mann continued [14]:

> If $J = 3/2^+$ is really right for these two cases, then our speculation might have some value and we should look for the last particle, say, Ω^- with $S = -3, I = 0$. At [a mass of] 1685 MeV it would be metastable and should decay by the weak interactions into $K^- + \Lambda$, $\pi^- + \Xi^0$, or $\pi^0 + \Xi^-$. Perhaps it would explain the old Eisenberg event [an unusual event found in emulsion exposed to cosmic radiation already in 1954 [16]]. A beam of K^- with momentum $\gtrsim 3.5$ GeV$/c$ would yield Ω^- by means of $K^- + p \to K^+ + K^0 + \Omega^-$.

This was a very definite prediction, comparable to that of missing elements by pointing out gaps in the Periodic Table by Moseley (Episode 24). The experiment was done by a group led by Samios[4] in Brookhaven. A hydrogen bubble chamber was exposed to a beam of negative K mesons produced by the Brookhaven Alternating Gradient Synchrotron and having a momentum of 5 GeV$/c$. In February 1964, an event as expected by Gell-Mann was reported

First observation of the creation and the decay of an Ω^- particle. The bubble-chamber photograph is shown on the left. On the right the relevant tracks seen in the chamber are redrawn as continuous lines and the trajectories of neutral particles (leaving no tracks) as broken lines. The production process is

$$K^- + p \to \Omega^- + K^0 + K^+ \quad .$$

The decay proceeds in three stages in each of which the strangeness (originally $S = -3$) is changed by one unit,

$$\Omega^- \to \Xi^0 + \pi^- \quad ,$$
$$\Xi^0 \to \Lambda^0 + \pi^0 \quad ,$$
$$\Lambda^0 \to p + \pi^- \quad .$$

Both photons from the decay of the short-lived neutral pion ($\pi^0 \to \gamma + \gamma$) produce an electron–positron pair and thus give rise to tracks recorded in the chamber. Figure reprinted with permission from [15]. Figure reprinted with permission from1964

---

[4] Nicholas Samios (born 1932)

[1]   Alston, M. et al., *Physical Review Letters*, **5** (1960) 520.

[2]   Anderson, H. L. et al., *Physical Review*, **85** (1952) 934, 936.

[3]   Alston, M. et al., *Physical Review Letters*, **6** (1961) 300.

[4]   Stonehill, D. et al., *Physical Review Letters*, **6** (1961) 624.

[5]   Erwin, A. R. et al., *Physical Review Letters*, **6** (1961) 628.

[6]   Maglic, B. C., Alvarez, L. W., Rosenfeld, A. H., and Stevenson, M. L., *Physical Review Letters*, **7** (1961) 178.

[7]   Johnson, G., *Strange Beauty – Murray Gell-Mann and the Revolution in Twentieth-Century Physics*. Knopf, New York, 1999.

[8]   Gell-Mann, M., *Physical Review*, **125** (1962) 1067.

[9]   Pevsner, A. et al., *Physical Review Letters*, **7** (1961) 421.

[10]   Kalbfleisch, G. R. et al., *Physical Review Letters*, **12** (1964) 527.

[11]   Bertanza, L. et al., *Physical Review Letters*, **9** (1962) 162.

[12]   Ne'eman, Y., *Nuclear Physics*, **26** (1961) 222.

[13]   Gell-Mann, M. and Ne'eman, Y. (eds.), *The Eightfold Way*. Benjamin, New York, 1964.

[14]   Gell-Mann, M. In Prentki, J. (ed.): *Proceedings of the 1962 Int. Conf. on High-Energy Physics at CERN*, p. 805. CERN, Geneva, 1962.

[15]   Barnes, V. E. et al., *Physical Review Letters*, **12** (1964) 204.

[16]   Eisenberg, Y., *Physical Review*, **96** (1954) 541.

[17]   Gell-Mann, M., *Physics Letters*, **8** (1964) 214.

[18]   Zweig, G., *An SU_3 Model for Strong Interaction Symmetry and its Breaking*. CERN–TH–401, Geneva, 1964.

[19]   Gell-Mann, M., *The Quark and the Jaguar*. Freeman, New York, 1994.

[20]   Blum, W., Brandt, S., Cocconi, V. T., et al., *Physical Review Letters*, **11** (1964) 353.

[15]. The photograph, arguably, is the best-known bubble-chamber picture ever taken.

The hadron multiplets belong to higher representations of $SU(3)$, not the simplest or *fundamental* representation with only three states. In January 1964, it was proposed that this fundamental triplet of states also represented particles and that they were the constituents of all known hadrons. The idea was announced independently by Gell-Mann [17] and by Zweig [18]. Zweig[5] had just earned his Ph.D. under Feynman at Caltech and was working on a post-doctoral position at CERN. The new particles had to have unusual properties. In particular, they had to have baryon number $B = 1/3$ and electric charges which were $2/3$ or $-1/3$ of the elementary charge (see the Table given below). They were named *quarks* by Gell-Mann after a cryptic sentence 'Three Quarks for Muster Mark' in the novel *Finnegan's Wake* by James Joyce; for details of Gell-Mann's motivation see [19]. The three quarks were distinguished by the letters u, d, s, now also referred to as *up, down*, and *strange*. Zweig called the fundamental particles *aces*, alluding to important playing cards. Gell-Mann submitted his paper [17] to the young European journal *Physics Letters*; it seems he feared that the term *quarks* might not be accepted by the referees of its already well-established American counterpart. Zweig's paper remained a *preprint*, i.e., a mimeographed typewritten manuscript which is already circulated while a paper is surveyed and printed. It never appeared in a regular journal.

Quarks					Antiquarks				
Name	Q	B	I_3	S	Name	Q	B	I_3	S
u	$2/3$	$1/3$	$1/2$	0	$\bar{u}$	$-2/3$	$-1/3$	$-1/2$	0
d	$-1/3$	$1/3$	$-1/2$	0	$\bar{d}$	$1/3$	$-1/3$	$1/2$	0
s	$-1/3$	$1/3$	0	-1	$\bar{s}$	$1/3$	$-1/3$	0	1

Electric charge Q (in units of the elementary charge), baryon number B, isospin quantum number I_3, and strangeness S of quarks and antiquarks.

All mesons in our diagrams are easily understood as quark–antiquark systems; their constituent quarks are given in parentheses, e.g., π^+ $(u\bar{d})$. Baryons are composed of three quarks, e.g., Ω^- (sss), and antibaryons of three antiquarks. One easily sees that the quantum numbers Q, B, I_3, S of hadrons are simply the sums of the corresponding numbers carried by the constituent quarks.

It was an open question, whether quarks existed or were only convenient mathematical constructs. In 1964 I was myself involved in an experiment at CERN in which we tried to produce quarks at the highest available accelerator energy and detect them by means of their fractional charge [20]. Up to now all such searches for free quarks were fruitless. But as a concept quarks gained more and more reality. As we shall see in Episodes 92 and 94, the modern theory of strong interaction relies on quarks and hard experimental facts can only be explained if fractionally charged quarks exist.

---

[5] George Zweig (born 1937)

CP – Another Symmetry Broken: The Peculiar System of the Neutral K Meson and Its Antiparticle (1964)

88

Before the 'Fall of Parity' in 1957 (Episode 77), it was taken for granted that the laws of physics were mirror symmetric. For every given process the mirror-inverted process should be possible. Also believed to hold was *time-inversion symmetry*, the symmetry between a process and the corresponding one with time running backwards. Of course, time cannot really be inverted but time inversion of a process is equivalent to viewing the process in a film with the projector running backwards. The symmetry says that the projected process should also be possible in reality. Another symmetry is that of *charge conjugation*. The name is a little misleading; what is implied is the replacement of every particle by its antiparticle and vice versa. For every process the one with this replacement performed should be possible, too.

All modern theories of particle interactions are so-called quantum field theories. Essentially, all conceivable theories of this type have the property that they are symmetric under the joint application of charge conjugation (C), space reflection (P), and time inversion (T). That was stated by Lüders in 1954 and by Pauli in 1955 as what is called the *CPT theorem*. Up to now it has not been challenged by experiment. When parity violation had been predicted by Lee and Yang but even before its experimental verification was announced in New York, Landau in Moscow submitted a paper [1] in which he proposed a new symmetry, which should be obeyed by all interactions. He called it *combined inversion*, charge conjugation combined with space reflection. Thus CP symmetry and T symmetry were expected to hold separately. In 1964, however, it was shown experimentally that this is not the case by a group from Princeton University, headed by Fitch[1] and Cronin[2] [2].

Ten years earlier, in October 1954, when parity violation was not yet an issue, Gell-Mann and Pais predicted that neutral kaons decay in different ways with very different lifetimes [3]. That followed from their common work on the classification of hadrons, in particular, from the fact that the neutral kaon K^0 was taken to be different from its antiparticle $\bar{K}^0$. Pais remembers [4] that the work was done in an apartment he had recently rented in Greenwich Village, New York City, in order to escape from Princeton to the metropolis over weekends. Gell-Mann, at that time, was working at Columbia University, on leave from the University of Chicago. We deviate here from the original reasoning of Gell-Mann and Pais by taking parity violation into account.

While parity and strangeness are conserved in the creation process of strange particles through *associated production*, in their decay there is no such conser-

Fitch

Cronin

---

[1] Val Logsdon Fitch (born 1923), Nobel Prize 1980   [2] James Watson Cronin (born 1931), Nobel Prize 1980

K^0–$\bar{K}^0$ Mixing – Three Pairs of Names for Neutral Kaons

Neutral K mesons can be seen in three different ways:

1. **Description with respect to their production through strong interaction**

   The strangeness quantum number S is conserved. The K^0 with $S = 1$ and its antiparticle $\bar{K}^0$ with $S = -1$ are clearly different. (In the reaction $\pi^- + p \rightarrow K^0 + \Lambda^0$ the produced meson is a kaon since the Λ^0 hyperon has $S = -1$.)

2. **Description with respect to their decay through weak interaction**

   Strangeness is not conserved. Both K^0 and $\bar{K}^0$ can decay into identical states, e.g., a pair of pions ($\pi^+\pi^-$ or $\pi^0\pi^0$). There can be transformations between the two, for instance, via the virtual process $K^0 \leftrightarrow \pi\pi \leftrightarrow \bar{K}^0$.

   It is often convenient to denote the quantum-mechanical wave function or *state* of a particle by simply placing the particle symbol in a Dirac *ket*. Since the weak interaction does not distinguish between the states* $\left|K^0\right\rangle$ and $\left|\bar{K}^0\right\rangle$, superpositions can be formed,

   $$\left|K_1^0\right\rangle = \frac{1}{\sqrt{2}}\left(\left|K^0\right\rangle - \left|\bar{K}^0\right\rangle\right) \quad , \qquad \left|K_2^0\right\rangle = \frac{1}{\sqrt{2}}\left(\left|K^0\right\rangle + \left|\bar{K}^0\right\rangle\right) \quad .$$

   (The factors $1/\sqrt{2}$, ensuring proper normalization of the states, are unimportant for our reasoning.) These states have definite transformation properties under CP. It is known from production experiments that the kaon has negative eigenparity, i.e., the parity operator P_{op} acting on the state K^0 (or $\left|\bar{K}^0\right\rangle$) yields the eigenvalue -1, $P_{\text{op}}\left|K^0\right\rangle = -\left|K^0\right\rangle$. By definition, the charge-conjugation operator C_{op} replaces the state of a particle by that of its antiparticle, $C_{\text{op}}\left|K^0\right\rangle = \left|\bar{K}^0\right\rangle$. With this it is easy to see that the combined CP parity of K_1^0 is positive and that of K_2^0 negative,

   $$C_{\text{op}}P_{\text{op}}\left|K_1^0\right\rangle = +\left|K_1^0\right\rangle \quad , \qquad C_{\text{op}}P_{\text{op}}\left|K_2^0\right\rangle = -\left|K_2^0\right\rangle \quad .$$

   A pair of two pions ($\pi^+\pi^-$ or $\pi^0\pi^0$) has positive CP parity while that of a triple ($\pi^+\pi^-\pi^0$ or $\pi^0\pi^0\pi^0$) is negative. Therefore, if CP is conserved in the decay as had to be assumed, two different kaons appear with the decay modes $K_1^0 \rightarrow \pi + \pi$ and $K_2^0 \rightarrow \pi + \pi + \pi$. For kinematic reasons the latter has a much longer lifetime than the former.

3. **Description with respect to their lifetime**

   However, CP is not strictly conserved; the long-lived meson was found to decay also into two pions. The states for the observed short-lived and long-lived kaons are therefore those of K_1^0 and K_2^0, respectively, with a small admixture ε of the other type,

   $$\left|K_{\text{S}}^0\right\rangle = a\left(\left|K_1^0\right\rangle + \varepsilon\left|K_2^0\right\rangle\right) \quad , \qquad \left|K_{\text{L}}^0\right\rangle = a\left(\varepsilon\left|K_1^0\right\rangle + \left|K_2^0\right\rangle\right) \quad ,$$

   where a is a normalization factor.

C and CP of pion systems

Under the action of C_{op}, the wave function of the photon, which is its own antiparticle, can at most change sign, $C_{\text{op}}\left|\gamma\right\rangle = C\left|\gamma\right\rangle$, with $C = \pm 1$. Since the photon is connected to the electromagnetic field and since the field changes sign if the charges causing it are inverted, the photon has odd C parity, $C(\gamma) = -1$. C like P is a multiplicative quantum number; the C parity of the π^0 which decays into two photons is even, $C(\pi^0) = +1$. All pions have odd parity P, thus $CP(\pi^0) = -1$. Angular momentum plays no role in kaon decay and therefore $CP(\pi\pi) = +1$ and $CP(\pi\pi\pi) = -1$.

* There is some freedom in the definition of the quantum-mechanical states. In the modern literature they are usually chosen such that $C_{\text{op}}P_{\text{op}}\left|K^0\right\rangle = \left|\bar{K}^0\right\rangle$. For historic reasons we stick to the convention $C_{\text{op}}\left|K^0\right\rangle = \left|\bar{K}^0\right\rangle$. As a result, some of our formulae are different from modern usage; the conclusions are not.

vation. Thus, when produced, K^0 and $\bar{K}^0$ are different. But that is no longer the case when they are left to themselves. Since they can both decay into pairs of pions, there is a virtual process, $K^0 \leftrightarrow \pi\pi \leftrightarrow \bar{K}^0$, transforming them from one to the other. If CP is taken to be conserved in the decay, it is reasonable to form kaon states with definite CP symmetries. Two states with opposite CP parity, named K_1^0 and K_2^0, can be constructed (see Box). CP conservation demands that the K_1^0 decays into two pions and the K_2^0 into three pions. Since in the latter decay nearly all of the kaon rest mass is needed to produce the rest mass of the three pions, the decay is inhibited with respect to the former: The K_2^0 was expected to have a much longer lifetime than that of the well-known neutral kaon decaying into two pions, which, apparently, was the K_1^0. Indeed, a particle fitting the description of the K_2^0 was found in 1956 by Lederman's group [6].

Already before the existence of the long-lived neutral kaon was thus verified, Pais, this time with Piccioni[3], predicted another astonishing effect: Long after the short-lived kaons had all decayed, such particles could be *regenerated* from the long-lived ones [5]. In their paper, they consider an experimental set-up (which they take to be placed in a cloud chamber) in which neutral kaons pass through four regions, in which they have to be understood as $(K^0, \bar{K}^0)$, as (K_1^0, K_2^0), again as $(K^0, \bar{K}^0)$, and yet again as (K_1^0, K_2^0). In the first region, a material target A, K^0 mesons are produced via strong interaction by high-energy pions. In the second region, the free space behind A, they can decay by weak interaction as K_1^0 (into two pions) and K_2^0 (into three pions). Because of their much longer lifetime only K_2^0 mesons reach the third region, target B. There strong interaction is enabled again. The particles appear as K^0 and $\bar{K}^0$. Since the latter are more readily absorbed in matter than the former, mainly K^0 leave B. In the fourth region, the space behind B, they decay again as K_1^0 and K_2^0. As net result there has been a *regeneration* of K_1^0 mesons.

In their paper [5] Pais and Piccioni discuss a related effect for which no material target is needed. The quantum-mechanical states are complex wave functions which oscillate in space and time. If one looks at the time dependence, the states K_1^0 and K_2^0, because of their different mean lives, correspond to damped oscillations with very different damping constants. If the masses of K_1^0 and K_2^0 are slightly different, then so are the oscillation frequencies. At any given time the amplitudes of $K^0, \bar{K}^0$ can be reconstructed from those of K_1^0, K_2^0 and, because of the different frequencies of the latter, a *beat* appears. The intensities of $K^0, \bar{K}^0$ are the absolute squares of the amplitudes. In figures we display here K_1^0, K_2^0 amplitudes (drawn not to scale with much reduced frequencies) and the $K^0, \bar{K}^0$ intensities as a function of time. In an originally pure K^0 beam the $\bar{K}^0$ intensity rises while the K^0 intensity falls. For some time the intensity ratio is even reversed; one speaks of K^0–$\bar{K}^0$ *oscillations*.

In the discussion, we had so far, on some of the beauties and intricacies of neutral K mesons we had no reasons to doubt the conservation of CP symmetry. Only in 1964 the Princeton group found that K_2^0 mesons, with a small but distinct probability of about 0.2%, decay into two charged pions, violating CP symmetry [2]. Fitch has told the history of that discovery at several occasions, for instance, in [7,8].

Fitch grew up in Gordon, Nebraska. His father, a farm owner but hindered by an accident to work in his farm, was in the insurance business. During

---

[3] Oreste Piccioni (1915–2002)

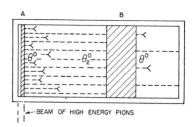

Illustration of K_1^0 regeneration by Pais and Piccioni. Although of the K^0 mesons (here still denoted by their original name θ^0) produced in target A only the long-lived K_2^0 reach target B, short-lived K_1^0 are observed behind it. Figure reprinted with permission from [5]. Copyright 1955 by the American Physical Society.

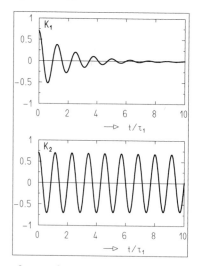

K_1^0 and K_2^0 amplitudes (shown is the real part) as a function of time in units of τ_1, the K_1^0 mean life. To make these graphs more readable the oscillation frequencies were vastly reduced.

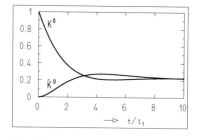

K^0–$\bar{K}^0$ oscillations: Intensities of K^0 and $\bar{K}^0$ computed from the amplitudes shown in the figure above this one.

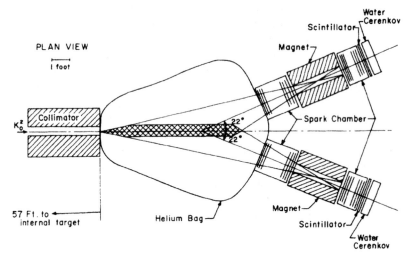

PLAN VIEW

1 foot

Water Cerenkov
Scintillator
Magnet
Spark Chamber
Collimator
K_0^z
22°
22°
57 Ft. to internal target
Helium Bag
Magnet
Scintillator
Water Cerenkov

Apparatus used in the discovery of CP violation. Long-lived neutral kaons enter a decay region through a collimator. Two magnetic spectrometers detect pairs of charged pions and measure their directions and momenta. Figure reprinted with permission from [2]. Copyright 1964 by the American Physical Society.

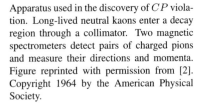
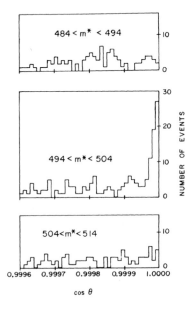

$484 < m^* < 494$

$494 < m^* < 504$

$504 < m^* < 514$

NUMBER OF EVENTS

0.9996   0.9997   0.9998   0.9999   1.0000

$\cos \theta$

Angular distribution for the system of two oppositely charged pions. Systems with an invariant mass very near that of the neutral kaon (497.6 MeV) were selected (middle) as well as systems with a mass slightly below (top) or above (bottom). The plot in the middle clearly shows an excess of events very near $\cos \theta = 1$, i.e., $\theta = 0$, which is the direction of the incident K_2^0 beam. Figure reprinted with permission from [2]. Copyright 1964 by the American Physical Society.

the war Fitch worked as soldier-turned-technician at Los Alamos, where the atomic bomb was being developed, and became fascinated by physics and, in particular, electronics. Only after the war he formally completed his physics studies at McGill University and took a Ph.D. at Columbia University in 1954 with work on mu-mesic atoms, i.e., atoms in which an electron is replaced by a muon. Fitch then moved on to Princeton University, becoming a full professor in 1960. From there he did experiments mainly at Brookhaven. Stimulated by Panofsky from Stanford, who was in Brookhaven for the summer of 1956, Fitch did his first experiment on neutral kaons. Together with Panofsky and others he confirmed the existence of the long-lived K_2^0 [9].

Cronin was born in Chicago. He studied in his home town and in 1955 obtained his Ph.D. at the University of Chicago with a thesis work on nuclear physics. Thereafter he joined Brookhaven National Laboratory to work at the recently completed 3-GeV Cosmotron accelerator and continued to experiment there also after he had moved to Princeton in 1958. In 1962 Cronin did an experiment at the Cosmotron on the production of the ρ meson which was discovered the year before (Episode 87). The neutral ρ meson decays immediately after its creation, predominantly into a positive and a negative pion. Cronin had two spectrometers, consisting of spark chambers and magnets to measure the directions and momenta of both pions. From this information the mass, the direction, and the momentum of the parent particle could be computed.

In the winter 1962/63 the results of a bubble-chamber experiment became known, which seemed to indicate an anomalously high regeneration of K_1^0 from K_2^0. Fitch remembers [7]: 'In view of my previous experience with K mesons I responded to these news like a fire horse to the sound of a bell. And my colleague Cronin and I had many discussions about this over lunch tables at Brookhaven.' The two decided that Cronin's apparatus was well suited to detect the decay $K_1^0 \rightarrow \pi^+ + \pi^-$. Together with Turlay[4], a visitor from the

---

[4] René Turlay (1932–2002)

high-energy research centre in Saclay, France, they wrote a two-page proposal (reproduced in [7]) to do a high-precision experiment on regeneration and also to obtain 'a much better limit for the partial rate $K_2^0 \to \pi^+ + \pi^-$'. The proposal was promptly accepted, the group was completed by Christenson, one of Cronin's graduate students, and Cronin's equipment was adapted and moved from the Cosmotron to the Alternating-Gradient Synchrotron. The experiment was done in June and July 1963. About half of the allotted 200 hours of accelerator time were used for the regeneration experiment which included material targets. In the other half, named the *CP Invariance Run* in the group's logbook, long-lived K mesons, produced far from the apparatus, could enter a helium-filled bag (a 'poor man's vacuum'). Charged particles from their decay were registered by the two spectrometers. The analysis of the spark-chamber data, recorded on film, followed after the actual running.

From the directions and momenta of the two pions, registered by the spectrometers, the properties of a parent particle which decayed into the two can be computed. In most cases the mass m^* of the 'parent particle', obtained in this way, was not the kaon mass, since the pions were part of the frequent decay $K_2^0 \to \pi^+ + \pi^- + \pi^0$. In some events, however, the kaon mass was found and within these there was a clear sample in which the parent particles had exactly the direction of the incoming kaons; these were events of the type $K_2^0 \to \pi^+ + \pi^-$, violating CP symmetry. After careful checking and re-checking the discovery was published in the summer of 1964 [2]. Results of the regeneration experiment and a first determination of the mass difference between K_2^0 and K_1^0 appeared a year later [10]. (The modern value for this tiny difference is 3.48×10^{-12} MeV.) For a beautiful experiment on K^0–$\bar{K}^0$ oscillations, done another ten years later, see [11].

Quite a number of theoretical models were developed to explain CP violation, including the existence of an additional type of interaction and the existence of more than three quarks. At present, after decades of precision experiments, it seems that the observed CP violation is consistent with the standard model of weak interaction, assuming six quarks.

It is quite possible that the seemingly tiny effect of CP violation has important consequences on a large scale. In 1967 the effect stimulated Sakharov[5] [12] to propose a model which could explain that the universe consists essentially of matter with very little antimatter, although equal amounts of both are probably created at a very early stage. His model assumes the observed non-conservation of CP, a hitherto unobserved non-conservation of the total baryon number B, and also phases of expansion of the universe without thermal equilibrium. In one of his talks, Fitch, obviously fond of Sakharov's idea, says [8]:

> Indeed, one might turn the question around and say that the first evidence, ever, for CP violation was the fact that we exist.

[1] Landau, L., *Nuclear Physics*, **3** (1957) 127.

[2] Christenson, J. H., Cronin, J. W., Fitch, V. L., and Turlay, R., *Physical Review Letters*, **13** (1964) 138.

[3] Gell-Mann, M. and Pais, A., *Physical Review*, **97** (1955) 1387.

[4] Pais, A., *A Tale of Two Continents – A Physicist's Life in a Turbulent World*. Oxford University Press, Oxford, 1997.

[5] Pais, A. and Piccioni, O., *Physical Review*, **100** (1955) 1487.

[6] Landé, K., Booth, E. T., Impeduglia, J., and Lederman, L. M., *Physical Review*, **103** (1956) 1901.

[7] Fitch, V. L., *A Personal View of the Discovery of CP Violation*. In Doncel, M. G. et al. (eds.): *Symmetries in Physics*. Universitat Autònoma de Barcelona, Bellaterra, 1987.

[8] Fitch, V. L., *Some Bits of the History of CP Violation*. In *Proceedings of the 27th SLAC Summer Institute on CP Violation*, Ch. 2. SLAC, Stanford, 1999.

[9] Panofsky, W. K. H., Fitch, V. L., Motley, R. M., and Chesnut, W. G., *Physical Review*, **109** (1958) 1353.

[10] Christenson, J. H., Cronin, J. W., Fitch, V. L., and Turlay, R., *Physical Review*, **140** (1965) B 74.

[11] Geweniger, C. et al., *Physics Letters*, **B 48** (1974) 487.

[12] Sakharov, A. D., *JETP Letters*, **5** (1967) 24.

---

[5] Andrei Dmitrievich Sakharov (1921–1989), Nobel Peace Prize 1979

Blackbody Radiation from the Early Universe (1965)

Wilson (left) and Penzias.

The foundations for *radio astronomy* were laid in 1933 in the Bell Telephone Laboratories in Holmdel, New Jersey. A young physicist from Oklahoma, Jansky[1], had been given the task to investigate the phenomenon called *static*, disturbing communication by short radio waves. Working with a rotatable antenna at a wavelength of 14.5 metres, he identified near and distant thunderstorms as sources of static but also the centre of our galaxy [1]. The development of radar for military use resulted in sensitive receivers for wavelengths of centimetres and below, which were put to scientific application after the Second World War. The interstellar gas, which contains atoms and molecules emitting radiation in this range, for instance, hydrogen H and cyan CN, thus became 'visible'. Radio astronomy was established as a new field of research. In 1965, also at the Bell Labs in Holmdel, a particular type of radiation from space was discovered by Penzias and Wilson, two physicists who had become radio astronomers.

Penzias[2] was born in Munich. He and his family nearly fell victims to the persecution of Jews. They were put on a train for deportation to Poland. A few days later they found themselves back in Munich and were allowed to leave to England in the spring of 1939. At the end of that year the family moved to New York. Penzias began to study chemical engineering at the City College of New York, but soon changed to physics. For his graduate studies he was accepted by Columbia University and he did his Ph.D. thesis work under Townes, building a maser as amplifier for radio-astronomical observations. In 1961 Penzias joined the Bell Labs, where he would eventually become vice president. Wilson[3], a Texan, studied physics at the Rice University in Houston. For his graduate studies he went to Caltech, for which recently the Owens Valley Radio Observatory had been erected, and soon found himself doing research in radio astronomy. Through his thesis work, which required masers, he came into contact with Bell Labs and became a researcher there in 1963.

Soon thereafter, Penzias and Wilson began to set up a particularly sensitive radio telescope working at a wavelength of 7.35 centimetres. Its largest component was a horn antenna with a 20-foot aperture. It had originally been built for trials of transatlantic telecommunication using an early satellite named *Echo*, which essentially was a large balloon with a metallized surface, reflecting the waves it received. Radiation, entering the antenna through the trapezoidal opening, fell on a parabolic mirror and was focussed into the neck of the horn. There, by an appropriate filter, a narrow frequency band was selected and its intensity measured by a *radiometer* and registered. The antenna was highly

[1] Karl Guthe Jansky (1905–1950) [2] Arno Allan Penzias (born 1933), Nobel Prize 1978 [3] Robert Woodrow Wilson (born 1936), Nobel Prize 1978

The 20-foot horn antenna at Holmdel used by Penzias and Wilson. Photo: NASA.

directional, strongly suppressing signals from directions other than intended. The radiometer was based on a maser amplifier, cooled to liquid-helium temperature (4.2 K). The output signal of the radiometer is given by the radiation received by the antenna plus *noise* generated in the detecting system itself, in particular, in the antenna and the radiometer. The wavelength or frequency spectrum of such noise is usually well described by Planck's radiation law (Episode 7) and therefore characterized by a temperature. Penzias and Wilson carefully calibrated their set-up with an external transmitter, compared it to the radiation of a 'black body' maintained at liquid-helium temperature and were then able to attribute effective noise temperatures, describing contributions to the output signal, to different parts of their system, to the ground, on which the antenna was mounted, and to the air above the antenna.

When pointing their telescope at the sky, Penzias and Wilson observed, in addition to the different noise contributions, an excess signal to which a temperature of (3.5 ± 1.0) K had to be attributed. The radiation which caused it was 'within the limits of our observations, isotropic, unpolarized, and free from seasonal variations' [2]. A result like this had been expected, though not by Penzias and Wilson, who had intended their system for the observation of localized astronomical objects. In fact, at Princeton University, only about 30 miles from Holmdel, a group of scientists led by Dicke had been preparing a similar experiment.

Dicke[4] had taken his Ph.D. at the University of Rochester in 1941 and had worked on radar during the war. In 1946, he was the first to use the combination of a (small) horn antenna and a radiometer (without laser amplifier, of course)

---

[4] Robert Henry Dicke (1916–1997)

to study the radiation from atmospheric water vapour and also from the sun and the moon and had given an upper limit of about 20 K on the temperature of 'radiation from cosmic matter at the radiometer wave-lengths', which were around 1 cm [3]. Also in 1946, Dicke went to Princeton. He became interested in cosmology and, in particular, the possibility of an oscillating universe.

The first to write down equations to describe the universe in the framework of general relativity, already in 1917, was Einstein himself [4]. His model is that of a homogeneous and isotropic universe, i.e., the mass it contains has the same density everywhere and there is no preferred direction. (These assumptions can describe the real world with its mass density averaged over sufficiently large distances.) Since there was no astronomical evidence for a change in time of the world on a very large scale, Einstein searched for a *static universe*, i.e., a time-independent solution of his equations. He found one, but at the expense of introducing in his equations an additional term, containing what is now called the *cosmological constant*.

In 1922 the Russian mathematician, meteorologist, and physicist Friedmann[5] [5] in Petrograd showed, that the equations (with and without cosmological constant) also have time-dependent solutions. In general relativity (Episode 26), the curvature of space is determined by the distribution of masses in space and, in turn, influences the motion of masses. Einstein's cosmological term, if present, also contributes to curvature. The curvature is constant throughout space but can change with time. In his paper [6], entitled *On the Curvature of Space*, Friedmann discussed the cases of a monotonically decreasing and of a periodic curvature. In the first case the curvature falls with time but stays positive (or approaches zero); the universe keeps expanding. Extrapolating back, a time corresponding to infinite curvature is found. Friedmann called this moment, mathematically described by a singularity, the *creation of the world*. In the second case the curvature is periodic, it falls up to a certain time, then rises towards the singularity, falls again, etc.; the universe periodically expands and contracts. The time to complete a full cycle was named *world period* by Friedmann. He pointed out that the astronomical knowledge was utterly insufficient to decide which case applied to the universe. But he did give a numerical example: 'If we assume [the cosmological constant to be] $\lambda = 0$ and [the mass of the universe to be] $M = 5 \times 10^{21}$ sun masses, then the world period becomes on the order of 10 billion [10^{10}] years.'

Einstein thought to have spotted a mistake in Friedmann's reasoning and, in a short note, claimed that the latter's conclusions were not valid. However, when Friedmann sent him a letter with details of his calculations, Einstein did not hesitate to retract his objections and called Friedmann's results 'correct and clarifying'. Friedmann died of typhoid fever at the age of only thirty-seven, four years before Hubble published his observation of an expanding universe (Episode 45). In the late 1920s and the 1930s cosmological models were studied, among others, by Einstein, by Eddington, and, in particular, by Lemaître[6], a Belgian astrophysicist who was also a catholic priest and a canon. Lemaître referred to Friedmann's singularity not as the 'creation of the world' but as the 'primeval atom' ('l'atome primitif' in his native French). About the

Friedmann

[5] Aleksandr Aleksandrovich Friedmann (1888–1925)   [6] Georges-Henri Lemaître (1894–1966)

present state of the universe, evolved from it, he writes [7]:

> The evolution of the world can be compared to a display of fireworks that just has ended: some few red wisps, ashes and smoke. Standing on a well-chilled cinder, we see the slow fading of the suns, and we try to recall the vanished origin of the worlds.

This, more or less, was Dicke's aim. Peebles[7] relates [8] that Dicke was impressed with the cosmology of Lemaître, specifically with the possibility of an oscillating universe. At the end of a contraction phase, the temperature was expected to be so high that all atoms and nuclei dissociated into their constituents. Electromagnetic radiation and the material particles (electrons and nucleons) would be in thermodynamic equilibrium, forming a very hot plasma. The radiation spectrum would be given by Planck's law (Episode 7) and determined by only one parameter, the temperature. Dicke considered it possible that remnants of that radiation, cooled off by the expansion of the universe, could be found experimentally. He asked Roll and Wilkinson[8] to construct an antenna plus radiometer and try to detect this radiation. In addition, he suggested that Peebles study the implications of detection or non-detection. Doing so, Peebles became aware of a quantitative prediction of what is now called the *cosmic background radiation* (CBR) or also *cosmic microwave background* (CMB).

The prediction goes back to work by Gamow, who had studied in Petrograd when Friedmann was a professor there and whom we met (Episode 50) when he applied quantum mechanics to nuclear physics. Gamow had returned to the Soviet Union in 1931 to become professor in Leningrad but soon found the atmosphere too oppressive. Together with his wife he made several unsuccessful attempts to escape; one was to try and paddle in a kayak across the Black Sea to Turkey. When he was allowed to participate in the 1933 Solvay Conference in Brussels, accompanied by his wife, he did not return but accepted a professorship at the George Washington University in Washington, DC. Gamow introduced nuclear physics into cosmology which, until then, had been based only on general relativity, i.e., on gravitation. Together with his Ph.D. student Alpher[9], he studied the formation of heavier nuclei from protons and neutrons in the early universe. In 1948, they found that the change of conditions due to the expansion of the universe was necessary to explain the observed abundance distribution of different nuclei. (It is now believed that only nuclei up to those of helium, lithium, and beryllium were formed so early and that heavier elements were composed in stars.) Gamow, who had a highly developed sense of humour, convinced Bethe to co-author the publication with Alpher, so that it could be remembered as the $\alpha\beta\gamma$ paper [9]. Also in 1948, Gamow presented a simple argument from which he obtained the temperature of the universe and its matter density at the time of deuteron formation [10,11]. Using these values and the known thermodynamic properties of the expanding universe, the present temperature of the cosmic microwave background can be computed from the present mass density (see Box).

Gamow did not take this last step. He used the information obtained on the early universe to explain the development of galaxies and express their size in terms of fundamental constants. But in 1949 Alpher and his colleague Her-

Lemaître

---

[7] Philip James Edwin Peebles (born 1935)   [8] Davis Todd Wilkinson (1935–2002)   [9] Ralph Asher Alpher (1921–2007)

Gamow Condition and CMB Temperature

Gamow considered the first step in nucleosynthesis, the formation of a deuteron d from a proton p and a neutron n, in which superfluous energy is released in the form of a γ quantum,

$$n + p \to d + \gamma \ . \tag{1}$$

The cross section σ for this reaction was known from nuclear physics experiments. The reaction becomes important at the temperature $T = 10^9$ K, corresponding to a nucleon velocity of $v \approx 5 \times 10^8$ cm s^{-1}. (At higher temperatures the deuterons dissociate in collisions with photons or nucleons.) Gamow found the condition $\sigma v \, \Delta t \, n_p \approx 1$, connecting the four quantities σ (known from experiment), v (known from the temperature), the time Δt, in which the bulk of the reaction (1) happens, and the proton density n_p (or, generally, the nucleon density n_{N}), see Figure. With a reasonable assumption for Δt the nucleon number density n_{N} and the mass density $\rho = m \, n_{\mathrm{N}}$ could be computed (m is the nucleon mass). One assumption for Δt is the half-life of the neutron [11], since the neutron has to interact before it decays. Gamow found at $T = 10^9$ K a matter density of $\rho = 10^{-6}$ g cm^{-3}, while the density of radiation (its energy density divided by c^2) was very much larger, about 10 g cm^{-3}. Therefore, as the universe expands, radiation cools, essentially, as if no matter was present. If R is a distance between two points, increasing during expansion, then the temperature is proportional to the inverse of R, $T \propto R^{-1}$; the density is inversely proportional to the third power of R, $\rho \propto R^{-3}$. This leads to the simple relation between two temperatures T_1, T_2 of the universe and the corresponding mass densities ρ_1, ρ_2,

$$\frac{T_1}{T_2} = \left(\frac{\rho_1}{\rho_2} \right)^{1/3} \ .$$

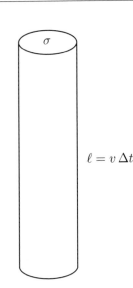

The cross section σ of a neutron moving with velocity v for a time Δt sweeps out a cylinder of volume $V = \sigma v \, \Delta t$. On the average one deuteron formation (1) will occur in Δt if there is one proton in the cylinder, i.e., if $V n_p = 1$, where n_p is the proton density.

man[10] [12], taking as the present mass density a value determined by Hubble, 10^{-30} g cm^{-3}, arrived at a temperature of around 5 K.

Since Dicke's group in Princeton did not expect any competition, Peebles gave a talk at the Johns Hopkins University in Baltimore about the cosmic radiation background and their hope to detect it. Word spread to Penzias and Wilson. The groups from Holmdel and Princeton met and agreed on the nature of the signal observed at the Bell Labs. Two papers were prepared to appear simultaneously in the *Astrophysical Journal*. The first, by the Princeton group [13], entitled *Cosmic Black-Body Radiation*, concentrates on cosmological considerations. The second, by Penzias and Wilson [2], bears the sober title *A Measurement of Excess Antenna Temperature at 4800 Mc/s* and sticks to the bare observations. Early in 1966, Roll and Wilkinson reported results obtained with their own instrument. At a wavelength of 3.2 cm they observed a signal corresponding to a temperature of (3.0 ± 0.5) K [14].

The conception of the expanding universe, evolving from a very dense and hot initial state, is now generally called the *Big Bang model*. It seems that the name *Big Bang* was coined by Hoyle in a transatlantic radio debate with Gamow, transmitted by BBC in 1949. In the discussion, which 'apparently became heated' [15], Hoyle was opposing Gamow's point of view.

---

[10] Robert C. Herman (1914–1997)

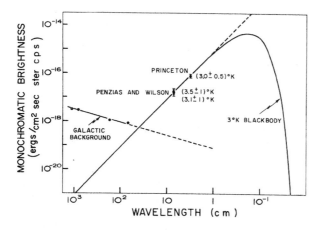

The measurement of Penzias and Wilson and that of Roll and Wilkinson (marked PRINCETON). The solid line correspond to a blackbody spectrum at a temperature of 3 K. Figure reprinted with permission from [14]. Copyright 1966 by the American Physical Society.

The cosmic background radiation, besides Hubble's law, became the main empirical pillar of the Big Bang model and was studied in great detail. A new level of precision was reached when a satellite was used as platform for radiation detectors. After fifteen years of preparation the COBE (Cosmic Background Explorer) satellite was launched in 1984. It carried three complementary instruments: the Diffuse Infrared Background Experiment DIRBE, the Far Infrared Absolute Spectrophotometer FIRAS, and the Differential Microwave Radiometers DMR [16]. More than a thousand persons worked on the COBE project. COBE project leader and also responsible for FIRAS was Mather[11] from NASA's Goddard Space Flight Center in Greenbelt, Maryland. Leaders of the DIRBE and DMR groups were Michael Hauser (also from Goddard) and Smoot[12] from Berkeley, respectively. The choice of the satellite orbit allowed the observation of the full sky. The instruments were protected by a shield from thermal radiation of earth and sun.

Mather

Smoot

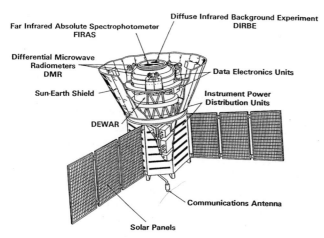

The COBE satellite with its major components. From [16]. Reprinted with permission of the American Astronomical Society.

---

[11] John C. Mather (born 1946), Nobel Prize 2006   [12] George F. Smoot (born 1945), Nobel Prize 2006

[1] Jansky, K. G., *Nature*, **132** (1933) 66.

[2] Penzias, A. A. and Wilson, R. W., *Astrophysical Journal*, **142** (1965) 419.

[3] Dicke, R. H. et al., *Physical Review*, **70** (1946) 340.

[4] Einstein, A., *Sitzungsber. Preuss. Akad. Wiss. (Berlin)*, (1911) 142.

[5] Tropp, E. A., Frenkel, V. Y., and Chernin, A. D., *Alexander A. Friedmann: The Man who Made the Universe Expand*. Cambridge University Press, Cambridge, 1993.

[6] Friedmann, A., *Zeitschrift für Physik*, **10** (1922) 377.

[7] Lemaître, G., *The Primeval Atom: An Essay in Cosmogony*. Van Nostrand, Toronto, 1950.

[8] Peebles, P. J. E., *Principles of Physical Cosmology*. Princeton University Press, Princeton, 1993.

[9] Alpher, R. A., Bethe, H., and Gamow, G., *Physical Review*, **73** (1948) 803.

[10] Gamow, G., *Physical Review*, **74** (1948) 505.

[11] Gamow, G., *Nature*, **162** (1948) 680.

[12] Alpher, R. A. and Herman, R. C., *Physical Review*, **75** (1949) 1089.

[13] Dicke, R. H., Peebles, P. J. E., Roll, P. G., and Wilkinson, D. T., *Astrophysical Journal*, **142** (1965) 414.

[14] Roll, P. G. and Wilkinson, D. T., *Physical Review Letters*, **16** (1966) 405.

[15] Alpher, R. A. and Herman, R., *Genesis of the Big Bang*. Oxford University Press, Oxford, 2001.

[16] Boggess, N. W. et al., *Astrophysical Journal*, **397** (1992) 420.

[17] Fixsen, D. J. et al., *Astrophysical Journal*, **473** (1996) 576.

[18] http://lambda.gsfc.nasa.gov.

[19] Bennett, C. L. et al., *Astrophysical Journal*, **464** (1996) L1.

FIRAS measured the intensity spectrum in the wavelength range between 0.1 mm and 10 mm. It compared the signal received through a horn antenna, subtending a 7° angle of the sky, with that of an on-board blackbody. The differences between Planck's radiation law and the observed one were less then fifty parts per million of the maximum value of the spectrum. The temperature of the cosmic background radiation was measured as $(2.728 \pm 0.004)\,\mathrm{K}$ [17].

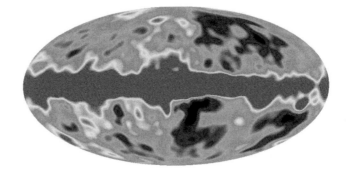

Temperature map of the universe as produced by the DMR instrument on the COBE satellite [18]. The map is in galactic coordinates; the Milky Way appears as the horizontal band along the equator. Outside it, fluctuations of the temperature, indicated in different grey shades, are clearly visible. Fluctuations typically are $\Delta T \approx 20\,\mu\mathrm{K}$. Figure: NASA.

DMR was designed to register possible tiny differences in temperature across the sky. Two specially designed 7° horn antennas looked at patches of the sky, which were 60° apart, and the difference of the two signals was measured, while the whole sky was swept. Minuscule temperature fluctuations were indeed observed [19], as expected by cosmologists. They are thought to give an impression of the beginning of matter clustering in the early universe which, eventually, led to the formation of galaxies. While the universe cooled off, there was still some interaction between radiation and matter as long as the temperature was high enough, about 10^5 K, for the ionization of hydrogen atoms. Matter fluctuations at that stage still find themselves conserved in temperature fluctuations of the cosmic background radiation.

Temperature map of the universe as measured by a later NASA satellite, the Wilkinson Microwave Anisotropy Probe (WMAP) [18]. In the representation, shown here, the local contribution from the Milky Way is subtracted. Figure: NASA.

Two Forces of Nature are Only One – Electroweak Interaction (1967)

The theory of *electroweak interaction*, also called *Glashow–Salam–Weinberg theory*, describes the electromagnetic and the weak force in a common framework; the two are no longer separate but two faces of one medal. It has been called a *unified theory*. The unification was likened to that of electricity and magnetism in the second half of the nineteenth century. Together with the modern theory of strong interactions (Episode 92), it is now usually referred to as the *standard model* [1], a rather modest name for a theory comprising most of physics and, so far, unchallenged by experiment.

The key to the electroweak theory is a property of quantum electrodynamics, which we mentioned in passing in Episode 71, the property of *gauge invariance*. The curious word *gauge* was introduced into theoretical physics in 1919 by the mathematician Weyl, whom we met briefly as colleague of Schrödinger in Episode 39, when he tried to construct a unified theory of classical electrodynamics and gravitation. The electric and the magnetic field can be defined as derivatives of a scalar and a vector potential, respectively. But there is some ambiguity here: A variety of potentials leads to the same fields. To use a particular gauge, for Weyl, meant to use a particular type of potential within this ambiguity. Physics, of course, had to be independent of that choice.

In quantum electrodynamics, the gauge concept appears in a new guise (see Box). The theory does not change if the electron wave function $\Psi(x)$ is replaced by $\Psi'(x) = \mathrm{e}^{\mathrm{i}\alpha}\Psi(x)$, where $\mathrm{e}^{\mathrm{i}\alpha} = \cos\alpha + \mathrm{i}\sin\alpha$ is a *phase factor* with the constant *phase* α. That is called *global* gauge invariance. In a *local* gauge transformation $\Psi(x)$ is multiplied by a phase factor $\mathrm{e}^{\mathrm{i}\alpha(x)}$, which depends on the coordinates x of space and time. Demanding local gauge invariance is equivalent to introducing the photon field. The photon, which transmits the electromagnetic interaction between charged particles, is the *gauge boson* of the theory. It has spin 1 (in units of $\hbar$) and, therefore, is a *vector boson*.

In 1954, two years before his collaboration with Lee on parity (Episode 39), Yang stayed as a visitor in Brookhaven, sharing the office with Mills[1]. Together, in a very influential work [2], they generalized the gauge principle. They introduced a fermion field comprising fermions with different electric charges (as an example, they took the proton and the neutron) and considered a local isospin transformation described by the group $SU(2)$ to perform a change from one charge to another. The theory contained three gauge bosons, electrically positive, negative, and neutral. Because of the *non-Abelian* nature of the group $SU(2)$ (see Box) the theory gives rise not only to fermion–boson but also to boson–boson interactions. This first *Yang–Mills theory* aimed at the strong interaction. The gauge bosons could not be pions, since those are not

Glashow (left) and Salam.

Weinberg lecturing

---

[1] Robert Lawrence Mills (1927–1999)

About Gauge Theories

Lagrangean. The model gauge theory is quantum electrodynamics, Episode 71. To extend it, usually the Lagrangean formalism is used, originally developed in the eighteenth century for classical mechanics. The Lagrange function or *Lagrangean L* of a system of particles is the difference $L = T - V$ of its kinetic energy T and its potential energy V. In general, it is a function of the positions and velocities of all particles and of the time. The system is completely described by Lagrange's equations, a set of differential equations in these variables. In field theory, the particle coordinates are replaced by fields, i.e., wave functions depending on space and time. The Lagrange function, originally an energy, becomes an energy density but is still called *Lagrangean*. Lagrange's equations, constructed with the corresponding replacements, are often referred to as *equations of motion* just as in classical mechanics. As an example, the Dirac equation of a free electron (see Box III in Episode 43) is obtained as equation of motion of a properly chosen Lagrangean.

Feynman rules. From the mathematical form of different terms appearing in the Lagrangean the Feynman rules, determining the theory, can be constructed. Graphically they are represented by Feynman diagrams.

Global gauge symmetry. Invariance of the Lagrangean reflects symmetries of the wave functions under transformations. A symmetry, usually, corresponds to a conservation law (see Episode 77). As an example, we consider the electron wave function $\Psi(x)$ occurring in the Dirac equation (technically a Dirac spinor which is a function of the four-vector x of spacetime.) A *global* gauge transformation is simply the multiplication of Ψ by a spacetime-independent phase factor, $\Psi(x) \rightarrow e^{i\alpha}\Psi(x)$. The Lagrangean of the Dirac equation is unchanged under this transformation; as a consequence the electric charge is a conserved quantity.

Local gauge symmetry, gauge-boson field. If the constant phase α is replaced by by an x-dependent one, $\alpha = \alpha(x)$, then the Lagrangean is no longer gauge invariant. The effects of such a *local* gauge transformation, however, can be exactly balanced by the introduction of an additional field, the *gauge field*, describing a *massless* boson. The photon is the gauge boson of electromagnetic interaction; by *demanding* gauge invariance the Dirac equation for a free electron is turned into an equation of motion for an electron interacting with a photon. In terms of group theory (see Box in Episode 87) the very simple transformation through multiplication by the phase factor $e^{i\alpha(x)}$ is described by the group $U(1)$. The result of two successive transformations does not depend on their order, since $\alpha_1(x) + \alpha_2(x) = \alpha_2(x) + \alpha_1(x)$; transformation groups with this property are called *Abelean*. Bosons of Abelean gauge groups only interact with fermions, not with gauge bosons.

Yang–Mills theory. Instead of a single fermion field (that of the electron in the case of quantum electrodynamics) several fields can appear in the Lagrangean which can be transformed into one another by transformations from the weak isospin group $SU(2)$. Local gauge invariance with respect to such transformations can be achieved by introducing a set of three massless gauge bosons with electric charges $+e, 0, -e$, where e is the elementary charge. The gauge group $SU(2)$ is *non-Abelean*; the gauge bosons not only interact with fermions but also with one another.

Spontaneous symmetry breaking. Higgs mechanism. The solutions of the equations of motion, in particular, the ground state, may lack the symmetry of the Lagrangean, see figures. In this case the gauge bosons can have mass. A 'wine-bottle' potential (taken as a simple model) has azimuthal symmetry, a local maximum at the origin, and a radial minimum off centre. Because of the independence of the potential of the azimuth, one boson seems to remain massless. That, however, is avoided by invoking the *Higgs mechanism*, see text.

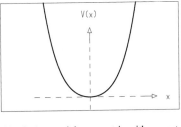

Classical potential, symmetric with respect to the single variable x, with minimum in the symmetry point.

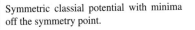

Symmetric classial potential with minima off the symmetry point.

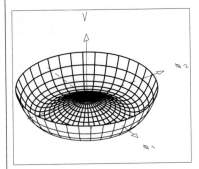

'Wine-bottle' potential as a function of two fields ϕ_1 and ϕ_2.

vector but scalar bosons, having spin 0. As to their physical mass, Yang and Mills did 'not have a satisfactory answer' [2].

Schwinger, in 1957, constructed an isospin triplet of the neutral and massless photon and two (positively and negatively) charged and massive vector bosons to transmit the electromagnetic and weak force, respectively. In his paper [3] he emphasizes that the long time scale (i.e., the very 'weakness') of the weak interaction 'becomes more comprehensible ..., if every observable process requires the virtual creation of a heavy particle'. This is because the massive boson appearing as an inner line, a propagator, of a Feynman diagram for low-energy processes such as the β decay is far off the mass shell (see Box III in Episode 71). In other words, the interaction appears to be weak because its carrier, the gauge boson, is heavy.

In 1960 Glashow, then in Copenhagen, proposed three massive bosons, positive, neutral, and negative, in addition to the photon. Glashow[2] was born in Manhattan. At the Bronx High School he became friends with other future distinguished physicists: Feinberg (Episode 85) and Weinberg. Together with Weinberg he studied at Cornell University, gaining his first degree in 1954. In 1958 Glashow completed his Ph.D. thesis entitled *The Vector Meson in Elementary Particle Decays* under Schwinger at Harvard University. Having won a postdoctoral fellowship, he wanted to work with Tamm in Moscow. But instead he did research at the Niels Bohr Institute in Copenhagen while waiting for the Soviet visa which never arrived. Glashow's theory of 1960 [4] already contains the four bosons of the present theory but its Lagrangean is not gauge invariant, since terms containing gauge-boson masses destroy the invariance. Accordingly, Glashow's paper carries the title *Partial-Symmetries of Weak Interactions*. A theory similar to Glashow's was published somewhat later by Salam and Ward[3] [5].

In the mid-1960s, the mass problem was tackled from a different angle by several physicists, in particular, by Higgs[4] from the University of Edinburgh, who connected gauge theories to *spontaneous symmetry breaking*, a phenomenon known from other domains of physics. It shows up if the Lagrangean is symmetric but the solutions of the equations of motion no longer display this symmetry. This is the classical example of such a case: A small ball can move along the x direction, its potential energy $V(x)$ is symmetric about the point $x = 0$. If $V(x)$ has a minimum at $x = 0$ and shows a continuous rise for increasing $|x|$, the ball will eventually end up at the point of symmetry, $x = 0$. (We assumed that there is some friction.) Let us change the potential a little. It is still symmetric, it still rises continuously with $|x|$ for large values of $|x|$, but it now has a maximum at $x = 0$ and, consequently, two minima situated symmetrically at $x = \pm a$. The ball will come to rest either at $x = +a$ or at $x = -a$, neither point being the centre of symmetry. In quantum field theory the potential becomes a function of fields (which themselves are functions of space and time coordinates). If the symmetry is spontaneously broken, the ground state acquires mass. In this way gauge bosons could become massive. It seemed, however, that even then one massless and spinless (or scalar) boson remained. In 1964 Higgs was able to show that the 'degree

Diagram for the muon decay

$$\mu^- \to e^- + \nu_\mu + \bar{\nu}_e$$

in the original theory without gauge boson.

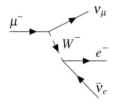

Diagram for the same decay with a propagator line corresponding to an *intermediate* charged gauge boson W^-.

Higgs

[2] Sheldon Lee Glashow (born 1932), Nobel Prize 1979   [3] John Clive Ward (1924–2000)   [4] Peter Ware Higgs (born 1929)

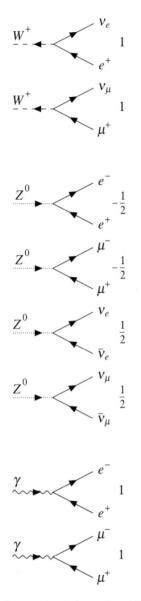

Basic diagrams for the interaction of W, Z, and γ bosons with leptons. The numbers indicate the relative weight of each diagram within its set. Particle names were written, assuming all particles moving to the right. (Additional diagrams are obtained by inverting the arrow directions and exchanging particle and antiparticle names. Note that neutral bosons are their own antiparticles.)

of freedom' offered by what apparently was an unwanted particle could be absorbed into the properties of the massive gauge boson of spin 1. The photon which has spin 1 but no mass can only be transversely polarized. Its polarization is described by components along two directions perpendicular to the photon direction of motion and to each other. A massive vector boson can also have longitudinal polarization. What had been interpreted as a scalar boson is indeed needed for the proper treatment of the polarization of the massive gauge bosons. This is now known as the *Higgs mechanism*.

Within a few weeks Higgs wrote two important papers. The first [6], published in *Physics Letters*, shows that the massless scalar boson does not necessarily appear. It ends announcing a subsequent note which was to contain a more specific application. That note, however, was rejected by the same journal. Higgs realized that this paper 'had been short of salestalk' [7]. He added two paragraphs on strong interactions, in which the spin-1 bosons are vector mesons, and sent the paper to *Physical Review Letters* who published it [8]. In the last paragraph attention is drawn to a characteristic additional scalar (spin-0) boson with mass, now called the *Higgs boson*. The work of Higgs allows the construction of a gauge theory with massive gauge bosons without the appearance of a massless scalar boson which, if it existed, should have been observed. The price to pay is to accept the existence of the Higgs boson. Also that has not been seen up to now. But there is no contradiction to experimental facts; the Higgs boson is believed to have a mass beyond the reach of present particle accelerators. Only three years after the work of Higgs his reasoning was applied to electroweak interactions by Weinberg and, independently, by Salam.

Weinberg[5], a New Yorker like Glashow, and, as we saw, his friend at school in Brooklyn and at Cornell University, went to Copenhagen for a year as graduate student, then took his Ph.D. at Princeton in 1957. He moved to Columbia University and from there in 1959 to Berkeley. His work on electroweak theory was done at MIT where, in 1967, he was a visitor on leave from Berkeley. He became professor at MIT in 1969 and later moved to Harvard and to the University of Texas at Austin.

Salam[6] was born in a small town near Lahore in British-ruled India, now in Pakistan. At the age of fourteen he was admitted to the University of Punjab and he took his M.A. there in 1946. He won a scholarship to St. John's College, Cambridge, and obtained a Ph.D. with a thesis on quantum electrodynamics in 1951. After two years in Pakistan, trying in vain to establish a school of modern research there, Salam returned to Cambridge as lecturer and, in 1957, became professor at Imperial College, London. In parallel to his work there he founded and led, from 1964 onwards, the International Centre of Theoretical Physics in Trieste, Italy, which offers research opportunities, in particular, for physicists from developing countries.

Weinberg entitles his paper of 1967 *A Model of Leptons* [9]. Accordingly, electroweak interaction of leptons only is described. To take into account the 'left-handedness' of the weak interaction he decomposes the electron wave function into a left-handed part e_{L}^{-} and a right-handed part e_{R}^{-}. The neutrino

[5] Steven Weinberg (born 1933), Nobel Prize 1979   [6] Abdus Salam (1926–1996), Nobel Prize 1979

ν_e, which is always left-handed, is placed in a charge doublet with e_L^-. A singlet remains for e_R^-. Concerning the gauge symmetry to be used, Weinberg is inspired by the isospin–strangeness or, rather, isospin–hypercharge symmetry of hadrons (Episode 74) and quarks (Episode 87). Quantum numbers now called *weak isospin* and *weak hypercharge* are introduced. The doublet has weak isospin $I^W = 1/2$, the singlet $I^W = 0$. Weak hypercharge is assigned such, that the Gell-Mann–Nishijima formula (relation (4) in Episode 74) for the charge Q, i.e., $Q = I_3^W + Y^W/2$, is fulfilled.

Up to this point all seems to be modelled after the case of the three quarks. However, the three wave functions do not transform under a common group $SU(3)$ since that would imply transformations from left-handed to right-handed states. Instead, Weinberg uses as gauge groups for the doublet the isospin group $SU(2)$ and for the singlet the trivial group $U(1)$. As result he obtains a triplet W^1, W^2, W^3 of gauge fields from the first group and a singlet B from the second. (We use names employed in the more recent literature.) From W^1, W^2 the fields of the charged gauge bosons W^+, W^- are constructed. The fields W^3 and B are 'mixed' (for an earlier example of mixing see that of neutral kaons in Episode 88),

$$A = B \cos \theta_W + W^3 \sin \theta_W \quad , \qquad Z = -B \sin \theta_W + W^3 \cos \theta_W \quad ,$$

to obtain the massless field A of the photon and the massive field of the neutral boson Z^0. Weinberg's theory needs only one coupling constant, the elementary charge e. But there is an additional parameter, the *weak mixing angle* (sometimes called the *Weinberg angle*) θ_W. Also the masses of the heavy gauge bosons are unknown, but the theory predicts the relation $M_W = M_Z \cos \theta_W$ between them. Although developed for the electron and the electron neutrino the theory, of course, also applies to the muon and its neutrino.

While the theory was developed by Weinberg at MIT, Salam and Ward, at Imperial College, London, were following very similar ideas which Salam presented in seminar talks. A formal publication appeared in 1968 [10].

Only gradually was the value of the new theory established. An indication for the existence of a heavy neutral gauge boson, the Z^0, was found in 1973 in the form of *weak neutral currents* (Episode 91). Free $W^\pm$ and Z^0 bosons were produced in 1983 (Episode 96). Difficulties were encountered when the theory was extended to quarks to describe the weak decay of hadrons. They were only removed by the bold conjecture of Glashow and collaborators in 1970 that there exist not three but four quarks. This turned out to be correct a few years later (Episode 93).

Theoretically, the theory of electroweak interactions was put on a firm footing in 1971 by t'Hooft[7] in Utrecht, who was able to show that, like quantum electrodynamics, Yang–Mills theories can be renormalized, both for the case of massless [11] and of massive [12] gauge bosons. Only if this is guaranteed reliable results can be obtained from perturbation series. In his famous proof, t'Hooft used methods developed by his teacher Veltman[8].

Lepton	Q	I^W	I_3^W	Y^W
ν_e	0	1/2	$+1/2$	-1
e_L^-	-1	1/2	$-1/2$	-1
e_R^-	-1	0	0	-2

Electric charge Q (in units of the elementary charge), weak isospin I^W, its third component I_3^W, and weak hypercharge Y^W of the doublet and the singlet of leptons as defined by Weinberg.

[1] Hoddeson, L., Brown, L., Riordan, M., and Dresden, M. (eds.), *The Rise of the Standard Model*. Cambridge University Press, Cambridge, 1997.

[2] Yang, C. N. and Mills, R., *Physical Review*, **96** (1954) 191.

[3] Schwinger, J., *Annals of Physics*, **2** (1957) 407.

[4] Glashow, S. L., *Nuclear Physics*, **22** (1961) 579.

[5] Salam, A. and Ward, J. C., *Physics Letters*, **13** (1964) 168.

[6] Higgs, P. W., *Physics Letters*, **12** (1964) 132.

[7] Higgs, P. In [1], page 506.

[8] Higgs, P. W., *Physical Review Letters*, **13** (1964) 508.

[9] Weinberg, S., *Physical Review Letters*, **19** (1967) 1264.

[10] Salam, A., *Weak and Electromagnetic Interactions*. In Svartholm, N. (ed.): *Elementary Particle Theory*, p. 367. Almqvist and Wiskell, Stockholm, 1968.

[11] t'Hooft, G., *Nuclear Physics*, **B 33** (1971) 173.

[12] t'Hooft, G., *Nuclear Physics*, **B 35** (1971) 167.

---

[7] Gerardus t'Hooft (born 1946), Nobel Prize 1999   [8] Martinus J. G. Veltman (born 1931), Nobel Prize 1999

Weak Neutral Currents – A Glimmer of Heavy Light (1973)

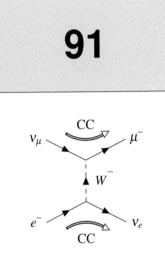

Diagram for charged-current (CC) interaction of a muon neutrino ν_μ with an electron.

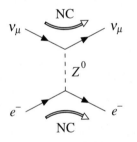

In this neutral-current (NC) interaction the nature of the incoming particles stays unchanged.

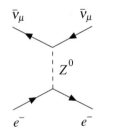

NC scattering of anti-muon neutrino and electron.

In Episode 90 we described how the unified theory of *electroweak interaction* gradually emerged. In the final theory there are four gauge bosons transmitting the interaction between fermions (leptons and quarks). Two of these bosons carry electric charge (the W^+ and the W^-), two are neutral (the Z^0 and the photon γ). While the photon is massless, the W and Z bosons carry mass. The theory does not provide numerical values for the masses M_W and M_Z, only the relation $M_W = M_Z \cos\theta_W$ between the two. But also the parameter θ_W, the *weak mixing angle*, was unknown. In 1983 the heavy bosons were directly observed (Episode 96) but in the 1970s they were out of reach for the accelerators of the time. Still, in 1973, a collaboration of physicists at CERN in Geneva found a first hint for the existence of the Z^0 and succeeded in a first measurement of θ_W. The Z^0 boson is a close relative of the photon; in the theory both originate as mixtures of two other fields. In some sense the Z^0 is just a heavy light quantum, hence the title of this episode.

The experiment was proposed by a theoretician. t'Hooft had shown in 1971 that electroweak theory was renormalizable and thereby had given it more credibility. Later that year he pointed out [1] that a theory with a Z^0 permits the elastic scattering of a muon neutrino or its antiparticle off an electron,

$$\nu_\mu + e^- \to \nu_\mu + e^- \quad , \qquad \bar{\nu}_\mu + e^- \to \bar{\nu}_\mu + e^- \quad , \tag{1}$$

while a theory without Z^0 does not. t'Hooft's point is best explained with the help of diagrams. The central point of the $V-A$ theory is the interaction of two charge-changing currents (see Box in Episode 79). Such a current is usually simply named *charged current* and abbreviated further to CC. Graphically the current corresponds to two connected fermion lines with a change of charge at the point of interaction. If the two pairs of fermion lines are linked by the propagator line of a boson transmitting the interaction, then that boson is necessarily charged to ensure charge conservation at the two vertices of the diagram. In the scattering of a muon neutrino off an electron, as described in the $V-A$ theory, the two initial leptons have to change charge; the final leptons are a muon and an electron neutrino. The new theory, however, allowed also for a neutral boson, the Z^0, corresponding to a *neutral current* (NC). In neutral-current scattering the initial leptons do not change their identity.

How do *neutral-current events* appear to the experimenter? If a beam of muon neutrinos falls on a target containing electrons (all matter does), then a neutrino can transmit momentum to an electron. Such a *knock-on electron* can be detected. A muon-neutrino beam also contains some electron neutrinos that can produce knock-on electrons in charged-current reactions and thereby imitate neutral-current events. It is therefore better to use a beam of anti-muon

430

neutrinos. These, as well, can knock on electrons. But anti-electron neutrinos cannot; they could knock on positrons but there are none in the target. (The admixture of electron neutrinos is negligible. We should add that also a charged-current reaction of an electron neutrino with a hadron can give rise to a fake event.) The ideal instrument for the experiment is a bubble chamber filled with a liquid of high density to provide enough target material for the scarce neutrino interactions. Since the chamber filling is not only target but also detector, it provides tracks of the knock-on particles beginning at the point of interaction. At the time of t'Hooft's proposal such a chamber, in fact the largest heavy-liquid bubble chamber ever built, was already running in a neutrino beam at CERN.

Lagarrigue[1] had entered the École Polytechnique in Paris in 1945, where he studied under Leprince-Ringuet[2]. He did his Ph.D. research on cosmic rays with cloud chambers under the guidance of Peyrou and worked at Berkeley at the time the Bevatron accelerator began operation. On returning to the École Polytechnique in 1955, he began to design and build bubble chambers. His chamber number 3, built at the laboratory of the Centre d'Énergie Atomique in Saclay, was the one used in the first round of CERN neutrino experiments. In 1964 Lagarrigue, together with Musset[3] and Rousset[4], proposed to build a large cylindrical bubble chamber, nearly 2 metres in diameter and nearly 5 metres long, filled with 18 tons of Freon (CF_3Br), which is 20 times heavier than liquid hydrogen and therefore suitable as neutrino target. When Leprince-Ringuet heard of the unwieldy instrument he called it *Gargamelle*, taking the name from the work *Gargantua and Pantagruel* by the French Renaissance writer Rabelais. There Gargamelle, in her eleventh month of pregnancy and after a heavy meal of tripe, gave birth to the giant Gargantua through her left ear. The giant chamber was again constructed at Saclay and moved to CERN, where it was mounted, tested, and finally run in the neutrino beam [2].

Even in the early 1960s neutrino experiments were run at CERN, confirming the existence of two neutrino species found at Brookhaven (Episode 85). In addition to spark chambers a heavy-liquid bubble chamber was employed. When, at Stanford, experiments with incident high-energy electrons gave insight into the substructure of the proton, a strong interest arose to repeat them with incident neutrinos. Accordingly, a facility was set up at CERN to produce intense beams of muon neutrinos or antineutrinos. Neutrinos originate from the decay of parent mesons. The CERN facility included a novel device, the *neutrino horn*, invented by van der Meer, through which the parent particles passed. It focussed mesons of one charge along the beam direction and removed those of the opposite charge. The decay of positive mesons yields neutrinos and that of negative ones antineutrinos.

Lagarrigue assembled a collaboration of physicists from seven laboratories (Aachen, Brussels, CERN, École Polytechnique, Milano, Orsay, and University College, London) to perform the neutrino experiment and analyse the results. The collaboration was already taking data when, in 1971, t'Hooft made his proposal to search for neutral currents. In practice, the search meant *scanning*, i.e., carefully inspecting, many thousands of photographs. The film rolls

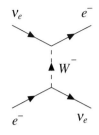

NC events of the type (1) can be imitated by such CC reactions if there are electron neutrinos in the beam.

Lagarrigue (left) and Rousset.

Faissner

---

[1] André Lagarrigue (1924–1975)   [2] Louis Leprince-Ringuet (1901–2000)   [3] Paul Musset (1935–1985)   [4] André Rousset (1930–2001)

The oblong cylindrical chamber body of Gargamelle within two coils which provide a strong magnetic field. Photo: CERN.

containing them had been distributed among the laboratories and chance had it that the first candidate of an event consisting of an isolated electron track was found in the group headed by Faissner[5] in Aachen. Perkins from Oxford who, at that time, was a member of the CERN group in the Gargamelle collaboration, remembers when and how he first heard about the event [3].

The first of the three leptonic neutral-current events observed in Gargamelle. The beam of anti-muon neutrinos ($\bar{\nu}_\mu$) enters from the left. Through the reaction $\bar{\nu}_\mu + e^- \rightarrow \bar{\nu}_\mu + e^-$ an electron (e^-) is accelerated and moves to the right, leaving a track which is slightly curved upwards by the magnetic field. The electron then loses most of its energy by radiating a photon which, in turn, produces an electron–positron pair and that process repeats itself further to the right. (The vertical track is not related to the event.) Photo: CERN.

On 30 December 1972, Faissner came to see him in Oxford to discuss the proposal of a future experiment and Perkins met him at London Airport. Right there Faissner showed him a photograph of the event. Perkins asked if it had been taken with the neutrino or the antineutrino beam; on being told the latter, he 'suggested to dash to the bar and celebrate'. (As pointed out above there is only a negligible background of electron neutrinos in the anti-muon-neutrino beam which could produce isolated electrons via charged-current interaction.) This first event and its analysis were published in September 1973 [4]. In the

---

[5] Helmut Faissner (1928–2007)

further course of the experiment two more such leptonic neutral-current events were found [5].

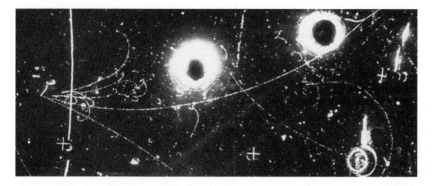

A hadronic neutral-current event observed in Gargamelle. Again the beam enters from the left. All particles produced in the interaction (in the left of the photograph) are identified as hadrons. (Again there is a vertical track, unrelated to the event.) Photo: CERN.

So far, we have discussed only interactions of muon neutrinos and antineutrinos with electrons, yielding the particularly clean but rare *leptonic events*. Much more numerous are *hadronic events*, resulting from collisions with nuclei or, rather, with quarks within nucleons within nuclei. They, too, can be classified as either CC (due to charged-current) or NC (neutral-current) interaction. A CC collision results in hadrons plus a lepton, which is determined by the type of the incident neutrino. An NC collision yields only hadrons, since the neutrino is not transformed into a charged lepton. Together with the first leptonic event the collaboration published their first results on hadronic events [6]; extensive analyses appeared later [7,8]. A historical account with emphasis on this part of the experiment is given in [9]. Essential for that work was the clear distinction between the tracks of leptons (electrons and muons) and hadrons (mesons and baryons) in Gargamelle. An electron, as seen in the picture of the first leptonic event, creates an electromagnetic shower in the heavy material of the bubble-chamber liquid. A muon only loses a little energy by ionization and leaves the chamber without strong interaction. Charged hadrons, on the average, interact more than once with nuclei in the liquid if their paths can be traced over a sufficiently long distance.

It was thus possible to determine the numbers of NC and CC events registered while Gargamelle was exposed to a muon-neutrino beam. The ratio $(N_{\mathrm{NC}}/N_{\mathrm{CC}})_{\nu_\mu}$ of these numbers carries information about the weak mixing angle, as does the corresponding ratio $(N_{\mathrm{NC}}/N_{\mathrm{CC}})_{\bar\nu_\mu}$ obtained with an antineutrino beam. Qualitatively, that is easy to see. The probability of an NC event depends on the mass of the exchanged Z^0 boson, that of a CC event on the mass of the $W^\pm$ boson. The cosine of the weak mixing angle is the ratio of the boson masses, $\cos\theta_{\mathrm{W}} = M_W/M_Z$, and therefore is related to the event-number ratio. It has become customary to quote the sine squared, $\sin^2\theta_{\mathrm{W}}$, of the weak mixing angle. The first value of this fundamental constant obtained by the Gargamelle collaboration was still crude. They stated $\sin^2\theta_{\mathrm{W}}$ to be in the range 0.3 to 0.4. Now, more than three decades later, the uncertainty is not in the first but in the fourth decimal: The current number is $0.231\,22\pm0.000\,15$.

[1] t'Hooft, G., *Physics Letters*, **37 B** (1971) 195.

[2] Rousset, A., *Gargamelle et les courants neutres*. École des Mines de Paris, Paris, 1996.

[3] Perkins, D. In [10], p. 428.

[4] Hasert, F. J. et al., *Physics Letters*, **46 B** (1973) 121.

[5] Blietschau, J. et al., *Nuclear Physics*, **B 114** (1976) 189.

[6] Hasert, F. J. et al., *Physics Letters*, **46 B** (1973) 138.

[7] Hasert, F. J. et al., *Nuclear Physics*, **B 73** (1974) 11.

[8] Blietschau, J. et al., *Nuclear Physics*, **B 118** (1977) 218.

[9] Haidt, D., *European Physical Journal*, **C 34** (2004) 25.

[10] Hoddeson, L., Brown, L., Riordan, M., and Dresden, M. (eds.), *The Rise of the Standard Model*. Cambridge University Press, Cambridge, 1997.

Quantum Chromodynamics (QCD) – The New Theory of Strong Interaction (1973)

92

The riddle of the strong interaction, i.e., of the force that keeps the nucleus together, was finally solved in the early 1970s in a rather unexpected way. In previous episodes we told of various attempts to describe nuclear forces, in particular, Heisenberg's introduction of isospin (Episode 51) and Yukawa's postulate of the meson as carrier of the strong force (Episode 57). While these theories led to important developments, they failed to reach their original goal. The present theory was arrived at in three steps. First, quarks were postulated to have an additional property which was to come in three different forms, soon called *colours*. Second, a Yang–Mills theory, similar to the one successfully constructed for the electroweak interaction (Episode 90), was formulated using as gauge symmetry an $SU(3)$ group providing transformations between the three colours. Finally, this type of theory was found to have an astonishing property called *asymptotic freedom*: The nearer two quarks come to each other, the smaller is the force between them. Small distances are reached only for very high momentum transfer between the two quarks; in such situations quarks approach the behaviour of free particles. On the other hand, the force increases with rising distance or falling momentum transfer. Let us now look at the developments in a little more detail.

The quark model, which so well explained the spectrum of hadrons (Episode 87), had one rather unpleasant defect. Consider the Δ^{++} baryon, which is a spin-3/2 fermion consisting of three u quarks. Its wave function is symmetric under the exchange of the coordinates of any two quarks instead of being antisymmetric as befits a fermion. One way out, proposed by Greenberg[1], was to assume that quarks did not obey Fermi–Dirac statistics but a *parastatistics* which would allow three identical quarks to share one state [1].

In 1965 Han[2] and Nambu[3] attributed a new, three-valued property to quarks [2]. (The three u quarks in the Δ^{++} could then each take a different value and thus follow Fermi–Dirac statistics.) Pais was the first to colloquially use colours to distinguish the three values of the new property [3]: 'If the baryon is made up of a red, a white, and a blue quark, they are all different fermions and you … reconcile Fermi statistics, symmetry of the spatial wave function, and the three-quark structure of the baryon.' The early choice of specific colours is generally replaced by *red*, *green*, and *blue*. Although, of course, the 'colour' property of quarks has nothing to do with visual colours, that choice is advantageous, since red, green, and blue mix to a colourless, white appearance. Similarly, a colour and its complementary colour, for instance, red and cyan –

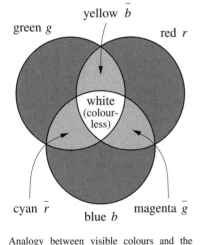

Analogy between visible colours and the colour quantum number attributed to quarks. Quarks q carry one of the three colours r, g, b and antiquarks $\bar{q}$ one of the three anticolours $\bar{r}, \bar{g}, \bar{b}$. A meson consists of a quark with some colour and an antiquark with the corresponding anticolour ($r\bar{r}, g\bar{g}, b\bar{b}$) and therefore is colourless. Likewise a baryon, consisting of three quarks with different colours (rgb), itself carries no net colour.

---

[1] Oscar Wallace Greenberg (born 1932)   [2] Moo-Young Han (born 1934)   [3] Yoichiro Nambu (born 1921)

particle physicists say red (r) and anti-red ($\bar{r}$), add to white. This will allow us to speak of the formation of colourless hadrons from coloured quarks. Han and Nambu used the mathematical group $SU(3)$ to describe symmetry transformations between colours, the same group that had proved useful to picture the relation between u, d, and s quarks (Episode 87). Thus there seemed to be two triplets of quark properties, a triplet of colours (r, g, b) and a triplet of *flavours* (u, d, s). While it is essential for QCD that there are three colours (the corresponding group is now called $SU(3)_c$), more flavours were found later (see Episode 93). As an aside, we note that Han and Nambu associated colour also with electric charge; they proposed three triplets with three quarks each, all quarks having integral electric charge. This reflects the general uneasiness that still prevailed with respect to fractionally charged fundamental particles.

In fact, at that time quarks played a role similar to that of atoms in most of the nineteenth century. Atoms had been useful to picture chemical reactions, quarks were a help in explaining the plenitude of hadrons; but did they really exist? Even Gell-Mann repeatedly spoke of quarks as 'fictitious', also in the work he did with Fritzsch in 1972. Fritzsch[4] grew up in East Germany and studied in Leipzig. Still a student, in 1967, he worked in a part of the library which was housed in the mediaeval university church and came across the work of Yang and Mills on non-Abelian gauge theories (Episode 90). Without much success, he tried to apply it to the interaction between quarks. Together with a friend, only a year later, Fritzsch succeeded to flee to the West travelling two hundred kilometres in a tiny folding boat across the Black Sea [4]. In Munich, working in Heisenberg's Max-Planck Institute for Physics and Astrophysics, he took his Ph.D. in 1971. That same year, in California, Fritzsch told Gell-Mann of his trials with Yang–Mills theories. The following year, when both were at CERN, they constructed such a theory with $SU(3)_c$ as gauge group [5]. They later named it *quantum chromodynamics* (QCD) in analogy to quantum electrodynamics (QED), alluding to the new colour force rather than to the well-known electromagnetic one (see abstract in [6]). In 1973 the collaboration was continued at Caltech in Pasadena, where they were joined by Leutwyler[5], then already a young professor at the university of his home town Berne, the Swiss capital [7].

QCD can be quite easily compared to and distinguished from QED in a few sentences. While there are two charges (the positive and the negative electric charge) in QED, there are six in QCD, the three colour charges r, g, b and the three anticolour charges $\bar{r}, \bar{g}, \bar{b}$. Quarks of any flavour can carry any colour; antiquarks carry anticolour. In QED, there is a single massless gauge boson, the uncharged photon. In QCD, there is an octet of gauge bosons; they carry colour and are called *gluons*. The name is, of course, derived from 'glue' and was used by Gell-Mann in a different context already in 1962 [8]. This time it stuck. Since the gauge group describes the transformation between quarks of different colour, a gluon carries both colour and anticolour, for instance, red and anti-blue, $r\bar{b}$. A red quark emitting such a gluon turns blue; an antiquark with the colour anti-red absorbing it becomes anti-blue. In this way a meson, for instance, a π^+ composed of a u quark and a $\bar{d}$ antiquark, stays colourless

Fritzsch (left) and Gell-Mann.

Leutwyler

[4] Harald Fritzsch (born 1943)   [5] Heinrich Leutwyler (born 1938)

The Colour Factor and the Ratio R

Consider the collision of an electron e^- and a positron e^+ at high energy. They annihilate to form a (virtual) photon, which can create another particle–antiparticle pair, for instance, a pair of $\mu^+\mu^-$ of muons or a pair $q\bar{q}$ of quarks. Both processes are described by the same Feynman diagram in QED. Their cross sections σ, proportional to the rate with which these processes occur, differ only because of the difference in charge between the muon and the quark. One has (for the production of a pair $q\bar{q}$ with specific colour and anticolour, e.g., $r\bar{r}$)

$$\frac{\sigma(e^+ + e^- \to q + \bar{q})}{\sigma(e^+ + e^- \to \mu^+ + \mu^-)} = Q_q^2 \quad,$$

where Q_q is the electric charge of the quark q in units of the elementary charge e. The numerical values are $Q_u^2 = (2/3)^2 = 4/9$ for the u quark and $Q_d^2 = Q_s^2 = 1/9$ for the d and the s quark. One assumes that production of hadrons in an electron–positron collision is initiated by a quark–antiquark pair production; the pair can be $u\bar{u}$, $d\bar{d}$, or $s\bar{s}$. Since every quark exists in three colours, one gets the ratio

$$R = \frac{\sigma(e^+ + e^- \to \text{hadrons})}{\sigma(e^+ + e^- \to \mu^+ + \mu^-)} = 3(Q_u^2 + Q_d^2 + Q_s^2) = 2 \quad.$$

The *colour factor* 3 would be missing if there were no colour.

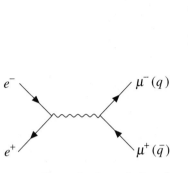

Feynman diagram for the production of a pair of muons (or quarks) in an e^+e^- collision.

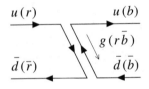

Example for the colour flow between two quarks forming a π^+ meson. A gluon $g(r\bar{b})$ is exchanged.

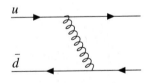

In a Feynman diagram the gluon is usually shown as a line similar to a spring.

while its constituents continuously change colour by exchanging gluons. This gluon exchange provides the binding between quark and antiquark. In much the same way the three quarks in a baryon are held together.

The Lagrangean of the new theory and with it, essentially, the complete structure of QCD was published in 1972 [5] and somewhat more explicitly in 1973 [7]. One might think that now the masses of all mesons and baryons had become easily calculable from the interaction between quarks, as had been the energy states of an atom when quantum mechanics was established. Alas, that is not the case as will become clear at the end of this episode. But the early work did predict an experimental result. In high-energy electron–positron collisions one can observe the rate for the production of hadrons and the rate for the creation of muon pairs and one can determine the ratio R of the two rates. That ratio was predicted to be three times as high than expected before (see Box). The prediction is based only on the existence of colour, not on QCD as a fully fledged theory.

In QED calculations take the form of a *perturbation series*, i.e., a sum of terms ordered in the power of the fine-structure constant α. Since $\alpha \approx 1/137$ is much smaller than 1, only the first or the first few terms of the series need be computed. The equivalent to α in QCD is called the *strong coupling constant* α_{s}. Obviously, it has to be quite a bit smaller than 1 to allow 'perturbative' calculations. In 1973 it was shown by Gross and Wilczek and by Politzer that this is the case, at least in some situations.

Before we discuss their work, let us look a little closer at the fine-structure constant α, which was introduced by Sommerfeld and which is proportional to the square of the elementary charge e (Episode 27). A charged particle, say, an electron of charge $-e$, is surrounded by 'virtual' electron–positron pairs.

These are created by the electric field, exist for the short time allowed by the uncertainty relation, and annihilate. The positrons are attracted by the original electron, the electrons are repelled. One speaks of *vacuum polarization* since the vacuum, filled with virtual pairs, develops a charge density under the influence of the original charge. For a test particle at some distance, the charge of the original electron seems reduced because the positrons diminish it more than the electrons add to it. The reduction is called *screening* of the original charge. Screening increases with distance. The nominal value $\alpha \approx 1/137$ corresponds to the 'large' distance typical of atomic physics. Small distances are reached if two particles collide with a large momentum difference or, more precisely, with a large four-momentum transfer between them; collisions are characterized by the (relativistically invariant) square q^2 of that transfer. At $q^2 \approx (100\,\mathrm{GeV}/c)^2$ the value of α is about 10% higher than at $q^2 \approx 0$. The electromagnetic interaction increases in strength with rising q^2 since the value of the elementary charge, and with it that of α, increases.

Experimentally, a completely different behaviour was found in the late 1960s for the strong interactions at the Stanford Linear Accelerator (SLAC) in California. There the production of hadrons in electron–proton collisions at very high energies was analysed. The data displayed a feature, dubbed 'scaling'; protons seemed to be made up of constituents (presumably quarks) which were the more loosely bound to each other the higher the relative q^2 between them was. Thus, while the electromagnetic interaction strengthens with rising q^2, the strong interaction appeared to weaken and eventually to vanish at infinite q^2. This property was later called *asymptotic freedom*. Many theoreticians therefore thought that no field theory of the strong interaction could be constructed and quite different approaches were tried.

Among them was Gross[6]. He was born in Washington, D.C., and grew up in nearby Arlington until, when he was eleven, his family moved to Israel. His father had become a member of an American advisory team and later joined the faculty of the Hebrew University in Jerusalem. Gross took his B.Sc. there before going to Berkeley for graduate studies. In 1966 he completed his Ph.D. thesis under Chew[7], moved on to Harvard, and became professor at Princeton in 1969. In 1972 he set himself the task to prove that no local gauge field theory could explain scaling. He decided first to show rigorously that asymptotic freedom is needed to explain scaling and, in a second step, that no local field theory displayed asymptotic freedom. The first step was achieved in collaboration with Callan[8], the second together with Coleman[9]; these studies, however, were restricted to Abelian gauge theories. In the extension to non-Abelian theories Gross engaged Wilczek, who had recently joined his group as a graduate student. Wilczek[10] was born in New York City and had attended high school in Queens before studying mathematics at Chicago. In Princeton, before working with Gross, he had intended to do research in mathematics. While the project of Gross and Wilczek was completed, Coleman was for an extended stay in Princeton. Meanwhile, his own graduate student Politzer, at Harvard University, worked on the same problem. Politzer[11], also a New Yorker, had attended

Gross

Wilczek

Politzer

[6] David J. Gross (born 1941), Nobel Prize 2004   [7] Geoffrey Chew (born 1924)   [8] Curtis G. Callan (born 1942)   [9] Sidney Richard Coleman (1937–2007)   [10] Frank Wilczek (born 1951), Nobel Prize 2004   [11] H. David Politzer (born 1949), Nobel Prize 2004

the Bronx High School of Science and had obtained his B.Sc. from the University of Michigan in 1969. Politzer tells [9] how one day he phoned Coleman in Princeton to report that, according to his calculations, non-Abelian theories could be asymptotically free. He did not use these words but exactly what they express resulted from a minus sign he had found for a certain function. Coleman suggested that he check again, since Gross and Wilczek had found the sign to be plus. When Politzer called a week later to confirm his result, Coleman said 'he knew because the Princeton team had found a mistake, corrected it, and already submitted a paper to *Physical Review Letters*'. Politzer made haste, his own paper appeared next to the one from Princeton [10,11].

A non-Abelian theory can show asymptotic freedom because its gauge bosons carry charge, i.e., colour in the case of QCD. This fact can completely change the effect of vacuum polarization we discussed above, since not only the fermions but also the bosons carry charge. Instead of screening one can have *anti-screening*, i.e., the colour force between two quarks increases with distance, acting like a spring when it is extended. In QCD this is the case because the number of different quark flavours is small; otherwise the effect of the gluons would be counteracted by that of too many quarks.

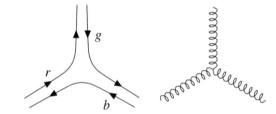

Interaction between gluons, shown as colour flow (left) and as Feynman diagram (right). It contributes decisively to the effect of anti-screening and to asymptotic freedom.

Thus, the strong coupling constant α_s, indeed, falls with q^2; in an energy region reached with modern particle accelerators its value is around 0.12. In this region perturbative calculations are possible and have been verified experimentally, as we shall see in Episode 94.

As q^2 decreases the value of α_s rises and very strongly binds quarks and gluons to one another. That, most probably, explains the *confinement* of quarks and gluons within hadrons, consistent with the fact that free quarks or gluons were never observed. Gluons, the gauge bosons of QCD, have no mass; still there is no conflict with experiment.

At low values of q^2 the method of perturbation series breaks down, since α_s becomes too large. The theory, QCD, still holds in that region; but we do not know how to perform calculations there. Or do we? Following pioneering work by Wilson[12] [12], computing techniques have been developed in which space and time are no longer treated as continuous but as a lattice of discrete points. There are good reasons to expect that in such a *lattice QCD* the formation of hadrons from quarks and the mass spectrum of hadrons can be obtained using powerful computer systems.

[1]   Greenberg, O. W., *Physical Review Letters*, **13** (1964) 598.

[2]   Han, M. Y. and Nambu, Y., *Physical Review*, **139** (1965) B 1006.

[3]   Pais, A. In Zichichi, A. (ed.): *Recent Developments in Particle Symmetries*, p. 406. Academic Press, New York, 1966.

[4]   Fritzsch, H., *Flucht aus Leipzig*. Piper, München, 1990.

[5]   Fritzsch, H. and Gell-Mann, M., *Current Algebra: Quarks and What Else?* In Jackson, J. D. and Roberts, A. (eds.): *Proceedings of the XVI International Conference of High Energy Physics, Chicago, 1972*. Vol. 2, p. 135. Also accessible as [6].

[6]   Fritzsch, H. and Gell-Mann, M. arXiv:hep-ph/0208010.

[7]   Fritzsch, H., Gell-Mann, M., and Leutwyler, H., *Physics Letters*, **47 B** (1973) 365.

[8]   Gell-Mann, M., *Physical Review*, **125** (1962) 1067.

[9]   Politzer, H. D., *Les Prix Nobel 2004*, p. 85. Nobel Foundation, Stockholm, 2005.

[10] Gross, D. J. and Wilczek, F., *Physical Review Letters*, **30** (1973) 1343.

[11] Politzer, H. D., *Physical Review Letters*, **30** (1973) 1346.

[12] Wilson, K. G., *Physical Review*, **D 10** (1974) 2445.

---

[12]   Kenneth Geddes Wilson (born 1936), Nobel Prize 1982

A Fourth Quark – Charm (1974)

At noon on 11 November 1974, in a seminar at the Stanford Linear Accelerator Center (SLAC), a discovery was announced which strongly influenced particle physics; for some time one even spoke of the *November Revolution*. Not only was the discovery important but it had been made independently by two groups, working on either coast of the United States, and not till that very morning had they heard of each other's success. These were the groups of Ting from MIT, experimenting with the proton synchrotron at Brookhaven, and a collaboration between Stanford and Berkeley and initiated by Richter, working at an electron–positron collider at Stanford. A meeting of a SLAC advisory committee, of which Ting was an external member, had been scheduled for the morning of 11 November. As Richter remembers [1], Ting said to him before the meeting: 'Burt, I have some interesting physics to tell you about.' His response was: 'Sam, I have some interesting physics to tell *you* about!' When they had told each other, the meeting was cut short and the noon-time seminar was announced.

Ting[1] was born in Ann Arbor, Michgan, while his parents – his father was professor of engineering, his mother professor of psychology – were visiting there. He grew up in China but returned to Michigan when he was twenty. He got degrees in both physics and mathematics and obtained his Ph.D. in 1962. One of his advisers was Perl, of whom we will hear more at the end of this episode. Having worked for two years at CERN in Geneva on a fellowship, Ting went to Columbia University, where he got interested in testing quantum electrodynamics by studying the production of electron–positron pairs in the collision of high-energy photons with nuclei for the special case that the directions of the electron and the positron formed a relatively large angle. That, incidentally had also been Richter's favourite subject in his early career. Ting accepted the offer to perform such an experiment at the Deutsches Elektronen-Synchrotron (DESY) in Hamburg. From then on, he remained fascinated with the pair production of electrons or muons. In his later experiments he would always see to it that such pairs would be measured with the highest precision attainable. In 1969 Ting became professor at MIT and in 1971, with his MIT group, he began the experiment at Brookhaven.

In his Nobel Lecture [2], Ting relates the history of that experiment. He wanted to search for the possible existence of heavy neutral vector mesons. Vector mesons have spin one and negative parity. Thus, if they are neutral, they have the quantum numbers of the photon and can decay into an e^+e^- pair. Three such mesons, $\rho^0, \omega^0, \varphi^0$, were known (Episode 87), the heaviest (the φ^0) having a mass of just over $1\,\mathrm{GeV}/c^2$. Ting proposed to use 28-GeV protons, provided by the Brookhaven Alternating-Gradient Synchrotron, for

Ting

---

[1] Samuel Chao Chung Ting (born 1936), Nobel Prize 1976

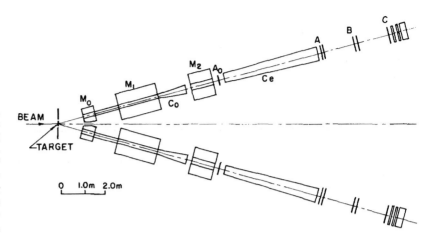

Ting's pair spectrometer. There are two identical arms, one for electrons, the other for positrons. The magnets M_0, M_1, M_2 bend the particle trajectories upwards. The multiwire proportional chambers A_0, A, B, C yield trajectory coordinates. Two Cherenkov counters C_0, C_e, and a lead-glass shower counter (the last element in each arm) make sure that only electrons (or positrons) are registered. Reprinted from [3], copyright 1975, with permission from Elsevier.

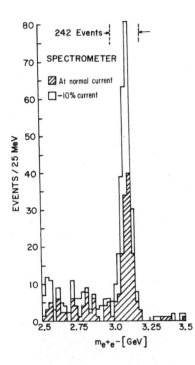

The J particle, observed by Ting's group as a sharp peak in the mass spectrum of e^+e^- pairs. A change in the magnet currents of the spectrometer, i.e., a change on particle trajectories, does not affect the peak position. Figure reprinted with permission from [4]. Copyright 1974 by the American Physical Society.

the production of possible heavier vector mesons but to identify them through their rare e^+e^- decay. This method would yield a clean signal of the new particles, provided electrons could be discriminated from hadrons with very high efficiency. His proposal was granted in May 1972. The group set up two spectrometers analysing electrons and positrons, respectively, which came from the beryllium target, bombarded by the accelerator protons. The momentum was determined by measuring the trajectory, bent in the field of three magnets.

That was achieved by *multiwire proportional chambers*. The proportional chamber with a single wire was invented by Rutherford and Geiger (Episode 13). In a multiwire chamber, a plane of many parallel wires is placed in a gas-filled room between two grounded electrodes. The wires carry positive high voltage and signals from all wires are registered separately and read out to a computer. A particle passing perpendicularly through the chambers induces a signal in one or a few wires; their position yields one coordinate on the trajectory. A second coordinate is provided by another chamber with wires oriented in a different way. The third coordinate is given by the position of the pair of chambers. As we shall see, such detection methods, which had been developed in several laboratories, were also used by the group at SLAC.

To distinguish electrons from hadrons the Cherenkov effect (Episode 56) was used and also the fact that only electrons give rise to electromagnetic showers in matter. Each spectrometer arm contained two Cherenkov counters, filled with hydrogen gas, which registered only electrons. Each arm was ended by a shower counter, consisting of blocks of lead glass in which electrons lost all their energy by forming showers; the resulting light signals were recorded by photomultipliers.

The apparatus functioned well with a proton beam early in the summer of 1974. The mass of a particle which decays into e^+ and e^- can be reconstructed from the momenta of these two particles. Each arm could register momenta in a certain range about a mean value defined by the magnet currents. That meant that for a given current setting a certain mass range was covered. Having begun in the range of 4 GeV to 5 GeV in which very few pairs were seen, the group moved to the region between 2.5 GeV and 4 GeV. Not only did they observe many clean pairs, but most were concentrated in a narrow mass interval around 3.1 GeV. The very narrow peak had a width of less than 5 MeV. A new particle had been found and one with an unexpectedly small width. Ting and his group decided to denote the particle by the capital letter J. (One reason, later given for this choice, was that unusual particles, like the intermediate W and Z bosons of the electroweak interaction, predicted but not observed at the time, were named by capital letters.) It was also decided that a number of detailed and thorough tests be done before publication. One was to somewhat change the magnet currents and thereby shift the particle trajectories to different regions of the chambers and counters. The peak stayed were it was. Early in November, Ting drafted a paper on the J particle. Then, on 11 November, he heard what had happened at SLAC.

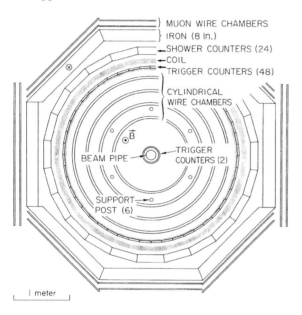

Cut, perpendicular to the beam pipe, through the Mark I detector at SPEAR. There is a large coil, producing a magnetic field **B** parallel to the beam. The magnetic flux is led back through an octagonal iron yoke near the outside. There is a set of four cylindrical multiwire proportional chambers to record the trajectories of charged particles. They are triggered by counters near the beam pipe and just inside the coil. Shower counters outside the coil register energy deposited by photons and electrons. Muons can traverse the iron and are detected in chambers on its outside. Figure reprinted with permission from [5]. Copyright 1975 by the American Physical Society.

While at Brookhaven the J particle had been produced via strong interaction and seen to decay into e^+e^-, at Stanford the reverse process had been observed: The same particle had been produced in e^+e^- collisions and detected by its strong decay into hadrons. It had been dubbed ψ. In the literature it is now referred to as the J/ψ.

We met Richter in Episode 83 as pioneer of particle colliders. After years of design work and efforts to get it funded, an electron–positron storage ring had finally been brought into operation in April 1972. It was called *SPEAR* for Stanford Positron Electron Accumulator Ring and could store and collide particles of up to 3.5 GeV energy. From the beginning Richter planned an ambitious experiment. The collision region was to be surrounded by a set of detectors which, themselves, were immersed in a magnetic field. Tracks of charged particles, bent by the magnetic field, could be reconstructed and the particle momentum vectors could thus be obtained. Complex events could be registered and analysed in a way that had been possible before only with bubble chambers. This goal was achieved in a collaborative effort; Richter had

Richter

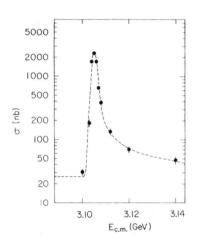

The ψ resonance curve as a function of the total e^+e^- energy. Shown are cross-section measurements for the production of more than two hadrons. Figure reprinted with permission from [7]. Copyright 1974 by the American Physical Society.

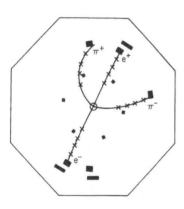

Event display showing four tracks appearing in the two-step decay

$$\psi' \to \psi + \pi^+ + \pi^-, \; \psi \to e^+ + e^-.$$

It is graphically similar to the letter ψ. Figure: SLAC.

asked three groups from Berkeley, led by Chinowsky, Goldhaber, and Trilling, and another group from SLAC, headed by Perl, to join in. The Stanford–Berkeley collaboration designed and built the *Mark I* detector which became the model for most later collider experiments. Several collaboration members had worked with bubble chambers before. That experience helped in the development of computer programs for track reconstruction and for the visual display of whole events.

When data taking began in 1973, the primary aim was to measure the cross section for hadron production as a function of the total e^+e^- energy in energy steps of 200 MeV. First results were presented by Richter at an international conference in London in July 1974. Goldhaber[2] tells what happened in the following months [6]. There was no pronounced energy dependence. However, two points, at 3.2 GeV and at 4.2 GeV, were a little too high for consistency with the others. Intermediate points were taken around these values, an irregularity at 3.1 GeV was found in October and a decision was taken in early November to remeasure that region before publication. On 9 November, the energy region around 3.1 GeV was scanned in 10-MeV steps and a jump up in cross section was seen at 3.12 GeV. Finally, on Sunday, 10 November, a scan in 2 MeV steps was done and the cross section was seen to grow by a factor of 100 within three such steps. The resonance curve was not symmetric; it had a long extension towards higher energy. This feature, the *radiative tail*, is caused by the simultaneous production of the J/ψ and a photon, the latter taking some of the energy. It was this effect that had caused the early high points. The width of the resonance, as observed, was that of the energy spread of the collider. It was later found that the natural width of the J/ψ is only 91 keV. For a short while, the group referred to the particle as SP(3105), where SP stood for SPEAR and the number for its mass in MeV. Then Goldhaber found out that the Greek letter ψ, its Latin transcription beginning with p and s, was not yet used for a particle. A later energy recalibration of SPEAR made the full name $\psi(3095)$.

Still on 11 November, physicists at the Italian National Laboratory in Frascati were informed of the discoveries. They had constructed ADONE, an e^+e^- collider designed for a maximum energy of 3 GeV. Two days later they began to look for the resonance signal, running the collider magnets at their limit, and found it on 15 November. Publications by the three groups, MIT [4], Stanford–Berkeley [7], and Frascati [8] appeared together in the first December issue of *Physical Review Letters*. Already the next issue, a week later, carried another letter from the Stanford–Berkeley group [9]. It announced the existence of a second narrow resonance. It was called $\psi(3695)$ and $\psi(3684)$ after recalibration. Usually it was referred to simply as ψ'. One decay mode of that particle would lead to particularly pretty events in the Mark I detector, which provided extra justification for the choice of the name ψ (see figure of event display).

For a while the nature of the new particles was debated. But soon it became clear that they were bound states ($c\bar{c}$) of a fourth quark c and its antiparticle $\bar{c}$. In the literature the possible existence of an additional quark had been discussed, but that conjecture had not influenced experimenters.

---

[2] Gerson Goldhaber (born 1924)

Arguments for four rather than three quarks had been given by several authors. The name *charm* for the fourth quark appears already in 1964, in a paper by Bjorken[3] and Glashow [10]. In 1970 Glashow, Iliopoulos[4], and Maiani[5] found that they could explain several serious discrepancies between theory and experiment, concerning the weak interaction of hadrons, if there was a fourth quark [11]. The charm quark c had the charge $2/3\,e$, where e is the elementary charge. It had its own flavour quantum number C, also called charm, which was analogous to strangeness. Quarks now came in two doublets, (u, d) and (c, s), the first quark in each doublet having charge $2/3\,e$, the second charge $-1/3\,e$. In this way symmetry was reached with the two doublets of leptons, the electron and the electron neutrino (e^-, ν_e) and the muon and its neutrino (μ^-, ν_μ). For the interaction with the W bosons of the weak interaction, the d and the s quark states appeared as mixed states. (We discussed the *mixing* of particle states already in Episodes 88 and 90.) The mixing angle is called *Cabibbo angle* θ_C. It had been introduced by Cabibbo[6] already in 1963 [12], before the advent of the quark hypothesis; Cabibbo had discussed hadronic charged currents similar to the leptonic charged currents we mentioned in Episode 90. By exactly defining the interactions of the four quarks with four bosons γ, Z^0, W^+, W^- Glashow, Iliopoulos, and Maiani completed the theory of electroweak interactions. The relative importance of basic Feynman diagrams involving the W boson is given by the cosine and the sine of the Cabibbo angle, respectively. Since $\theta_C \approx 13°$, the former are called 'Cabibbo favoured', the latter 'Cabibbo suppressed'. The discrepancies we mentioned were removed because in the new theory with four quarks the contributions from some diagrams cancelled. That welcome effect is referred to in the literature as *GIM mechanism*, the letters standing for the authors of paper [11].

In spite of this success the charm model was hardly perceived by experimentalists, possibly because it predicted the existence of many more hadrons than had been observed so far. With the additional charm quantum number the triangle of quarks in the I_3, S plane became a tetrahedron in I_3, S, C space, and the hexagonal or triangular hadron multiplets became polyhedrons with many free places, of which we show only the one for vector mesons. In April 1974, Glashow lectured in Boston on the 4th International Conference on Experimental Meson Spectroscopy. His appeal to the experimentalists present was clad as prediction for the next meeting, scheduled for 1976:

> There are just three possibilities:
> 1. Charm is not found, and I eat my hat.
> 2. Charm is found by hadron spectroscopy and we celebrate.
> 3. Charm is found by outlanders, and you eat your hats!

A kind of handbook, entitled *Search for Charm*, was compiled in the first half of 1974 and distributed to research groups as a 'preprint' in August; but it appeared in print only in 1975, appended with a long *note added in proof* about the November discoveries [13]. It contained many properties of the expected new particles, in particular, the distinction between one or several narrow states at lower energy which were to be followed by broader resonances of higher energy.

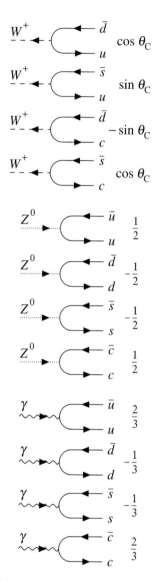

Basic diagrams for the interaction of W, Z, and γ bosons with quarks. The expressions or numbers on the right indicate the relative weight of each diagram within its set. The diagrams complement those given for leptons on page 428.

---

[3] James Daniel Bjorken (born 1934)   [4] John Iliopoulos (born 1940)   [5] Luciano Maiani (born 1941)   [6] Nicola Cabibbo (born 1935)

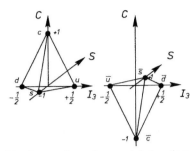

The four quarks u, d, s, c and their antiparticles as points in a three-dimensional space spanned by axes representing the quantum numbers I_3, S, C of isospin, strangeness, and charm, respectively.

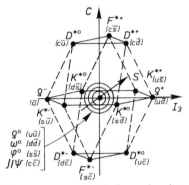

Vector mesons composed of a quark and an antiquark. The plane $C = 0$ contains only one additional meson compared to the original three-quark model, the J/ψ. There are two planes ($C = \pm 1$) which contain three mesons each with open charm. (The states ρ^0 and ω^0 are superpositions, see page 410.)

The J/ψ and the ψ' were identified as these narrow states. To let the J/ψ (or ψ') decay, the c and $\bar{c}$, forming it, have to annihilate. In the decay lighter quarks and, finally, lighter hadrons, are formed. This type of decay takes relatively long. By the uncertainty relation the width in energy (and mass) of the J/ψ is small. The J/ψ and ψ' have the net charm quantum number $C = 0$ and are therefore said to have *hidden charm*. Broad structures appear at energies of more than twice the c-quark mass, when the decay into heavier mesons, containing a c quark and said to have *open charm*, is possible.

The c quark is considerably heavier than the three quarks known previously and the binding energy of the c and $\bar{c}$ in the J/ψ is small in comparison. This situation allowed the construction of a simple quantum-mechanical model of $(c\bar{c})$ states. The two quarks were taken as ordinary particles and the force between them was assumed to increase in a simple way with the distance between c and $\bar{c}$. In the model the two quarks form an 'atom', called *charmonium*, and it is a relatively simple exercise to compute the energy levels of this atom. The charmonium spectrum contains states with various angular momenta, $J = 0, 1, 2, \ldots$. The J/ψ and the ψ' are the two lowest states of $J = 1$; states with $J = 0$ and $J = 2$ were observed in later experiments.

The success of the charmonium model was taken as convincing evidence for the reality of quarks. It was realized that high-energy electron–positron colliders opened a particularly clean way to study the strong interaction: The primary result of an e^+e^- annihilation is simply a quark–antiquark pair of high energy. Its strong interaction can be studied to test quantum chromodynamics. We shall learn in Episode 94 how that was done and how clear experimental evidence was obtained for the existence of the gluon, the carrier of the strong interaction.

With the discovery of the charm quark two doublets of leptons and two doublets of quarks were firmly established. The electronic leptons (e^-, ν_e) together with the light quarks (u, d) became to be called the *first generation* of fundamental fermions. The matter surrounding us, including ourselves, is

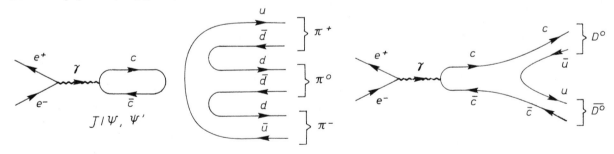

Hidden charm (left): Production and decay of the J/ψ (or ψ') meson into lighter mesons. Open charm (right): If produced with enough energy, a $(c\bar{c})$ state can give rise to the creation of one or several quark–antiquark pairs. The result is a final state with mesons containing a charm (or anti-charm) quark, in this example a D^0 and a $\bar{D}^0$.

composed of it. The *second generation* consists of the muonic leptons (μ^-, ν_μ) and the heavier quarks (c, s). Bound states containing a second-generation fermion are short-lived, and it takes accelerator or cosmic-ray energies to create them. Are there more generations? The answer is yes. Perl[7] and his group who analysed more data taken with the Mark I detector at SPEAR [14], already in 1975, found a new charged heavy lepton, partner to the electron and

---

[7] Martin Lewis Perl (born 1927), Nobel Prize 1995

the muon. The particle was named τ; it was produced in pairs in e^+e^- collisions, and it decayed into three particles, its own neutrino ν_τ and a pair of lighter leptons, either a muon and its neutrino or an electron and its neutrino. The 'signature' of τ-pair production was the observation of a pair of 'unlike charged leptons' (e^-, μ^+ or e^+, μ^-) and of 'missing energy' and 'missing momentum' in the detector, carried away by the neutrinos.

Perl

Now that there were three lepton generations, there had to be a third generation of quarks, too. The fifth quark, called b for *bottom* (sometimes also *beauty*) was observed in 1977 by a group led by Lederman [15] at the Fermi National Accelerator Laboratory, better known as *Fermilab* in Batavia, Illinois, near Chicago. The experiment had similarity to Ting's. It used a two-arm spectrometer which, this time, was constructed for muon pairs. The mass spectrum of such pairs, produced in collisions of protons from Fermilab's 400-GeV synchrotron with target nuclei, showed two maxima near 9.5 GeV and 10 GeV. These resonances are $(b\bar{b})$ bound states; they were called Υ and Υ'. The sixths quark, the *top t*, was found only in 1995 [16,17]. It has an extraordinarily large mass and could be produced only after the Fermilab synchrotron had been converted into a proton–antiproton collider.

Leptons		Quarks			
e	$0.511\,\mathrm{MeV}/c^2$	u	$\sim 2\,\mathrm{MeV}/c^2$	d	$\sim 5\,\mathrm{MeV}/c^2$
μ	$106\,\mathrm{MeV}/c^2$	c	$\sim 1.25\,\mathrm{GeV}/c^2$	s	$\sim 95\,\mathrm{MeV}/c^2$
τ	$1.777\,\mathrm{GeV}/c^2$	t	$\sim 174\,\mathrm{GeV}/c^2$	b	$\sim 4.2\,\mathrm{GeV}/c^2$

Masses of charged leptons and quarks.

The question, of course, arises: How many generations of leptons and quarks are there? We shall present an answer in Episode 98.

[1] Richter, B., *SLAC Beam Line*, **November** (1976) 3.

[2] Ting, S. C. C., *Nobel Lectures, Physics 1971–1980*, p. 316. World Scientific, Singapore, 1992.

[3] Aubert, J. J. et al., *Nuclear Physics*, **B 89** (1975) 1.

[4] Aubert, J. J. et al., *Physical Review Letters*, **33** (1974) 1404.

[5] Augustin, J. E. et al., *Physical Review Letters*, **34** (1975) 233.

[6] Goldhaber, G., *From Psi to Charmed Mesons*. In Hoddeson, L., Brown, L., Riordan, M., and Dresden, M. (eds.): *The Rise of the Standard Model*, p. 57. Cambridge University Press, Cambridge, 1997.

[7] Augustin, J. E. et al., *Physical Review Letters*, **33** (1974) 1406.

[8] Bacci, C. et al., *Physical Review Letters*, **33** (1974) 1408.

[9] Abrams, G. S. et al., *Physical Review Letters*, **33** (1974) 1453.

[10] Bjørken, B. J. and Glashow, S. L., *Physics Letters*, **11** (1964) 255.

[11] Glashow, S. L., Iliopoulos, J., and Maiani, L., *Physical Review D*, **2** (1970) 1285.

[12] Cabibbo, N., *Physical Review Letters*, **10** (1963) 531.

[13] Gaillard, M. K., Lee, B. W., and Rosner, J. L., *Reviews of Modern Physics*, **47** (1975) 277.

[14] Perl, M. L. et al., *Physical Review Letters*, **35** (1975) 1489.

[15] Herb, S. W. et al., *Physical Review Letters*, **39** (1977) 252.

[16] Abe, F. et al., *Physical Review Letters*, **74** (1995) 2632.

[17] Abachi, S. et al., *Physical Review Letters*, **74** (1995) 2626.

The Discovery of the Gluon (1979)

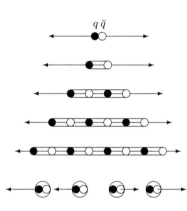

$q\,\bar{q}$

Hadronization. When a quark q and an antiquark $\bar{q}$ fly apart, in the colour field between them new $q\bar{q}$ pairs are created, forming jets of hadrons.

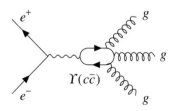

e^+

g

g

$\Upsilon(c\bar{c})$

g

e^-

Diagram showing the production of the Υ meson and its decay into three gluons.

---

[1] Gail Gulledge Hanson (born 1947)

In Episode 14 we witnessed Perrin uncover 'la réalité moléculaire', the reality of atoms and molecules, in a time when these objects were considered still far from real by many. By experimental analysis he showed that a *visible phenomenon*, the Brownian motion of tiny spheres, was caused by the movement of molecules, although the latter remained invisible. The experiments discussed now prove the 'reality of quarks and gluons', although, as mentioned in Episode 92, these particles are confined inside hadrons and do not exist as separate entities. Perrin observed his little spheres instead of the invisible molecules; particle physicists study *jets* of hadrons instead of the confined quarks and gluons. The first to do so was Hanson[1] as a member of the Stanford–Berkeley group which had discovered the charm quark with the Mark I detector at the SPEAR collider (Episode 93). In data taken with the highest total e^+e^- energy reached by SPEAR, 7.4 GeV, she observed that the hadrons produced in the electron–positron annihilation appeared as two bundles of particles, called jets, which flew into opposite directions [1]. These directions were oriented in space, with respect to the line of flight of the incident electron and positron, as could be expected of a pair of muons from the reaction $e^+ + e^- \rightarrow \mu^+ + \mu^-$. Since quarks, like muons, are spin-1/2 particles, the same angular distribution was also expected for quarks and antiquarks from the reaction $e^+ + e^- \rightarrow q + \bar{q}$.

A simple model of *hadronization* explains why jets carry information about quarks. The quark q and the antiquark $\bar{q}$ have very high momenta. While they fly apart, a colour field, due to the confining strong interaction, develops between them (Episode 92); part of the kinetic energy of the $q\bar{q}$ pair is converted to potential energy of the field. Once that is high enough, it can create additional quarks and antiquarks which build up fields between themselves, etc. Finally, neighbouring quarks and antiquarks regroup to form hadrons and these appear as jets. Jets become pronounced if they have high enough energy. Their total momentum then gets near to the original quark momentum and the momentum components of individual hadrons transverse to the jet direction remain small.

In a similar way the model suggested that a gluon of high energy causes a directed region with a strong colour field and a jet of hadrons is emitted in that direction. The existence of the gluon could therefore be established by observing jets in reactions in which gluons of sufficiently high energy appeared. Quantum chromodynamics predicted, in particular, two such reactions [2]:

1. The decay of the Υ meson into three gluons,

$$\Upsilon \rightarrow g + g + g \quad .$$

446

Characterization of Jet Events

We consider the momentum vectors $\mathbf{p}_i$ of all N final-state particles in the centre-of-mass system of a collision (which, in the collider experiments discussed here, coincides with the laboratory system).

- **Two-Jet Events** *Jet Axis.* Although apparently obvious, the jet axis can be defined in different ways, which need not yield the same result. The *principal axis* [3] or *thrust axis* is the one for which the sum of the (absolute) momentum components, parallel to it, is a maximum. For the *sphericity axis* [4] the sum of the squared momentum components perpendicular to it is a minimum.

  Jet Slenderness. The principal axis is found by trying all possible groupings of the particle momenta in two classes C_1, C_2. For each grouping the momentum vectors in each class are summed up to $\mathbf{P}(C_1)$ and $\mathbf{P}(C_2)$, respectively. Their moduli are added and that grouping C_1^*, C_2^* is chosen for which the sum of moduli is largest. The momenta $\mathbf{P}(C_1^*), \mathbf{P}(C_2^*)$ have the direction of the jet axis. Division of the maximum, just described, by the sum of all momentum moduli yields the quantity *thrust* [5],

$$T = \{|\mathbf{P}(C_1^*)| + |\mathbf{P}(C_2^*)|\} \big/ \sum_{i=1}^{N} |\mathbf{p}_i| \quad .$$

  It varies between $T = 1$ for perfectly aligned two-jet events and $T = 1/2$ for spherical events. Another measure is *sphericity* [4],

$$S = (3/2) \sum_{i=1}^{N} \mathbf{p}_{i\perp}^2 \big/ 2 \sum_{i=1}^{N} \mathbf{p}_i^2 \quad ,$$

  where the $\mathbf{p}_{i\perp}$ contain only components perpendicular to the sphericity axis. Perfect two-jet events have $S = 0$, spherical events $S = 1$.

- **Three-Jet Events** Particle momenta are now grouped in three classes. For one set of classes C_1^*, C_2^*, C_3^* the sum $|\mathbf{P}(C_1^*)| + |\mathbf{P}(C_2^*)| + |\mathbf{P}(C_3^*)|$ is largest; the directions of $\mathbf{P}(C_1^*), \mathbf{P}(C_2^*), \mathbf{P}(C_3^*)$ define three *jet directions*. The quantity *triplicity* [6],

$$T_3 = \{|\mathbf{P}(C_1^*)| + |\mathbf{P}(C_2^*)| + |\mathbf{P}(C_3^*)|\} \big/ \sum_{i=1}^{N} |\mathbf{p}_i| \quad ,$$

  varies between $T_3 = 1$ for events with three perfectly slim jets and $T_3 = 0.65$ for completely spherical events. Events with three well-separated slim jets have triplicity $T_3 \approx 1$ but thrust T considerably smaller than 1. Another quantity characterizing three-jet events, *tri-jettiness*, was developed by extending the concept of sphericity [7].

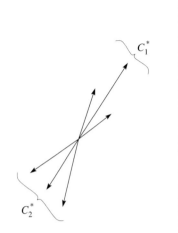

Grouping particle momenta into two classes C_1^*, C_2^* defining two jets.

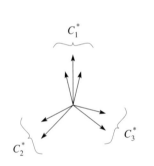

Grouping particle momenta into three classes C_1^*, C_2^*, C_3^* defining three jets.

2. The emission of *gluon bremsstrahlung* by one member of a $q\bar{q}$ pair, produced in an e^+e^- collision,

$$e^+ + e^- \rightarrow q + \bar{q} + g \quad .$$

The term is taken from the terminology used for X rays (page 81). Here a quark is decelerated in the colour field of the other one and radiates a gluon.

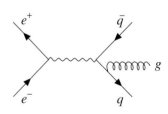

Diagram showing the production of a quark–antiquark pair and subsequent gluon bremsstrahlung.

By the *November Revolution*, the discovery of charm, the interest of particle physicists became focussed on e^+e^- colliders. The planning for two much larger colliders, *PETRA* at DESY in Hamburg and *PEP* at SLAC in Stanford, began immediately. PETRA began to operate in July 1978 nearly two years before PEP; that is why the gluon was first observed in Hamburg.

The history of DESY goes back to 1955 when Jentschke was offered a physics chair at the University of Hamburg. Jentschke[2], a Viennese, studied in his home town and, already in 1939, published an important paper on the fission products of uranium. After the war he was invited to the United States and became director of the cyclotron laboratory of the University of Illinois in Urbana. In Germany he won support for his idea to establish a national accelerator laboratory. It was decided to construct an electron synchrotron in order to be complementary to the physics program of CERN which was based on a large proton machine. The 7.5-GeV Deutsches Elektronen Synchrotron DESY began operation in 1964. It also gave the laboratory its name. An e^+e^- collider named *DORIS* with a total collision energy of up to 10.2 GeV was completed in 1973. When in 1970 Jentschke became Director General at CERN, he was succeeded by Paul who accepted to lead DESY for an interim period. Paul convinced Voss[3], an outstanding accelerator scientist, to return to DESY from the United States as leader of the accelerator department and Schopper[4] from the Technical University of Karlsruhe to become director of DESY. Still in the month of the November Revolution, Schopper proposed the construction of PETRA to the Federal Government in Bonn which, fortunately, had recently announced an economic reflation program and unexpectedly soon granted the necessary funds. PETRA was designed and built by Voss and his team and first produced collisions in July 1978. The discovery of the gluon is the result of the excellent work of DESY as a whole and of four international collaborations working in friendly competition. Only to add personal flavour to this episode do I, in the following, sometimes mention my own involvement.

Voss

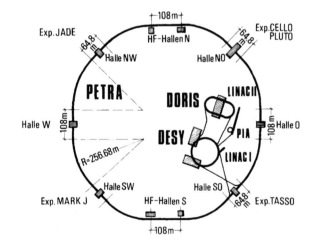

Schematic map of accelerators and experiments at DESY in 1979. Figure: DESY.

Before PETRA the largest e^+e^- collider was DORIS. Directly after the discovery of charm it was used successfully in the spectroscopy of charmonium. When the Υ meson, a bound state $(b\bar{b})$ of the b quark and its antiquark, was found in 1977 (see end of Episode 93), DORIS was the only collider with

---

[2] Willibald Jentschke (1911–2002)   [3] Gustav-Adolf Voss (born 1929)   [4] Herwig Schopper (born 1924)

which it could be produced. My group from the University of Siegen was a member of a collaboration which ran the *PLUTO* detector at DORIS. PLUTO was similar in design to MARK I, its smaller size was compensated by its higher magnetic field, produced by a superconducting coil.

The collimation of hadrons and the name *jets* was known from cosmic-ray experiments since the 1950s. In the centre-of-mass system of a hadron–hadron collision at high energy, particles are emitted in two back-to-back jets taking approximately the directions of the incident hadrons. When I worked with bubble chambers at CERN in 1964, I was the most junior member of a small group which developed a method to define and measure the axis of such a two-jet event [3]. We called it the *principal axis*. My co-authors were Peyrou[5], the leader of the CERN Track Chamber Division, as well as Sosnowski[6] and Wróblewski[7], two research fellows and good friends from Warsaw. After the appearance of jets in e^+e^- collisions the method was re-invented by Farhi and extended to yield not only the axis but also a measure for the slenderness of two-jet events, called *thrust* [5]. When events with three jets were predicted in the framework of QCD, my colleague Dahmen[8] at Siegen University and I characterized those events by a quantity we called *triplicity* [6]. Only a little later, Sau Lan Wu and Zobernig presented a somewhat different measure, *tri-jettiness* [7]. For more details on these methods, see Box.

At the end of June 1979, I spoke at a conference in Geneva about an analysis of PLUTO data on the Υ decay by my graduate student Meyer[9] and myself [8]. We had found that the data, analysed in terms of thrust and triplicity, were better described by a model with three jets than by a structureless ('phase-space') model and much better than by a two-jet model, which did fit the data well below the Υ resonance. However, the jet structure was not clearly pronounced and that did not surprise us, since the average energy of each of the three jets was only 3 GeV, the value at which it had barely been observable by Hanson [1] for two jets.

Only two months later, at the end of August 1979, another international conference was held at Fermilab. In the conference proceedings one finds reports on the observation of three-jet events of considerably higher energy by four collaborations working with the PETRA collider [9–12]. PETRA had four interaction regions; each contained a large detector, built and run by an international collaboration. One was PLUTO, which had been upgraded and moved from DORIS to PETRA. The other three were newly designed. TASSO was a detector in a large solenoid magnet with two external arms specially instrumented for particle identification, hence the name which stood for Two Arm Spectrometer Solenoid. JADE was another solenoid containing high-resolution track chambers. The detector of the Mark-J group, led by Ting, was optimized for the precision measurement of electrons and muons. It did detect hadrons but had no magnetic field in the central region. At a later time, PLUTO was replaced by CELLO. The latter, like PLUTO, had a superconducting magnet but was much larger. Except for PLUTO the track detectors of all experiments were *drift chambers*. This type of detector, pioneered by

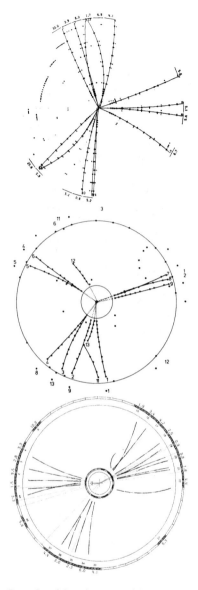

Examples of three-jet events of the type

$$e^+ + e^- \rightarrow q + \bar{q} + g$$

as observed at PETRA in the detectors TASSO (top), PLUTO (middle), and JADE (bottom). The arcs are trajectories of charged particles. Symbols in the outer regions indicate energies of neutral particles. Figures: DESY.

---

[5] Charles Peyrou (1918–2003)   [6] Ryszard Sosnowski (born 1932)   [7] Andrzej Kajetan Wróblewski (born 1933)   [8] Hans Dieter Dahmen (born 1936)   [9] Hans Jürgen Meyer (born 1946)

[1]   Hanson, G. et al., *Physical Review Letters*, **35** (1975) 1609.

[2]   de Rujúla, A., Ellis, J., Floratos, E. G., and Gaillard, M. K., *Nuclear Physics B*, **138** (1978) 387.

[3]   Brandt, S., Peyrou, C., Sosnowski, R., and Wróblewski, A., *Physics Letters*, **12** (1964) 57.

[4]   Bjorken, J. D. and Brodsky, S. J., *Physical Review D*, **1** (1970) 1416.

[5]   Farhi, E., *Physical Review Letters*, **39** (1977) 1587.

[6]   Brandt, S. and Dahmen, H. D., *Zeitschrift für Physik C*, **1** (1979) 61.

[7]   Wu, S. L. and Zobernig, G., *Zeitschrift für Physik C*, **2** (1979) 107.

[8]   Brandt, S., *Experimental Search for Υ Decay into 3 Gluons*. In *Intern. Conf. on High-Energy Physics*, p. 338. CERN, Geneva, 1979.

[9]   Mark-J: Newman, H. In [18], p. 3.

[10] PLUTO: Berger, C. In [18], p. 19.

[11] TASSO: Wolf, G. In [18], p. 34.

[12] JADE: Orito, S. In [18], p. 52.

[13] Wu, S. L., *Physics Reports*, **107** (1984) 60.

[14] TASSO Collab.: Brandelik, R. et al., *Physics Letters*, **86 B** (1979) 243.

[15] Mark-J Collab.: Barber, D. P. et al., *Physical Review Letters*, **43** (1979) 830.

[16] PLUTO Collab.: Berger, C. et al., *Physics Letters*, **86 B** (1979) 418.

[17] JADE Collab.: Bartel, W. et al., *Physics Letters*, **91 B** (1980) 142.

[18] Kirk, T. B. W. and Abarbanel, H. D. I. (eds.), *Proceedings of the 1979 International Symposium on Lepton and Photon Interactions*. Fermilab, Batavia, 1979.

Charpak[10] at CERN, is similar to the multiwire proportional chamber (page 440) but has a higher spatial resolution. (Electrons left by an ionizing particle along its track take some time to drift to the wire. That time is measured and converted into the distance between track and wire.) For a detailed description of all PETRA experiments and their results in the first five years see [13].

The energy of an e^+e^- collider is limited by synchrotron radiation. The energy radiated by electrons and positrons has to be resupplied by radio-frequency cavities distributed around the machine. By installing more radio-frequency power the total energy of PETRA was raised step by step. By the time of the Fermilab conference first data were available for energies between 13 GeV and 32 GeV. That allowed to study the appearance of three-jet events as a function of energy. The first collaboration to submit a full paper to a journal was TASSO [14], followed only two days later by Mark-J [15], and, another two weeks later, by PLUTO [16]. JADE [17], unfortunately, needed a few more months. After a mishap, detector repairs and more running time were needed. All groups performed similar analyses, differing in details. At low energy the events consisted of two symmetric jets. With increasing energy in part of the events one of the jets extended transversally, it became 'fatter' than the other. As the energy increased further, it became clear that this extension was confined to a plane; the events became planar and, finally, the fat jet resolved into two. The event plane was spanned by the momentum vectors of three jets.

Three-jet structure of a high-energy event in PLUTO. Shown are the momentum vectors projected onto the triplicity plane (top left), onto a perpendicular plane normal to the fastest jet (top right), and onto a plane containing the direction of the fastest jet (bottom). Solid and dotted lines correspond to charged and neutral particles, respectively. The directions of the jet axes are indicated as fat bars near the margins of the figures. Reprinted from [16], copyright 1979, with permission from Elsevier.

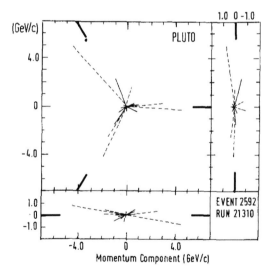

Together with Meyer and Barreiro[11] I analysed the PLUTO data in terms of thrust T and triplicity T_3. Genuine three-jet events have large triplicity but small thrust. We took $T_3 > 0.9, T < 0.8$. At high energy (27 GeV to 32 GeV) the number of events observed in this region could only be explained as observation of gluon bremsstrahlung, $e^+ + e^- \rightarrow q + \bar{q} + g$, the gluon giving rise to its proper jet [16]. Since all four PETRA collaborations had clearly observed that reaction, the existence of the gluon was firmly established.

---

[10] Georges Charpak (born 1924), Nobel Prize 1992   [11] Fernando Barreiro (born 1949)

The Quantum Hall Effect (1980)

Superconductivity (Episode 18) and superfluidity (Episode 58), both discovered unexpectedly, were later shown to be quantum phenomena on a macroscopic scale, not on the scale of atoms, nuclei, or particles. Another such macroscopic phenomenon, the *quantum Hall effect*, was observed in the small hours of 5 February 1980, by von Klitzing. It, too, came as a complete surprise.

Von Klitzing[1] was born in Schroda (now Środa, Poland). His family, belonging to the many refugees caused by the Second World War, settled in Lower Saxony. Here von Klitzing attended primary school and the Gymnasium. He studied physics at the Technical University in Brunswick, obtaining his diploma in 1969, and moved to Würzburg for graduate studies. There, in 1972, he got his Ph.D. with a dissertation, entitled *Galvanomagnetic Properties of Tellurium in Strong Magnetic Fields*. Von Klitzing became Privatdozent at the University of Würzburg in 1978 and stayed there until after the discovery of the quantum Hall effect. In 1980 he became professor at the Technical University in Munich and, in 1985, director at the Max-Planck Institute for Solid-State Research in Stuttgart.

In the vocabulary of von Klitzing's Ph.D. thesis, his discovery of 1980 could be called 'galvanomagnetic properties of a two-dimensional electron gas in strong magnetic fields'. This special electron gas, as von Klitzing used it, exists in a carefully prepared metal–oxide–silicon field-effect transistor or *MOSFET*. This type of transistor is the workhorse of modern electronics, because very many can be placed on a small flat silicon crystal, a 'chip', which needs to be structured on one side only. In one kind of MOSFET the crystal itself, the *substrate*, is acceptor-doped, i.e., *p*-type. There are two *n*-type islands, called *source* and *drain*, connected to leads. Except for these connections the crystal is covered with insulating silicon oxide. Over the thin oxide region between source and drain a metallic *gate* electrode is placed. Normally, there is no current between source and drain because the electrons, charge carriers in those *n*-type islands, cannot pass through the *p*-type region. However, if a voltage is applied to the gate, which is positive with respect to the substrate, then a thin layer of the crystal, directly below the oxide, becomes conducting for electrons. Through the voltage the conduction band of the substrate is strongly lowered in energy; it forms a triangular well at the interface to the oxide in which a two-dimensional electron gas (see page 406) can exist at very low temperatures. Two colleagues, Dorda[2] from the Siemens Research Laboratories in Munich and Pepper[3] from the Cavendish Laboratory in Cambridge, prepared special MOSFET devices of high quality with additional electrodes, with which von Klitzing studied the Hall effect in the two-dimensional electron

von Klitzing

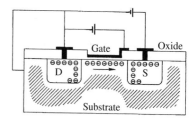

Section through a MOSFET with *p*-type substrate and *n*-type source (S) and drain (D). For positive gate voltage, a thin layer below the oxide cover becomes conducting for electrons.

---

[1] Klaus von Klitzing (born 1943), Nobel Prize 1985   [2] Gerhard Dorda (born 1932)   [3] (Sir) Michael Pepper (born 1942)

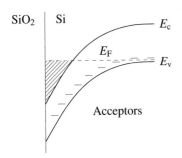

Lowering of the band edges E_c and E_v at the silicon-oxide interface by a positive gate voltage. A two-dimensional electron gas can form in the hatched triangular region, see figure below.

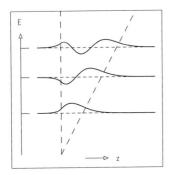

The three lowest eigenstates of an electron in a triangular potential well. The long-dashed broken line indicates the potential as function of the position z. The horizontal bars at the energy scale are the energy eigenvalues. The wave functions for the different eigenvalues are shown as continuous lines. At very low temperatures only the lowest state is occupied. An electron in this state is confined in z but free to move in the xy plane.

The integer quantum Hall effect as first observed by von Klitzing. At a constant field ($B = 18$ T) and with the current I between source and drain held fixed at $1\,\mu$A, the Hall voltage U_H and the longitudinal voltage U (denoted U_{pp} in the figure and measured between the potential probes indicated in the inset) were recorded as a function of the gate voltage V_g. As V_g, and with it the electron density n_e increases, more Landau levels are filled. The first three plateaus of U_H in this figure correspond to the filling factors $\nu = 2, 3, 4$. Figure reprinted with permission from [1]. Copyright 1980 by the American Physical Society.

gas.

If a current I passes through a long, flat bar and if the bar is in a magnetic field B perpendicular to its large, flat surface, then a voltage U_H appears across the bar, perpendicular to I and B. The *Hall voltage* U_H is easily computed by considering the electric and magnetic forces on the charge carriers in the bar. Writing $U_H = R_H I$, the *Hall resistance* R_H is found to be $R_H = B/(en_e)$, i.e., proportional to the magnetic field B and inversely proportional to the density n_e of the carriers, each having charge e.

Von Klitzing decided to measure the Hall effect under the extreme conditions of high magnetic field and low temperature that could be obtained in the High-Field Magnet Laboratory, operated in Grenoble jointly by the German Max-Planck Society and the French research organization CNRS. The discovery of 5 February 1980 was made at a temperature of 1.5 Kelvin with a field of 18 Tesla [1].

As he slowly varied the gate voltage V_g of his MOSFET but kept the current I fixed, von Klitzing recorded the longitudinal voltage drop U (designated U_{pp} in the figure) and the Hall voltage U_H as a function of V_g. Increasing V_g meant increasing the carrier density n_e. Oscillations of U with varying n_e were already known. Totally new was the behaviour of U_H and, with it, of $R_H = U_H/I$. Rather than to fall smoothly like $1/n_e$, these quantities displayed pronounced 'plateaus', regions in which they stayed constant. Doing a little calculation on the recorder paper, carrying the curves $U_H = U_H(V_g)$ and $U_{pp} = U_{pp}(V_g)$ (reproduced in [2]), von Klitzing identified these special values of the Hall resistance as given directly by two fundamental constants,

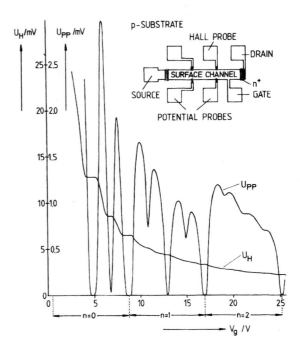

The Hall Effects – Basic Concepts

Definitions: An external circuit ensures that a constant current I flows in x direction through a so-called Hall probe, which is placed in the xy plane and exposed to a magnetic field B in z direction. Two voltages are measured, the voltage drop U over the distance L_x and the *Hall voltage* U_H over L_y. We can write

$$U = RI \quad , \qquad U_H = R_H I \quad . \tag{1}$$

The first formula is Ohm's law with R being the conventional Ohmian resistance. The second formula with the Hall resistance R_H is written in analogy to the first. We introduce the electric fields E (in x) and E_H (in y) and, taking advantage of the flatness of the probe, the current density j (per unit length), and the resistivity ρ,

$$E = U/L_x \quad , \qquad E_H = U_H/L_y \quad , \qquad j = I/L_y \quad , \qquad \rho = RL_y/L_x \quad . \tag{2}$$

Note that both R and ρ have the same dimension and can be expressed in Ohms. Substituting (2) in (1), one obtains

$$E = \rho j \quad , \qquad E_H = \rho_H j \quad , \qquad \text{where} \quad \rho_H = R_H \quad . \tag{3}$$

Actually, we should have written more complicated equations, in which the electric field and the current density are vectors and the resistivity is a matrix. Therefore, equations for j cannot be simply derived from (3), see, for instance, [2]. They read (for $\rho \ll \rho_H$):

$$j = \sigma E \quad , \qquad j_H = \sigma_H E_H \quad , \qquad \sigma = \rho/\rho_H^2 \quad , \qquad \sigma_H = 1/\rho_H \quad . \tag{4}$$

Note that $\sigma = 0$ for $\rho = 0$ (again because of the underlying matrix equation).
Conventional Hall Effect: Classical and quantum-mechanical calculations yield

$$R_H = R_H^0 = \rho_H^0 = \frac{B}{qn_e} \quad , \qquad \sigma_H^0 = \frac{qn_e}{B} \quad , \tag{5}$$

where q is the charge of a carrier ($q = -e$ for an electron and $q = e$ for a hole; e is the elementary charge). B is the magnetic field and n_e the carrier density, i.e., the number of electrons per unit area. The Hall resistance R_H^0 varies smoothly with n_e and with B.

Quantum Hall Effects: Under special conditions preferred values of R_H are observed,

$$R_H = R_H^{(\nu)} = \rho_H^{(\nu)} = \frac{1}{\nu}\frac{h}{e^2} \quad , \qquad \sigma_H^{(\nu)} = \nu\frac{e^2}{h} \quad , \tag{6}$$

where h is Planck's constant and ν is a number. For the integer quantum Hall effect ν is a natural number ($\nu = 1, 2, \ldots$). For the fractional quantum Hall effect it is a simple fraction of natural numbers like $1/3$ or $2/3$.

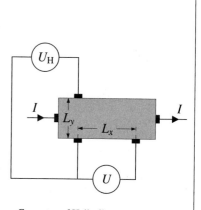

Geometry of Hall-effect measurements.

Filling Factor: The magnetic flux through an area A of the Hall probe is $\Phi = BA$. Since the magnetic flux quantum is $\Phi_0 = h/e$, the total number of flux quanta in A is $N_{\Phi_0} = BA/(h/e)$. The number of electrons in the area A is $N_e = n_e A$. The ratio

$$\eta = \frac{N_e}{N_{\Phi_0}} = \frac{n_e h}{eB} \quad , \tag{7}$$

i.e., the number of electrons per flux quantum, is called *filling factor*. We can now write the conventional Hall resistance (5) as

$$R_H = \frac{1}{\eta}\frac{h}{e^2} \quad . \tag{8}$$

It coincides with the quantum Hall resistance $R_H^{(\nu)}$ in (6), if the filling factor η assumes the integer or fractional values ν.

namely, Planck's constant h and the elementary charge e,

$$R_H = R_H^{(\nu)} = \frac{1}{\nu}\frac{h}{e^2} \quad ,$$

where ν is a natural number, $\nu = 1, 2, \ldots$. This discovery is now known as the *integer quantum Hall effect*.

Quite apart from fundamental physics von Klitzing's finding became important for *metrology*, the science of precision measurement. The quantum Hall

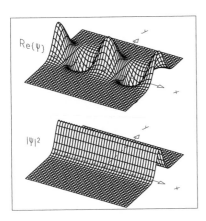

Wave function (top) and probability density of a single electron (in the lowest Landau level) in a two-dimensional Hall configuration. In a completely filled, undisturbed Landau level, states like this are densely stacked side by side.

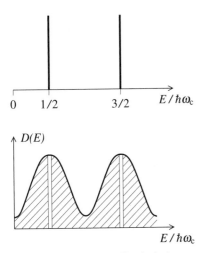

Landau levels of an undisturbed electron (top) and the density of states $D(E)$ for electrons in a real crystal. The hatched energy regions correspond to localized states in y.

effect serves as the most precise representation of the *ohm*, the unit of resistance. The quantity $R_K = h/e^2$ was given the name *von Klitzing constant*. Quantum Hall measurements of R_K, performed in a standardized procedure by various institutions, such as the Physikalisch-Technische Bundesanstalt in Brunswick or the National Institute of Standards and Technology (the former National Bureau of Standards) in Washington, typically differ by less than 2 parts in 10^9. In 1990 the International Committee for Weights and Measures introduced a new, practical representation of the ohm, based on the quantum Hall effect and defined a conventional, i.e., adopted, value for the constant: $R_{K-90} = 25\,812.807\,\Omega$.

All explanations of the quantum Hall effect start out from Landau's quantum-mechanical treatment of the conduction electrons in a metal, which is placed in a magnetic field [3]. It was published in 1930 when Landau was 22 years old. In classical description, an electron (charge $-e$, mass m), in the plane perpendicular to the field B, orbits on a circular path with the *cyclotron frequency* $\omega_c = eB/m$ (Episode 47). In quantum mechanics the possible energy eigenstates are those of a harmonic oscillator of circular frequency ω_c. (See page 355 for an illustration of the oscillator energies.) These energies are called *Landau levels*. Their spacing is $\Delta E = \hbar\omega_c$ and increases with B. Since the electrons are spread out over the plane, there can be many electrons in one Landau level. Their maximum number in an area A is (neglecting spin orientations) equal to the number N_{Φ_0} of elementary flux quanta Φ_0 (see page 365), which make up the total magnetic flux $\Phi = BA$ in that area. At very low temperature only the lowest energy states are occupied. The ratio $\eta = N_e/N_{\Phi_0}$ of the number of electrons and the number of flux quanta in the area is called the *filling factor* (see Box). Thus for an integer filling factor, $\eta = 1, 2, \ldots$, the η lowest Landau levels are exactly filled, whereas the others are empty. The wave function of a single electron in the two-dimensional Hall configuration is simple. The corresponding probability density is well concentrated in y (the direction along which the Hall voltage is measured), but independent of x (the direction of the current), see figure. (The momentary average position is the centre of the cyclotron orbit and that drifts along the x direction.) For a completely filled Landau level the different energy states are placed, side by side, at slightly different values of y. This picture will help us to understand Laughlin's thought experiment, described below.

For an integer filling factor the conventional Hall resistance coincides with the integer quantum Hall resistance. But why does the latter stay constant, if the filling factor is lowered or increased by quite a bit?

No crystal is completely perfect. In a real crystal the sharp Landau levels are broadened; one obtains a continuous density of states, $D(E)$, as a function of energy with maxima located at the original Landau energies. At energies, which are more than a little different from these maxima, electrons are trapped at irregularities of the crystal. Such *localized states* can be filled, but they do not contribute to the current. Let us imagine to gradually fill one broadened Landau level, i.e., to increase the filling factor from some integer value η to $\eta+1$. As long as η is in the region of localized states, there is no conductivity σ in longitudinal direction; the Hall conductivity σ_H and the Hall resistance R_H stay constant at one of their plateau levels. Only in the region of mobile states, is $\sigma \neq 0$. Here σ_H and R_H quickly change to the next plateau. The regions of mobile states are similar to energy bands in a conductor or semiconductor (Episode 44), regions between them to the gaps between bands. The latter are called *mobility gaps*. Localized states, taken by themselves, explain Hall plateaus, not their extreme flatness. But a symmetry argument does, which was put forward by Laughlin in 1981 [4].

The magnetic field can be defined as spatial derivative of a vector potential. There is some freedom of choice in the vector potential without changing the field. Laughlin used this *gauge symmetry* in a thought experiment, performed in its first stage for an ideal, undisturbed two-dimensional electron gas with ν Landau levels filled and all others empty. He changed the geometry, winding up the longish rectangular Hall probe to become a narrow circular ribbon; the original x coordinate denotes the position along the ribbon and y that across it. The field B, everywhere, is perpendicular to its surface. The ribbon carries the current I; between its edges the Hall voltage U_H appears. Landau had considered his metal bar to extend infinitely and this, effectively, is realized in the new geometry. The quantum-mechanical states on the ribbon are those discussed and depicted above. Now, still in the thought experiment, a narrow tube is placed on the axis of the ring, far away from the ribbon. It carries a magnetic flux independent of the field B on the ribbon. However, a change in that flux corresponds to a change in the vector potential of B. Laughlin showed that the quantum-mechanical state is unchanged if that change in flux is exactly one flux quantum Φ_0; this change corresponds to a permissible gauge transformation. On the other hand, the change in flux, by Faraday's law, leads to induction in the ring and an increase of the system's energy. Laughlin showed that this energy is equal to that needed to move one electron across the width of the ribbon against the Hall voltage U_H, provided the relation of the integer quantum Hall effect between U_H and I holds, $U_H = R_H^{(\nu)} I$. The thought experiment's net result is the following: The quantum-mechanical state is unchanged; but ν electrons are shifted across the width of the Hall probe. One can also say that, for each of the ν Landau levels, every electron is shifted up by one position in y. The state, originally at the upper edge in y, is somehow removed. The state, vacated at the lower edge, is filled.

In the second stage of his thought experiment, Laughlin considered the more realistic case with mobile and localized states. With η in a mobility gap, all mobile states in the Landau levels below it are filled, all others empty. The quantum-mechanical states are now quite different from those of the undisturbed case, but the gauge-symmetry argument (a gauge transformation cannot change a state) still holds and leads necessarily to the integer quantum Hall relation $U_H = R_H^{(\nu)} I$. The removed mobile state at high y becomes a localized state. At low y a localized state is turned mobile. Thus, Laughlin's gauge argument not only places the Hall plateaus in the mobility gap but also explains their constancy.

Laughlin[4], a Californian, began to study mechanical engineering in Berkeley but soon changed to physics. After graduation and two years of military service he entered graduate school at MIT in 1974, working on theoretical solid state

[4] Robert Betts Laughlin (born 1950), Nobel Prize 1998

Formation of quantum Hall plateaus. In the region of localized states the longitudinal conductivity σ vanishes; the Hall conductivity σ_H stays constant. In the region of mobile states $\sigma \neq 0$ and σ_H transfers from one plateau to another.

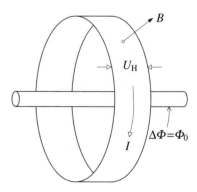

Geometry of Laughlin's thought experiment.

Laughlin

Tsui

Störmer

physics and earning his Ph.D. in 1979. The faculty at MIT recommended him for a position at the Bell Laboratories in Murray Hill, New Jersey. It is there that he did his thought experiment, inspired by contacts at Bell Labs with Störmer and Tsui who were then repeating von Klitzing's work with more advanced Hall probes. Soon after writing the now-famous paper [4], Laughlin left and took a position at the Lawrence Livermore National Laboratory in California.

Tsui[5], until he was twelve, grew up in a remote village in the province of Henan in central China. His formal schooling began only when he was sent to Hong Kong by his parents. In 1958, soon after graduating from school, Tsui was granted a scholarship to Augusta College in Rock Island, Illinois. He moved there and later took his Ph.D. at the University of Chicago in experimental solid state physics. In 1968 he began to work at the Bell Laboratories.

Störmer[6] grew up in a small town near Frankfurt. After a short spell as a student of architecture in Darmstadt he studied physics in Frankfurt. When he had taken his diploma in 1974, a friend persuaded him to move to Grenoble and to do research in the high-field laboratory, where von Klitzing was to make his discovery a few years later. Störmer's thesis adviser was Queisser[7] from the Max-Planck Institute and the University in Stuttgart. He urged him to finish his thesis on 'electron–hole droplets' in high magnetic fields without delay and move on to the United States, recommending the Bell Labs, where he had been working himself. Störmer arrived at Murray Hill in the summer of 1977.

In the following year, Störmer participated in the realization of a two-dimensional electron gas in a double heterostructure [5] and, in 1979, in a single heterojunction [6]. This work was briefly described at the end of Episode 86. In 1981 a sample containing an AlGaAs–GaAs heterojunction was prepared by Arthur Gossard and William Wiegmann, also at Bell Labs, and it was structured to make it serve as a Hall probe. The electron density n_e was quite small. Tsui and Störmer took the sample to the Francis Bitter Magnet Laboratory of MIT at Cambridge. Because of its small electron density n_e even the lowest Landau level would be only partially filled in a high magnetic field. The aim was to try and find signs for a two-dimensional electron crystal, that had been predicted by Wigner.

Hall measurements began on 7 October 1981. Since, under these conditions, n_e was fixed, the magnetic field B was varied between 0 and more than 20 Tesla. The filling factor is inversely proportional to B, and the curve for the conventional Hall resistance as a function of B is a straight line, not a hyperbola as in von Klitzing's plot, in which it is a function of V_g. At the temperature of boiling helium, $T = 4.2\,\text{K}$, a straight line was indeed observed and an indication, at low field, of von Klitzing's plateaus at integer filling factors. As the temperature was lowered, these became very pronounced. Unexpectedly, a plateau also developed at the filling factor of $\nu = 1/3$. Tsui, Störmer, and Gossard published their discovery of the *fractional quantum Hall effect* early in 1982 [7], mentioning that it was at variance with Laughlin's gauge argument, since that would require the existence of quasiparticles of charge $e/3$. (In the thought experiment the charge, transported across the ribbon, is νe.)

---

[5] Daniel Chee Tsui (born 1939), Nobel Prize 1998   [6] Horst Ludwig Störmer (born 1949), Nobel Prize 1998   [7] Hans-Joachim Queisser (born 1931)

Laughlin rose to the challenge. In previous theories, including his own thought experiment, electrons had been taken as independent, except for the Pauli principle. In 1983 he wrote down a wave function, depending on the coordinates of all electrons [8]. It is constructed such that the Pauli principle is automatically fulfilled and that it is antisymmetric under the exchange of any two electrons, as required for fermions. In the probability density every electron 'sees' a triple zero at the positions of all other electrons. The system takes the lowest energy, if there are three flux quanta at each of these positions; it can be considered to consist of *composite particles*, each made up of an electron and three flux quanta. Its energy defines a Landau sub-level, separated from others by a mobility gap. If one electron is missing from the system, there are three spare flux quanta, each mimicking a *quasiparticle* of charge $-e/3$. Likewise a spare electron mimics three quasiparticles of charge $e/3$. Compelling experimental evidence for the existence of such objects (not to be confused with real, elementary particles) was found in 1997 [9,10].

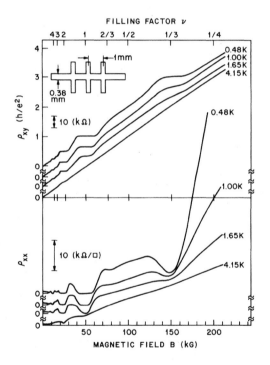

Observation of the fractional quantum Hall effect by Tsui, Störmer, and Gossard. Shown are the Hall resistance R_H (denoted ρ_{xy}) and the longitudinal resistance R (denoted ρ_{xx}) as function of the magnetic field B. At very low temperature a new plateau in R_H appears at the fractional filling factor $\nu = 1/3$. Figure reprinted with permission from [7]. Copyright 1982 by the American Physical Society.

The quantum Hall effect became a wide field of research. As the sample quality was further improved, more and more fractional plateaus were found and theoretical activity flourished. In a review on 25 years of quantum Hall effect, by von Klitzing and two colleagues [11], we find that, in 2005, there was about one publication on the subject per day.

[1] von Klitzing, K., Dorda, G., and Pepper, M., *Physical Review Letters*, **45** (1980) 494.

[2] Yu, P. and Cardona, M., *Fundamentals of Semiconductors*. Springer, Berlin, 1996.

[3] Landau, L., *Zeitschrift für Physik*, **64** (1930) 629. Engl. transl. in [12], p. 31.

[4] Laughlin, R. B., *Physical Review*, **B 23** (1981) 5632.

[5] Dingle, R., Störmer, H. L., Gossard, A. C., and Wiegmann, W., *Applied Physics Letters*, **33** (1978) 665.

[6] Störmer, H. L., Dingle, R., Gossard, A. C., Wiegmann, W., and Sturge, M. D., *Solid State Communications*, **29** (1979) 705.

[7] Tsui, D. C., Störmer, H. L., and Gossard, A. C., *Physical Review Letters*, **48** (1982) 1559.

[8] Laughlin, R. B., *Physical Review Letters*, **50** (1983) 1395.

[9] de Picciotto, R. et al., *Nature*, **389** (1997) 162.

[10] Saminadayar, L. et al., *Physical Review Letters*, **79** (1997) 2526.

[11] von Klitzing, K., Gerhardts, R., and Weis, J., *Physik Journal*, **4 (6)** (2005) 37.

[12] Landau, L., *Collected Papers*. Gordon and Breach, New York, 1965.

96 *W* and *Z* Boson Discovered (1983)

Rubbia

van der Meer

With the discovery of the heavy gauge bosons W^+, W^-, and Z^0 in 1983 the theory of electroweak interactions (Episode 90) was firmly established. Since these particles are partners of the massless photon, the discovery can be likened to that of the Compton effect (Episode 31) which irrefutably demonstrates the photon's existence. But here the similarity ends. While Compton's experiment was relatively simple, the methods developed and the apparatus needed to produce and detect the heavy bosons were of unprecedented complexity. Although the theory did not predict the masses of the W and Z bosons, experiments at low energy suggested them to lie in the range between 50 GeV and 100 GeV. With the success of electron–positron colliders and, in particular, with the discovery of charm (Episode 93) it became clear that such a collider of sufficiently high energy would be the ideal instrument to produce the Z^0 and to create the W^+ and W^- in pairs. Consequently, in 1976, the report of a study group was published by CERN in which the physics potential of such a collider was summarized. The machine, called *LEP* for large electron–positron collider, was eventually funded and constructed; it began to operate in 1989 (Episode 98). Also in 1976, at an international conference in Aachen, Rubbia presented an idea he had worked out with McIntyre and Cline [1]. They proposed to convert an existing proton synchrotron into a proton–antiproton collider which could then produce heavy bosons. Within less than seven years the ambitious project was indeed realized [2].

Rubbia[1] was born in Gorizia (also Görz or Gurize), a town in the north-east of Italy, which for centuries had been influenced by Austrians, Italians, and Slovenians and, in 1945, became part of Yugoslavia. Rubbia's family – his father was an electrical engineer, his mother a school teacher – fled to Venice and then to Udine. He studied at the prestigious Scuola Normale Superiore in Pisa, Fermi's alma mater, and obtained his doctorate under Conversi[2]. In 1958 Rubbia went to Columbia University. It is there that he began to do experiments on weak-interaction processes, which were to become his primary interest. At the time of the 1976 proposal, Rubbia had already worked at CERN and also at several laboratories in the United States and was professor at Harvard. He knew well the experimental facilities on both sides of the Atlantic Ocean. Both at CERN and at Fermilab large synchrotrons had just been completed, which could accelerate protons to energies of up to several hundred GeV.

The history of CERN – its full name is *European Organization for Nuclear Research* – can be traced back to 1949, when Louis de Broglie, at the European Cultural Conference in Lausanne, proposed the creation of a European science laboratory. The organization was officially founded by 11 European

---

[1] Carlo Rubbia (born 1934), Nobel Prize 1984   [2] Marcello Conversi (1917–1988)

states in 1954. From the beginning the scientific emphasis was on nuclear and subnuclear physics with high-energy accelerators. In 1960 the 28-GeV proton synchrotron (PS), for years the workhorse of CERN, was completed by a team under Adams[3]. In the late 1960s a much larger machine, called the super proton synchrotron (SPS), was planned. In view of this large project Adams was appointed Executive Director General of CERN. In 1976 van Hove[4] became Director General of Research. The SPS is housed in an underground ring tunnel, 2200 metres in diameter, on both sides of the Swiss–French border north of Geneva. It was completed in May 1976 and scheduled to serve a large number of fixed-target experiments. Only a few months later Rubbia, in his talk in Aachen, suggested to turn the SPS (or, alternatively, an accelerator at Fermilab) into a collider and, at least for some years, to completely change its scientific programme.

Fermilab did not take the risk. But at CERN the chances were seen, although Adams was very reluctant. Richter tells of a rather dramatic meeting with the two Directors General [3] at which van Hove threatened to resign unless Adams would support the collider proposal. Finally, agreement was reached on a two-step process. The first step was to demonstrate on a large scale that a sufficient number of antiprotons could be produced and stored ready for injection into the SPS. In the case of success, the conversion of that machine into a proton–antiproton collider, dubbed S$\bar{p}p$S, would follow.

Antiprotons can be produced if protons of sufficient energy fall on a material target (Episode 75). The 26-GeV protons from the PS could be used but the antiprotons had a wide range of energies and directions. A synchrotron, on the other hand, accepts only particles within a very limited range. An antiproton accumulator ring (AA), a special small synchrotron of relatively large acceptance, was designed to capture and store antiprotons of about 3.5 GeV. Rather few antiprotons could actually be caught in the AA (about one for a million protons falling on the target). Before these could be handed over to other machines for acceleration they had to be 'cooled' in the AA to reduce the spread of position and momentum vectors. Fortunately, there was an accelerator scientist at CERN, van der Meer, who had invented the method of 'stochastic cooling', which was the key for obtaining the antiproton intensities required.

Van der Meer[5] was born in The Hague. He studied technical physics in Delft and specialized in measurement and regulation technology. After obtaining his engineering degree in 1952 van der Meer worked at the Phillips Research Laboratory at Eindhoven before joining CERN in 1956. There, he became one of the builders of the PS and contributed to other accelerators and to experiments; in Episode 91 we mentioned his 'neutrino horn'.

In an electron storage ring cooling happens automatically. Electrons lose some of their energy by synchrotron radiation and the focussing forces of the machine see to it that, after some time, all particles circulate very near to the nominal machine orbit with a momentum very near the nominal value. Because of their much larger mass synchrotron radiation is negligible for protons and antiprotons; their positions and momenta have to be measured and then

Rubbia and van der Meer at CERN in October 1984 after learning that they had been awarded the Nobel Prize.

---

[3] (Sir) John Adams (1920–1984) [4] Léon van Hove (1924–1990) [5] Simon van der Meer (born 1925), Nobel Prize 1984

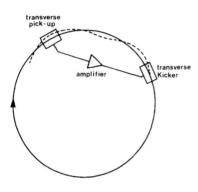

Block diagram illustrating the stochastic cooling of the transverse betatron oscillations. Reprinted from [4], copyright 1980, with permission from Elsevier.

actively corrected. That cannot, of course, be done for every particle individually but only on a statistical basis; hence the name *stochastic cooling* [4]. As an example, we sketch the reduction of the transverse deviation from the nominal orbit. In a synchrotron particles perform so-called betatron oscillations about the nominal orbit (see Box II in Episode 64). The aim is to decrease their amplitude. This is reached by making the particles pass a sensor or 'pick-up' measuring the average beam position and, a little later, a fast 'kicker' magnet correcting it. The action of the magnet is steered by the signal from the pick-up. Since the particles travel practically with the speed of light and the signal processing and kicker activation takes time, the signal has to take a short-cut with respect to the particles: It proceeds on a straight line rather than on an arc. It needs very many passages through this and similar systems to sufficiently cool the antiproton beam.

Once cooled, the antiprotons could be transferred back to the PS, accelerated to 26 GeV, then injected into the SPS, accelerated to 270 GeV, and be brought to collision with counter-rotating protons. The first collisions were observed in 1981 [5].

The proton–antiproton accelerator complex at CERN. Protons p are accelerated to 26 GeV in the proton synchrotron PS, then ejected onto a target. Antiprotons $\bar{p}$ produced there with energies around 3.5 GeV and a certain angular spread are collected in the antiproton accumulator ring AA. The process is repeated every 2.4 seconds for about a day until about 10^{11} antiprotons are collected and 'cooled'. They are then transferred to the PS, accelerated to 26 GeV, and injected into the super proton synchrotron SPS. Just before, about 10^{12} protons of the same energy from the PS were injected into the SPS in the opposite direction. In that machine protons and antiprotons are accelerated to up to 270 GeV and made to collide. (Also the intersecting storage rings ISR can be filled with protons and antiprotons. Moreover, antiprotons can be delivered to the low-energy antiproton ring LEAR.) Reprinted from [5], copyright 1980, with permission from Elsevier.

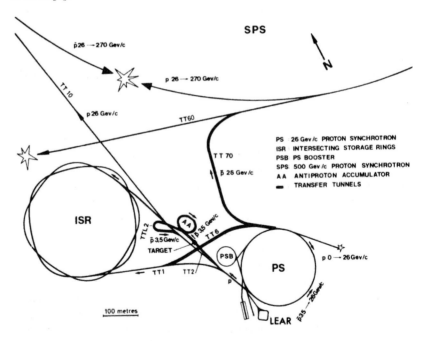

Since the ring tunnel of the SPS was deep underground, two large caverns had to be excavated to accommodate the experiments. One of them, called *UA1* (UA stands for underground area), was conceived to measure with high precision practically all particles produced an a $p\bar{p}$ interaction. The apparatus was designed and built by a collaboration of about 130 physicists from 11 laboratories led by Rubbia, as one of his colleagues put it [6], in the spirit of 'enlightened absolutism'. The second large experiment, *UA2*, was more modest in scope and concentrated on measuring the flow of energy emitted by an interaction, rather than individual particles. About 60 scientists from six institutes formed the UA2 collaboration with Darriulat[6] as spokesman.

The formation of a heavy boson takes place in the collision of a quark from a proton and an antiproton, e.g., $u\bar{d} \to W^+$, $d\bar{u} \to W^-$, $u\bar{u} \to Z^0$ (see basic diagrams on page 443). The remaining two quarks and two antiquarks are called 'spectators'. They give rise to jets of hadrons emitted mainly in the directions of the incident proton and antiproton. The heavy bosons can decay again into a pair of quarks but also into a pair of leptons (see basic diagrams on page 428). Quarks produce jets and are difficult to distinguish from jets originating from strong interactions between the colliding quark and antiquark. A clear signature of heavy-boson production is provided by the leptons from their decay. The heavy bosons are, in general, not at rest in the laboratory but they have no notable momentum transverse to the direction of the colliding beams. The two leptons, therefore, have equal and opposite transverse momenta. This makes it relatively easy to detect the Z^0; one has to detect a pair of oppositely charged electrons or muons with large and opposite transverse momenta. For the $W^\pm$, however, only one lepton can be observed, the other being a neutrino. But the neutrino momentum can be obtained indirectly. It is the total momentum 'missing' if that of all other particles is measured.

Essential for the experiments was a type of detector, called *calorimeter*. The name, indicating the measurement of heat, is misleading. What is measured is the total energy carried by a particle or a characteristic fraction from which the total is computed. To this end the particle is made to lose all its energy in matter. A *sampling calorimeter* is a sandwich of target matter interspersed with detectors. The particle triggers a cascade of interactions in the target material and the total signal (number of track elements in track chambers of total light observed in scintillators) is a measure for the particle energy. Calorimeters are sensitive not only to charged but also to neutral hadrons because charged particles are formed in the cascade. For *electromagnetic calorimeters* optimized for electrons and photons, a target material of high nuclear charge such as lead is used. *Hadron calorimeters* only require a material with many nucleons per unit volume; iron is most commonly taken. Muons have no strong and only very little electromagnetic interaction. Therefore, particles detected behind a totally absorbing hadron calorimeter are identified as muons.

The UA1 detector combined all detector techniques of its day. A large drift chamber in a transverse magnetic field surrounded the beam pipe. It was enclosed in an electromagnetic calorimeter. The iron yoke of the magnet acted as target material of the hadron calorimeter. Its outside was covered with chambers detecting muons. A particular property of UA1 was its *hermeticity*; only particles with an angle smaller than $0.2°$ with respect to the beam line could escape undetected. The UA2 detector had a magnetic field only in a limited region and relied wholly on calorimetry.

Of interest in this episode are three running periods of the collider, each following with much increased performance with respect to the foregoing. The first run, in December 1981, was devoted to the detection of jets from the strong interaction of a quark with an antiquark. We show a spectacular event regis-

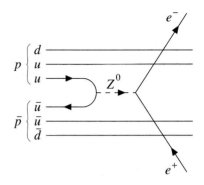

Production of the Z^0 boson in a proton–antiproton collision and its subsequent decay into an electron–positron pair.

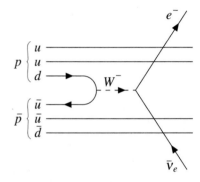

Production and leptonic decay of a W^- boson.

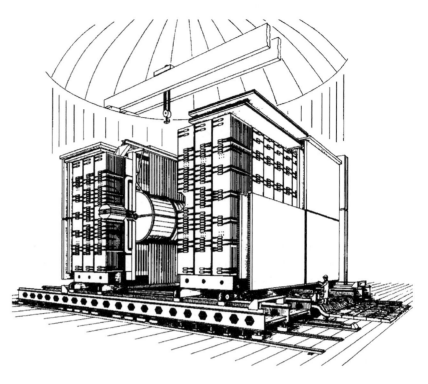

Schematic view of the UA1 detector in its underground cavern. The large cylinder houses the central track detector. The two yokes of the magnet are moved apart to allow access to the central detector. The iron yokes are segmented and form part of the hadron spectrometer. The large flat structures outside it are muon chambers. The electromagnetic calorimeter (not visible in the figure) is located between the central track detector and the hadron calorimeter. Reprinted from [7], copyright 1980, with permission from Elsevier.

tered by UA2 with two jets emitted in opposite directions nearly perpendicular to the beam line [8].

In the run of November and December 1982, a clear signal of the $W^{\pm}$ was observed both by UA1 and UA2 [9,10]. By careful selection criteria events were found with a large amount of energy registered in the electromagnetic

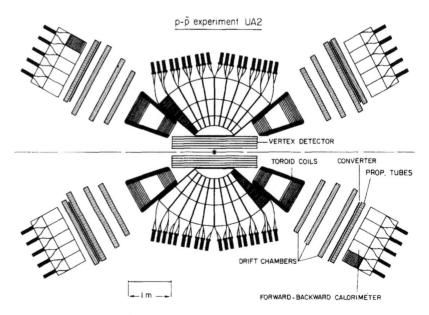

Cut through the UA2 detector. It has a cylindrical symmetry about the beam direction. There is a small track detector for the reconstruction of the interaction vertex. The main emphasis is on calorimetry. The central calorimeter covers the angular region between $40°$ and $140°$ with respect to the beam axis. There is a magnetic field and there are tracking chambers in front of the forward calorimeters ($20°$–$40°$ and $140°$–$160°$). Electromagnetic and hadronic cascades are distinguished by the different type of pattern they leave in the calorimeters. Reprinted from [8], copyright 1982, with permission from Elsevier.

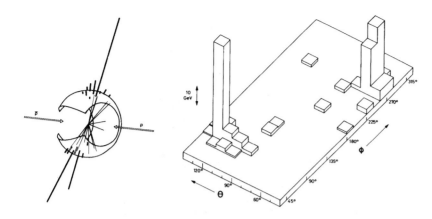

Two-jet event as observed in the UA2 detector. *Left:* The lengths of the different bars represent the energy registered in different sections of the central calorimeter. *Right:* The same information projected onto a plane (θ and ϕ are the polar and azimuthal angle, respectively, with respect to the direction of the incident proton). Reprinted from [8], copyright 1982, with permission from Elsevier.

calorimeter and a similar balancing 'missing' energy, which could be attributed to a neutrino. UA1 observed five such events and UA2 found four; first estimates of the W mass were given.

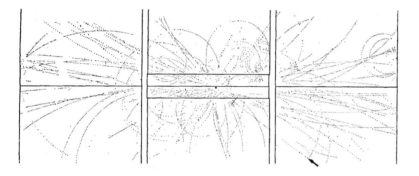

An event with 65 tracks in the central detector of UA1. The track marked by an arrow is identified to belong to an electron, since it points to a cell in the electromagnetic calorimeter registering the energy of 43 GeV. The momentum transverse to the beam is (24 ± 0.6) GeV/c. It is well balanced by the transverse momentum 'missing' in the detector of (24.4 ± 4.6) GeV/c. The electron is therefore interpreted as originating in the decay of a W boson. Reprinted from [9], copyright 1983, with permission from Elsevier.

In the run of April/May 1983 many more events were observed. This allowed to measure the angular distributions of the electrons from W^- and of the positrons from W^+ decay. They were found to be fully compatible with the predictions of the $V - A$ theory of weak interaction. The most important result of that run was the discovery of the Z^0 boson.

The UA1 collaboration reported four events with a Z^0 boson decaying into an electron–positron pair and one event in which the boson decayed into a pair

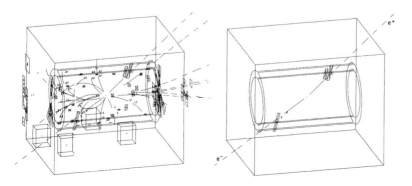

A Z^0 boson observed in the UA1 detector to decay into an electron e^- and a positron e^+. All tracks observed in the central detector and energies observed in the calorimeters are shown on the event display on the left. The display on the right is restricted to objects with a transverse momentum of more than 2 GeV/c. Reprinted from [11], copyright 1983, with permission from Elsevier.

of oppositely charged muons [11]. A little later the UA2 group published the observation of eight events of the type $Z^0 \to e^+ + e^-$ [12].

[1]   Rubbia, C., McIntyre, P., and Cline, D., *Producing Massive Neutral Intermediate Vector Bosons with Existing Accelerators.* In Faissner, H., Reithler, H., and Zerwas, P. (eds.): *Proceedings of the International Neutrino Conference in Aachen*, p. 683. Vieweg, Braunschweig, 1977.

[2]   Watkins, P., *Story of the W and Z.* Cambridge University Press, Cambridge, 1986.

[3]   Richter, B., *The Rise of Colliding Beams.* In Hoddeson, L., Brown, L., Riordan, M., and Dresden, M. (eds.): *The Rise of the Standard Model*, p. 261. Cambridge University Press, Cambridge, 1997.

[4]   Möhl, D., Petrucci, G., Thorndahl, L., and van der Meer, S., *Physics Reports*, **58** (1980) 73.

[5]   The Staff of the CERN Proton–Antiproton Project, *Physics Letters*, **107 B** (1981) 306.

[6]   Denegri, D., *CERN Courier*, **43**(4) (2003) 26.

[7]   UA1 Collaboration: Barranco Luque, M. et al., *Nuclear Instruments and Methods*, **176** (1980) 175.

[8]   UA2 Collaboration: Banner, M. et al., *Physics Letters*, **118 B** (1982) 203.

[9]   UA1 Collaboration: Arnison, G. et al., *Physics Letters*, **122 B** (1983) 103.

[10] UA2 Collaboration: Banner, M. et al., *Physics Letters*, **129 B** (1983) 130.

[11] UA1 Collaboration: Arnison, G. et al., *Physics Letters*, **126 B** (1983) 398.

[12] UA2 Collaboration: Bagnaia, P. et al., *Physics Letters*, **122 B** (1983) 476.

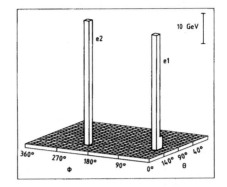

A decay $Z^0 \to e^+ + e^-$ observed in the calorimeter of the UA2 experiment [12]. The isolated amounts $e1, e2$ of energy were deposited by the electron and the positron. Reprinted from [11], copyright 1983, with permission from Elsevier.

The efforts of CERN and many European laboratories were soon recognized. Already in 1984 the Nobel Prize was awarded to Rubbia and van der Meer 'for their decisive contributions to the large project, which led to the discovery of the field particles W and Z, communicators of weak interaction'. Experiments at the S$\bar{p}p$S collider continued until 1990. Among many other results they yielded accurate measurements of the $W^\pm$ and Z^0 masses. These, in turn, were much improved with the arrival of the e^+e^- collider LEP. The present values are $m_W = (81.403 \pm 0.029)\,\mathrm{GeV}/c^2$ and $m_Z = (91.1876 \pm 0.0021)\,\mathrm{GeV}/c^2$.

In 1985 the first proton–antiproton collisions were registered at Fermilab. Following CERN's example, the laboratory had converted its very large proton synchrotron, called *Tevatron* for its proton energy of $1\,\mathrm{TeV} = 1000\,\mathrm{GeV}$, into a $p\bar{p}$ collider. The top quark t was discovered with the help of this machine in 1995 (see end of Episode 93).

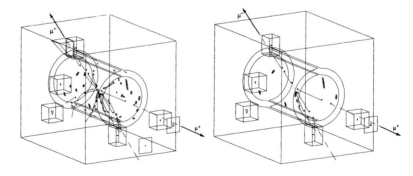

The decay $Z^0 \to \mu^+ + \mu^-$ as seen in the UA1 detector. *Left:* All registered information. *Right:* Tracks and calorimeter information with small transverse momenta removed. Reprinted from [11], copyright 1983, with permission from Elsevier.

Cooling and Trapping Neutral Atoms (1985)

Atomic beams, pioneered by Stern (Episodes 30 and 52), opened the way to study non-interacting atoms which, essentially, have only one component in velocity, namely, that in beam direction. Ions could even be confined in a trap (Episode 80) with small values of all three velocity components. We shall see now how that was achieved also for neutral atoms. A measure of quality for a gas of trapped atoms is its temperature, i.e., the average kinetic energy of the atoms expressed in Kelvin. By carefully exploiting the inner structure of atoms it became possible, with the help of laser light and magnetic fields, to cool and trap small samples of highly dilute atomic gases, reaching temperatures as low as 180 nK $(180 \times 10^{-9}$ K$)$.

A first important step was the proposal of an experimental technique, laid down in 1974 at Stanford University in a two-page letter [1] by Hänsch[1] and Schawlow. The experiments were done by three groups working in friendly competition: the group of Chu at the Bell Labs in Holmdel, New Jersey, and later at Stanford University, the group of Cohen-Tannoudji at the Collège de France and the École Normale Supérieure in Paris, and the group of Phillips at a laboratory of the National Bureau of Standards (now National Institute of Standards and Technology) in Gaithersburg, Maryland.

Chu[2] grew up in Garden City, New York. His father was a professor at the nearby Brooklyn Polytechnic. Both his parents had come from China to the United States for their university studies in the 1940s. He took his B.Sc. at the University of Rochester in 1970 and his Ph.D. from the University of California in Berkeley in 1976. The research he did for his thesis and continued to do afterwards in Berkeley was high-precision laser spectroscopy in thallium, revealing effects of the unified theory of electromagnetic and weak interaction (Episode 90) even at the low energies of atomic physics. Chu joined the Bell Labs in 1978 and stayed there until 1987 when he accepted a professorship at Stanford University. In 2004, he was appointed director of the Lawrence Berkeley National Laboratory.

Cohen-Tannoudji[3] was born in Constantine, Algeria (then part of France) as descendant of a Sephardic family that had fled Spain in the time of the Inquisition and had come to Algeria via Tangiers, hence the name which means Cohen from Tangiers. He went to primary and secondary school in Algiers. In 1953, after a competitive exam, he entered the prestigious École Normale Supérieure. Initially Cohen-Tannoudji was more interested in mathematics but, under the influence of lectures by Kastler[4], he soon opted for physics. After his first exam in 1958 he did his military service, lasting for more than two years,

Energy, Velocity, Temperature

Consider a gas of atoms of mass m and mean squared velocity v^2. These have the average kinetic energy

$$E_{\text{kin}} = mv^2/2 \quad,$$

which is related to the temperature T by

$$E_{\text{kin}} = (3/2)kT \quad,$$

where k is Boltzmann's constant. To give a numerical example: Sodium atoms at room temperature ($T = 300$ K) have the average velocity $v = 570$ m/s.

Chu

---

[1] Theodor Wolfgang Hänsch (born 1941), Nobel Prize 2005   [2] Steven Chu (born 1948), Nobel Prize 1997   [3] Claude Cohen-Tannoudji (born 1933), Nobel Prize 1997   [4] Alfred Kastler (1902–1984), Nobel Prize 1966

Cohen-Tannoudji

Phillips

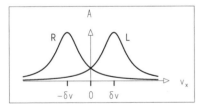

Probability A for an atom with velocity v_x in x direction to absorb a photon from the laser beams shining to the left (L) and to the right (R). Both beams are tuned below resonance absorption for an atom at rest. The detuning is compensated by the Doppler effect for the velocities δv (L beam) and $-\delta v$ (R beam).

and then returned to Paris for doctoral research under Kastler. As a Ph.D. student he developed the quantum theory of optical pumping and predicted new sublevels of the Zeeman effect, brought about by irradiation of atoms with light. This effect, jocularly, was named 'lamp shift' by Kastler, who alluded to the Lamb shift (Episode 67), and is now known as *light shift*. With this work he obtained his Ph.D. in 1962. In 1964 he joined the faculty of the University of Paris and in 1973 he was appointed professor at the venerable Collège de France, founded in 1530. Cohen-Tannoudji's research on laser cooling and trapping was performed with a group he set up in 1984, including Aspect[5] as associate director and Dalibard[6] as graduate student.

Phillips[7] was born in Wilkes-Barre, Pennsylvania. Both his parents were social workers and had studied at Juniata College in Huntingdon, also in Pennsylvania. Phillips graduated from the same small school in 1970 and then moved on to MIT for graduate studies. For his thesis he performed a high-precision measurement of the magnetic moment of the proton in water and also studied collisions of laser-excited atoms. He obtained his Ph.D. in 1976, stayed at MIT for another two years with a fellowship, and was engaged in early attempts to achieve Bose–Einstein condensation of polarized hydrogen atoms. In 1978 he joined the National Bureau of Standards. There, at first, laser cooling was his spare-time activity.

We now return to the proposal by Hänsch and Schawlow [1], which is based on *resonant fluorescence*. If an atom absorbs a photon of a characteristic frequency, one of its electrons acquires a state of higher energy; the atom returns to the original state by emitting a photon of the same frequency. The research, described here, was mostly done with sodium atoms; for the resonance the D_1 line was used, one of the two closely spaced yellow sodium lines. Consider an atom, moving in x direction, and a beam of laser light, in resonance with the atom and directed in $-x$ direction. In each absorption process not only the energy of the photon but also its momentum is transferred to the atom, which is thus slowed down. Re-emission is a random process without preferential direction; averaged out over many individual events it does not contribute to the velocity of the atom. The net effect of the laser light is that the atom is slowed down very effectively.

To achieve resonance in an atom which moves against the laser light, the laser frequency has to be reduced because some energy is provided by the atom's motion. This *Doppler shift* has to be taken into account as long as it is significantly larger than the natural width of the resonance energy. That can be done by continuously adjusting the laser frequency (a technique called 'chirping') during the slowing-down process. Once low velocities have been reached, they can be maintained by a system of six laser beams, shining in the directions $x, -x, y, -y, z, -z$ and all tuned slightly below the resonance frequency of the atom at rest. We consider the motion in x. The detuning, on the order of the natural resonance width, is compensated by the Doppler effect for an atom moving (to the right) with velocity δv. For such an atom the probability to absorb a photon from the light shining to the left is at a maximum. Likewise an atom with velocity $-\delta v$ has high probability to absorb

[5] Alain Aspect (born 1947) [6] Jean Dalibard (born 1958) [7] William Daniel Phillips (born 1948), Nobel Prize 1997

a photon from the light shining to the right. By the deceleration mechanism the absolute value of the velocity is therefore kept around or below δv. With six laser beams this statement holds in all three dimensions. The theoretical limit, called *Doppler limit*, of this type of cooling is half the natural energy width of the resonance transition used. For sodium atoms it is $240\,\mu K$.

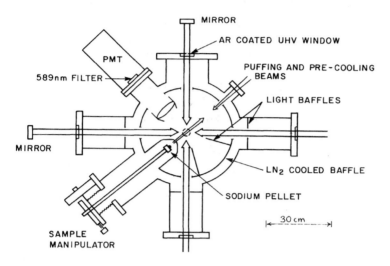

The set-up of 'optical molasses' by Chu and his group. Six laser beams are directed onto the centre of the ultra-high vacuum (UHV) vessel (four in the plane of the figure, two perpendicular to it). Additional laser beams are used to 'puff' sodium atoms off a sample of solid sodium and to pre-cool them before they reach the centre. Light from the cooled sodium gas in the central region is registered by a photomultiplier tube (PMT). The baffle, cooled by liquid nitrogen (LN$_2$), facilitates the maintenance of ultra-high vacuum. Figure reprinted with permission from [2]. Copyright 1985 by the American Physical Society.

More than ten years after it was proposed, the system of six pairwise counter-propagating, orthogonal laser beams was realized by Chu's group at Holmdel [2]. The experiment was done in a vacuum chamber, fitted with six windows for the laser beams and additional ports. One contained a sample holder, carrying metallic sodium. A short burst of laser light, entering through an opposite port, could produce free sodium atoms by evaporation, a process called 'puffing'. Atoms, travelling towards the central intersection region of the six laser beams, were pre-cooled by light from a 'chirped' laser opposing them. A photomultiplier registered light emitted by the cooled gas. Its intensity was a measure for the number of atoms in the central region.

Because of its strong slowing-down effect on atoms Chu called his system *optical molasses*, alluding to the saying 'slow as molasses in January'. It was operated in a repetition cycle of 0.1 seconds. Each cycle began with a puffing pulse of 10 nanoseconds, in which atoms of about $1000\,K$ were produced. They were pre-cooled within 0.5 milliseconds on a distance of less than 5 cm and then drifted to the central region, where about 10^6 atoms per cm^3 were further cooled and confined by the molasses. The fluorescence of the gas was visible to the eye. Its temperature was measured by letting the gas expand ballistically: The light of the molasses was shut off for a few milliseconds and the photomultiplier signal was registered before and after that time interval. The decrease of the light intensity was a measure for the number of atoms which left the central region. From it their average velocity and thus their temperature could be inferred. Within experimental errors it turned out to be equal to the predicted Doppler limit of $240\,\mu K$.

The atoms stayed for some time in the optical molasses because they moved with very small velocities but they were not actually trapped. The first trap

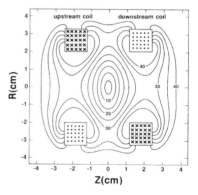

Field configuration of the first magnetic trap for neutral atoms. It is rotationally symmetric about the z axis. The current, flowing in opposite directions in the two coils, produces a field which is indicated by lines of constant magnitude (given in milliTesla). Figure reprinted with permission from [3]. Copyright 1985 by the American Physical Society.

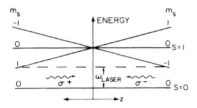

Principle of the magneto-optical trap (MOT). The resonance energy for transitions from the $s = 0$ ground state to the $s = 1, m_s = 1$ excited state rises with the position coordinate z, while it falls for transitions to the state $s = 1, m_s = -1$. The former type of transition is brought about by σ^+ photons (preferentially in the region $z < 0$), the latter by σ^- photons (mainly for $z > 0$). The momenta of these photons, on the average, push the atoms towards $z = 0$. Figure reprinted with permission from [5]. Copyright 1987 by the American Physical Society.

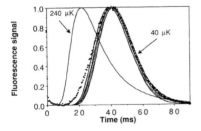

Time-of-flight measurements (shown as dots) which revealed cooling of a gas of sodium atoms to $40\,\mu$K, far below the Doppler limit of $240\,\mu$K. The curves correspond to distributions expected for $240\,\mu$K and $40\,\mu$K, respectively. Figure reprinted with permission from [6]. Figure reprinted with permission from 1988

for cold atoms was built by Phillips and his group. They had cooled atoms already a little before Chu (albeit only in one dimension) by shining a laser beam against a beam of sodium atoms emanating from an oven [4]. Rather than to change the laser frequency by chirping they had varied the resonance frequency of sodium by applying a magnetic field, decreasing along the deceleration path. The Zeeman effect in this field led to a shift in the ground state of the resonance transition and thus of the resonance energy. The group now added a magnetic trap to their set-up [3]. Sodium atoms with their single electron in the outer shell have a magnetic moment analogous to that of silver atoms we discussed for the Stern–Gerlach effect in Episode 30. In an inhomogeneous magnetic field the atoms therefore experience a force in the direction of rising or falling field if the magnetic moment is parallel or antiparallel to the field direction, respectively. If the field has a maximum or minimum in space, then atoms are attracted to it and can be trapped in its neighbourhood. Since the forces are small, the atoms have to be quite cold for the trap to function. Phillips and collaborators used two coils, carrying identical but counter-rotating currents, and were able to trap sodium atoms for about a second in a volume of a few cubic centimetres. The trap had a *depth* of 17 mK, i.e., atoms of that temperature or less were confined in it.

Dalibard had the idea to combine the magnetic trap with a special version of optical molasses and thus to create a *magneto-optical trap* (MOT) which would be considerably deeper, i.e., it could confine atoms up to higher temperatures than the ordinary magnetic trap. That idea was realized at Bell Labs in 1987 [5]. The operating principle is best exemplified for a hypothetical atom with a ground state of spin quantum number $s = 0$ and magnetic spin quantum number (spin component in z direction) $m_s = 0$. The excited state has $s = 1$; $m_s = -1, 0, +1$. In a magnetic field, changing linearly with z, i.e., like $B(z) = bz$, near $z = 0$ (as the field in the magnetic trap does), the three possible m_s states behave differently as a function of z. The state $m_s = 1$ rises in energy as z increases and the state $m_s = -1$ falls, while the state $m_s = 0$ is unaffected. Only a so-called σ^+ photon (a photon of right-handed circular polarization with respect to its direction of motion) can excite the atom from the ground state to the $m_s = 1$ state. Likewise, the excitation to $m_s = -1$ requires a σ^- photon. Laser light, σ^+ in z direction and σ^- opposite to it, tuned below the resonance energy, has the following effect: More σ^+ photons are absorbed at $z < 0$ because there the transition energy is lowered and the detuning is less effective. At $z > 0$ more σ^- photons are absorbed. As a net effect, the atoms experience a force pushing them towards $z = 0$. This trapping principle can be applied in all three dimensions. In reality the states to be considered are more complicated but the principle is the same. The first such trap had a depth of 0.4 K and could trap 10^7 atoms for 2 minutes [5].

In 1988 Phillips and his group found a surprising result: Doppler cooling works better than expected theoretically. A sample of sodium gas, cooled in optical molasses, was prepared and then the cooling lasers were turned off. Another laser beam was directed at a point 1.1 cm below the centre of the molasses, and the fluorescence in the sodium atoms, caused by that beam, was recorded as a function of the time after shut-off. From the intensity distribution, caused by the sample falling through the laser beam, the sample temperature was determined to only $40\,\mu$K.

An explanation of this welcome effect was given by Dalibard and Cohen-Tannoudji [7] and by Chu's group [8]. It is based on the *light-shift* effect, which Cohen-Tannoudji had discovered and described in his Ph.D. thesis. If two opposing laser beams are linearly polarized in orthogonal directions, then, by interference, a standing light wave results, which is circularly polarized and in which the sense of polarization (σ^+, σ^-) alternates every quarter wavelength. The ground state of the sodium D_1 resonance can have two spin orientations, denoted by the quantum numbers $m_s = 1/2, -1/2$, which, as a result of the light shift, become space-dependent. It could be shown, that in one cycle from a ground state to an excited state and back to a ground state, on average, slightly more energy is re-emitted than previously absorbed. The extra energy is taken from the kinetic energy of the atom. Cohen-Tannoudji called this type of cooling the *Sisyphus effect*, because the atom seems to move more often 'uphill' than 'downhill'. (Sisyphus is the character in Greek mythology, cursed to perpetually roll a huge boulder up a hill.)

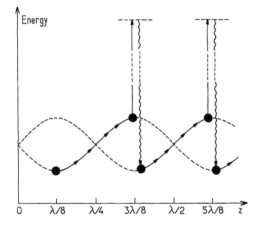

Illustration of the Sisyphus effect. The $m_s = \pm 1/2$ levels of the ground state become position-dependent. On the horizontal axis, the position is shown in units of the wavelength λ of the laser beams. On the average, an atom is slowed, because it moves more often 'uphill' (indicated by arrows) than 'downhill'. From [7]. Reprinted with permission of the Optical Society of America.

While the Doppler limit was now removed, a new barrier appeared. The photon, which is re-emitted after a resonance-absorption process, imparts a recoil momentum on the atom, leading to a random motion that is again equivalent to a temperature. For sodium atoms this *recoil limit* is $1.2\,\mu$K, i.e., 200 times lower than the Doppler limit. This limit was undercut by the Paris group, who succeeded to prepare cold atoms in a particular state, in which they would neither absorb nor re-emit photons. The experiment was done with helium atoms, possessing a resonance line with a ground state that can exist in the form two Zeeman states, each described by its own wave function. But the ground state can also be a (coherent) superposition of these two wave functions, in which case resonance transitions are inhibited. Atoms with such a ground state can be accumulated with time. The new method, called *velocity-selective coherent population trapping*, was demonstrated in optical molasses in one [9], two [10], and, finally, in three [11] dimensions. A temperature of $180\,$nK was reached, a value 22 times smaller than the recoil limit for helium. At this temperature helium atoms 'crawl' with velocities around $1\,\text{cm/s}$.

[1] Hänsch, T. W. and Schawlow, A. L., *Optics Communications*, **13** (1975) 68.

[2] Chu, S., Hollberg, L., Bjorkholm, J. E., Cable, A., and Ashkin, A., *Physical Review Letters*, **55** (1985) 48.

[3] Migdall, A., Prodan, J., and Phillips, W. D., *Physical Review Letters*, **54** (1985) 2596.

[4] Prodan, J., Migdall, A., Phillips, W. D., So, I., Metcalf, H., and Dalibard, J., *Physical Review Letters*, **54** (1985) 992.

[5] Raab, E. L., Prentiss, M., Cable, A., Chu, S., and Pritchard, D. E., *Physical Review Letters*, **59** (1987) 2631.

[6] Lett, P. D., Watts, R. N., Westbrook, C. I., Phillips, W. D., Gould, P. L., and Metcalf, H. J., *Physical Review Letters*, **61** (1988) 169.

[7] Dalibard, J. and Cohen-Tannoudji, C., *Journal of the Optical Society of America*, **B 6** (1989) 2023.

[8] Ungar, P. J., Weiss, D. S., Riis, E., and Chu, S., *Journal of the Optical Society of America*, **B 6** (1989) 2058.

[9] Aspect, A., Arimondo, A., Kaiser, R., Vansteenkiste, N., and Cohen-Tannoudji, C., *Physical Review Letters*, **61** (1988) 826.

[10] Lawall, J., Bardou, F., Saubamea, B., Leduc, M., and Cohen-Tannoudji, C., *Physical Review Letters*, **73** (1994) 1915.

[11] Lawall, J., Coulin, S., Saubamea, B., Bigelow, M., Leduc, M., and Cohen-Tannoudji, C., *Physical Review Letters*, **75** (1995) 4194.

There are Just Three Generations (1989)

In Episode 96, we mentioned that CERN was already planning a large electron–positron collider, named *LEP*, to study electroweak interactions and, in particular, the properties of the $W^{\pm}$ and Z^0 bosons, when a shortcut was proposed by which these bosons were discovered years earlier. Nevertheless, the LEP project was continued because it did offer the possibility of precision measurements which allowed thorough testing of the *standard model* of particle physics and the determination of its basic parameters. In this episode we concentrate on a single question: What is the number of lepton *generations*, i.e., are there more than the three known generations $(e, \nu_e), (\mu, \nu_\mu), (\tau, \nu_\tau)$, mentioned at end of Episode 93, and, if so, how many? Because of the symmetry between leptons and quarks, presumably, there are as many quark generations as there are lepton generations.

LEP was designed to operate in two stages, first at a total $e^+ e^-$ energy of around 100 GeV to produce the Z^0 $(e^+ + e^- \to Z^0)$ and later at about twice that energy to produce $W^{\pm}$ bosons in pairs $(e^+ + e^- \to W^+ + W^-)$. The machine had to be very large in order to keep the synchrotron radiation manageable, since the latter is inversely proportional to the radius but directly proportional to the fourth power of the energy. LEP consists of curved and straight sections; accelerating equipment and the experiments are placed in the latter. The ring tunnel, housing the machine, has a circumference of nearly 27 kilometres. The plane of the ring is not quite horizontal but follows the general

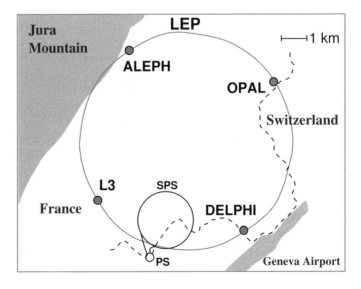

Schematic map of the LEP collider with its four experiments. (The two accelerators PS and SPS, which serve to pre-accelerate electrons and positrons, were described in Episode 96.) Reprinted from [1], copyright 2006, with permission from Elsevier.

inclination of the topography between Lake Geneva and the Jura Mountains. Schopper, under whose direction of DESY the PETRA collider had been built in record time, became Director General of CERN before construction works began. The LEP project proper was led by Picasso[1].

Four large international collaborations were formed to design and build the experimental set-ups and to prepare the analysis of the expected data. They named themselves *ALEPH, DELPHI, L3*, and *OPAL*. A collaboration, on average, comprised about 420 scientists from somewhat more than 30 universities and research institutes. (My own group from the University of Siegen was part of the ALEPH collaboration.) The detector designs were inspired by earlier collider experiments but the detectors were more precise and they allowed to better distinguish the types of particles produced. Differing in many details, they followed the main general principle: There were various shells of sub-detectors enclosing in their centre the LEP beam pipe in which the interactions took place. Also, there was a large magnetic field parallel to the beam. Close to the interaction region there was a central track detector, registering the momentum vectors of charged particles. All used the *drift-chamber* principle but were very different in actual design. Outside the track detector was an electromagnetic calorimeter, a hadron calorimeter, and finally a detection system for muons (compare also the design of UA1, Episode 96). Within each collaboration, different institutes usually worked on different sub-detectors, on the electronic and computer means to control them and to collect the data, and later on different of the very many physics topics, which could be explored by analysing these data.

The ALEPH detector, opened up and seen along the beam direction. This allows to distinguish the larger subdetectors. The beam pipe in the centre is surrounded by a large special drift chamber, recognizable by the characteristic pattern of its end faces. Outside it is the electromagnetic calorimeter, segmented in 12 modules. The ring-like structure is the cryostat housing the superconducting coil which generates the magnetic field. It is enclosed by the iron yoke which also serves as hadron calorimeter. On the very outside (in this picture only partially mounted near the top) are muon chambers. Steinberger, founder and spokesman of the ALEPH collaboration, stands in front of the detector, kindly giving an impression of its size.

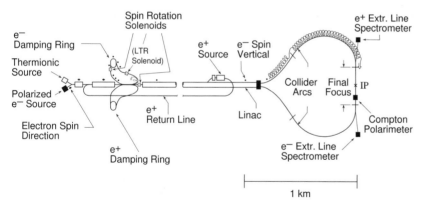

Schematic map of the Stanford Linear Collider SLC. The detector (not shown) encloses the interaction point (IP). Reprinted from [1], copyright 2006, with permission from Elsevier.

Competing with LEP was an ambitious project at Stanford, the *Stanford Linear Collider* (SLC). It was based on the existing two-mile long Stanford Linear Accelerator for electrons. The machine was adapted also to produce and accelerate positrons. A packet of electrons and one of positrons were given an energy of about 46.5 GeV. Next, guided by magnets, electrons took a curved path, called 'arc', and positrons a similar, mirror-symmetric path. Then the two packets were made to collide. The arcs had a strong curvature and about 1 GeV of particle energy was lost in each by synchrotron radiation. Particles which did not collide were lost. Because of this, there are much less colli-

---

[1] Emilio Picasso (born 1927)

The Line Shape of the Z^0 Resonance

The production of the Z^0 boson by an e^+e^- collider and its subsequent decay into a fermion–antifermion pair $f\bar{f}$ can be written as

$$e^+ + e^- \to f + \bar{f} \quad . \tag{1}$$

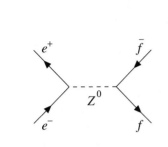

Diagram for the production of a Z^0 boson in an electron–positron collision and its subsequent decay into a fermion–antifermion pair.

All fundamental fermions, i.e., quarks, charged leptons, and neutrinos can appear in pairs on the right-hand side of (1), provided the energy suffices for the production of such a pair. The final-state particles are hadrons if f is one of the five quarks u, d, s, c, b; the t quark is too heavy to be produced. The three known charged leptons e, μ, τ are produced as well as their partner neutrinos ν_e, ν_μ, ν_τ, which, essentially, are massless. Should there exist one or more charged leptons $L, L', \ldots$ with masses larger than half the Z^0 mass, then these cannot be produced. The corresponding neutrinos $\nu_L, \nu_{L'}, \ldots$, however, presumably are also very light and are produced. Although they cannot be detected, they leave their mark on the Z^0 resonance curve.

We denote by E_{cm} the total e^+e^- energy in the centre-of-mass system of the collision and by $s = E_{\mathrm{cm}}^2$ its square. The total cross section for the production of hadrons in reaction (1) is (with $c = 1$)

$$\sigma_{\mathrm{had}}(s) = \sigma_{\mathrm{had}}^0 \frac{s\Gamma_Z^2}{(s - M_Z^2)^2 + s^2\Gamma_Z^2/M_Z^2}[1 + \delta_{\mathrm{rad}}(s)] \quad . \tag{2}$$

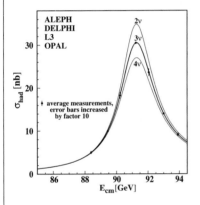

The line shape of the Z^0 resonance computed for 2, 3, and 4 light neutrino generations. The points are the combined measurements by all four experiments taken in their complete running period at LEP. Reprinted from [1], copyright 2006, with permission from Elsevier.

M_Z is the mass of the Z^0 boson and Γ_Z the width of the resonance curve, also called *total width* of the Z^0. The term $\delta_{\mathrm{rad}}(s)$ describes the fact that the initial electron or positron in (1) can radiate a photon and thus effectively reduce s. This 'radiative correction' is quantitatively important. But since it can be precisely computed in quantum electrodynamics, we do not discuss it further. The expression for the peak cross section,

$$\sigma_{\mathrm{had}}^0 = 12\pi \frac{\Gamma_{ee}\Gamma_{\mathrm{had}}}{M_Z^2\Gamma_Z^2} \quad , \tag{3}$$

contains the *partial widths* Γ_{ee} (describing the production vertex in the diagram on the margin) and Γ_{had} (describing the decay vertex). The total width,

$$\Gamma_Z = \Gamma_{\mathrm{had}} + \Gamma_{\mathrm{charged\ leptons}} + \Gamma_{\mathrm{neutrinos}}$$

$$= \sum_{q=u,d,s,c,b} \Gamma_{q\bar{q}} + \sum_{\ell=e,\mu,\tau} \Gamma_{\ell\bar{\ell}} + \sum_{\nu=\nu_e,\nu_\mu,\nu_\tau,\ldots} \Gamma_{\nu\bar{\nu}} \quad , \tag{4}$$

is the sum over the partial widths for all possible fermions. All partial widths can be calculated in the theory of electroweak interactions. (Except for $\Gamma_{\nu\bar{\nu}}$ they can also be directly measured.) The only unknown in the form (2) of the cross-section curve or in the peak cross section (3) is the number of different neutrino species participating in the Z^0 decay.

sions in a linear collider than in a storage ring like LEP, unless the particle packets are made tiny in lateral extension (about 1 micrometre or less) and are aimed at each other with utmost precision. On the other hand, a linear collider, consisting of two linear accelerators opposing one another, is not at all limited by synchrotron radiation and is considered to be the e^+e^- collider of the future. The *Mark II* detector, which had served well at the PEP storage ring at Stanford, was moved to the SLC. It was later replaced by a new detector, named *SLD*. The SLC, with Mark II installed, began to operate in 1989, several months before LEP. An initial measurement of the Z^0 parameters was published in August [2]. There were, however, not enough data to give a clear-cut answer to our question about the number of generations.

That this question can be answered by measuring the Z^0 resonance is owed to Heisenberg's uncertainty principle (Episode 42). Written as $\Delta E\,\Delta t = \hbar/2$, it states that the uncertainty or *width* in energy E (or mass M_Z) is the larger, the smaller is the uncertainty in the time t for which the Z^0 boson exists. That time is simply the average lifetime τ_Z of the Z^0. If we denote by Γ_Z the total width in energy of the Z^0 resonance, the uncertainty relation reads $\tau_Z = \hbar/(2\Gamma_Z)$. The width Γ_Z is equal to the sum of the so-called partial widths of all possible Z^0-decay processes (see Box). Every decay process increases the width and thus shortens the lifetime. That is also true for decays in which only a pair of neutrinos is produced, e.g., $Z^0 \to \nu_e + \bar{\nu}_e$, which leave no trace in a detector. All partial widths can be precisely computed in the theory of electroweak interactions. They can be added up for all decay channels and compared with the experimental width, obtained by measuring the total and partial cross sections (which are proportional to the number of events per e^+e^- collision) as a function of the total e^+e^- energy. Possible are all decays into a fermion f and its antiparticle $\bar{f}$, as long as they are energetically allowed, i.e., as long as the mass of the fermion is smaller than half the Z^0 mass. Known fermions contributing are the quarks u, d, s, c, b (which give rise to hadrons), the charged leptons e^-, μ^-, τ^-, and their neutrinos ν_e, ν_μ, ν_τ. Suppose there exists a further lepton L^- which is too heavy to appear in the Z^0 decay. Then its neutrino ν_L, in all probability, is light and contributes to the width Γ_Z. A precise measurement of the Z^0 resonance curve therefore reveals the existence of new light neutrinos, 'light' signifying a mass of less than $45\,\mathrm{GeV}/c^2$. Most sensitive to the number of neutrino species is the measurement of the maximum or peak cross section for the production of hadrons in Z^0 decays. Comparison of theory and measurement yields the number N_ν of generations as a decimal fraction with a measurement error. The true value is, of course, an integer.

On 13 October 1989, the four LEP collaborations presented their results in the CERN amphitheatre. The Mark II group had been a little faster and given a public report at Stanford the day before. Taking the five experiments together, an additional fourth neutrino type was excluded already then. A few months later, with more data, all groups published results with much higher precision. There was definitely no additional generation, at least not a conventional one with a light neutrino. (In our Box we show the Z^0 resonance curve, computed for 2, 3, and 4 neutrino generations as well as the complete and combined data of all LEP collaborations.)

LEP and its four detectors ran highly successfully until the end of 2000. In spite of its smaller luminosity, SLC could compete in selected questions, since it could provide electrons and positrons with their spins aligned parallel or anti-parallel to their direction of motion, thus facilitating to probe the 'left-handedness' of the weak interaction (Episode 79). A proton–proton collider, called *LHC* for large hadron collider, was then installed in the LEP tunnel. Not limited by synchrotron radiation it will reach a total energy of $14\,000\,\mathrm{GeV} = 14\,\mathrm{TeV}$ and is expected to start up while this book goes to print.

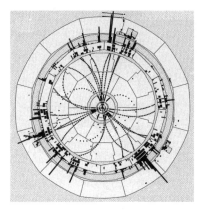

The decay of a Z^0 boson into hadrons as registered in the ALEPH detector. Trajectories of charged particles are seen in the centre. The inner and outer ring of bar diagrams show the energy distributions observed in the electromagnetic and the hadron calorimeter, respectively. Events, in which the Z^0 decays into a pair of charged leptons, are very different and easily distinguished from hadronic events. Figure: ALEPH Collaboration.

Group	N_ν	Reference
Mark II	2.8 ± 0.6	[3]
ALEPH	3.27 ± 0.3	[4]
DELPHI	2.4 ± 0.64	[5]
L3	3.42 ± 0.48	[6]
OPAL	3.12 ± 0.42	[7]

The first measurements of number N_ν of neutrino generations.

[1] ALEPH, DELPHI, L3, OPAL, and SLD Collaborations; LEP Electroweak Working Group; SLD Electroweak and Heavy Flavour Groups, *Physics Reports*, **427** (2006) 257.

[2] Abrams, G. S. et al., *Physical Review Letters*, **63** (1989) 724.

[3] Abrams, G. S. et al., *Physical Review Letters*, **63** (1989) 2173.

[4] ALEPH Collab.: Decamp, D. et al., *Physics Letters*, **B 231** (1989) 519.

[5] DELPHI Collab.: Aarino, P. et al., *Physics Letters*, **B 231** (1989) 539.

[6] L3 Collab.: Adeva, B. et al., *Physics Letters*, **B 231** (1989) 509.

[7] OPAL Collab.: Akrawy, M. Z. et al., *Physics Letters*, **B 231** (1989) 530.

Bose–Einstein Condensation of Atoms (1995)

The phenomenon, which is now called Bose–Einstein condensation, was predicted by Einstein in 1925 for an ideal gas of indistinguishable particles with rest mass on the basis of a new statistics introduced by Bose (Episode 33). It was shown later (Episode 41) that these particles have to have integer spin $(0, \hbar, 2\hbar, \ldots)$; they were called bosons. As the temperature of a gas of bosons drops below a critical temperature T_c, which depends on the mass of the atoms and their density in space, a fraction of all bosons (rising as the temperature falls further) accumulates in the state of lowest energy. The remaining ones still follow a statistical distribution. Einstein put it like this [1]:

> A separation set in; a part "condenses", the rest remains a "saturated" ideal gas.

Einstein used double quotes on two words, because in the concept of an *ideal gas* there is no interaction between its constituents, whereas condensation and saturation are properties of a *real gas* of interacting atoms or molecules. The 'interaction' bringing about Bose–Einstein condensation is the symmetry requirement for multi-particle wave functions of bosons, sometimes called *exchange force* (see page 176). For this to become effective, the wave functions of individual bosons have to overlap, i.e., the average (or *thermal*) de Broglie wavelength of the bosons has to be on the order of or larger than the average distance between them. In other words, a Bose–Einstein condensate is expected to form if the ratio of wavelength and distance exceeds a certain value. (The third power of this ratio is called phase-space density, see margin.) At very low temperature the average distance can be quite large; interactions between bosons become unimportant and the situation of an ideal gas is well approximated. We saw in Episode 97 how such temperatures for samples of neutral atoms were reached. These methods, greatly simplified and extended, were the basis for the achievement of Bose–Einstein condensates by the groups of Wieman and Cornell [2] and of Ketterle [3] in 1995, seventy years after Einstein's prediction.

There is no stable elementary boson with mass. The photon is massless and therefore does not qualify for Bose–Einstein condensation. Laser light, a coherent state of many photons, though showing some similarities, is not a Bose–Einstein condensate, having neither a well-defined number of particles nor a temperature. Composite bosons are therefore used in experiments, in particular, atoms of alkali metals. These have one electron (spin $\hbar/2$) in the outer shell; the spins of all other electrons cancel. For isotopes with a half-integer nuclear spin, i.e., an odd atomic number A, the total spin of the atom is integer: The atom as a whole is a boson, albeit composed of fermions. The very first condensate was made of rubidium atoms, soon followed by a sodium

Condition for Bose–Einstein condensation

A gas of identical bosons is characterized by its density n (number of particles per unit volume), the particle mass m, and the temperature T. The *thermal de Broglie wavelength* is defined as

$$\lambda_{\mathrm{dB}} = h/\sqrt{2\pi m k T} \quad,$$

where h and k are Planck's and Boltzmann's constants, respectively.

The average distance between bosons is $d = n^{-1/3}$. The wave functions of bosons overlap if $\lambda_{\mathrm{dB}} \approx d$. The exact condition for Bose–Einstein condensation is

$$\rho_{\mathrm{ps}} = n\lambda_{\mathrm{dB}}^3 > 2.612 \quad.$$

The quantity ρ_{ps} is called *dimensionless phase-space density*.

condensate. Originally, hydrogen seemed the obvious choice. But, not lending itself to laser cooling, it was condensed only in 1998 [4].

A decisive, technological step was taken by Wieman in 1990. Wieman[1] was born in Corvallis, a small university town in Oregon. His father worked in the lumber industry and the family lived far out in the country before moving into Corvallis. After high school Wieman was accepted to study at MIT. Already as an undergraduate he spent much time in the laboratories of Kleppner[2] and of David Pritchard, who was then a postdoctoral fellow. Both were to contribute significantly to the field of Bose–Einstein condensation. Pritchard took part in the realization of the magneto-optical trap (MOT), Episode 97, and it was Kleppner who produced the first condensate of hydrogen. Wieman's activities included the construction of a recently invented device, a dye laser with tunable frequency. He was so much absorbed by his work that he even gave up his dormitory room and lived in the lab for half a year. After receiving his B.Sc. in 1973, Wieman went to Stanford University. For his Ph.D., obtained under Hänsch in 1977, he built new dye lasers and did laser spectroscopy on hydrogen, including Lamb-shift measurements. Later Wieman worked for several years at the University of Michigan on the detection of parity violation, i.e., effects of the electroweak interaction, in atomic physics. He found it promising to use caesium atoms and pursued this idea when he became professor at the University of Colorado in Boulder, where he would work at the *Joint Institute for Laboratory Astrophysics* (JILA), run by the university together with the National Institute of Standards and Technology (NIST). Within a year the difficult experiment was done successfully in cooperation with his young wife.

Wieman (left) and Cornell.

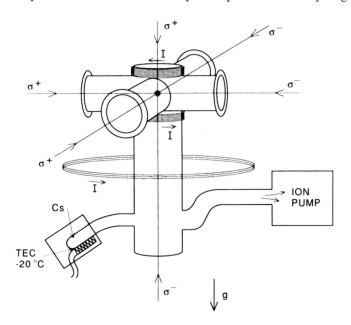

The glass vapour cell for laser cooling and trapping of caesium atoms developed by Wieman's group. The pressure of caesium atoms is defined by the vapour pressure of caesium at the temperature maintained by a thermoelectric cooler (TEC). The central feature is the magneto-optical trap (MOT) with its two coils, wound directly on the glass body, and its six laser beams, see also Episode 97. The symbols σ^+, σ^- stand for right-handed and left-handed circular polarized laser light, respectively. Figure reprinted with permission from [5]. Copyright 1990 by the American Physical Society.

Wieman's technological revolution of laser cooling and trapping at first was simply a by-product of the caesium work. His low-cost design, realized with a

---

[1] Carl Edwin Wieman (born 1951), Nobel Prize 2001   [2] Daniel Kleppner (born 1932)

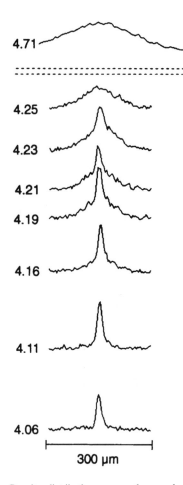

4.71

4.25

4.23

4.21

4.19

4.16

4.11

4.06

|———————————|
300 µm

Density distributions across the sample of cold atoms as observed by the group of Wieman and Cornell. The curves can also be seen as distributions of a velocity component, see text. The numbers on the left give the frequency (in MHz) determining the depth of the evaporative trap and are a measure for the temperature of the gas of rubidium atoms. In the figure the temperature decreases from top to bottom. The curve at 4.25 MHz corresponds to a temperature of 170 nK, a density of 2.6×10^{12} cm^{-3}, and a total number of 2×10^4 atoms. At this temperature and below a separation becomes apparent: One fraction, growing as the temperature falls, displays a bell-shaped (Gaussian) density similar to that of the ground state in a harmonic oscillator; the other corresponds to higher energies and velocities. From [2]. Reprinted with permission from AAAS.

small group, soon became a new standard. It replaced the bulky stainless-steel vessel with a simple glass cell and tunable dye lasers with vastly cheaper solid-state lasers. A 'cold finger' contained caesium, kept at a temperature of about $-20°$C by a thermo-electric cooler (TEC). The very low vapour pressure at that temperature defined the pressure of caesium atoms in the cell. Optical molasses, created by laser beams, collected atoms directly from that vapour. The coils for the magneto-optical trap were a few turns of wire wound around the cell. In this cell trapped atoms at a temperature of 1 µK were recorded in 1990 [5], two orders of magnitude lower than achieved by others. This was the time when Cornell joined Wieman's group as a postdoctoral fellow.

Cornell[3] was born in Palo Alto, California, where his parents were completing their graduate studies. He grew up in Cambridge, Massachusetts; his father had become professor of civil engineering at MIT. Cornell followed undergraduate studies at Stanford, interrupted by a year of travel and language perfection in Taiwan, Hong Kong, and mainland China. For graduate studies he went to MIT in 1985, doing his doctoral research under Pritchard on precision measurements on single, trapped ions. Cornell obtained his Ph.D. in 1990. A little earlier, while looking for a postdoctoral position, he had visited Wieman, who showed him the vapour cell and accepted him into his group. At MIT, Kleppner was already deeply involved in his attempts to achieve Bose–Einstein condensation of hydrogen. It was therefore quite natural for Cornell to think of a condensate of alkali atoms in a vapour cell.

To reach that goal further cooling was needed. At MIT the method of *evaporative cooling* had been pioneered in 1986 [6]. Atoms were loaded into a magnetic trap. By allowing those with the highest energy to leave, the average energy, i.e., the temperature of the remaining ones sank. The process could be influenced by disturbing the trap field with an external radio-frequency field, which was equivalent to a gradual decrease in trap depth. Cornell built a new vapour cell for rubidium atoms, in which he installed a magneto-optical trap with evaporative cooling. It worked as expected: The cooled sample of atoms contracted near the 'bottom' of the trap, where there is no magnetic field, but was then lost. (We saw in Episode 97 that the field is needed to align the magnetic moment, so that a well-defined force can act on it.) To solve this problem Cornell superimposed a time-dependent magnetic field, which moved the field minimum around and affected the atoms only in its time-averaged form. Using this improved trap with its time-averaged, orbiting potential (TOP) he reached vastly larger phase-space densities and storage times on the order of 100 seconds [7]. The TOP potential is that of a harmonic oscillator, dear to every student of quantum mechanics, with its well-known bound states and wave functions (see figure on page 355).

In June 1995 Bose–Einstein condensation of rubidium atoms (^{87}Rb) was observed in Boulder [2]. In a well-timed series of consecutive steps, atoms were collected, laser cooled, trapped in a MOT, and cooled further. The laser light was then removed and a TOP constructed around the sample, which was then cooled by evaporation. Finally, the trap field was removed and the sample of atoms was allowed to expand freely for 60 milliseconds, after which

---

[3] Eric Allin Cornell (born 1961), Nobel Prize 2001

time the spatial density of the cloud of atoms was measured. This was done by producing a shadow image in resonant light. A transverse scan through the image yields a density distribution. As mentioned the TOP potential is that of a harmonic oscillator. The probability density of its ground state has a Gaussian, bell-shaped form both in spatial and momentum (or velocity) coordinates. Such a distribution, the more pronounced the lower the temperature, was indeed observed. The experimental procedure was automated, yielding well-reproducible results and allowing quantitative measurements under different conditions, although, in the final step, the sample was destroyed by the illumination with resonant light. At temperatures below 170 nK, the separation of the gas of bosons, expected by Einstein, did set in. Indeed, a part showed the properties similar to the ground state in a harmonic-oscillator potential; the rest had higher energies (see figure). This was a clear signature of Bose–Einstein condensation – clearer, in fact, than had been expected by physicists before the experiment.

A few months later much larger samples of atoms were condensed by the group of Ketterle at MIT. Ketterle[4] was born in Heidelberg and began to study physics at the old university in his home town. After the first exam, he changed to the Technical University in Munich. Having obtained his diploma there in 1982 with a thesis on theoretical solid state physics, Ketterle did experimental work for his doctoral dissertation under Walther at the Max-Planck-Institut für Quantenoptik in Garching. With research on laser spectroscopy of triatomic hydrogen (H_3) and helium hydride (HeH) he received his Ph.D. in 1986. Ketterle returned to Heidelberg for postdoctoral research in applied physics and, in 1990, joined Pritchard's group at MIT as a research fellow.

Ketterle

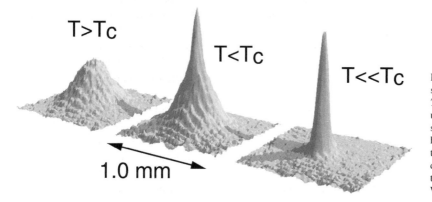

T>Tc   T<Tc   T<<Tc

1.0 mm

Formation of a Bose–Einstein condensate of sodium ions below the critical temperature T_c as observed by Ketterle's group. The figure was generated by computer from a density recording in two dimensions. The two horizontal dimensions indicate the position in the horizontal plane, the vertical dimension is density. A very similar figure is shown by the group of Wieman and Cornell. Figure: W. Ketterle, MIT.

Inspired by Wieman's work, Ketterle built a vapour cell for sodium atoms, developed a new type of trap, and became assistant professor. In 1993 Pritchard made him the generous offer to hand over to him the project to condense sodium, including laboratory and funding, if he would stay at MIT. Ketterle accepted and Bose–Einstein condensation was reached two years later [3]. For this work the hole in the magneto-optical trap was not plugged by another magnetic field but by an intensive very thin laser beam directed through the centre of the trap. The light of higher frequency than the sodium resonance induced

---

[4] Wolfgang Ketterle (born 1957), Nobel Prize 2001

electric dipole moments in the atoms, which were then pushed away from the centre by the electric field of the laser light.

Once Bose–Einstein condensates had been produced, it became possible to study their properties in detail. The all-important feature is *coherence*, i.e., the existence of a well-defined quantum-mechanical state of the ensemble of atoms. In fact, not all atoms participating in the formation of a Bose–Einstein condensate are in the ground state but about ten states are involved, each occupied by very many atoms. That they, together, form a coherent state was shown in an interference experiment by Ketterle's group [8]. A trap was constructed, in which two condensates could be formed side by side. The fields, forming the trap, were then removed and two condensates were allowed to fall freely, to expand, and thus to overlap. The density in the overlap region was measured by non-destructive imaging with non-resonant light. It showed pronounced interference fringes, completely analogous to those observed in the intensity pattern of two overlapping, coherent light beams.

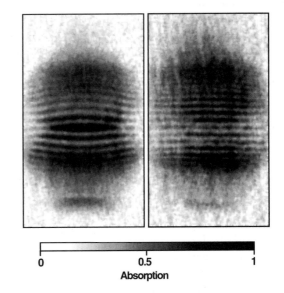

Interference pattern of two Bose–Einstein condensates of sodium atoms. The fields of view are 1.1 mm horizontally by 0.5 mm vertically. The two images correspond to somewhat different initial separations and expansion times for the condensates. From [8]. Reprinted with permission from AAAS.

A source of coherent matter waves, comprising large numbers of atoms, has been called an *atom laser* [9]. The name stuck despite the principal difference between light and matter, mentioned earlier, and even though there is, of course, no stimulated emission, i.e., creation of atoms, as the name might imply. Pulsed atom lasers were built at MIT [10] in the form of traps releasing 'blobs' of Bose–Einstein condensate for further study. In 1997 the group of Hänsch, now at the Max-Planck-Institut für Quantenoptik in Garching, built a continuous atom laser, in which a collimated coherent beam of rubidium atoms was extracted from a trap over a period of 100 milliseconds [11].

[1]   Einstein, A., *Sitzungsber. Preuss. Akad. Wiss. (Berlin)*, (1925) 3.

[2]   Anderson, M. H., Ensher, J. R., M. R, M., Wieman, C. E., and Cornell, E. A., *Science*, **269** (1995) 198.

[3]   Davis, K. B., Mewes, M. O., Andrews, M. R., van Druten, N. J., Durfee, D. S., Kurn, D. M., and Ketterle, W., *Physical Review Letters*, **75** (1995) 3969.

[4]   Fried, D. G., Kilian, T. C., Willmann, L., Landhuis, D., Moss, S. C., Kleppner, D., and Greytag, T. J., *Physical Review Letters*, **81** (1998) 3811.

[5]   Monroe, C., Swann, W., Robinson, H., and Wieman, C., *Physical Review Letters*, **65** (1990) 1571.

[6]   Hess, H. F., *Physical Review*, **B 34** (1986) 3476.

[7]   Petrich, W., Anderson, M. H., Ensher, J. R., and Cornell, E. A., *Physical Review Letters*, **74** (1995) 3352.

[8]   Andrews, M. R., Townsend, C. G., Miesner, H.-J., Durfee, D. S., Kurn, D. M., and Ketterle, W., *Science*, **275** (1997) 637.

[9]   Wiseman, H. M., *Physical Review*, **A 56** (1997) 2068.

[10] Mewes, M. O., Andrews, M. R., Kurn, D. M., Durfee, D. S., Townsend, C. G., and Ketterle, W., *Physical Review Letters*, **78** (1997) 582.

[11] Bloch, I., Hänsch, T. W., and Esslinger, T., *Physical Review Letters*, **82** (1999) 3008.

Neutrinos Have Mass (1998, 2001)

In the early times of particle physics, discoveries were made in cosmic rays; but soon accelerators took over as a source of primary particles. The most elusive of particles, the neutrino, was finally observed, because low-energy neutrinos are copiously produced in a nuclear reactor (Episode 76). Accelerators were used for experiments with high-energy neutrinos, which more readily interact with matter (Episodes 85 and 91). But eventually neutrino physics turned to cosmic sources with a surprising result: Neutrinos have mass. In his first printed remarks on the neutrino (quoted on page 202) Pauli stated that its mass had to be small and might be equal to zero. The latter possibility came to be widely accepted, in particular, since it simplified the theoretical description (Episode 79). In principle, the neutrino mass can be measured in a β-decay process. But elaborate experiments yielded only an upper limit: The mass of the (electron) neutrino ν_e (or, to be more precise, of the $\bar{\nu}_e$) is less than $2\,\mathrm{eV}/c^2$.

Davis[1] is the pioneer of experiments with neutrinos from the sun. He was born in Washington as son of a photographer at the National Bureau of Standards, studied chemistry at the University of Maryland, and received his Ph.D. in physical chemistry from Yale in 1942. Until the end of the war he worked on weapon tests as a reserve officer and later in the laboratories of a chemical company. In 1948, after a few years in the chemical industry, Davis joined the newly founded Brookhaven National Laboratory. There, the chairman of the chemistry department suggested that he choose a research topic that appealed to him. Gratefully, Davis spent some time in the library and, through a recent review article, became intrigued with the possibility to search for the neutrino. He began with experiments on the recoil of nuclei in β decay, which furnished indirect information on the neutrino. In 1951, Davis started his work on the neutrino-induced reaction $\nu_e + {}^{37}\mathrm{Cl} \rightarrow {}^{37}\mathrm{Ar} + e^-$, which had been discussed by Pontecorvo already in 1946. In 1948, Alvarez had suggested to build a large tank, fill it with a solution of sodium chloride (kitchen salt), and detect solar neutrinos by finding some ${}^{37}\mathrm{Ar}$ atoms. Neither Alvarez nor anybody else had actually performed a chlorine–argon experiment. Davis decided to try it. Rather than a salt solution he used a cleaning liquid, carbon tetra chloride $\mathrm{CCl_4}$, and developed methods to 'wash out', collect, and detect the argon. ${}^{37}\mathrm{Ar}$ is radioactive, it decays back into ${}^{37}\mathrm{Cl}$ by *K capture* (see end of Episode 53) and the ${}^{37}\mathrm{Cl}$ atom emits an X ray of characteristic energy; the half-life is 35 days. Davis performed several experiments without, however, detecting a signal from solar neutrinos. His detector simply was not large enough.

In parallel to the work by Davis, a detailed model of nuclear reactions in

Davis

---

[1] Raymond Davis Jr. (1914–2006), Nobel Prize 2002

The chlorine detector of Davis in the Homestake Gold Mine. Photo: Brookhaven National Laboratory.

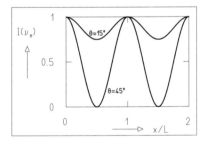

Illustration of neutrino oscillations for the case of only two neutrino species. Electron neutrinos ν_e of fixed energy travel in x direction. Since they can change into muon neutrinos ν_μ and back into ν_e, the intensity $I(\nu_e)$ of electron neutrinos decreases and increases periodically. For complete mixing ($\theta = 45°$, lower curve) it can pass through zero.

the sun, leading to the production of neutrinos, was developed by Bahcall[2] in Pasadena. We do not discuss here the complicated energy spectrum of neutrinos from the sun but only mention that the chlorine method is only sensitive to solar neutrinos of the highest energy. In March 1964, two papers appeared back-to-back in *Physical Review Letters*, one by Bahcall [1] describing his solar model, the other by Davis [2] proposing an experiment sensitive enough to test it. Bahcall predicted the probability per second for a single ^{37}Cl atom on the earth to be transformed into ^{37}Ar to be about 4×10^{-35}. When more data on nuclear reactions in the laboratory became available and were entered into the solar model, that value sank by quite a bit. A neutrino flux corresponding to the transformation of 1 nucleus out of 10^{36} per second (^{37}Cl in the case of chlorine experiments) is now named 1 *solar neutrino unit* or, for short, 1 SNU. Davis concluded that he needed a detector of about 100 000 gallons of liquid C_2Cl_4, another cleaning liquid, to register a signal in a reasonable time. The detector had to be installed deep below ground to be shielded from cosmic particles other than solar neutrinos.

Four years later, Davis and his collaborators had indeed constructed a detector of that size, filled with 390 cubic metres of C_2Cl_4, and installed in a depth of nearly 1500 metres in the Homestake Gold Mine in Lead, South Dakota. The group's first publication [3] was negative. No signal from solar neutrinos was found but the sensitivity of the experiment allowed them to derive 3 SNU as upper limit for the neutrino flux. The newest version of the solar model, published in a companion paper [4], predicted (7.3 ± 3) SNU. Thirty years later the discrepancy still persisted, and it was known to much higher precision. Davis and his group measured a flux of (2.56 ± 0.23) SNU [5] and Bahcall's group predicted (7.6 ± 1.2) SNU [6].

In 1968, Pontecorvo, who contributed so many ideas to neutrino physics, together with Gribov[3], offered an explanation for the phenomenon of seemingly missing solar neutrinos [7]. It was based on the idea of *mixing* (which we discussed in Episode 88 for the $K^0 \bar{K}^0$ system) and required two rather radical assumptions: First, there are neutrinos of different, non-zero mass; we call them ν_1 and ν_2. Second, the lepton numbers are no longer strictly conserved; the electron neutrino ν_e and the muon neutrino ν_μ appear as superpositions of ν_1 and ν_2. If, for brevity, we use the particle symbols to denote wave functions, then the mixing can be expressed by

$$\nu_e = \nu_1 \cos\theta + \nu_2 \sin\theta \quad , \qquad \nu_\mu = -\nu_1 \sin\theta + \nu_2 \cos\theta \quad ,$$

where θ is the mixing angle. Just as in the $K^0 \bar{K}^0$ system, ν_e and ν_μ change into one another because ν_1 and ν_2 have slightly different oscillation frequencies. But, in contrast to the $K^0 \bar{K}^0$ system, the oscillation between ν_e and ν_μ goes on indefinitely, because the ν_1 and ν_2 do not decay. Consider a ν_e of energy E_ν to travel from $x = 0$ along the x direction. The probability for it to be a ν_e at position x is

$$P_{\nu_e}(x) = 1 - \sin^2(2\theta)\sin^2(\pi x/L) \quad , \qquad L = \frac{4\pi\hbar c E_\nu}{\Delta m^2} \quad .$$

---

[2] John Norris Bahcall (1934–2005)   [3] Vladimir Naumovich Gribov (1930–1997)

Here L is the *oscillation length* and $\Delta m^2 = m_1^2 - m_2^2$ is the difference between the squared masses of ν_1 and ν_2. (For more than two neutrino species the formulae become more complicated but the principle remains the same.)

While the assumption of *neutrino oscillations* could explain a diminished flux of solar electron neutrinos on the earth, most physicists expected that either the solar model or the detection methods had to be improved. But neither experiments, using other neutrino-induced nuclear reactions, nor further work on the solar model removed the discrepancy between theory and experiment. The matter was resolved with a new experimental technique, pioneered in Japan under the leadership of Koshiba.

Koshiba[4] was born in Toyohashi. He studied at the University of Tokyo and obtained his Ph.D. in 1955 from the University of Rochester with a thesis on cosmic rays. In the following years he worked at the universities of Chicago and Tokyo, from 1963 onwards at Tokyo. In the 1970s a unified theory of strong and electroweak interactions (based on the group $SU(5)$) was proposed, which implied a finite lifetime of the proton. The proton was expected to have a mean life of about 4.5×10^{29} years and to decay, for instance, into a positron and a neutral pion, $p \rightarrow e^+ + \pi^0 \rightarrow e^+ + \gamma + \gamma$. As an experimental test Koshiba proposed to install a large, water-filled tank deep underground and to look for signals from γ rays, accompanying the possible decay of a proton from the water. The walls of the tank were to be fitted with many highly sensitive photomultipliers to detect these signals. Special photomultipliers with a sensitive surface of 20 inch diameter were developed together with the company Hamamatsu [9] and a detector containing 3000 cubic metres of water was installed in a mine in Kamioka, 1000 metres below ground. It was called *Kamiokande*, where the last three letters stood for *nucleon decay experiment*. When the proton was found to be stable (or, more accurately, to have a lifetime at least a hundred times higher than predicted by $SU(5)$) these letters were re-interpreted as *neutrino detection experiment*. As we shall see, Kamiokande, which was largely a Japanese experiment, gave important results. It was succeeded by a large international effort, named Super-Kamiokande, a detector of 50 000 cubic metres, looked at by about 11 000 photomultipliers.

In 1988 Kamiokande reported what came to be known as the *atmospheric neutrino anomaly*. The principal source of high-energy neutrinos is the decay of charged pions, produced by cosmic radiation in the earth's atmosphere. The pions decay into muons and their neutrinos,

$$\pi^+ \rightarrow \mu^+ + \nu_\mu \quad , \qquad \pi^- \rightarrow \mu^- + \bar{\nu}_\mu \quad ,$$

and the muons into electrons and neutrinos of either type,

$$\mu^+ \rightarrow e^+ + \nu_e + \bar{\nu}_\mu \quad , \qquad \mu^- \rightarrow e^- + \bar{\nu}_e + \nu_\mu \quad .$$

Therefore, the number of atmospheric muon neutrinos is expected to be twice that of atmospheric electron neutrinos. By collisions in the detector the neutrinos can produce charged leptons. Electron neutrinos $(\nu_e, \bar{\nu}_e)$ yield electrons (e^-, e^+), muon neutrinos $(\nu_\mu, \bar{\nu}_\mu)$ muons (μ^-, μ^+). Electrons and muons are detected by their Cherenkov effect in water. Cherenkov light is emitted in a

---

[4] Masatoshi Koshiba (born 1926), Nobel Prize 2002

Koshiba

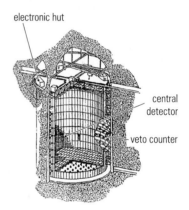

Super-Kamiokande. The detector is 41.4 m high and 39.3 m in diameter. Reprinted from [8], copyright 1992, with permission from Elsevier.

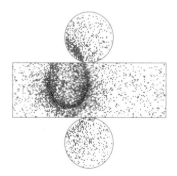

The pattern created by an electron in Super-Kamiokande. Barrel, bottom, and lid of the cylindrical detector are shown with the photomultier signals registered on them. Figure reprinted with permission from [10]. Copyright 2001 by the American Physical Society.

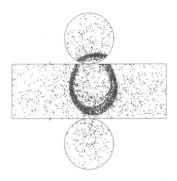

The pattern due to a muon. Figure reprinted with permission from [10]. Copyright 2001 by the American Physical Society.

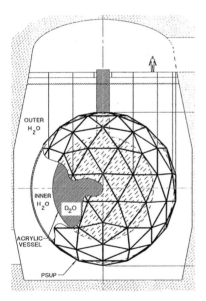

The Sudbury Neutrino Observatory SNO. The photomultiplier support structure PSUP has a diameter of 17.8 m. The spherical acrylic vessel (hatched area) contains $1000 \, \text{m}^3$ of heavy water D_2O and is surrounded by light water H_2O. Reprinted from [14], copyright 2000, with permission from Elsevier.

cone defined by the particle direction and velocity and registered as a ring in the photomultipliers on the detector surface. An electron initiates an electromagnetic shower in the water. The directions of the shower electrons scatter around the direction of the original electron. Therefore, the Cherenkov ring, caused by an electron, is somewhat fuzzy; electrons and muons can quite easily be distinguished. In a careful analysis of 277 events in Kamiokande it was concluded that there was a deficiency of atmospheric muon neutrinos: only $(59 \pm 7)\%$ of the expected number was observed [11].

In 1989 Kamiokande published on solar neutrinos [12], which, because of their lower energy, are more difficult to detect than atmospheric ones. The flux of neutrinos from the sun was found to agree with that measured by Davis. Moreover, for the first time, these neutrinos were actually found to come from the direction of the sun. That was possible because an electron, produced in the detector, practically follows the direction of the incident neutrino, which therefore is the axis of the Cherenkov cone.

Thus, at the end of the 1980s, not only solar electron neutrinos but also atmospheric muon neutrinos were missing. In 1998 Super-Kamiokande presented convincing evidence that the latter effect was caused by neutrino oscillations. The primary cosmic radiation, consisting mainly of high-energy protons, is isotropic, i.e., it has no preferred direction. The same holds for the neutrinos produced by it in the atmosphere. Because of their tiny interaction probability the earth is no obstacle for neutrinos; their directions, recorded in the detector, should also be isotropic. A striking anisotropy, however, was observed for muon neutrinos [13]; less came from below than from above. A muon neutrino ν_μ, travelling upwards through the earth, has a much larger chance to change into a neutrino of another kind than a neutrino travelling downwards from the atmosphere to the detector. If it changed into a tau neutrino ν_τ, then it may create a τ^- lepton. Because of the latter's large mass that is highly improbable and tau neutrinos go largely undetected. The complete analysis of the Super-Kamiokande data in terms of $(\nu_\mu \to \nu_\tau)$ oscillations was consistent with maximal mixing $(\theta = 45°)$ and gave a value for the difference between the squared neutrino masses, obtained from atmospheric neutrinos, of $\Delta m_{\text{atm}}^2 \approx 2 \times 10^{-3} \, \text{eV}^2/c^4$ and a lower limit for the mass of the heavier neutrino,

$$m_{\nu_1} \geq \sqrt{\Delta m_{\text{atm}}^2} \approx 0.045 \, \text{eV}/c^2 \quad .$$

Thirty years after they had been proposed, neutrino oscillations had now been observed. The bearing of this discovery on the original problem, the riddle of missing solar electron neutrinos, however, was still indirect since the observation had been in the (ν_μ, ν_τ) system. This still open question was finally answered by an elegant experiment with which it was not only possible to detect electron neutrinos from the sun but also muon and tau neutrinos. The experiment was set up by a large international collaboration in a mine near Sudbury in Ontario, Canada, slightly more than 2000 metres below the surface and named *Sudbury Neutrino Observatory* or, simply, SNO [14]. It is a water-filled Cherenkov detector like Kamiokande and Super-Kamiokande but in its centre there is an acrylic, transparent sphere filled with 1000 tons of heavy water. With the SNO detector it is possible to distinguish two different

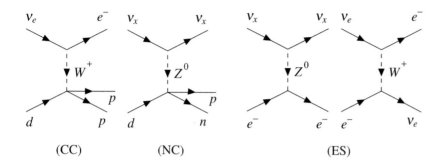

Diagrams for the reactions charged-current (CC) and neutral-current (NC) neutrino–deuteron scattering and for elastic neutrino–electron scattering (ES).

reactions between a neutrino from the sun and a deuteron d, the nucleus of deuterium, i.e., of heavy hydrogen, which is composed of a proton p and a neutron n. A third reaction, the elastic scattering of a solar neutrino by an electron, is also detected and appears different from the first two. The three reactions are

$$\nu_e + d \to p + p + e^- \quad \text{(CC)} \quad ,$$
$$\nu_x + d \to p + n + \nu_x \quad \text{(NC)} \quad ,$$
$$\nu_x + e^- \to \nu_x + e^- \quad \text{(ES)} \quad .$$

The first one, marked (CC) for *charged current*, i.e., W exchange (see Episode 91), can be brought about only by an electron neutrino, since only that can change into the electron on the right-hand side. The negative charge of the electron is compensated by the change of the neutron in the deuteron into a proton. Reaction (CC) is used by SNO to measure the flux $\phi(\nu_e)$ of solar electron neutrinos. The *neutral-current* reaction (NC) proceeds via Z^0 exchange and is independent of the neutrino type; ν_x can stand for ν_e, ν_μ, and ν_τ. It is used to measure the flux $\phi(\nu_e) + \phi(\nu_\mu, \nu_\tau)$ independent of the neutrino flavour. The *elastic-scattering* reaction (ES) takes place by Z^0 exchange for all neutrinos ν_x and, in addition, by W exchange for ν_e. It measures $\phi(\nu_e) + \phi(\nu_\mu, \nu_\tau)/6.5$. The SNO collaboration found that the flux of electron neutrinos is only one third of the total registered flux of solar neutrinos [15]. This was consistent with previous measurements, but the SNO experiment clearly demonstrated that not only did electron neutrinos disappear but other neutrino flavours appeared instead.

A second value for $\Delta m_\odot^2$ for the difference in squared masses is obtained from solar-neutrino oscillations. At the time of writing (2007) the experimental situation is consistent with the existence of three neutrino mass states ν_1, ν_2, ν_3, each being a mixture of the flavour states ν_e, ν_μ, ν_τ [16].

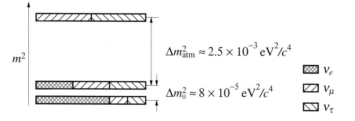

Spectrum of neutrino mass states. The fractions of ν_e, ν_μ, ν_τ are indicated by hatching.

[1] Bahcall, J. N., *Physical Review Letters*, **11** (1964) 300.

[2] Davis, R., *Physical Review Letters*, **11** (1964) 303.

[3] Davis, R., Harmer, D. S., and Hoffman, K. C., *Physical Review Letters*, **20** (1964) 1205.

[4] Bahcall, J. N., Bahcall, N. A., and Shariv, G., *Physical Review Letters*, **20** (1968) 1209.

[5] Cleveland, B. T. et al., *Astrophysical Journal*, **496** (1998) 505.

[6] Bahcall, J. N., Pinsonneault, M. H., and Basu, S., *Astrophysical Journal*, **555** (2001) 909.

[7] Gribov, V. and Pontecorvo, B., *Physics Letters*, **B 28** (1968) 493.

[8] Koshiba, M., *Physics Reports*, **220** (1992) 229.

[9] Kume, H. et al., *Nuclear Instruments and Methods*, **205** (1983) 443.

[10] Kajita, T. and Totsuka, Y., *Reviews of Modern Physics*, **73** (2001) 85.

[11] Hirata, K. S. et al., *Physics Letters*, **B 205** (1988) 416.

[12] Hirata, K. S. et al., *Physical Review Letters*, **63** (1989) 16.

[13] Fukuda, Y. et al., *Physical Review Letters*, **81** (1998) 1562.

[14] Boger, J. et al., *Nuclear Instruments and Methods*, **A 449** (2000) 172.

[15] Ahmad, Q. R. et al., *Physical Review Letters*, **87** (2001) 071301.

[16] Particle Data Group: Yao, W.-M. et al., *Journal of Physics*, **G 33** (2006) 1.

Epilogue – What Have We Learnt?
What is to Come?

Practically all of our hundred episodes deal with one of the two great themes of the twentieth century: the structure of space and time and the structure of matter. Progress in the former was achieved almost single-handedly by Einstein with his creation of the special and the general theory of relativity. Understanding of the latter was brought about by generations of experimentalists and theoreticians, one being Einstein.

The quest for the structure of matter may be seen as a triple task: finding the constituents of matter, identifying the forces between them, and understanding the general dynamical laws, applying to them. In the course of time more and more layers of constituents were found. Gradually, it became clear that, besides the gravitational force of Newton and the electromagnetic force understood in the nineteenth century, there are two more forces which were simply called strong and weak, respectively. Our present knowledge can be summarized like this. The forces are transmitted by fundamental bosons, namely, the electromagnetic force by the massless photon γ, the weak force by the very massive $W^{\pm}$ and Z^0 bosons, and the strong force by the gluon g. It is assumed that gravitation is also transmitted by a fundamental boson, the graviton. The fundamental fermions are the leptons and the quarks, each coming in three generations of pairs – leptons as $(e^-, \nu_e), (\mu^-, \nu_\mu), (\tau^-, \nu_\tau)$, quarks as $(u, d), (c, s), (t, b)$, as well as their antiparticles. Hadrons, like the proton and the neutron, are composed of quarks, which are bound together by gluons. Nuclei are made of protons and neutrons, atoms of nuclei and electrons, and macroscopic matter is made of atoms. The general dynamical laws for the constituents of matter are those of quantum mechanics. This theory, which can be seen as an ingenious extension of Newton's laws of motion, went through several stages of development, pushed along by new and more precise experiments: the quantum theory of Planck, Einstein, Bohr, and Sommerfeld, the quantum mechanics of Heisenberg, Dirac, and Schrödinger, the relativistic quantum mechanics of Dirac, and, finally, the quantum field theory. The latter was originally developed for the electromagnetic interaction and given the name quantum electrodynamics, after the inclusion of weak interactions it became the unified electroweak field theory. Applied to the strong interactions of quarks and gluons it is called quantum chromodynamics.

With the help of quantum mechanics mysterious and unexpected phenomena in macroscopic matters were explained, some long after they were discovered, such as superconductivity, superfluidity, the behaviour of conductors and semiconductors, and the quantum Hall effects. Others would scarcely have been observed, had they not been predicted by quantum mechanics, for instance, maser and laser action or Bose–Einstein condensation.

Progress often required novel instruments and creations of such instruments are discoveries in their own right. Prominent examples are the cloud chamber, the Geiger counter, the different particle accelerators, the nuclear reactor, the bubble chamber, the various traps for ions and atoms, and the laser. Of these the reactor, the transistor, and the laser have gained enormous industrial and practical importance.

Four episodes tell of work concerning a very large, cosmic scale. Three (the bending of light, Hubble's law, and the cosmic background radiation) are directly connected to general relativity. The fourth (energy production in stars) deals with nuclear physics and, of course, special relativity through the conversion of mass into energy.

Having squeezed a century of physics into a few sentences, we now ask: What will come next? Since predictions are difficult, the answer will be brief and simply list some open questions. In particle physics, the burning question is about the masses of fundamental particles. Why are they different? How can their values be explained? In Episode 90 we mentioned the Higgs mechanism, which might be responsible for the mass of particles. At the time of writing experiments are under way to search for the Higgs boson and the mechanism will gain credibility if this boson is found. But why are there three generations of leptons and quarks with different masses? Is there yet another substructure? Are these particles, after all, not fundamental?

General relativity, on a microscopic scale, has to be formulated in terms of quantum mechanics because this governs the dynamics on that scale. So far, these formulations, called quantum gravity theories, are not fully satisfactory. The graviton (corresponding to the photon in the quantum theory of the electromagnetic field) has not been found. Even gravitational waves on a macroscopic scale (corresponding to Hertzian waves in classical electrodynamics) have not yet been observed, but several large experiments are being set up with the aim to find them.

Details in the rotation of galaxies show that they contain a sizeable fraction of non-luminous 'dark' matter. It is considered probable that this is not simply composed of ordinary dust but is of some other origin. If so, what is it? Moreover, recent extensions of Hubble's work on the motion of very distant objects seem to indicate that in the cosmological models (Episode 89) a non-vanishing value for Einstein's cosmological constant is needed. This would imply the existence of some kind of repulsion between masses, in addition to the familiar gravitational attraction. What is the source of this mysterious repulsion, which is referred to as 'dark energy'?

The good understanding of quantum effects in macroscopic matter will certainly continue to provide us with new materials and new electronic and optical devices. But we may also be in for surprises as in the case of the quantum Hall effects.

Last, but certainly not least, physicists hope for the discovery of phenomena far beyond the expected, which they simply call 'new physics'. There was plenty of new physics in the twentieth century and there is bound to be more in the twenty-first!

Photo Credits

The following is a list of credits for portraits and photographs of scientists. All portraits, not appearing here, have been ascertained or are judged to be in the public domain. The author would appreciate to be informed (brandt@physik.uni-siegen.de) of cases, in which this judgement might be erroneous. (For photographs of technical nature the credit line is included in the figure caption.)

Episode 1 Portrait Goldstein *Archiv der Max-Planck-Gesellschaft, Berlin-Dahlem.*

Episode 2 Portrait Becquerel *Archiv der Max-Planck-Gesellschaft, Berlin-Dahlem.*

Episode 3 Photographs Zeeman, Lorentz *Niels Bohr Archive.*

Episode 4 Photograph Thomson *Archiv der Max-Planck-Gesellschaft, Berlin-Dahlem.*

Episode 5 Photograph Pierre and Marie Curie *ACJC–Curie and Joliot–Curie fund.*

Episode 6 Portrait and photograph of Rutherford *Archiv der Max-Planck-Gesellschaft, Berlin-Dahlem.*

Episode 7 Portraits Planck *Archiv der Max-Planck-Gesellschaft, Berlin-Dahlem.* Portrait Wien *Courtesy Prof. Karl Wien, Darmstadt.*

Episode 9 Portrait Soddy *AIP Gallery of Member Society Presidents.*

Episode 10 Portrait Einstein *Deutsches Museum, München.*

Episode 11 Portrait Einstein *Archiv der Max-Planck-Gesellschaft, Berlin-Dahlem.*

Episode 12 Portrait Nernst and photograph of Nernst lecturing *Archiv der Max-Planck-Gesellschaft, Berlin-Dahlem.* Photograph Lindemann, Nernst *Photograph by Francis Simon, courtesy AIP Emilio Segrè Visual Archives, Francis Simon Collection.*

Episode 13 Photograph Geiger, Rutherford and group photograph *Archiv der Max-Planck-Gesellschaft, Berlin-Dahlem.*

Episode 14 Portrait Perrin *Burndy Library, courtesy AIP Emilio Segrè Visual Archives.*

Episode 15 Photograph Millikan *Lawrence Berkeley National Laboratory (LBL News, Vol. 6, No. 3, Fall 1981).* Group photograph *Archiv der Max-Planck-Gesellschaft, Berlin-Dahlem.*

Episode 17 Photograph Wilson *AIP Emilio Segrè Visual Archives.*

Episode 18 Group photograph *Niels Bohr Archive.*

Episode 19 Portrait Hess and photograph of Hess in balloon *Courtesy William J. Breisky.* Photograph Hess with instrument *Archiv der Universität Innsbruck.*

Episode 20 Portrait Laue *Archiv der Max-Planck-Gesellschaft, Berlin-Dahlem.* Laue in uniform *Deutsches Museum, München.*

Episode 21 Portrait W. H. Bragg *AIP Emilio Segrè Visual Archives, Physics Today Collection.* Portrait W. L. Bragg *AIP Emilio Segrè Visual Archives, Weber Collection.* Group photo-graph *AIP Emilio Segrè Visual Archives, Fankuchen Collection.*

Episode 22 Portrait Thomson *Archiv der Max-Planck-Gesellschaft, Berlin-Dahlem.* Portrait Aston *AIP Emilio Segrè Visual Archives, W. F. Meggers Gallery of Nobel Laureates.*

Episode 23 Two portraits of Bohr *Niels Bohr Archive.* Photograph Bohr, Planck *Archiv der Max-Planck-Gesellschaft, Berlin-Dahlem.*

Episode 24 Photograph Moseley *University of Oxford, Museum of the History of Science, courtesy AIP Emilio Segrè Visual Archives, Physics Today Collection.*

Episode 25 Portrait Franck *Niels Bohr Archive.* Portrait Hertz *Universitätsarchiv Leipzig (UAL), Signatur FS N 151.*

Episode 26 Photographs *Archiv der Max-Planck-Gesellschaft, Berlin-Dahlem.*

Episode 27 Portrait Sommerfeld *Deutsches Museum, München.* Photograph Sommerfeld lecturing *Archiv der Max-Planck-Gesellschaft, Berlin-Dahlem.*

Episode 28 Portrait Rutherford *Archiv der Max-Planck-Gesellschaft, Berlin-Dahlem.* Portrait Blackett *Photograph by Lotte Meitner-Graf, London, courtesy AIP Emilio Segrè Visual Archives.*

Episode 29 Portrait Eddington *AIP Emilio Segrè Visual Archives.*

Episode 30 Portrait Stern *Photograph by Francis Simon, courtesy AIP Emilio Segrè Visual Archives, Francis Simon Collection.* Portrait Gerlach *Archiv der Max-Planck-Gesellschaft, Berlin-Dahlem.*

Episode 31 Photograph Compton *Lawrence Berkeley National Laboratory.*

Episode 32 Portrait de Broglie *AIP Emilio Segrè Visual Archives, Brittle Books Collection.*

Episode 33 Portrait Bose *Indian National Council of Science Museums, courtesy AIP Emilio Segrè Visual Archives.*

Episode 34 Portrait Bothe *Archiv der Max-Planck-Gesellschaft, Berlin-Dahlem.*

Episode 35 Photographs *CERN Pauli Collection.*

Episode 36 Photograph Kronig and Pauli *CERN Pauli Collection.* Photograph Uhlenbeck, Kramers, and Goudsmit *AIP Emilio Segrè Visual Archives, Goudsmit Collection.*

Episode 37 Portrait Heisenberg and photograph Born *Archiv der Max-Planck-Gesellschaft, Berlin-Dahlem.* Portrait Jordan *Courtesy Prof. Jürgen Ehlers, Potsdam.*

Episode 38 Photograph and portrait *Niels Bohr Archive.*

Episode 39 Portrait Schrödinger *Niels Bohr Archive.* Schrödinger in Alpbach *Deutsches Museum, München.*

Episode 40 Group photograph *CERN Pauli Collection.*

Episode 41 Photographs *CERN Pauli Collection.*

Episode 42 Photograph Heisenberg, Bohr *CERN Pauli Collection.* Photograph Bohr, Heisenberg in Tyrol *Niels Bohr Archive.*

Episode 43 Photograph Dirac *Niels Bohr Archive.* Photograph Dirac, Pauli *CERN Pauli Collection.*

Episode 44 Portrait Bloch *Niels Bohr Archive.* Group photograph *AIP Emilio Segrè Visual Archives, Peierls Collection.*

Episode 45 Photograph Hubble *Huntigton Library.*

Episode 46 Photographs *CERN Pauli Collection.*

Episode 47 Photographs Lawrence and Livingston *Lawrence Berkeley National Laboratory.*

Episode 48 Portrait Chadwick *Niels Bohr Archive.*

Episode 49 Photograph Anderson *Lawrence Berkeley National Laboratory (LBL News, Vol. 6, No. 3, Fall 1981).* Photograph Occhialini and Blackett *Amaldi Archives, Physics Department, Sapienza – University of Rome.*

Episode 50 Portrait Gamow *Niels Bohr Archive.* Group photograph *AIP Emilio Segrè Visual Archives.*

Episode 51 Photograph Heisenberg, Landau *AIP Emilio Segrè Visual Archives, Physics Today Collection.*

Episode 52 Photograph Stern *Niels Bohr Archive.* Photograph Stern in the Alps *CERN Pauli Collection.*

Episode 53 Photograph Fermi *Lawrence Berkeley National Laboratory (LBL News, Vol. 6, No. 3, Fall 1981).* Group photograph *AIP Emilio Segrè Visual Archives, Segrè Collection.*

Episode 54 Group photographs *ACJC–Curie and Joliot–Curie fund.* Portrait Gentner *Archiv der Max-Planck-Gesellschaft, Berlin-Dahlem.* Gentner lecturing *Deutsches Museum, München.*

Episode 55 Group photograph *Amaldi Archives, Physics Department, Sapienza – University of Rome.*

Episode 56 Portrait Vavilov *Russian State Archive for Cinema and Photo Documents, Krasnogorsk, courtesy AIP Emilio Segrè Visual Archives.* Portraits Cherenkov, Tamm, Frank *Physics Department, Lomonosov Moscow State University.*

Episode 57 Self-portrait *AIP Emilio Segrè Visual Archives, Yukawa Collection.* Portrait *Archiv der Max-Planck-Gesellschaft, Berlin-Dahlem.*

Episode 58 Portrait Kapitza *Archiv der Max-Planck-Gesellschaft, Berlin-Dahlem.*

Episode 59 Portrait Weizsäcker *Archiv der Max-Planck-Gesellschaft, Berlin-Dahlem.* Portrait Bethe *Courtesy Cornell – LEPP Laboratory.* Photograph Bethe lecturing *CERN.*

Episode 60 Photograph Hahn, photograph Meitner, photograph Hahn and Meitner in the labboratory *Hahn–Meitner-Institut, Berlin.* Group photograph with Rutherford, photograph Hahn and Meitner 1938, portrait Strassmann *Archiv der Max-Planck-Gesellschaft, Berlin-Dahlem.* Portrait Frisch *Niels Bohr Archive.*

Episode 61 Photographs McMillan and Seaborg *Lawrence Berkeley National Laboratory.*

Episode 62 Portrait Landau in Copenahgen *Niels Bohr Archive.* Portrait Landau in Russia *Physics Department, Lomonosov Moscow State University.*

Episode 63 Group photograph *Archival Photofiles [apf3-00232], Special Collections Research Center, University of Chicago Library.*

Episode 64 Photograph McMillan *Lawrence Berkeley National Laboratory.* Portrait Veksler *Joint Institute of Nuclear Research, Dubna.* Photograph Lawrence and others in magnet yoke *Lawrence Berkeley National Laboratory.* Photograph Courant and others *Courtesy Prof. Ernest D. Courant, New York.*

Episode 65 Portrait Bloch *CERN.* Portrait Rabi *Niels Bohr Archive.* Portrait Purcell *AIP Emilio Segrè Visual Archives.*

Episode 66 Photograph Occhialini, Powell *AIP Emilio Segrè Visual Archives, Segrè Collection.* Photograph Lattes *Istituto Nazionale di Fisica Nucleare.*

Episode 67 Photograph Lamb *National Archives and Records Administration, courtesy AIP Emilio Segrè Visual Archives.*

Episode 68 Portrait Rochester *Courtesy Prof. Arnold Wolfendale, Durham.* Portrait Butler *Courtesy Prof. Peter Dornan, Imperial College.*

Episode 69 Portrait Schottky *Deutsches Museum, München.* Group photograph *Alcatel-Lucent/Bell Labs.*

Episode 70 Portrait Beck *Niels Bohr Archive.* Photograph Goeppert-Mayer, Fermi *Photograph by Samuel Goudsmit, courtesy AIP Emilio Segrè Visual Archives, Goudsmit Collection.* Photographs of Jensen and of Goeppert-Mayer with Jensen *Archiv der Max-Planck-Gesellschaft, Berlin-Dahlem.*

Episode 71 Photograph Feynman *American Institute of Physics.* Photograph Schwinger *Harvard University, courtesy AIP Emilio Segrè Visual Archives, Weber Collection.* Photograph Tomonaga *AIP Emilio Segrè Visual Archives, Yukawa Collection.* Photograph Weisskopf, Dyson *CERN Pauli Collection.*

Episode 72 Photograph Glaser *Lawrence Berkeley National Laboratory.* Group photograph *in the author's possession.* Photographs of Alvarez *Lawrence Berkeley National Laboratory.*

Episode 73 Portrait Townes *Lawrence Berkeley National Laboratory (Magnet, Vol. 14, Nos. 10–11, October–November 1970, p. 2).* Photograph Basov and Prokhorov *AIP Emilio Segrè Visual Archives, Physics Today Collection.* Portrait Bloembergen *Harvard University, courtesy AIP Emilio Segrè Visual Archives.*

Episode 74 Portrait Pais *Photograph by Ulli Steltzer, courtesy AIP Emilio Segrè Visual Archives.* Portrait Gell-Mann *AIP Emilio Segrè Visual Archives.*

Episode 75 Portraits Segrè, Chamberlain and group photograph *Lawrence Berkeley National Laboratory.*

Episode 76 Photograph Reines *AIP Emilio Segrè Visual Archives.*

Episode 77 Photograph Lee, Yang *AIP Emilio Segrè Visual Archives*. Portrait Wu *AIP Emilio Segrè Visual Archives*.
Episode 78 Portraits Meissner, Ochsenfeld *Physikalisch-Technische Bundesanstalt (PTB)*. Photograph Heinz and Fritz London *Photo by K. Mendelssohn, Courtesy Prof. Horst Meyer, Duke University*. Portrait Bardeen *AIP Emilio Segrè Visual Archives, Segrè Collection*. Portrait Cooper *AIP Emilio Segrè Visual Archives*. Portrait Schrieffer *AIP Emilio Segrè Visual Archives, Physics Today Collection*.
Episode 79 Portrait Pontecorvo *Joint Institute of Nuclear Research, Dubna*. Photograph Gell-Mann, Feynman *AIP Emilio Segrè Visual Archives, Marshak Collection*.
Episode 80 Portrait Paul *AIP Emilio Segrè Visual Archives, W. F. Meggers Gallery of Nobel Laureates*. Snapshot of Paul *in the author's possession*. Photograph Dehmelt *Photograph by Mary Levin, University of Washington, courtesy AIP Emilio Segrè Visual Archives*.
Episode 81 Portrait Mössbauer *Archiv der Max-Planck-Gesellschaft, Berlin-Dahlem*.
Episode 82 Photograph Schawlow *Bell Laboratories, courtesy AIP Emilio Segrè Visual Archives, Hecht Collection*.
Episode 83 Portrait Panofsky *Photo Courtesy of Peter Ginter*. Photograph Richter *Stanford Linear Accelerator Center*. Photograph Touschek *INFN-LNF/SIS Photovideo*.
Episode 84 Photograph Franken *AIP Emilio Segrè Visual Archives*.
Episode 85 Group photograph *CERN*. Photograph Schwartz with chamber *Brookhaven National Laboratory*.
Episode 86 Portrait Kroemer *Tony Masters / UCSB*. Portrait Alferov © *Hannah Förster, PIK, Potsdam*.
Episode 87 Portrait Gell-Mann *AIP Emilio Segrè Visual Archives, Physics Today Collection*.
Episode 88 Portrait Fitch *Photograph by Bill Saunders, courtesy AIP Emilio Segrè Visual Archives*. Portrait Cronin *AIP Emilio Segrè Visual Archives, Segrè Collection*.
Episode 89 Photograph Wilson, Penzias *AIP Emilio Segrè Visual Archives*. Portrait Smoot *CERN*. Portrait Mather *NASA*.
Episode 90 Photograph Glashow, Salam *Abdus Salam International Centre for Theoretical Physics*. Photograph Weinberg *CERN*. Portrait Higgs *Peter Tuffy, Edinburgh University*.
Episode 91 All photographs *CERN*.
Episode 92 Photograph Fritzsch, Gell-Mann *Courtesy Prof. Harald Fritzsch, Munich*. Portrait Leutwyler *Courtesy Prof. Heinrich Leutwyler, Berne*. Photograph Gross *Robert P. Matthews, Princeton University, courtesy AIP Emilio Segrè Visual Archives, Physics Today Collection*. Portrait Wilczek *CERN*. Portrait Politzer *California Institute of Technology*.
Episode 93 Portrait Ting *CERN*. Portrait Richter *Stanford Linear Accelerator Center*. Portrait Perl *AIP Emilio Segrè Visual Archives, Physics Today Collection*.
Episode 94 Portrait Voss *DESY*.
Episode 95 Portrait von Klitzing *Archiv der Max-Planck-Gesellschaft, Berlin-Dahlem*. Portrait Laughlin *AIP Emilio Segrè Visual Archives, W. F. Meggers Gallery of Nobel Laureates*. Portrait Tsui *Courtesy Prof. Daniel Tsui, Princeton*. Portrait Störmer *AIP Emilio Segrè Visual Archives, Physics Today Collection*.
Episode 96 Portraits and Photograph Rubbia, van der Meer *CERN*.
Episode 97 Portraits Chu and Cohen-Tannoudji *CERN*. Portrait Phillips *NIST*.
Episode 98 Photograph Steinberger in front of the ALEPH detector *CERN*.
Episode 99 Photograph Wieman and Cornell © *Ken Abbot*. Portrait Ketterle *Photo Justin Knight*.
Episode 100 Photograph Davis *Brookhaven National Laboratory*. Portrait Koshiba *CERN*.

Name Index

A page marked by an asterisk (∗) contains a footnote with biographical data; one marked by a frame (□) carries a portrait or a group photograph.

Subject Index